# Diet and Microbiome in Health and Aging

# Diet and Microbiome in Health and Aging

Editors

**Sonia González**
**Nuria Salazar**
**Silvia Arboleya**

MDPI • Basel • Beijing • Wuhan • Barcelona • Belgrade • Manchester • Tokyo • Cluj • Tianjin

*Editors*

Sonia González
Instituto de Investigación
Sanitaria del Principado de
Asturias (ISPA)
University of Oviedo
Spain

Nuria Salazar
Instituto de Investigación
Sanitaria del Principado de
Asturias (ISPA)
University of Oviedo
Spain

Silvia Arboleya
Instituto de Investigación
Sanitaria del Principado de
Asturias (ISPA)
University of Oviedo
Spain

*Editorial Office*

MDPI
St. Alban-Anlage 66
4052 Basel, Switzerland

This is a reprint of articles from the Special Issue published online in the open access journal *Nutrients* (ISSN 2072-6643) (available at: https://www.mdpi.com/journal/nutrients/special_issues/Microbiome_Aging).

For citation purposes, cite each article independently as indicated on the article page online and as indicated below:

LastName, A.A.; LastName, B.B.; LastName, C.C. Article Title. *Journal Name* **Year**, *Volume Number*, Page Range.

**ISBN 978-3-0365-5363-4 (Hbk)**
**ISBN 978-3-0365-5364-1 (PDF)**

# Contents

*nutrients*

*Editorial*

# Diet and Microbiome in Health and Aging

**Silvia Arboleya [1,2,*], Sonia González [2,3,*] and Nuria Salazar [1,2,*]**

1 Department of Microbiology and Biochemistry of Dairy Products, Instituto de Productos Lácteos de Asturias (IPLA-CSIC), 33300 Villaviciosa, Spain
2 Diet, Microbiota and Health Group, Instituto de Investigación Sanitaria del Principado de Asturias (ISPA), 33011 Oviedo, Spain
3 Department of Functional Biology, University of Oviedo, 33006 Oviedo, Spain
* Correspondence: silvia.arboleya@ipla.csic.es (S.A.); soniagsolares@uniovi.es (S.G.); nuriasg@ipla.csic.es (N.S.); Tel.: +34-985-89-2131 (S.A. and N.S.)

After several years of research, sufficient evidence has been found supporting that diet is one of the main factors able to modulate both composition and activity of the intestinal microbiota, thus positioning it as a cornerstone in the host-microbiota interface. The gut microbiota plays a crucial role in the maintenance of normal host physiology. The rapid development of next-generation sequencing methods for nucleic acids, in the last decade, has facilitated in-depth studies of gut microbiome composition and function.

The articles collected in this Special Issue of Nutrients journal are intended to contribute to the progress of knowledge in the field as well as the basis for putative dietary interventions aimed at counteracting microbiota dysbiosis. These novel papers deal with the study of the relationship of diet on the intestinal microbiota from the early stages of life, deepening in certain pathologies, particularly relevant in this period of life, such as allergies, autism or overweight, up to adulthood and senescence. In addition, comprehensive review papers on hot topics such as the gut-brain axis, or the potential benefits of probiotics and prebiotics in the diet for allergy modulation were included. By providing updated and contrasted data, the authors propose several hypotheses that will be addressed in future research, which will undoubtedly arouse the interest of Nutrients journal readers.

The correct establishment of the gut microbiota at early life is known to be a milestone process for the later health of humans. Exponential studies during the last years have correlated aberrant gut microbiota colonization at the beginning of life with impairment on the intestinal, immune or nervous systems development [1]. Overweight, allergic diseases or neurodevelopmental disorders, like autism spectrum disorder (ASD), have been associated with gut microbiota alterations. Therefore, studying the composition of the gut microbiota at early life to be used as a predictor or to be target for modulation, is of great interest to prevent possible future diseases. In this context, Gonzalez et al. [2] evaluated the link between gut microbes and infant weight gain in the course of the first year of life in a cohort of full-term one-month aged neonates. They found significant associations between specific microbial groups and higher weight at 6 and 12 months, albeit being differently in vaginally and C-section delivered babies. Those gut microbes could be considered as potential microbial predictors for later weight gain.

The study of the connection between gut microbes, their metabolites and brain is currently favorable. Recent studies provide a close correlation of gut microbiota with different behavioral and cognitive traits, becoming a key stimulus during the first stages of neurodevelopment [3]. The exhaustive review by Johnson et al. [4] summarized the putative mechanisms implicated in the microbiome-brain interaction in the context of ASD. Genetic, environmental and epigenetic factors take part in an etiology puzzle that is not yet fully understood, in which the gut microbiome but also the mother's vaginal and oral microbiomes are playing a role. The authors highlighted diet and probiotics as gut microbiome modulators promising breakthrough interventions in the direction to get

**Citation:** Arboleya, S.; González, S.; Salazar, N. Diet and Microbiome in Health and Aging. *Nutrients* **2022**, *14*, 3250. https://doi.org/10.3390/nu14163250

Received: 26 July 2022
Accepted: 2 August 2022
Published: 9 August 2022

more individualized treatment approaches with lower side effects to guarantee the best clinical outcomes. Atopic diseases like asthma and allergic rhinitis often begin in early childhood when intestinal microbiota is underdeveloped. Evidence clearly supports a role for gut colonization in promoting and maintaining a balanced immune response in early life, thus this period could be considered as a window of opportunity [5]. Meirlaen et al. [6] aimed in their review to investigate if prevention and/or treatment of those atopic diseases could be accomplished by targeting gut microbiome. They performed an up to date search including both animal and clinical studies where probiotics, prebiotics or synbiotics were administered for the prevention or treatment of asthma and allergic rhinitis. The authors concluded that the current evidence is not enough to make recommendations of the use in children mainly due to the large heterogeneities derived from clinical study designs, but highlighted the benefits arisen from controlled pre-clinical studies. In concordance, they pointed out the need of well-designed and standardized studies to further clarify the action of those compounds on atopic diseases.

Diet has been identified as one of the main factors influencing gut microbiota modulation from early life, with breastfeeding as the greatest influencer at this time. In adulthood, solid evidence supports that long-term diets modulate the composition of the major microbial communities inhabiting the colon [7]. However, nowadays no reliable tool for calculating the healthiest dietary pattern in terms of microbiota has been identified for different diseases or adult life stages. We have a broad understanding of the impact of diet on the gut microbiota but formulating meaningful targeted dietary strategies remains a key challenge. In this sense, the work reported by Ruiz-Saavedra et al. [8] compared people over 50 years of age in different dietary indices, widely used in the literature, evaluating their potential for predicting the composition of the intestinal microbiota together with several other indicators of inflammatory state and oxidative stress, which are of special relevance in the aging process. On the other hand, some dietary components as well as isolated foods, have also demonstrated the capacity to modulate the intestinal microbiota at different levels. In this direction, the consumption of coffee in the regular diet, in a sample of healthy people aged between 19 and 95 years, has been associated with intestinal microbiota composition by González et al. [9]. In this descriptive and observational work, novel hypotheses have been proposed for the modulation of the *Bacteroides-Prevotella-Porphyromonas* group, associated in several studies with improved metabolic health, through coffee or the polyphenols contained in this beverage.

The impact of functional foods including probiotics, prebiotics and other bioactive compounds in the gut microbiota and host health has been evaluated in this Special issue through two *in vivo* murine models and a clinical trial. Massot-Cladera et al. [10] demonstrated that multivitamin and mineral supplementation together with prebiotic fibers (inulin and acacia gum) for 4 weeks differentially modulated gut microbiota composition, mineral absorption, and some immune and metabolic biomarkers in Wistar adult rats. Intestinal immune enhancement was reported in inulin-enriched supplement whereas acacia fiber supplement had stronger prebiotic activity, which may favor increasing mineral absorption. In another preclinical trial employing also diabetic type 2 Wistar rats, Toejing et al. [11] assessed the potential antidiabetic properties of the strain *Lactobacillus paracasei* HII01 isolated from the fermentation of northern Thai pickle. The strain was tested alone or in combination with the first-line drug antidiabetic drug metformin during a 12 weeks' period and potential beneficial effects were observed. The authors demonstrated that *L. paracasei* HII01 enhanced glycemic parameters including improvement in glucose intolerance, insulin, leptin and lipids levels, insulin-signaling proteins including from skeletal muscle that are involved in insulin-stimulated glucose uptake. This strain in diabetic rats also modulated the rat's gut microbiota reducing the plasma endotoxemia and systemic inflammation and increased caecum short chain fatty acids levels. The results suggested that there were no synergistic effects of metformin and probiotic *L. paracasei* HII01 but the data pointed out that this strain could be considered as a complementary supplement dietary strategy for type 2 diabetic patients.

The importance of the correct vaginal microbiota composition in vaginal health and success in pregnancy was also assessed by Fernández et al. [12]. The authors reported differences in vaginal parameters (pH, Nugent score, microbiota composition and soluble immune factor levels) between women with reproductive failure and fertile women. The lowest vaginal pH values and Nugent scores were associated with vaginal communities dominated by lactobacilli, while those with the highest pH values and Nugent scores were associated with a depletion of lactobacilli. Moreover, for the first time an antibiotic-associated depletion of vaginal lactobacilli was associated with long-term health, infertility and lower pregnancy success rates. The administration of the strain *Ligilactobacillus salivarius* CECT5713 for 6 months was also tested by the first time to women with reproductive failure and resulted in improved reproductive success by the modulation of the gut microbiota and it also induced several changes in biochemical and immunological parameters in women who got pregnant. These results demonstrated that the assessment of the microbial profiles in the reproductive tract should be evaluated in cases of reproductive failure of unknown cause or origin and the administration of *L. salivarus* CECT5713 is a novel and promising strategy to modulate the reproductive tract microbiome in order to increase the success of pregnancy. Moles et al. [13] in a comprehensive review, also assessed the dietary changes across human history and the evolution of the gut microbiota as result to these changes. They disclose the power of diet over one-off treatments, such as probiotics or prebiotics, on the gut microbiota modulation and highlighted the need to unravel the diet-host-microbiota interaction to achieve a preventive and personalized medicine.

Demographic aging is a global challenge. Through its impact on various levels such as the immune system, digestive tract or cognitive impairment, the intestinal microbiota is a potential target for enhancing life quality throughout old age. With this aim, van Soest et al. [14] have studied the effect of the administration of a Mediterranean diet, rich in fresh fruits and vegetables, on inflammatory status and cognitive decline in European individuals over 65 years of age belonging to the NU-AGE cohort. While confirming a positive association between the consumption of a pro-inflammatory diet rich in animal products with a more pro-inflammatory microbiota, no impact on the cognitive decline of the participants was observed. Undoubtedly, slowing down cognitive decline along with the prevention of conditions such as Alzheimer's disease is one of the major challenges of the nutrition field in the elderly in the last decade. The comprehensive review by Megur et al. [15] analyzed clinical and experimental studies highlighting the key role of gut microbiota dysbiosis in the development of Alzheimer's disease. Several mechanisms of action are proposed through which the microbiota could act as a communicator between the gut and the brain.

In another study, the supplementation of isolated polyphenol rich fractions from blueberry (BB) employing in vitro fecal batch fermentations demonstrated the differential effects of the blueberry ingredients on the fecal microbiota composition in the artificial colon model [16]. Moreover, the same authors in a pilot clinical study reported that freeze-dried whole BB consumption by healthy female volunteers in two age groups (young and older) for 6 weeks changed the gut microbiota composition. The BB consumption produced higher effects in microbiota diversity in older women and its modulation was associated with antioxidant activity in healthy adults. These results support the idea that BB consumption is related with beneficial effects by both the polyphenolic and fiber content of this fruit and could be potentially used for a healthy ageing.

Some recent research has also suggested that physical activity, independent of diet, may impact positively on the composition of the microbiome, however this is not yet elucidated at the extremes of life. On the basis of evidence indicating that physically active seniors had better gastrointestinal health [17]. Fart et al. [18] explored gut microbiota composition and diversity in elderly people, according to their physical activity. Results showed significant reductions in the proportion of some microorganisms such as *Parasutterella excrementihominis* and *Bilophila wadsworthia* associated with a beneficial effect on gastrointestinal health.

The collection of articles included in this Special Issue evidenced some of the current progress on the knowledge about the effects of diet on host health through the gut microbiota modulation. Understanding the complex and dynamic interaction between dietary exposures and gut microbiota throughout lifespan can help to elucidate their potential role in different pathologies and to guide future strategies for the prevention and treatment of diseases.

**Author Contributions:** Conceptualization, S.A., S.G., N.S.; writing—original draft preparation S.A., S.G., N.S.; writing—review and editing, S.A., S.G., N.S. All authors have read and agreed to the published version of the manuscript.

**Conflicts of Interest:** The authors declare no conflict of interest.

## References

1. Salazar, N.; Arboleya, S.; Valdés, L.; Stanton, C.; Ross, P.; Ruiz, L.; Gueimonde, M.; de los Reyes-Gavilán, C.G. The human intestinal microbiome at extreme ages of life. Dietary intervention as a way to counteract alterations. *Front. Genet.* **2014**, *5*, 406. [CrossRef] [PubMed]
2. González, S.; Selma-Royo, M.; Arboleya, S.; Martínez-Costa, C.; Solís, G.; Suárez, M.; Fernández, N.; de los Reyes-Gavilán, C.G.; Díaz-Coto, S.; Martínez-Camblor, P.; et al. Levels of Predominant Intestinal Microorganisms in 1 Month-Old Full-Term Babies and Weight Gain during the First Year of Life. *Nutrients* **2021**, *13*, 2412. [CrossRef] [PubMed]
3. Borre, Y.E.; O'Keeffe, G.W.; Clarke, G.; Stanton, C.; Dinan, T.G.; Cryan, J.F. Microbiota and Neurodevelopmental Windows: Implications for Brain Disorders. *Trends Mol. Med.* **2014**, *20*, 509–518. [CrossRef] [PubMed]
4. Johnson, D.; Letchumanan, V.; Thurairajasingam, S.; Lee, L.-H. A Revolutionizing Approach to Autism Spectrum Disorder Using the Microbiome. *Nutrients* **2020**, *12*, 1983. [CrossRef] [PubMed]
5. Kang, Y.B.; Cai, Y.; Zhang, H. Gut Microbiota and Allergy/Asthma: From Pathogenesis to New Therapeutic Strategies. *Allergol. Immunopathol.* **2017**, *45*, 305–309. [CrossRef] [PubMed]
6. Meirlaen, L.; Levy, E.I.; Vandenplas, Y. Prevention and Management with Pro-, Pre and Synbiotics in Children with Asthma and Allergic Rhinitis: A Narrative Review. *Nutrients* **2021**, *13*, 934. [CrossRef] [PubMed]
7. Wu, G.D.; Chen, J.; Hoffmann, C.; Bittinger, K.; Chen, Y.-Y.; Keilbaugh, S.A.; Bewtra, M.; Knights, D.; Walters, W.A.; Knight, R.; et al. Linking Long-Term Dietary Patterns with Gut Microbial Enterotypes. *Science* **2011**, *334*, 105–108. [CrossRef] [PubMed]
8. Ruiz-Saavedra, S.; Salazar, N.; Suarez, A.; de los Reyes-Gavilan, C.G.; Gueimonde, M.; Gonzalez, S. Comparison of Different Dietary Indices as Predictors of Inflammation, Oxidative Stress and Intestinal Microbiota in Middle-Aged and Elderly Subjects. *Nutrients* **2020**, *12*, 3828. [CrossRef] [PubMed]
9. González, S.; Salazar, N.; Ruiz-Saavedra, S.; Gómez-Martín, M.; de los Reyes-Gavilán, C.G.; Gueimonde, M. Long-Term Coffee Consumption is Associated with Fecal Microbial Composition in Humans. *Nutrients* **2020**, *12*, 1287. [CrossRef] [PubMed]
10. Massot-Cladera, M.; Azagra-Boronat, I.; Franch, À.; Castell, M.; Rodríguez-Lagunas, M.J.; Pérez-Cano, F.J. Gut Health-Promoting Benefits of a Dietary Supplement of Vitamins with Inulin and Acacia Fibers in Rats. *Nutrients* **2020**, *12*, 2196. [CrossRef] [PubMed]
11. Toejing, P.; Khat-Udomkiri, N.; Intakhad, J.; Sirilun, S.; Chaiyasut, C.; Lailerd, N. Putative Mechanisms Responsible for the Antihyperglycemic Action of Lactobacillus Paracasei Hii01 in Experimental Type 2 Diabetic Rats. *Nutrients* **2020**, *12*, 3015. [CrossRef] [PubMed]
12. Fernández, L.; Castro, I.; Arroyo, R.; Alba, C.; Beltrán, D.; Rodríguez, J.M. Application of Ligilactobacillus Salivarius CECT5713 to Achieve Term Pregnancies in Women with Repetitive Abortion or Infertility of Unknown Origin by Microbiological and Immunological Modulation of the Vaginal Ecosystem. *Nutrients* **2021**, *13*, 162. [CrossRef] [PubMed]
13. Moles, L.; Otaegui, D. The Impact of Diet on Microbiota Evolution and Human Health. Is Diet an Adequate Tool for Microbiota Modulation? *Nutrients* **2020**, *12*, 1654. [CrossRef] [PubMed]
14. Van Soest, A.P.M.; Hermes, G.D.A.; Berendsen, A.A.M.; van de Rest, O.; Zoetendal, E.G.; Fuentes, S.; Santoro, A.; Franceschi, C.; de Groot, L.C.P.G.M.; de Vos, W.M. Associations between Pro- and Anti-Inflammatory Gastro-Intestinal Microbiota, Diet, and Cognitive Functioning in Dutch Healthy Older Adults: The NU-AGE Study. *Nutrients* **2020**, *12*, 3471. [CrossRef] [PubMed]
15. Megur, A.; Baltriukienė, D.; Bukelskienė, V.; Burokas, A. The Microbiota–Gut–Brain Axis and Alzheimer's Disease: Neuroinflammation Is to Blame? *Nutrients* **2021**, *13*, 37. [CrossRef] [PubMed]
16. Ntemiri, A.; Ghosh, T.S.; Gheller, M.E.; Tran, T.T.T.; Blum, J.E.; Pellanda, P.; Vlckova, K.; Neto, M.C.; Howell, A.; Thalacker-Mercer, A.; et al. Whole Blueberry and Isolated Polyphenol-Rich Fractions Modulate Specific Gut Microbes in an in Vitro Colon Model and in a Pilot Study in Human Consumers. *Nutrients* **2020**, *12*, 2800. [CrossRef] [PubMed]
17. Algilani, S.; Östlund-Lagerström, L.; Kihlgren, A.; Blomberg, K.; Brummer, R.J.; Schoultz, I. Exploring the Concept of Optimal Functionality in Old Age. *J. Multidiscip. Healthc.* **2014**, *7*, 69–79. [CrossRef] [PubMed]
18. Fart, F.; Rajan, S.K.; Wall, R.; Rangel, I.; Ganda-Mall, J.P.; Tingö, L.; Brummer, R.J.; Repsilber, D.; Schoultz, I.; Lindqvist, C.M. Differences in Gut Microbiome Composition between Senior Orienteering Athletes and Community-Dwelling Older Adults. *Nutrients* **2020**, *12*, 2610. [CrossRef] [PubMed]

nutrients

MDPI

Article

# Long-Term Coffee Consumption is Associated with Fecal Microbial Composition in Humans

**Sonia González [1,2], Nuria Salazar [2,3], Sergio Ruiz-Saavedra [2,3], María Gómez-Martín [1,2], Clara G. de los Reyes-Gavilán [2,3] and Miguel Gueimonde [2,3,*]**

1 Department of Functional Biology, University of Oviedo, 33006 Oviedo, Spain; soniagsolares@uniovi.es (S.G.); mariagomart@gmail.com (M.G.-M.)

2 Diet, Microbiota and Health Group, Instituto de Investigación Sanitaria del Principado de Asturias (ISPA), 33011 Oviedo, Spain; nuriasg@ipla.csic.es (N.S.); sergioruizsa3@gmail.com (S.R.-S.); greyes_gavilan@ipla.csic.es (C.G.d.l.R.-G.)

3 Department of Microbiology and Biochemistry of Dairy Products, Instituto de Productos Lácteos de Asturias (IPLA-CSIC), 33300 Villaviciosa, Asturias, Spain

* Correspondence: mgueimonde@ipla.csic.es; Tel.: +34-985892131

Received: 14 April 2020; Accepted: 28 April 2020; Published: 1 May 2020

**Abstract:** Coffee consumption has been related to a preventive effect against several non-transmissible pathologies. Due to the content of this beverage in phytochemicals and minerals, it has been proposed that its impact on health may partly depend on gut microbiota modulation. Our aim was to explore the interaction among gut microbiota, fecal short chain fatty acids, and health-related parameters in 147 healthy subjects classified according to coffee consumption, to deepen the association of the role of the (poly)phenol and alkaloid content of this beverage. Food daily intake was assessed by an annual food frequency questionnaire (FFQ). Coffee consumption was categorized into three groups: non-coffee-consumers (0–3 mL/day), moderate consumers (3–45 mL/day) and high-coffee consumers (45–500 mL/day). Some relevant groups of the gut microbiota were determined by qPCR, and concentration of fecal short chain fatty acids by gas chromatography. Serum health related biomarkers were determined by standardized methods. Interestingly, a higher level of *Bacteroides–Prevotella–Porphyromonas* was observed in the high consumers of coffee, who also had lower levels of lipoperoxidation. Two groups of coffee-derived (poly)phenol, methoxyphenols and alkylphenols, and caffeine, among alkaloids, were directly associated with *Bacteroides* group levels. Thus, regular consumption of coffee appears to be associated with changes in some intestinal microbiota groups in which dietary (poly)phenol and caffeine may play a role.

**Keywords:** coffee; (poly)phenol; gut microbiota; *Bacteroides*

## 1. Introduction

Coffee is one of the most consumed non-alcoholic beverages worldwide and it may exert different effects at a physiological level [1]. Although it has traditionally been considered as a beverage with very low nutritional value, epidemiological evidence suggests that moderate coffee consumption may reduce the risk of chronic diseases such as metabolic syndrome, obesity, type 2 diabetes [2], cardiovascular diseases [3], or some types of cancer [4–6]. Coffee may impact directly on the host gastrointestinal physiology by increasing intestinal motility and reducing intestinal transit time [7,8]. Some of these widely described benefits of coffee have been attributed to its high content in non-nutritional compounds such as phenolic compounds, fibers, minerals, and caffeine [9], which may also influence host metabolic pathways related to health maintenance. From these compounds, caffeine, (poly)phenols, and fibers are able to reach and exert some of their effects in the large intestine, being fermented by the gut microbiota [10]. Thus, given the pivotal role that microbiota plays on human nutrition and health [11], it is possible that some of the beneficial effects of the coffee components may be related with the participation of the gut microbiota in the metabolism of such compounds. Interventional studies analyzing the impact of a moderate coffee intake during three weeks in a healthy population have reported an increase of *Bifidobacterium* [9], sometimes also linked to a decrease of *Clostridium* and *Escherichia coli* [9,12–14]. Regarding other bacterial groups, such as *Bacteroides*, the results in the literature remain controversial [9,13,15,16]. Among the possible mechanisms to explain these associations, data from in vitro studies pointed to a direct relationship between chlorogenic acids and selective changes on the *Blautia coccoides–Eubacterium rectale* group [10] and between caffeine and the abundance of the *Lactobacillus* species [17]. Based on previous evidences indicating that theobromine, an alkaloid present in coffee, can enhance (poly)phenol absorption in the intestine [18], a synergistic effect for phenolic compounds and alkaloids on the intestinal microbiota at this location may be plausible. To date, most of the studies analyzing the impact of coffee on the composition of the intestinal microbiota come from in vitro, animal, or intervention studies. However, to the best of our knowledge, no observational studies are currently available analyzing the impact of regular coffee consumption on fecal microbiota, taking into consideration the influence that the content of this beverage in caffeine and phenolic compounds may exert on the microbiota. This information would contribute to expand the existing knowledge about the impact of coffee on gastrointestinal physiology and therefore, on health maintenance.

## 2. Subjects and Methods

The study included 147 participants, with ages ranging from 19 to 95 years and body mass index (BMI) scores between 19.0 and 39.0 kg/m$^2$ who were recruited in Asturias (Atlantic coast of Spain). Volunteers were cited individually, informed about the study, and gave informed written consent before enrolment. Inclusion criteria were the absence of diagnosed immune or digestive related pathologies as well as non-consumption of corticoids, immunosuppressive drugs, monoclonal antibodies, antibiotics, or immunotherapy, and not having consumed probiotics or prebiotics as dietary supplements during the previous month.

The study was approved by the Bioethics Committee of CSIC (Consejo Superior de Investigaciones Científicas) and the Regional Ethics Committee for Clinical Research (Servicio de Salud del Principado de Asturias n 13/2010).

### 2.1. Nutritional Assessment

Participants were instructed to maintain their usual dietary pattern before the study. Regular food intake was assessed by trained personnel in a personal interview of approximately 1 h duration, using an annual semi-quantitative food frequency questionnaire (FFQ), previously validated [19,20]. Methodological issues about dietary assessment were published elsewhere [18]. Food consumption was transformed into energy and macronutrients intake using the food composition tables of CESNID (Centro de Enseñanza Superior de Nutrición Humana y Dietética) [21]. Caffeine intake was estimated from the United States Department of Agriculture (USDA) food composition database [22]. The polyphenols content in foods was completed using the Phenol Explorer database that compiled detailed information from over 400 foods and beverages, including coffee [23] and fiber components, and were ascertained using the Marlett et al. food composition tables [24].

At the time of carrying out the blood extraction, height and weight were taken by standardized protocols previously described [25] in order to calculate the BMI by the formula: weight (kg)/height (m$^2$).

### 2.2. Blood Biochemical Analyses

Fasting blood samples were drawn by venipuncture and centrifuged (1000× g, 15 min). Plasma and serum aliquots were kept at −20 °C until analyses. Plasma glucose, cholesterol, and triglycerides were determined by standard methods. Serum C-reactive protein (CRP) levels were determined by ELISA (CRP Human Instant ELISA, Ebioscience, San Diego, CA, USA), and malondialdehyde (MDA) by using the Byoxytech LPO-586 assay (Oxis International S.A., Paris, France) [26]. Serum leptin was determined by using the Human Leptin ELISA Development Kit 900-K90 (PeproTech Inc., Rocky Hill, NJ, USA) according to the manufacturer's instructions.

### 2.3. Fecal Collection and Microbial Analysis

Participants were provided with a sterile container for fecal sample collection; after deposition samples were immediately frozen at −20 °C and transported to the laboratory. For analyses, samples were melted at room temperature (24 ± 2 °C), weighed, diluted 1/10 in sterile PBS, and homogenized (LabBlender 400 Stomacher, Seward Medical, London, UK) for 4 min; the DNA was extracted using the QIAamp DNA stool mini kit (Qiagen, Hilden, Germany) as described elsewhere [27]. Quantification of different bacterial populations, covering the major bacterial groups present in the human gut, was achieved in a 7500 Fast Real-Time PCR System (Applied Biosystems, Foster City, CA, USA) using SYBR Green PCR Master Mix (Applied Biosystems) [27] (Table 1). One microliter of template fecal DNA (~5 ng) and 0.2 μM of each primer were added to the 25 μL reaction mixture. PCR cycling consisted of an initial cycle of 95 °C 10 min, followed by 40 cycles of 95 °C 15 s, and 1 min at the appropriate primer–pair temperature. The number of cells was determined by comparing the Ct values obtained from a standard curve as previously described [27]. Fecal DNA extracts were analyzed and the mean quantity per gram of fecal wet weight was calculated for each bacterial group.

Major short chain fatty acid (SCFA), acetate, propionate, and butyrate were analyzed by gas chromatography from the supernatants of 1 mL of the homogenized feces as previously indicated [28]. A chromatograph 6890N (Agilent Technologies Inc., Palo Alto, CA, USA) connected to a mass spectrometry detector (MS) 5973N (Agilent Technologies) and a flame ionization detector (FID) was used for identification and quantification of SCFA, respectively, as described previously [29].

**Table 1.** Primers and annealing temperatures used for the quantification of intestinal microbial groups by qPCR.

| Microbial Group | Primer Sequence (5′-3′) | Tm. (°C) |
|---|---|---|
| *Akkermansia* | F: CAGCACGTGAAGGTGGGGAC<br>R: CCTTGCGGTTGGCTTCAGAT | 60 |
| *Bacteroides-Prevotella-Porphyromonas* | F: GAGAGGAAGGTCCCCCAC<br>R: CGCKACTTGGCTGGTTCAG | 60 |
| *Bifidobacterium* | F: GATTCTGGCTCAGGATGAACGC<br>R: CTGATAGGACGCGACCCCAT | 60 |
| *Clostridia* cluster *XIVa* group | F: CGGTACCTGACTAAGAAGC<br>R: AGTTTYATTCTTGCGAACG | 55 |
| *Faecalibacterium prausnitzii* | F: GGAGGAAGAAGGTCTTCGG<br>R: AATTCCGCCTACCTCTGCACT | 60 |
| *Lactobacillus* group | F: AGCAGTAGGGAATCTTCCA<br>R: CATGGAGTTCCACTGTCCTC | 60 |

Adapted from Reference [28].

### 2.4. Statistical Analyses

Statistical analysis was performed using the IBM SPSS program version 24.0 (IBM SPSS, Inc., Chicago, IL, USA). Goodness of fit to the normal distribution was analyzed by means of the Kolmogorov–Smirnov test. When the distribution of variables was skewed, the natural logarithm of each value was used in the statistical test. For descriptive purposes, mean values are presented on untransformed variables. Differences in general and anthropometric characteristics, blood parameters, gut microbial groups, and fecal SCFA were assessed in accordance to tertiles of coffee intake through multivariate analyses adjusted by age, gender, BMI, and energy, based on the strong evidences linking these factors with human microbial composition. Pearson correlation was conducted to elucidate the interplay between caffeine and polyphenols from coffee and intestinal microbiota. The conventional probability value (0.05) for significance was used in the interpretation of results. Results obtained from the sample analysis were plotted using Microsoft Excel Software version 2016 (Microsoft Corporation, Redmon, Washington, USA). Data resulting from the Pearson correlation tests were plotted using GraphPad Prism version 8 for Windows (GraphPad Software, San Diego, CA, USA).

## 3. Results

The general characteristics of the sample are described in Table 2 according to the coffee tertiles. No statistically significant differences were found based on coffee consumption for any of the variables evaluated, with the exception of age, which was lower in subjects with the highest consumption of coffee (tertile 3).

**Table 2.** General characteristics of the study sample according to coffee consumption tertiles.

| Characteristic | Coffee (mL/day) | | |
|---|---|---|---|
| | T1 (0–3) (*n* = 49) | T2 (>3–45) (*n* = 49) | T3 (>45–500) (*n* = 49) |
| Age (years) | 58.8 ± 18.62 $_a$ | 67.57 ± 14.77 $_b$ | 47.10 ± 10.86 $_c$ |
| Gender (% female) | 69% | 71% | 71% |
| BMI (kg/m$^2$) | 28.08 ± 4.52 $_a$ | 27.32 ± 3.55 $_a$ | 26.73 ± 5.16 $_a$ |
| Sleep duration (h/day) | 6.78 ± 1.06 $_a$ | 6.73 ± 1.07 $_a$ | 7.00 ± 1.31 $_a$ |
| Energy intake (Kcal/day) | 1906.93 ± 494.27 $_a$ | 1776.89 ± 531.64 $_a$ | 2040.23 ± 622.47 $_a$ |
| Coffee consumption (mL/day) | 0.15 ± 0.59 $_a$ | 27.53 ± 11.14 $_b$ | 151.84 ± 92.10 $_c$ |
| Tobacco user (%) | 25% | 28% | 25% |
| Depositions (n°/week) | 8.89 ± 6.26 $_a$ | 6.49 ± 2.68 $_b$ | 7.74 ± 3.84 $_{a,b}$ |

All values are shown as mean ± standard deviation (SD). Values in the same row showing a different subscript present a statistically significant difference ($p \leq 0.05$). Tobacco user refers people with smoking-habit at the time of the study.

The possible existence of a dietary pattern linked to coffee consumption was explored, as shown in Figure 1.

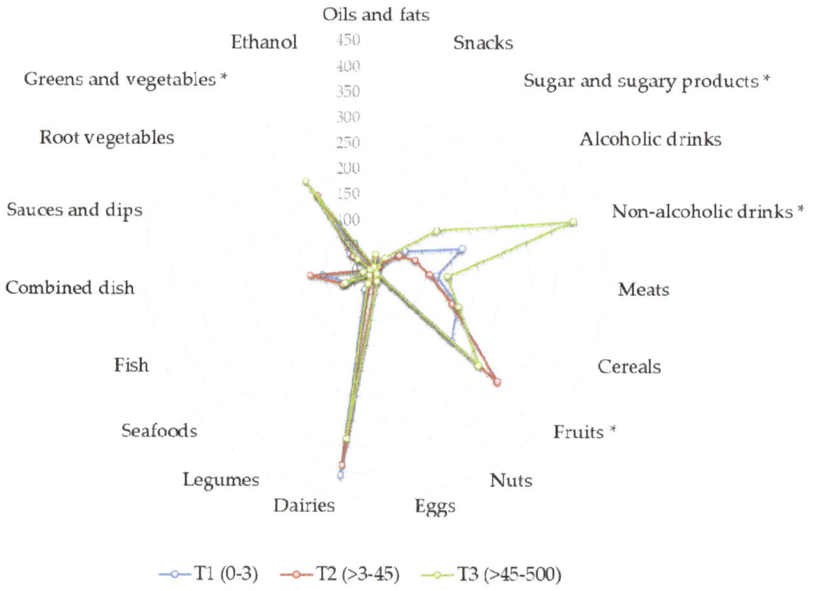


**Figure 1.** A radar plot representing differences in dietary patterns according to coffee consumption (mL/day) tertiles. * $p \leq 0.05$.

From the 19 items analyzed, only a moderate increase in the consumption of greens and vegetables was found across coffee tertiles, this being higher in tertile 3. As expected, since coffee is usually consumed with sugar, significant differences were also observed in the intake of non-alcoholic beverages and sugar products. In spite of this, the scarce differences found do not allow for a differential dietary pattern in the high-consumers group to be defined. When the average counts of the major gut microbial groups were analyzed, based on coffee consumption tertiles (Table 3), the sole difference observed was a significantly higher level of *Bacteroides-Prevotella-Porphyromonas* in tertile 3. Nevertheless, no differences were detected in fecal levels of SCFA according to coffee consumption, neither in the studied serum biomarkers, with the exception of MDA, an indirect biomarker of lipid peroxidation, whose concentration was lower in tertile 3.

**Table 3.** Differences in gut microbiota composition, fecal short chain fatty acids concentration (SCFA), and serum markers according to coffee consumption tertiles.

| | Coffee (mL/day) | | |
|---|---|---|---|
| | **T1 (0–3)** | **T2 (>3–45)** | **T3(>45–500)** |
| **Model 1. Microbial group (Log n cells/gram feces) (n, 138) *** | | | |
| *Akkermansia* | $5.70 \pm 0.40_{a}$ | $5.76 \pm 0.33_{a}$ | $6.30 \pm 0.36_{a}$ |
| *Bacteroides-Prevotella-Porphyromonas* | $8.03 \pm 0.27_{a}$ | $8.74 \pm 0.27_{a,b}$ | $9.14 \pm 0.30_{b}$ |
| *Bifidobacterium* | $7.61 \pm 0.26_{a}$ | $7.73 \pm 0.26_{a}$ | $8.19 \pm 0.28_{a}$ |
| *Clostridia* cluster XIVa group | $7.48 \pm 0.29_{a}$ | $7.49 \pm 0.29_{a}$ | $7.50 \pm 0.31_{a}$ |
| *Lactobacillus* group | $6.28 \pm 0.27_{a}$ | $5.97 \pm 0.27_{a}$ | $5.97 \pm 0.29_{a}$ |
| *Faecalibacterium prausnitzii* | $7.10 \pm 0.17_{a}$ | $7.29 \pm 0.17_{a}$ | $7.51 \pm 0.19_{a}$ |
| **Model 2. Fecal SCFA concentration (mM) (n, 132) *** | | | |
| Acetic acid | $36.77 \pm 2.51_{a}$ | $36.80 \pm 2.48_{a}$ | $33.99 \pm 2.58_{a}$ |
| Propionic acid | $12.55 \pm 1.09_{a}$ | $13.97 \pm 1.08_{a}$ | $12.56 \pm 1.12_{a}$ |
| Butyric acid | $10.17 \pm 1.18_{a}$ | $10.90 \pm 1.17_{a}$ | $10.10 \pm 1.21_{a}$ |
| **Model 3. Blood parameters *** | | | |
| Serum MDA (µM) (n,102) | $2.51 \pm 0.11_{a}$ | $2.28 \pm 0.07_{a,b}$ | $1.89 \pm 0.20_{b}$ |
| C reactive protein (mg/L) (n,108) | $1.37 \pm 0.24_{a}$ | $1.27 \pm 0.17_{a}$ | $0.69 \pm 0.46_{a}$ |
| Leptin (ng/mL) (n,102) | $11.05 \pm 1.21_{a}$ | $11.15 \pm 0.85_{a}$ | $8.34 \pm 2.34_{a}$ |
| LDL–HDL ratio (n,125) | $2.51 \pm 0.18_{a}$ | $2.86 \pm 0.13_{a}$ | $2.85 \pm 0.35_{a}$ |
| Triglycerides (mg/dL) (n,125) | $121.70 \pm 11.12_{a}$ | $122.08 \pm 7.80_{a}$ | $103.14 \pm 21.44_{a}$ |
| Glucose (mg/dL) (n,125) | $100.12 \pm 5.62_{a}$ | $103.05 \pm 3.94_{a}$ | $103.73 \pm 10.84_{a}$ |
| Antioxidant capacity (mM) (n,72) | $0.36 \pm 0.02_{a}$ | $0.34 \pm 0.01_{a}$ | $0.34 \pm 0.04_{a}$ |

* Results obtained from multivariate analyses adjusted by age, gender, BMI, and energy. Values in the same row showing different subscripts present a statistically significant difference; ($p \leq 0.05$). MDA, malondialdehyde; LDL, low density lipoprotein; HDL, high density lipoprotein.

Coffee is a dietary source of various bioactive compounds including (poly)phenols and alkaloids (Figure 2). At the compound level, the major phenolic compounds provided by coffee were caffeoylquinic and feruloylquinic acids among hydroxycinnamics, and guaiacol from methoxyphenols (Figure 2A). As shown in Figure 2B, coffee was the major contributor to the intake of caffeine in the sample, explaining more than 90% of its consumption.

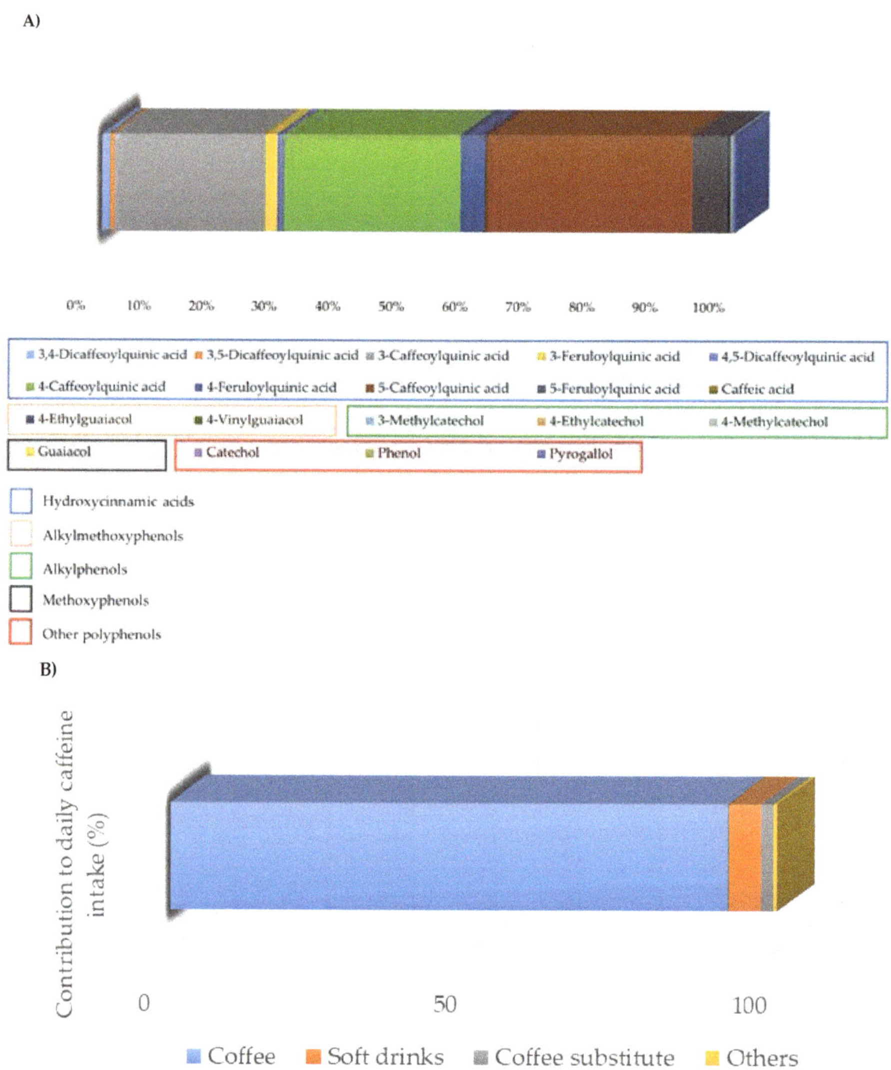


**Figure 2.** Representation of (**A**) the contribution of each coffee phenolic compound in the sample and (**B**) the dietary caffeine sources in the sample.

Furthermore, the linear relationships between coffee derived dietary components and the microbiota was estimated through Pearson's correlation test and are presented graphically in the heatmap of Figure 3. From the different phenolic compounds analyzed, those derived from coffee have shown the highest correlation with intestinal microbial groups together with caffeine. While metoxyphenols and alkylmethoxyphenols were correlated with the levels of the *Bacteroides–Prevotella–Porphyromonas* group with a $r = 0.177$ and 0.182, respectively, caffeine intake was directly associated with fecal *Bacteroides* levels ($r = 0.200$).

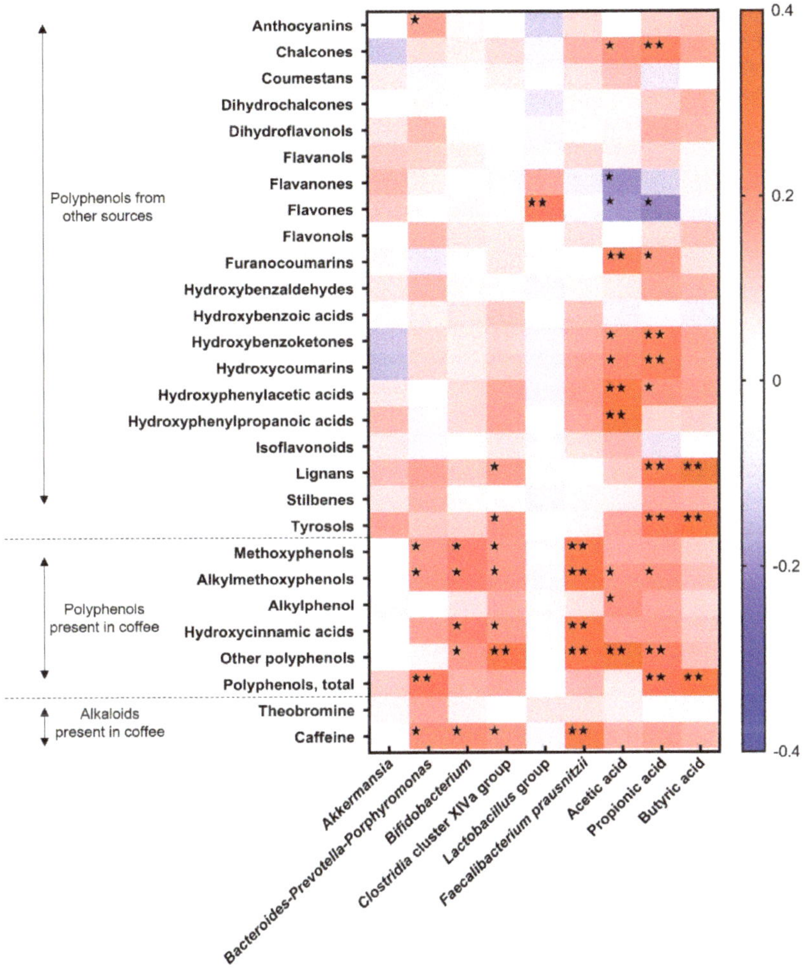


**Figure 3.** A heatmap showing Pearson correlations among intestinal microbial groups (Log n cells/gram feces), fecal short chain fatty acids (mM), polyphenol groups (mg/day), and alkaloids (mg/day), from coffee and other dietary sources. Columns correspond to major intestinal microbial groups and fecal SCFA; rows correspond to dietary polyphenols and alkaloids. Blue and red colors denote negative and positive association, respectively. The intensity of the colors represents the degree of association between variables. Asterisks indicate significant associations: * $p \leq 0.05$; ** $p \leq 0.01$.

## 4. Discussion

Our results represent a first step in broadening the knowledge of the association between the regular intake of coffee and fecal microbiota in an apparently healthy human population, suggesting a possible implication of coffee phenolic compounds and caffeine in this relationship.

The mean consumption of coffee is highly variable in the study sample, ranging between 0 and 500 mL/day, in a similar way to that observed in other European countries with a Mediterranean-type dietary pattern, such as Italy or Greece [30]. Given the absence of a reference value to establish coffee consumption levels, tertiles have been used to categorize the sample. The defined cut-off points are coherent from a methodological point of view, since they group non-coffee-consumers (0–3 mL/day) in tertile 1, moderate consumers (3–45 mL/day) in tertile 2, which could correspond to those subjects consuming a little cup of coffee per day of the so-called Italian coffee, and the tertile 3 of high consumers. Still, it must be taken into account that this tertile 3 has a lower mean intake of coffee than what has been reported in other countries, such as Germany [30]; therefore, our data may not be directly extrapolated to other countries with different trends in coffee consumption. It is also important to note that important differences in the coffee preparation procedures exist among different consumers and different countries. In our case, the mean coffee intake in tertile 3 is slightly lower than the range of 400–600 mL [9], associated with a protective effect against various pathologies [3,31]. Interestingly, in some studies the long-term consumption of seven cups of coffee per day has been associated with a reduction in the risk of metabolic syndrome, obesity, and type 2 diabetes [32,33]. However, since the coffee cup volume could vary from 150 mL to 300 mL among the studies included in the meta-analysis of Grosso et al. [32] indicated just before, extrapolations to our results are limited.

In some impaired health conditions, alterations in the intestinal microbiota have been described, mainly a decrease in the abundance of *Bacteroides* and/or an increase in the Firmicutes/ Bacteroidetes ratio [34–36]. In line with this, we have found higher fecal levels of the *Bacteroides-Prevotella-Porphyromonas* group in the high coffee consumers, supporting previous evidences in humans and animal models having higher fecal levels of these microorganisms in groups of better metabolic status [34–36]. It is tempting to speculate about this potential association; however, given the descriptive nature of this cross-sectional study, we cannot establish cause–effect relationships or directionality. Members of the phylum Bacteroidetes have been hypothesized to reduce intracellular oxygen levels, thus favoring the growing of anaerobic species which could promote the maintenance of intestinal balance, and they are identified as key glycan degrading bacteria [37,38] being more able to metabolize polyphenols than other groups such as Firmicutes. In this sense, coffee polyphenols explained a 20% of total polyphenol intake in the subjects with the highest consumption (tertile 3). Therefore, we hypothesized that at an equivalent total intake of polyphenols, the physiological effect of these compounds may differ among subjects depending on their dietary origin. It has been demonstrated that coffee-derived polyphenols were able to interact with intestinal bacteria in a bidirectional way: polyphenols may modify the gut environment [39], and/or they can be catabolized by intestinal bacteria converting them into a large variety of compounds with greater antioxidant activity than the compound of origin [40]. Despite some coffee-derived phenolic compounds, such as chlorogenic acid, having been associated in in vitro studies with important antioxidant and anti-inflammatory effects [41–44], we did not find differences in serum antioxidant capacity or in C-reactive protein, depending on the coffee consumption levels. Several factors may explain these results. Firstly, it is possible that the amount of coffee consumed in this sample was insufficient to observe a differential effect among tertiles, and secondly, subjects with low coffee consumption received polyphenols through other foodstuffs, thus ultimately achieving a similar polyphenol intake to the high coffee consumers (mean intake of 1604.4 and 1487.7 mg/day in T2 and T3 respectively, $p = 0.533$). Considering the high impact of phenolic compounds on gut microbial modulation, future human intervention studies analyzing the impact of coffee on fecal microbiota [9] should evaluate the intake of dietary and specific coffee polyphenols.

Moreover, there is evidence from animal research showing that caffeine administration counteracts shifts in the ratio, Firmicutes/Bacteroides, resulting from a western diet [45]. Although we cannot attribute the observed differences in fecal microbial composition to a single compound, caffeine has been positively correlated in this work with most of the gut microbial groups analyzed. In turn, evidences of lower MDA concentrations in high coffee consumers may be in consonance with data in the literature describing the down-regulation effect of caffeine on lipid binding proteins and consequently in lipogenesis [46–49].

At the time of interpreting these results, the limited sample size should be considered. Nonetheless, we have found associations whose consistency and strength justify further research. Human experimentation in healthy subjects is limited by the logistical problems associated with carrying out direct measurements. Fecal SCFA accounts for only 5% to 10% of SCFA production that is not absorbed in the colon [50].

The FFQ is one of the most valid dietary tools in order to describe subjects' regular dietary intake. However, to accurately quantify the coffee derived polyphenols, it would be advisable to register more detailed information about coffee such as the variety, the amount of powder used, ground grain size, and the final volume obtained [30]. Since we did not have individuals with daily intakes greater than 500 mL, it would be desirable in the future to be able to expand this group to deepen the association between this beverage and intestinal microbiota and oxidative stress, and to determine whether this would be dose dependent.

## 5. Conclusions

The interaction between coffee consumption, a modifiable factor, and intestinal bacteria may be useful for the development of dietary strategies in humans focused on diverse pathologies where the concentration of the *Bacteroides* group was altered.

**Author Contributions:** The authors' responsibilities were as follows: S.G. and M.G. designed the experimental work. S.R.-S. and M.G.-M. carried out the nutritional and anthropometric determinations. S.G., N.S., C.G.d.l.R.-G., and M.G. obtained the biological samples. N.S., C.G.d.l.R.-G., and M.G. were involved in microbiota and SCFA analysis, meanwhile S.R.-S. and M.G.-M. performed the statistical analysis. S.G. wrote the original draft. All authors have read and agreed to the published version of the manuscript.

**Funding:** This research was funded by Plan Estatal de I+D+I through projects AGL2017-83653-R (AEI/FEDER, UE) and RTI2018-098288-B-I00 (MCIU/AEI/FEDER, UE), and by contracts with Biopolis SL (Valencia, Spain), CAUCE Foundation (Oviedo, Spain) and Alimerka Foundation (Llanera, Spain). N.S. was granted a postdoctoral contract awarded by the Fundación para la Investigación Biosanitaria de Asturias (FINBA); M.G-M. was supported by a FPU (FPU18/03393) predoctoral grant from the Spanish Ministry of Science, Innovation and Universities; and S.R.-S. is the recipient of a Research Training contract awarded under project RTI2018-098288-B-I00.

**Acknowledgments:** We show our greatest gratitude to all the volunteers participating in the study.

**Conflicts of Interest:** The authors declare no conflict of interest.

## References

1. Mubarak, A.; Bondonno, C.P.; Liu, A.H.; Considine, M.J.; Rich, L.; Mas, E.; Croft, K.D.; Hodgson, J.M. Acute effects of chlorogenic acid on nitric oxide status, endothelial function, and blood pressure in healthy volunteers: A randomized trial. *J. Agric. Food Chem.* **2012**, *60*, 9130–9136. [CrossRef] [PubMed]
2. Farias-Pereira, R.; Park, C.-S.; Park, Y. Mechanisms of action of coffee bioactive components on lipid metabolism. *Food Sci. Biotechnol.* **2019**, *28*, 1287–1296. [CrossRef] [PubMed]
3. O'Keefe, J.H.; DiNicolantonio, J.J.; Lavie, C.J. Coffee for cardioprotection and longevity. *Prog. Cardiovasc. Dis.* **2018**, *61*, 38–42. [CrossRef] [PubMed]
4. Grosso, G.; Godos, J.; Galvano, F.; Giovannucci, E.L. Coffee, Caffeine, and health outcomes: An umbrella review. *Annu. Rev. Nutr.* **2017**, *37*, 131–156. [CrossRef] [PubMed]
5. Dórea, J.G.; da Costa, T.H.M. Is coffee a functional food? *Br. J. Nutr.* **2005**, *93*, 773–782. [CrossRef]
6. Higdon, J.V.; Frei, B. Coffee and Health: A review of recent human research. *Crit. Rev. Food Sci. Nutr.* **2006**, *46*, 101–123. [CrossRef]
7. Poole, R.; Kennedy, O.J.; Roderick, P.; Fallowfield, J.A.; Hayes, P.C.; Parkes, J. Coffee consumption and health: Umbrella review of meta-analyses of multiple health outcomes. *BMJ* **2017**, *359*, j5024. [CrossRef]
8. Brown, S.R.; Cann, P.A.; Read, N.W. Effect of coffee on distal colon function. *Gut* **1990**, *31*, 450–453. [CrossRef]
9. Jaquet, M.; Rochat, I.; Moulin, J.; Cavin, C.; Bibiloni, R. Impact of coffee consumption on the gut microbiota: A human volunteer study. *Int. J. Food Microbiol.* **2009**, *130*, 117–121. [CrossRef]
10. Mills, C.E.; Tzounis, X.; Oruna-Concha, M.-J.; Mottram, D.S.; Gibson, G.R.; Spencer, J.P.E. In vitro colonic metabolism of coffee and chlorogenic acid results in selective changes in human faecal microbiota growth. *Br. J. Nutr.* **2015**, *113*, 1220–1227. [CrossRef]
11. Salvucci, E. The human-microbiome superorganism and its modulation to restore health. *Int. J. Food Sci. Nutr.* **2019**, *70*, 781–795. [CrossRef] [PubMed]
12. Forsyth, C.B.; Shannon, K.M.; Kordower, J.H.; Voigt, R.M.; Shaikh, M.; Jaglin, J.A.; Estes, J.D.; Dodiya, H.B.; Keshavarzian, A. Increased intestinal permeability correlates with sigmoid mucosa alpha-synuclein staining and endotoxin exposure markers in early Parkinson's disease. *PLoS ONE* **2011**, *6*, e28032. [CrossRef] [PubMed]

13. Nakayama, T.; Oishi, K. Influence of coffee ( Coffea arabica ) and galacto-oligosaccharide consumption on intestinal microbiota and the host responses. *FEMS Microbiol. Lett.* **2013**, *343*, 161–168. [CrossRef] [PubMed]
14. Khokhlova, E.V.; Smeianov, V.V.; Efimov, B.A.; Kafarskaia, L.I.; Pavlova, S.I.; Shkoporov, A.N. Anti-inflammatory properties of intestinal *Bifidobacterium* strains isolated from healthy infants. *Microbiol. Immunol.* **2012**, *56*, 27–39. [CrossRef] [PubMed]
15. Gniechwitz, D.; Reichardt, N.; Blaut, M.; Steinhart, H.; Bunzel, M. Dietary fiber from coffee beverage: Degradation by human fecal microbiota. *J. Agric. Food Chem.* **2007**, *55*, 6989–6996. [CrossRef]
16. Faust, K.; Sathirapongsasuti, J.F.; Izard, J.; Segata, N.; Gevers, D.; Raes, J.; Huttenhower, C. Microbial co-occurrence relationships in the human microbiome. *PLoS Comput. Biol.* **2012**, *8*, e1002606. [CrossRef]
17. Kleber Silveira, A.; Moresco, K.S.; Mautone Gomes, H.; da Silva Morrone, M.; Kich Grun, L.; Pens Gelain, D.; de Mattos Pereira, L.; Giongo, A.; Rodrigues De Oliveira, R.; Fonseca Moreira, J.C. Guarana (Paullinia cupana Mart.) alters gut microbiota and modulates redox status, partially via caffeine in Wistar rats. *Phyther. Res.* **2018**, *32*, 2466–2474. [CrossRef]
18. Yamamoto, T.; Takahashi, H.; Suzuki, K.; Hirano, A.; Kamei, M.; Goto, T.; Takahashi, N.; Kawada, T. Theobromine enhances absorption of cacao polyphenol in rats. *Biosci. Biotechnol. Biochem.* **2014**, *78*, 2059–2063. [CrossRef]
19. Cuervo, A.; Valdés, L.; Salazar, N.; de los Reyes-Gavilán, C.G.; Ruas-Madiedo, P.; Gueimonde, M.; González, S. Pilot study of diet and microbiota: Interactive associations of fibers and polyphenols with human intestinal bacteria. *J. Agric. Food Chem.* **2014**, *62*, 5330–5336. [CrossRef]
20. Cuervo, A.; de los Reyes-Gavilán, C.G.; Ruas-Madiedo, P.; Lopez, P.; Suarez, A.; Gueimonde, M.; González, S. Red wine consumption is associated with fecal microbiota and malondialdehyde in a human population. *J. Am. Coll. Nutr.* **2015**, *34*, 135–141. [CrossRef]
21. Centro de Ensenanza Superior de Nutricion Humana y Dietetica (CESNID). *Tablas de Composicion de Alimentos por Medidas Caseras de Consumo Habitual en España*; McGrawHill, Ed.; Publicaciones y Ediciones de la Universidad de Barcelona: Barcelona, Spain, 2008.
22. U.S. Department of Agriculture Agricultural Research Service. FoodData Central. Available online: fdc.nal.usda.gov (accessed on 12 November 2019).
23. Neveu, V.; Perez-Jimenez, J.; Vos, F.; Crespy, V.; du Chaffaut, L.; Mennen, L.; Knox, C.; Eisner, R.; Cruz, J.; Wishart, D.; et al. Phenol-Explorer: An online comprehensive database on polyphenol contents in foods. *Database* **2010**, *2010*, bap024. [CrossRef] [PubMed]
24. Marlett, J.A.; Cheung, T.-F. Database and quick methods of assessing typical dietary fiber intakes using data for 228 commonly consumed foods. *J. Am. Diet. Assoc.* **1997**, *97*, 1139–1151. [CrossRef]
25. Fernández-Navarro, T.; Díaz, I.; Gutiérrez-Díaz, I.; Rodríguez-Carrio, J.; Suárez, A.; de los Reyes-Gavilán, C.G.; Gueimonde, M.; Salazar, N.; González, S. Exploring the interactions between serum free fatty acids and fecal microbiota in obesity through a machine learning algorithm. *Food Res. Int.* **2019**, *121*, 533–541. [CrossRef] [PubMed]
26. Gérard-Monnier, D.; Erdelmeier, I.; Régnard, K.; Moze-Henry, N.; Yadan, J.-C.; Chaudière, J. Reactions of 1-Methyl-2-phenylindole with Malondialdehyde and 4-Hydroxyalkenals. Analytical applications to a colorimetric assay of lipid peroxidation. *Chem. Res. Toxicol.* **1998**, *11*, 1176–1183. [CrossRef]
27. Nogacka, A.M.; Salazar, N.; Arboleya, S.; Ruas-Madiedo, P.; Mancabelli, L.; Suarez, A.; Martinez-Faedo, C.; Ventura, M.; Tochio, T.; Hirano, K.; et al. In vitro evaluation of different prebiotics on the modulation of gut microbiota composition and function in morbid obese and normal-weight subjects. *Int. J. Mol. Sci.* **2020**, *21*, 906. [CrossRef]
28. Valdes, L.; Salazar, N.; Gonzalez, S.; Arboleya, S.; Rios-Covian, D.; Genoves, S.; Ramon, D.; los Reyes-Gavilán, C.G.; Ruas-Madiedo, P.; Gueimonde, M. Selections of potential probiotic bifidobacteria and prebiotics for elderly by using in vitro fecal batch cultures. *Eur. Food. Res. Technol.* **2017**, *243*, 157–185. [CrossRef]
29. Salazar, N.; Gueimonde, M.; Hernandez-Barranco, A.M.; Ruas-Madiedo, P.; de los Reyes-Gavilan, C.G. Exopolysaccharides produced by intestinal bifidobacterium strains act as fermentable substrates for human intestinal bacteria. *Appl. Environ. Microbiol.* **2008**, *74*, 4737–4745. [CrossRef]
30. Rothwell, J.A.; Keski-Rahkonen, P.; Robinot, N.; Assi, N.; Casagrande, C.; Jenab, M.; Ferrari, P.; Boutron-Ruault, M.; Mahamat-Saleh, Y.; Mancini, F.R.; et al. A metabolomic study of biomarkers of habitual coffee intake in four european countries. *Mol. Nutr. Food Res.* **2019**, *63*, 1900659. [CrossRef]

31. Chrysant, S.G. The impact of coffee consumption on blood pressure, cardiovascular disease and diabetes mellitus. *Expert Rev. Cardiovasc. Ther.* **2017**, *15*, 151–156. [CrossRef]
32. Grosso, G.; Micek, A.; Godos, J.; Pajak, A.; Sciacca, S.; Bes-Rastrollo, M.; Galvano, F.; Martinez-Gonzalez, M. Long-term coffee consumption is associated with decreased incidence of new-onset hypertension: A dose–response meta-analysis. *Nutrients* **2017**, *9*, 890. [CrossRef]
33. Santos, R.M.M.; Lima, D.R.A. Coffee consumption, obesity and type 2 diabetes: A mini-review. *Eur. J. Nutr.* **2016**, *55*, 1345–1358. [CrossRef] [PubMed]
34. Ley, R.E.; Backhed, F.; Turnbaugh, P.; Lozupone, C.A.; Knight, R.D.; Gordon, J.I. Obesity alters gut microbial ecology. *Proc. Natl. Acad. Sci. USA* **2005**, *102*, 11070–11075. [CrossRef] [PubMed]
35. Ley, R.E.; Turnbaugh, P.J.; Klein, S.; Gordon, J.I. Human gut microbes associated with obesity. *Nature* **2006**, *444*, 1022–1023. [CrossRef] [PubMed]
36. Bhandarkar, N.S.; Mouatt, P.; Goncalves, P.; Thomas, T.; Brown, L.; Panchal, S.K. Modulation of gut microbiota by spent coffee grounds attenuates diet-induced metabolic syndrome in rats. *FASEB J.* **2020**, *34*, 4783–4797. [CrossRef] [PubMed]
37. Williamson, G.; Clifford, M.N. Colonic metabolites of berry polyphenols: The missing link to biological activity? *Br. J. Nutr.* **2010**, *104*, S48–S66. [CrossRef]
38. Aura, A.-M.; Martin-Lopez, P.; O'Leary, K.A.; Williamson, G.; Oksman-Caldentey, K.-M.; Poutanen, K.; Santos-Buelga, C. In vitro metabolism of anthocyanins by human gut microflora. *Eur. J. Nutr.* **2005**, *44*, 133–142. [CrossRef]
39. Aura, A.-M. Microbial metabolism of dietary phenolic compounds in the colon. *Phytochem. Rev.* **2008**, *7*, 407–429. [CrossRef]
40. Couteau, D.; McCartney, A.L.; Gibson, G.R.; Williamson, G.; Faulds, C.B. Isolation and characterization of human colonic bacteria able to hydrolyse chlorogenic acid. *J. Appl. Microbiol.* **2001**, *90*, 873–881. [CrossRef]
41. Godos, J.; Pluchinotta, F.R.; Marventano, S.; Buscemi, S.; Li Volti, G.; Galvano, F.; Grosso, G. Coffee components and cardiovascular risk: Beneficial and detrimental effects. *Int. J. Food Sci. Nutr.* **2014**, *65*, 925–936. [CrossRef]
42. Martini, D.; Del Bo', C.; Tassotti, M.; Riso, P.; Del Rio, D.; Brighenti, F.; Porrini, M. Coffee consumption and oxidative stress: A review of human intervention studies. *Molecules* **2016**, *21*, 979. [CrossRef]
43. Moreira, A.S.P.; Nunes, F.M.; Domingues, M.R.; Coimbra, M.A. Coffee melanoidins: Structures, mechanisms of formation and potential health impacts. *Food Funct.* **2012**, *3*, 903. [CrossRef] [PubMed]
44. Mulak, A. Brain-gut-microbiota axis in Parkinson's disease. *World J. Gastroenterol.* **2015**, *21*, 10609. [CrossRef] [PubMed]
45. Cowan, T.E.; Palmnäs, M.S.A.; Yang, J.; Bomhof, M.R.; Ardell, K.L.; Reimer, R.A.; Vogel, H.J.; Shearer, J. Chronic coffee consumption in the diet-induced obese rat: Impact on gut microbiota and serum metabolomics. *J. Nutr. Biochem.* **2014**, *25*, 489–495. [CrossRef] [PubMed]
46. Baek, J.-H.; Kim, N.-J.; Song, J.-K.; Chun, K.-H. Kahweol inhibits lipid accumulation and induces Glucose-uptake through activation of AMP-activated protein kinase (AMPK). *BMB Rep.* **2017**, *50*, 566–571. [CrossRef] [PubMed]
47. Farias-Pereira, R.; Oshiro, J.; Kim, K.-H.; Park, Y. Green coffee bean extract and 5-O-caffeoylquinic acid regulate fat metabolism in caenorhabditis elegans. *J. Funct. Foods* **2018**, *48*, 586–593. [CrossRef]
48. Lally, J.S.V.; Jain, S.S.; Han, X.X.; Snook, L.A.; Glatz, J.F.C.; Luiken, J.J.F.P.; McFarlan, J.; Holloway, G.P.; Bonen, A. Caffeine-stimulated fatty acid oxidation is blunted in CD36 null mice. *Acta Physiol.* **2012**, *205*, 71–81. [CrossRef]
49. Su, S.-H.; Shyu, H.-W.; Yeh, Y.-T.; Chen, K.-M.; Yeh, H.; Su, S.-J. Caffeine inhibits adipogenic differentiation of primary adipose-derived stem cells and bone marrow stromal cells. *Toxicol. In Vitro* **2013**, *27*, 1830–1837. [CrossRef]
50. Topping, D.L.; Clifton, P.M. Short-chain fatty acids and human colonic function: Roles of resistant starch and nonstarch polysaccharides. *Physiol. Rev.* **2001**, *81*, 1031–1064. [CrossRef]

nutrients

MDPI

*Review*

# The Impact of Diet on Microbiota Evolution and Human Health. Is Diet an Adequate Tool for Microbiota Modulation?

**Laura Moles * and David Otaegui**

Multiple Sclerosis Group, Neurosciences Area, Biodonostia Health Research Institute, C. P. 20014 Donostia-San Sebastián, Spain; david.otaegui@biodonostia.org
* Correspondence: laura.moles@biodonostia.org; Tel.: +34-943-00-61-33

Received: 29 April 2020; Accepted: 31 May 2020; Published: 2 June 2020

**Abstract:** The human microbiome is emerging as an interesting field in research into the prevention of health problems and recovery from illness in humans. The complex ecosystem formed by the microbiota is continuously interacting with its host and the environment. Diet could be assumed to be one of the most prominent factors influencing the microbiota composition. Nevertheless, and in spite of numerous strategies proposed to modulate the human microbiota through dietary means, guidelines to achieve this goal have yet to be established. This review assesses the correlation between social and dietary changes over the course of human evolution and the adaptation of the human microbiota to those changes. In addition, it discusses the main dietary strategies for modulating the microbiota and the difficulties of putting them properly into practice.

**Keywords:** gut microbiota; Western diet; chronic disease; prebiotic; probiotic

---

## 1. Introduction

The human gut microbiota is a complex ecosystem formed by thousands of microorganisms that play an important role in human immune and metabolic functions, among others. It is estimated that more than 1000 species and $3 \times 10^{13}$ microbial cells live in or on us, being similar in number to human cells [1–3]. In terms of complexity and richness, the microbiota is even larger considering its genome (the microbiome). Specifically, the human microbiome has at least 100-fold more genes than the human genome; besides this, only 10% of the microbiome is shared between individuals. Therefore, the human microbiome is a unique fingerprint, and its richness and variability may explain its ability to adapt fast to environmental conditions [4–6].

The human microbiome is a relatively new field, but in recent years research into it has been increasing exponentially. The importance of the microbiota was underlined by it starting to be considered an organ in itself [4,5,7–9]. As the so-called "forgotten organ" [10] and considering its wide-ranging interaction with the host, research in this field could contribute to our understanding of many health problems that, so far, have proven difficult to tackle. In this context, the association of the human microbiota with health and disease is being intensively studied, and every day evidence emerges relating dysbiosis in the microbiota to more health problems, including diverse gastrointestinal and neurological disorders such as colitis, obesity, irritable bowel syndrome, Alzheimer's disease, autism, or multiple sclerosis; allergies; and some types of cancer (Figure 1) [11–17].

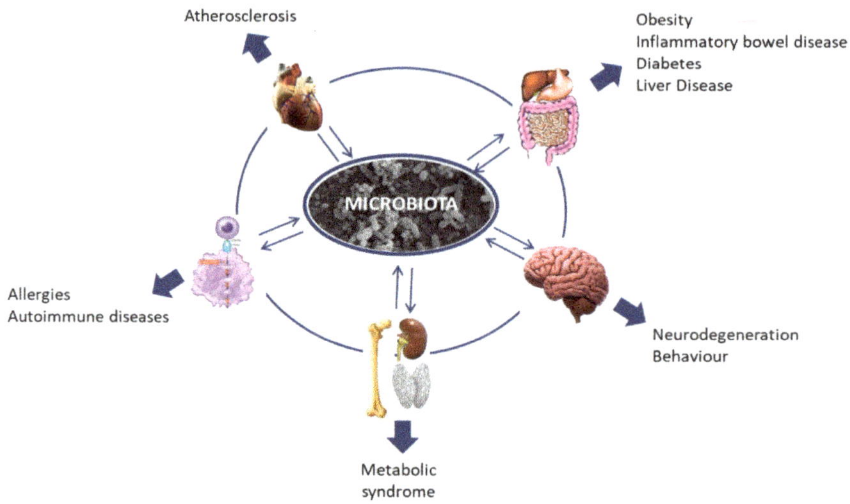


**Figure 1.** Host microbiota interactions and their relationship with disease.

The gut microbiota is especially moldable during infancy and notably stable in adulthood [18]. The limited microbiota present at birth undergoes dramatic changes before reaching the relative equilibrium that is characteristic of adulthood [7–9,19–21]. It is precisely in infancy when factors modulating the microbiota have the most marked influence [8,18,22]. Diet has been recognized as one of the strongest modulators of infant microbiota. In fact, numerous studies have described differences in the gut microbiota of breastfed and formula-fed infants [23,24]. It is believed that once the microbiota reaches an equilibrium (at 2–3 years of life), it is much more difficult to restore and modulate its composition. Once in adulthood, the gut microbiota remains relatively stable but diet continues to determine its composition. Studies carried out evaluating the microbiota associated with different diets in adulthood agree on the dominant presence of *Prevotella* in the gut of vegetarians, while levels of the genus *Bacteroides* and overall levels of the phylum Firmicutes are higher in people following diets high in protein and animal fats [25–27].

Consequently, it seems evident that our diet has the potential to modulate our microbiota. In this context, the aim of this study is to outline the challenges in dietary modulation of the microbiota, reviewing evolutionary changes in both diet and gut microbiota, their potential relationship, and consequences for human health and subsequently examining the strategies available for modulating the microbiota.

## 2. Dietary Changes across Human Evolution

Nutrition is one of the basic needs for a living being to survive and grow. Without a doubt, the human diet has dramatically changed from the time of first hominids to the present, and the changes have been especially fast and marked over the last 100 years [28]. Here, we describe the main features that have characterized these modifications in human diet.

### 2.1. Ancient Diet

The first hominids based their diet on plants gathered and animals hunted in the wild. Though relatively little is known about this time, it is believed that plants were the main foods eaten, while meat was limited to days on which hunting was successful. It is important to highlight that the ratios of plant to animal contributions of the hunter-gatherer diet are still controversial [29,30]; nevertheless, data from the current hunter-gatherer populations around the world suggest the predominance of plant food [22,31–37]. The development of the ability to control fire had a great impact on many aspects of human life. It provided protection from predators and warmth and light and was a determinant factor in the development of cooking. Cooking contributes to food energy accessibility by the efficient denaturing of proteins and starch gelatinization; it also preserves foods for longer periods by substantially reducing foodborne pathogens [38–40].

Another feature that caused marked dietary changes was the domestication of plants and animals. Agriculture allowed the availability of food to, more or less, meet the demand, and is considered a key element in the emergence of community life and civilization [38]. The spread of agriculture had other consequences, however; in particular, it led to a reduction in nutritional intake diversity.

### 2.2. First Civilization's Diet

Populations from the first civilizations were able to produce their own food to meet their energy requirements. Dietary patterns were characterized by the development of the first fermentable foods, such as bread, beer, yoghurt, and wine. Furthermore, for centuries—though there were differences between civilizations and cultures—dietary habits were generally based on the consumption of carbohydrate-dominant foods (such as potato, rice, maize, wheat, and vegetables), probably because these were the most easily accessible [41,42].

Protein intake was primarily from legumes, as proteins of animal origin were only consumed occasionally [38,41–43]. Animal domestication facilitated access to meat and animal-derived products; nevertheless, cattle slaughter was commonly carried out only once or twice a year, and the meat obtained was used to supply whole families. On the other hand, the techniques used to preserve meat and fish were still quite limited; epidemics, famines, and wars marked civilizations for long periods, restricting access to valuable products, including animal-derived foods. Therefore, animal proteins remained a minor component of diets [38] (Figure 2).

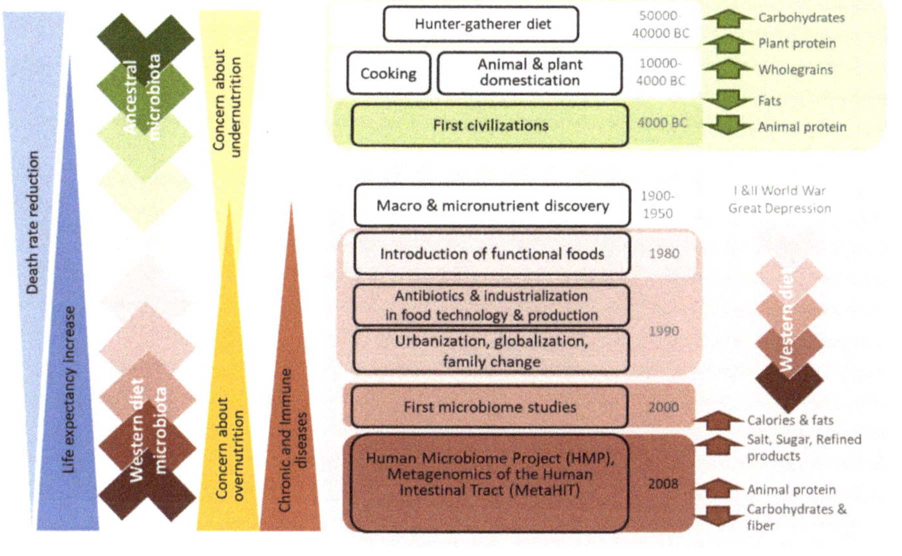


**Figure 2.** Drivers of dietary trends and their relation to microbiota composition and changes in human health.

### 2.3. Modern Diet

Demographic and lifestyle changes, such as urbanization, the abandonment of rural areas, and the increase in women working outside the home, have marked current populations. Through the 1990s, the worldwide growth in the use of antibiotics and industrialization in ranching and agro industries led to a new revolution in food technology and production. Together, these changes had a huge impact on food production and dietary habits [38].

On the one hand, globalization and advances in agriculture have nearly eliminated the seasonality of foods in developed countries [44]. Indeed, food availability is such that individuals have a wide choice of what to eat. On the other hand, the adaptation of consumer behavior to modern lifestyles has led to the demand for safer and longer-lasting food; this, in turn, has driven industry to increase the use of additives and develop new preservation technologies. Cooking has become a secondary concern, as the consumption of and demand for pre-cooked and ready-to-eat products has exponentially increased. These products must be tasty as well as easy to prepare and store; in this context, the addition of fats, sugar, and salt is imperative in meeting these requirements [45]. In addition, competitive markets force producers to use cheap ingredients in processed foods, and these are hardly ever the healthiest ones [46]. All these changes underlie the food industry's transition from the natural products consumed by our ancestors to the processed products currently available, which tend to be high in artificial and added ingredients such as preservatives, colorants, fats, sugar, and salt (Figure 2) [44,47,48].

### 2.4. What the Numbers Say

Global industrialization has facilitated these changes in diet and, notably, similar changes are observed in many countries despite differences in culture, lifestyle, and culinary traditions. According to the annual food balance sheets published by the Food and Agriculture Organization (FAOSTAT database; http://www.fao.org/faostat/en/#home), calorie intake has been increasing over recent decades. From the 1960s to the present (last data from 2013), the world's average energy intake has increased by nearly 500 kcal per capita per day (Figure 3). The origin of this calorie increase is slightly different in developed and developing countries. While the consumption of meat, sugars, and vegetable oils has increased in developing countries, developed countries have seen rises in meat and fat intake [44]. The Food and Agriculture Organization data show that the increase in energy intake in European Union countries is approximately twice that observed in developing countries (Table 1).

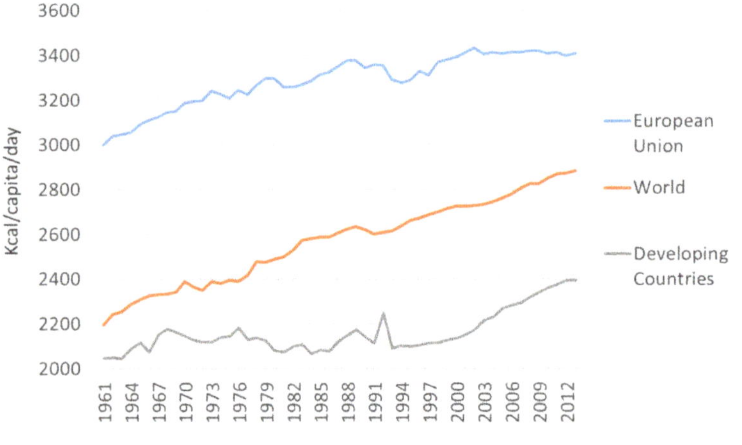


**Figure 3.** Increase in calorie intake in different regions over the last 50 years.

**Table 1.** Energy intakes in the last five decades (kcal/capita/day).

| Area | Item | 1961–1970 | 1971–1980 | 1981–1990 | 1991–2000 | 2001–2010 | Variation from the 60s to 2010 |
|---|---|---|---|---|---|---|---|
| European Union | **TOTAL** | 3094.9 (59.4) | 3241.3 (37) | 3316.9 (46.2) | 3336.5 (41) | 3415.1 (8) | **320.2** |
| | **Plant-based products** | 2223.7 (23.7) | 2257.2 (14.5) | 2272.6 (35.7) | 2321.6 (32.8) | 2407.5 (11.6) | **138.8** |
| | **Animal products** | 871.2 (37.5) | 984.2 (31.2) | 1044.4 (13) | 1014.8 (17.7) | 1007.7 (11.2) | **136.5** |
| | **Sugar and sweeteners** | 339.7 (18.9) | 378.5 (7.8) | 365 (6.1) | 364.3 (2.8) | 373.6 (9.3) | **33.9** |
| | **Vegetable oils** | 273 (18.4) | 322.9 (11.6) | 373.6 (22) | 431.1 (13.2) | 456.3 (15.6) | **183.3** |
| | **Meat** | 282.3 (20.5) | 355.4 (19.6) | 388.2 (12.1) | 393.8 (7.7) | 385.3 (4.8) | **103** |
| | **Animal fats** | 235.4 (6.7) | 242.6 (3.5) | 242.3 (6.5) | 214.6 (6.9) | 196.9 (6.4) | **−38.5** |
| | **Milk** | 270.6 (7.3) | 290.5 (7.3) | 312.2 (6.6) | 305.7 (5.7) | 317.4 (2.3) | **46.8** |
| Developing Countries | Total | 2105.6 (51.3) | 2130.7 (25.8) | 2109.4 (35.7) | 2125 (45.2) | 2262.9 (69.4) | **157.3** |
| | **Plant-based products** | 1914.3 (44.2) | 1933.8 (28.2) | 1918.9 (39.9) | 1886.1 (31.7) | 2011.8 (54.9) | **97.5** |
| | **Animal products** | 191.2 (8.3) | 196.8 (5.7) | 190.4 (11) | 239 (26) | 251.1 (15.1) | **59.9** |
| | **Sugar and sweeteners** | 71.2 (9.9) | 78.9 (4.9) | 90.6 (3.3) | 103.2 (7.8) | 107.8 (5.7) | **36.6** |
| | **Vegetable oils** | 52.7 (4.1) | 66.2 (5.7) | 86.9 (11) | 114.8 (5.9) | 137.5 (12.4) | **84.8** |
| | **Meat** | 79.2 (3.6) | 81.1 (3.5) | 84.1 (3.4) | 100.1 (10.3) | 98.5 (5.3) | **19.3** |
| | **Animal fats** | 27.8 (2.5) | 29.5 (1.2) | 27.2 (2.5) | 28 (5) | 25.6 (1.7) | **−2.2** |
| | **Milk** | 67.4 (1.9) | 66.4 (1.4) | 59.9 (5.2) | 91.7 (12.8) | 106 (6.4) | **38.6** |

Values expressed as mean and standard deviation.

The dietary pattern involving a high intake of saturated fats and sucrose and a low intake of fiber is commonly known as a "Western diet". Diet is one of the strongest modulators of chronic inflammation and Western diets represent a growing health risk, contributing to higher rates of metabolic diseases and inflammation [31,47,48].

## 3. Human Gut Microbiota Evolution

The human microbiota has been structured by its biological interaction with its host, and the resultant ecosystem is the consequence of thousands of years of evolution [49]. Furthermore, microbes have impressive abilities to spread, interact, and adapt to the environment; hence, microbial communities should not be considered in isolation, but rather as part of an interacting community [50]. The knowledge in this relatively new field is still quite limited. In fact, it is difficult to know the extent to which the human microbiome has been shaped by the selective pressure of modern diet, hygiene, antibiotic exposure, built environment, and lifestyle [51]. Despite these limitations, the following paragraphs attempt to outline the current knowledge on gut microbiota composition and its evolution.

### *3.1. Defining "Healthy Microbiota"*

In general, it is accepted that only four bacterial phyla dominate the human microbiota (Firmicutes, Proteobacteria, Actinobacteria, and Bacteroidetes), while others (Chlamydiae, Cyanobacteria, Deferribacteres, Deinococcus–Thermus, Fusobacteria, Spirochaetes, or Verrucomicrobia) may be found at lower abundances [52–55]. Strict anaerobes, mainly represented by members of the phyla Bacteroidetes and Firmicutes, dominate the gut, outnumbering aerobe microorganisms by 100- to 1000-fold [56–60]. Facultative anaerobes account for less than 1% of the microbiota and are mainly represented by the family *Enterobacteriaceae* and the genera *Enterococcus* and *Lactobacillus* [19].

Despite research efforts, there is no consensus on a detailed description of a "normal" or "healthy" microbiota. The enormous complexity and inter-individual variability in the microbiota make this goal very hard to achieve with current tools. Microbial richness (number of species) and diversity (variety and relative abundance of the species in a niche) are global parameters associated with health. Stability has also been considered as a key feature of a healthy microbiota, and this is related to the concepts of resistance (ability of a community to resist change in the context of ecological stress) and resilience (its ability to return to an equilibrium state following a stress-related perturbation).

Nevertheless, the idea that there is an ideal composition of the microbiota seems too simplistic. In fact, it minimizes the importance of microbiota–host interactions, individual genomic differences, and variations in susceptibility to disease, all of which probably play a determining role in shaping the microbiota. An alternative concept consists of characterizing the collection of genes and metabolic pathways provided by the microbiome rather than just the microbiota composition. This approach is probably more appropriate, but also requires a greater in-depth knowledge of the human microbiome [59].

In any case, it should be a priority to reach a consensus on the definition of a healthy microbiota in order to clarify the goal of strategies for microbiota modulation.

### *3.2. Clustering Individuals According to Their Microbiota Composition*

As a step towards defining the composition of a healthy gut microbiota, an interesting publication clustered the fecal microbiota of a healthy cohort into three so-called "enterotypes". Each cluster was characterized by the presence of some highly abundant genera that defined the group and many less abundant genera. Enterotype 1 was enriched in the genus *Bacteroides* and enterotype 2 in *Prevotella*, while enterotype 3 was dominated by *Ruminococcus*. The dominant genera tended to be observed together with other minority ones (*Parabacteroides*, *Desulfovibrio*, and *Akkermansia*, respectively) that, despite their low abundance, performed specialized functions beneficial to the host and are important for defining the enterotype. Though each enterotype preferred certain routes for generating energy, which suggests a specialization to their ecological niches [54], the data available support the idea that the gut microbiota is characterized by a high functional redundancy. In fact, 25% to 43% of the enzymatic functions of the microbiota have been found to be shared, regardless of the enterotype to which the microbiota belonged [21,60].

Some years later, another publication associated these enterotypes with long-term diets. The *Bacteroides* enterotype was strongly associated with a variety of amino acids from animal proteins and saturated fats, and therefore with Western diets. In contrast, the *Prevotella* enterotype was closely associated with carbohydrates and simple sugars, indicating an association with typical diets of agrarian societies [25].

A recent publication uses a metagenomic approach to classify individuals according to the number of gut microbiota-encoding genes as "low gene count" (LGC) or "high gene count" (HGC), depending on whether their microbiota harbor fewer or more than 480,000 genes, respectively. This approach is based on the functionality of the microbiota and its relation with the microbiota composition. The difference in the mean number of encoding genes between groups is notably high, reaching some 40%, and is related to the microbial richness. Broadly, LGC individuals have a less rich microbiota, dominated by *Bacteroides*, *Parabacteroides*, *Ruminococcus*, *Campylobacter*, *Dialister*, *Porphyromonas*, *Staphylococcus*, *Anaerostipes* and most members of the phylum Bacteroidetes. In contrast, the phylum Firmicutes and the genera *Faecalibacterium*, *Bifidobacterium*, *Lactobacillus*, *Butyrivibrio*, *Alistipes*, *Akkermansia*, *Coprococcus*, and *Methanobrevibacter* are associated with HGC individuals [61,62].

### 3.3. Ancient Microbiota

Regarding the changes in our gut ecosystem over the course of human evolution, some studies suggest that these are both pronounced and worrying. The study of the ancient microbiota is not easy due to the low number of available ancestral biological samples. The gut microbiota of ancestral specimens was evaluated in mummies, revealing the predominance of species of the genera *Clostridium* and *Bacteroides* in the larger intestine [63–65]. These studies provided valuable information, however, in addition to the small sampling size, the storage conditions and the possible post-mortem alterations in the bacterial communities should be considered in the interpretation of the results. In this context, current hunter-gatherer populations are also being studied. Research with uncontacted Amerindians who continue to live a seminomadic hunter-gatherer lifestyle revealed that their fecal microbiota is the most diverse ever reported in humans, and the proportion of shared microbiota between them is also much higher than in other human populations [32]. This high microbial biodiversity was also observed in studies carried out in other hunter-gatherer populations, such as the Matses from the Peruvian Amazon [66], the Hadza from Tanzania [33], or indigenous ethnic groups from the Central African Republic [36]. The Ameridians' microbiota seems to be characterized by a high abundance of the phyla Verrucomicrobia and Mollicutes; the families *Aeromonadaceae*, *Oxalobacteraceae*, and *Methanomassiliicoccaceae;* and the genus *Prevotella*, while the abundance of the genus *Bacteroides* is lower [32]. The microbiota of the Matses is characterized by the abundance of the genera *Clostridium*, *Catenibacterium*, *Eubacterium*, *Lachnospira*, and *Treponema* [66]. The Hadza population presented a microbiota enriched in *Prevotella*, *Succinivibrio*, *Treponema*, and *Eubacterium* and impoverished in *Bacteroides*, *Blautia*, and *Dorea* genera [33]. The Central African Republic hunter-gatherer population's microbiota is characterized by the predominance of *Prevotella* and *Treponema* [36]. Despite the differences, it is worth noting that the abundance of *Prevotella*, *Treponema*, and *Eubacterium* and the scarcity of *Bacteroides* in the microbiota of these populations may be common characteristics of the ancestral microbiota.

Metagenomic approaches allow us to analyze the genetic composition and function of complex communities. The application of these tools to the ancient microbiota provide further evidence to support the view that it has a higher functional diversity, characterized by increased metabolic pathways involving amino acid metabolism; glycosyltransferases; and the biosynthesis of lipopolysaccharides, terpenoid-quinones, and vitamins [32]. These findings suggest that not only is microbial diversity being lost, but also some of the functionality of gut microbials. As a consequence, it is not surprising that there is growing interest in ancient microbiome research and recovery [22,51,67].

### 3.4. Western Diet Microbiota and Its Consequences

Evidence suggests that lifestyle changes, including poor diet, urbanization, scarce physical activity, built environment, wide-spread antibiotic exposure, and better hygiene, have impacted the composition of our microbiota and also the emergence of the so-called diseases of modern civilization. These changes are included in the concept of "Westernalization" and contribute to microbiota alteration and disease [68]. Even if all aspects of Western lifestyle should be considered in this process, diet is accepted as one of the most potent ones shaping microbial communities [68–70].

There is a tendency to lose the overall diversity of the gut microbiota in people following Western diets. The gut microbiota composition has also undergone specific changes, characterized by an increase in the abundance of the phylum Firmicutes and the family *Enterobacteriaceae* and a decrease in the phylum Actinobacteria and the genus *Prevotella*. The presence of some bacterial species associated with anti-inflammatory conditions and the capacity to produce beneficial metabolites is diminishing in our guts. These species include *Akkermansia muciniphila, Faecalibacterium prausnitzii, Roseburia spp., Eubacterium hallii, Clostridium clusters XIVa* and *IV,* and *Ruminococcus*, among others. Indeed, some research has revealed the extinction of several bacterial groups from the guts of people following Western diets. It is difficult to assess the significance of that loss, but we are probably witnessing just the beginning of its consequences [12,31,47,71].

Furthermore, the gut microbiome's circadian rhythm is influenced by factors such as light–dark cycles, sunlight exposure, sleep, and dietary patterns; some of them are common stressors of the modern lifestyle [72]. The consumption of food in an undisturbed daily rhythm coincides with the light phase of the light–dark cycle and the activity phase of the day, which has consequences on the regulation of the hosts' intestinal cell transcription, the rhythms of the circulating metabolites, and the gut microbiota composition and function [72,73]. Some studies evidenced exacerbated effects on the gut microbiota of people with the circadian disruption of high sugar and fat diets; these effects were characterized by a drastic reduction in bacterial diversity and the Firmicutes/Bacteroidetes ratio [74,75].

The aforementioned changes in microbiota composition have been associated with a greater tendency to develop inflammation and, in turn, with a higher incidence of obesity; diabetes; allergies; cardiovascular disease; and metabolic, gut, and neurological disorders. Microbiota dysbiosis in these diseases may be involved in the alteration of certain specific microbial groups; nevertheless, in most cases the overall loss of microbial biodiversity is an important factor defining the dysbiosis [76]. The misalignment of the rhythms that control our energy metabolism also increases the risks of suffering diseases such as metabolic syndrome, including type 2 diabetes mellitus and obesity [77,78]. The growing incidence of these diseases in contemporary, industrialized populations over recent decades is believed to be associated, among other factors, with a lack of adaptation of our metabolism to the rapid dietary and lifestyle changes that have occurred over the course of human evolution [22,32,51,79–81].

It is likely that several different factors are contributing to the changes in microbiota composition and the increased prevalence of associated diseases. In any case, the impact of these changes on human health underlines the urgent need to find effective tools to halt this trend.

## 4. Modulation of Human Gut Microbiota with Diet

While it has already been described that geographical localization, culture, and genetic background all affect the microbiota composition, some authors consider that diet is responsible for more than 50% of the variability in the microbiota [82,83]. Even though it is difficult to determine this value accurately, there is evidence that dietary interventions, with significant changes in content, are able to exert modulatory effects on microbiota composition that may be seen within 1–4 days and are strong enough to shift the enterotype [84,85]. Nevertheless, dietary modulatory effects are diluted over time when the diet is discontinued, and there is a tendency to return to the original state [82].

The capacity of microbiota to recover its original status has also been observed after a course of antibiotics. Several studies have evidenced that some weeks after the use of antibiotics (one of the treatments that most dramatically alter the microbiota), the microbiota has nearly completely returned to its original composition, though this recovery is treatment- and age-dependent [86–88]. The frequent use of antibiotics or the requirement for prolonged treatments has a more marked effect on the microbiota composition [89,90]. Similarly, the modulatory effect of probiotics (live microorganisms which, when administered in adequate amounts, confer a health benefit to the host [70]; dead microbes, microbial products, or microbial components do not come under the probiotic classification [91]) is believed to disappear progressively together with the loss of the beneficial strains.

All this evidence supports the idea that treatments to modulate gut microbiota must be maintained over time. In line with this, the use of diet as a modulatory tool could be ideal whenever diet is considered as a long-lasting change in everyday habits.

### *4.1. Strategies for Modulating the Microbiota: Prebiotics*

Possibly the most widely explored strategy for modulating the microbiota is the use of prebiotics. Prebiotics are defined as "substrate that is selectively utilized by host microorganisms conferring a health benefit" [92]—that is, nutrients resistant to gastric acid secretion and digestive enzymes that once in the gut stimulate the growth of beneficial microbes or their activity. Certain dietary components such as inulin, fructooligosaccharides (FOS), galactooligosaccharides (GOS), and resistant starch (RS) have been studied as prebiotics, and their efficacy is commonly indirectly measured by the production of short chain fatty acids (SCFAs) or the decrease in intestinal pH [76,93]. Inulin is a fructan carbohydrate that may vary its polymerization degree and whose fructose chains ranges from 2 to 60 monomers [94]. Inulin stimulates the growth of lactobacilli and bifidobacteria; besides this, an increase in *F. prausnitzii* and *A. muciniphila* populations in the gut has been described and seems to produce early satiety by modulating the gut endocrine function. Nevertheless, it is still difficult to determine the mechanisms underlying these effects [93].

The FOS are oligosaccharides of glucose and fructose that differ from inulin in their polymerization degree that is under 10; whereas, GOS are oligosaccharides of glucose and galactose with a polymerization degree of 2 to 8. As typical prebiotics, FOS and GOS have been used to stimulate the growth of the beneficial bacteria, bifidobacteria and lactobacilli. The administration of FOS in a culture-dependent study resulted in an increase in *Bifidobacterium* and *F. prausnitzii*, while culture-independent studies based on high-throughput sequencing have revealed changes in more than 100 bacterial taxa. The most marked changes in abundance were an increase in *Bifidobacterium;* reductions in the genera *Phascolarctobacterium*, *Enterobacter*, *Turicibacter*, *Coprococcus*, and *Salmonella*; an overall increase in Bacteroidetes; and a decrease in the phylum Firmicutes [95]. Other genera that could be increased by FOS administration are *Lactobacillus* and the butyrate producers *Faecalibacterium*, *Ruminococcus*, and *Oscillospira* [96]. On the other hand, GOS administration resulted in increases in *Bifidobacterium* levels and decreases in the levels of *Ruminococcus*, *Dehalobacterium*, *Synergistes*, and *Holdemania* [95]. The effect of GOS on the gut microbiota could also improve the butyrate production and the presence of butyrate producers, such as *Eubacterium rectale* [97].

RS is defined as the total amount of starch and the products of starch degradation that resists digestion and has been shown to be composed of a linear molecule of α-1, 4-D-glucan, derived from the retrograded amylose fraction. Various classifications have been proposed for RS based on four or five types, and the content of RS in foods is influenced by the physical form of the food, the size and composition of the starch granules (amylose–amylopectin ratio), and the food processing methods and conditions [98,99]. Notably, it has been found that an increment in RS in the diet is associated with colonization by higher levels of the phylum Bacteroidetes and the genera *Bifidobacterium*, *Akkermansia*, and *Allobactum* [98].

### *4.2. Strategies for Modulating the Microbiota: Probiotics*

The field of probiotics—which, as stated above, are live microorganisms which when administered in adequate amounts confer a health benefit to the host [70,91]—has notably grown in recent years. The microorganisms commonly used as probiotics are the yeast *Saccharomyces cerevisiae* and members of the bacterial genera *Lactobacillus* and *Bifidobacterium*, though some formulations may also include some *Streptococcus*, *Enterococcus*, *Pediococcus*, *Propionibacterium*, *Bacillus*, or *Escherichia* strains. Most *Lactobacillus* and *Bifidobacterium* species have been assigned "Generally Recognized As Safe" status by the US Food and Drug Administration and "Qualified presumption of safety" status by the European Food Safety Authority, facilitating their preferential use as probiotics. On the other hand, their long history of use as probiotics means that there is a substantial body of evidence for a wide range of beneficial properties [100], though we must recall that probiotic properties are strain-specific—that is, they are not a characteristic of a species [76].

Nevertheless, it can be expected that, in the near future, other species will be used as probiotics—ones that are more commonly found in the human gut and play important functions in mitigating intestinal inflammation, inducing immune regulation, or enhancing the intestinal barrier function. These are likely to include species with anti-inflammatory properties (*A. muciniphila*, *F. prausnitzii*) and butyrate-producing bacteria [91,101].

Currently, probiotics are used in a wide range of contexts, normally being indicated for healthy people in a special situation (e.g., during infancy, pregnancy, breastfeeding, and old age) and for preventing or treating several specific health problems. As a consequence, the assessment of the safety of probiotics must pay attention not only to the selected strain, manner and frequency of administration, and dose and treatment duration but also to the potential vulnerability of the consumer and the physiological function that the strain may play in the host [102].

Despite the heterogeneity of clinical studies making it difficult to determine the most suitable strains and therapeutic guidelines, there is clear evidence of probiotic effectiveness in the prevention or treatment of diseases such as necrotizing enterocolitis, antibiotic-associated diarrhea, colitis, or acute gastroenteritis [103,104]. Nonetheless, to date there is a lack of evidence of the effectiveness of prebiotics or probiotics in achieving long-term changes in the microbiota.

### *4.3. Strategies for Modulating the Microbiota: Controlling the Gut Environment*

The concept of gut environment modulation considers not only the composition of the microbiota but also its function and interaction with the host. Nowadays, it is known that fat-rich diets enhance bile secretion to facilitate lipid digestion. Bile acids have antimicrobial properties; their detergent effect produces damage to bacterial membranes and exerts a strong selective pressure on microbiota composition. Studies carried out in rats reveal the strong resistance of the phylum Firmicutes to bile acids, in particular, the classes *Clostridia* and *Erysipelotrichia* and the family *Enterobacteriaceae* [105]. These findings are in agreement with the predominant presence of Firmicutes in people following Western diets [79].

On the other hand, intestinal pH may oscillate between 5 and 7 under normal physiological conditions, depending on the fermentation products present (such as SCFAs) and the metabolite absorption by host epithelial cells and their level of bicarbonate secretion. Intestinal pH affects not only the microbiota composition but also its metabolism. Some Firmicutes species, especially those belonging to *Clostridium cluster XIVa*, are tolerant of a low pH; however, many Bacteroidetes and Actinobacteria members are more sensitive to pH changes [12,106].

The Mediterranean diet is associated with a higher production of SCFAs in the gut [107,108]. These substances play important roles maintaining the integrity of the large bowel and small intestinal barrier, providing energy to epithelial cells and reducing inflammation [108] and support higher microbial richness in the gut [107,109].

The modulation of the gut microbiota through the induction of environmental changes has been less explored; but the gut environment remains closely related to dietary pattern and therefore should not be disregarded in future dietary interventions.

### 4.4. Challenges in Microbiota Modulation: Interindividual Variability

The complexity of gut microbiota modulation lies in interpersonal variability not only in the microbiota composition and functioning but also in differences in lifestyle and genetic predisposition to disease. It has been described that changes in fiber consumption lead to various changes in the gut microbiota. In general, the abundance of the family *Lachnospiraceae* increases with the incorporation of insoluble fiber or wheat bran to the diet, while enrichment in RS causes an increase in *Ruminococcaceae*, but changes are individual specific [62]. Along the same lines, a recent study revealed that changes in oral glucose tolerance and the microbiota induced by a given prebiotic intervention vary between individuals. Notably, a close positive correlation was found between glucose tolerance and the presence of some butyrate-producing bacteria [95].

These data indicate the enormous importance of knowledge of the microbiota composition to predict the effects of modulatory treatments. The available data suggest that parameters such as age, diet, and lifestyle, body mass composition, and the presence of specific diseases could help to define some key features of the microbiota composition, but treatments to modulate the microbiota are still based on estimates and it is likely that some effects are being masked.

### 4.5. Challenges in Microbiota Modulation: Microbial Metabolic Redundancy

To estimate the effects of a modulating treatment, in addition to microbiota composition there is a need to consider metabolic redundancy. The use of next-generation sequencing and metagenomics is greatly helping to advance our understanding of this factor. In this context, recent publications describe the metabolic capacity of some bacterial groups. It seems that members of the genus *Bacteroides* have a wide metabolic arsenal allowing them to utilize different polysaccharides, whereas Firmicutes members have less metabolic diversity for polysaccharide degradation and greater nutritional specialization [62].

Metabolic analyses of LGC and HGC individuals suggest a great capacity to handle oxidative stress but also a higher production of detrimental metabolites and a predisposition to inflammation in LGC individuals. On the other hand, HGC individuals' microbiota has a greater capacity to produce organic acids, including SCFAs. Further, it seems that there is a higher incidence of obesity and metabolic syndrome in LGC individuals and dietary interventions for weight loss increase microbiota diversity and the overall gene counts [61,62,110].

In accordance with this, the analysis of the metabolic specialization of the enterotypes indicated that individuals with the *Bacteroides* enterotype are better able to digest lipids, proteins and carbohydrates of animal origin, while the *Prevotella* enterotype showed a plant fiber hydrolysis specialization. In contrast, the *Ruminococcus* enterotype has no such marked specialization, but has been linked to a higher microbiota diversity and a lower inflammatory status in the host [84].

The aforementioned results suggest a healthy microbiota should be closer to that of HGC individuals and the *Ruminococcus* enterotype. Nonetheless, no consensus has yet been reached on the definition of a healthy microbiota. In fact, all enterotypes have been associated with some types of disease and it is believed that each enterotype has a different susceptibility to given illnesses. In any case, a greater knowledge of individuals' microbiota could be a good tool to guide treatment decisions and estimate patient response.

**Table 2.** Microorganisms that produce short chain fatty acids and other fermentation products.

| Microorganism | | Short Chain Fatty Acids and Other Fermentation Products | | | | | | | |
|---|---|---|---|---|---|---|---|---|---|
| **Phyla** | **Species** | **Acetate** | **Propionate** | **Butyrate** | **Ethanol** | **Formate** | **Lactate** | **Butanol** | **Succinate** |
| Firmicutes | *Eubacterium rectale* | | | X | | X | X | | |
| | *Roseburia inulinivorans* | | X | X | | X | X | | |
| | *Eubacterium hallii* | | | X | | X | | X | |
| | *Anaerostipes hadrus* | | | X | | | | | |
| | *Coprococcus catus* | | X | X | | | | | |
| | *Ruminococcus obeum* | X | | | | | X | | |
| | *Blautia wexlerae* | X | | | | | | | X |
| | *Faecalibacterium prausnitzii* | | | X | | X | X | | |
| | *Ruminococcus bromii* | X | | | X | | | | |
| Bacteroidetes | *Bacteroides thetaiotaomicron* | X | X | | | | | | X |
| | *Bacteroides vulgatus* | X | X | | | | | | X |
| Actinobacteria | *Bifidobacterium adolescentis* | X | | | | X | X | | |
| | *Collinsella aerofaciens* | X | | | | | X | | |

Adapted from "Links between diet, gut microbiota composition and gut metabolism" by Flint HJ et al. [60].

An alternative more affordable approach to exploring the microbiome function without resorting to metagenomics is the measurement of SCFA production. These metabolites are considered important for their influence on intestinal homeostasis, pH, and preventing the growth of potentially pathogenic bacteria. Moreover, SCFAs have anti-inflammatory and anti-apoptotic properties, contribute to the gut barrier integrity, and are key elements of the gut–brain axis. These fatty acids seem to be necessary for the proper maturation and functioning of microglia, also known as brain macrophages, that participate in brain physiology and homeostasis have phagocytic capacity and also participate in the integrity of the blood–brain barrier [53,111–113].

Microbiota-derived metabolites such as SCFAs depend directly on diet; mainly on its content of non-digestible carbohydrates, lipids, and proteins and the metabolic pathways available in the microbiome [12,62]. Some common microorganisms of the gut microbiota and the SCFAs that they produce are listed in Table 2. Individual dietary choices influence not only gut microbiota composition but its function, but we still lack an understanding of how the microbiome interacts with nutritional and host-genomic axes to confer predisposition to disease.

## 5. Conclusions

Diet and lifestyle habits have undergone dramatic changes since the origin of humanity. The first hominid's diet was based on the consumption of raw vegetables collected from the wild and a low intake of protein of animal origin. After thousands of years of evolution, the modern diet, influenced by globalization and consumerism, is characterized by an excessive lipid and energy intake and the introduction of processed and refined foods that are rich in lipids, sugars, salt, and preservatives. These foods have facilitated the accelerated pace of today's life, but it is likely that their negative consequences for human health are just beginning to be noticed.

Despite the relative youth of microbiome research, microbiota dysbiosis has already been associated with diverse health problems. Numerous international projects are focusing their research efforts on the study of human microbiome in different contexts, including the Human Microbiome Project (HMP: https://www.hmpdacc.org/hmp/), the International Cancer Microbiome Consortium (ICMC: https://www.icmconsortium.org/), the International Multiple Sclerosis Microbiome Study (iMSMS: http://imsms.org/home/), and the Inflammatory Arthritis Microbiome Consortium (IAMC); more are likely to be launched in the near future. Diet has been identified as one of the factors with the greatest influence on microbiota acquisition (in newborns) and modulation. Therefore, detailed studies of the complex interactions that occur between diet and microbiota are necessary to effectively direct desirable changes in human microbiota or alter its abnormal composition in disease. So far, making simple predictions of dietary effects remains extremely difficult.

Microorganisms, especially prokaryotes, are characterized by their amazing capacity to adapt to the environment, as evidenced by their ability to remain stable in an "altered microbiota". Human beings seem to be much less efficient at environmental adaptation. Certainly, the growing rate of diseases related to compromised immune or nervous system function and alterations in metabolism may be a response to a lack of adaptation to microbiota dysbiosis.

Most of the strategies designed to modulate the microbiota are based on one-off treatments that could be effective during their administration, but do not seem to produce stable changes in the microbiota, maybe with the exception of a fecal transplant (although there is still challenges with this practice [114]). In this context, diet will probably be the most powerful tool for microbiota modulation, but to achieve that a better understanding of diet–host–microbiota interactions is necessary. Learning how to use diet to generate a healthy microbiota must be a priority for society, and arguably represents one of the key steps to achieving real preventive and personalized medicine.

**Author Contributions:** Both authors contributed equally to this manuscript. All authors have read and agreed to the published version of the manuscript.

**Funding:** This research received no external funding

**Acknowledgments:** We would like to acknowledge the BIOEF (Basque Foundation for Health Innovation and Research) language service for the thorough revision of the text.

**Conflicts of Interest:** The authors declare that the research was conducted in the absence of any commercial or financial relationships that could be construed as a potential conflict of interest.

## References

1. Zoetendal, E.G.; Rajilic-Stojanovic, M.; De Vos, W.M. High-throughput diversity and functionality analysis of the gastrointestinal tract microbiota. *Gut* **2008**, *57*, 1605–1615. [CrossRef] [PubMed]
2. Pasolli, E.; Asnicar, F.; Manara, S.; Zolfo, M.; Karcher, N.; Armanini, F.; Beghini, F.; Manghi, P.; Tett, A.; Ghensi, P.; et al. Extensive Unexplored Human Microbiome Diversity Revealed by Over 150,000 Genomes From Metagenomes Spanning Age, Geography, and Lifestyle. *Cell* **2019**, *176*, 649–662. [CrossRef] [PubMed]
3. Sender, R.; Fuchs, S.; Milo, R. Revised Estimates for the Number of Human and Bacteria Cells in the Body. *PLoS Biol.* **2016**, *14*, e1002533. [CrossRef] [PubMed]
4. Macpherson, A.J.; Harris, N.L. Interactions between commensal intestinal bacteria and the immune system. *Nat. Rev. Immunol.* **2004**, *4*, 478–485. [CrossRef] [PubMed]
5. Schéle, E.; Grahnemo, L.; Anesten, F.; Hallén, A.; Bäckhed, F.; Jansson, J.O. Regulation of body fat mass by the gut microbiota: Possible mediation by the brain. *Peptides* **2016**, *77*, 54–59. [CrossRef] [PubMed]
6. Turnbaugh, P.J.; Ley, R.E.; Hamady, M.; Fraser, C.M.; Knight, R.; Gordon, J.I. The Human Microbiome Project. *Nature* **2007**, *449*, 804–810. [CrossRef] [PubMed]
7. Matamoros, S.; Gras-Leguen, C.; Le Vacon, F.; Potel, G.; De La Cochetiere, M.F. Development of intestinal microbiota in infants and its impact on health. *Trends Microbiol.* **2013**, *21*, 167–173. [CrossRef]
8. Scholtens, P.A.M.J.; Oozeer, R.; Martín, R.; Ben Amor, K.; Knol, J. The Early Settlers: Intestinal Microbiology in Early Life. *Annu. Rev. Food Sci. Technol.* **2012**, *3*, 425–447. [CrossRef]
9. Sharon, I.; Morowitz, M.J.; Thomas, B.C.; Costello, E.K.; Relman, D.A.; Banfield, J.F. Time series community genomics analysis reveals rapid shifts in bacterial species, strains, and phage during infant gut colonization. *Genome Res.* **2013**, *23*, 111–120. [CrossRef]
10. O'Hara, A.M.; Shanahan, F. The Gut Flora as a Forgotten Organ. *EMBO Rep.* **2006**, *7*, 688–693. [CrossRef]
11. Gopalakrishnan, V.; Helmink, B.A.; Spencer, C.N.; Reuben, A.; Wargo, J.A. The Influence of the Gut Microbiome on Cancer, Immunity, and Cancer Immunotherapy. *Cancer Cell* **2018**, *33*, 570–580. [CrossRef] [PubMed]
12. Jeffery, I.B.; O'Toole, P.W. Diet-microbiota interactions and their implications for healthy living. *Nutrients* **2013**, *5*, 234–252. [CrossRef] [PubMed]
13. Kang, C.S.; Ban, M.; Choi, E.J.; Moon, H.G.; Jeon, J.S.; Kim, D.K.; Park, S.K.; Jeon, S.G.; Roh, T.Y.; Myung, S.J.; et al. Extracellular Vesicles Derived from Gut Microbiota, Especially *Akkermansia muciniphila*, Protect the Progression of Dextran Sulfate Sodium-Induced Colitis. *PLoS ONE* **2013**, *8*, e76520. [CrossRef] [PubMed]
14. Pisa, D.; Alonso, R.; Fernández-Fernández, A.M.; Rábano, A.; Carrasco, L. Polymicrobial Infections in Brain Tissue from Alzheimer's Disease Patients. *Sci. Rep.* **2017**, *7*, 1–14. [CrossRef] [PubMed]
15. Rea, D.; Coppola, G.; Palma, G.; Barbieri, A.; Luciano, A.; Del Prete, P.; Rossetti, S.; Berretta, M.; Facchini, G.; Perdoná, S.; et al. Microbiota effects on cancer: From risks to therapies. *Oncotarget* **2018**, *9*, 17915–17927. [CrossRef] [PubMed]
16. Xu, M.; Xu, X.; Li, J.; Li, F. Association Between Gut Microbiota and Autism Spectrum Disorder: A Systematic Review and Meta-Analysis. *Front. Psychiatry* **2019**, *10*, 473. [CrossRef]
17. Hanski, I.; Von Hertzen, L.; Fyhrquist, N.T.; Koskinen, K.; Torppa, K.A.; Laatikainen, T.; Karisola, P.; Auvinen, P.; Paulin, L.; Mäkelä, M.J.; et al. Environmental biodiversity, human microbiota, and allergy are interrelated. *Proc. Natl. Acad. Sci. USA* **2012**, *109*, 8334–8339. [CrossRef]
18. Ottman, N.; Smidt, H.; de Vos, W.M.; Belzer, C. The function of our microbiota: Who is out there and what do they do? *Front. Cell. Infect. Microbiol.* **2012**, *2*, 1–11. [CrossRef]
19. Adlerberth, I.; Wold, A.E. Establishment of the gut microbiota in Western infants. *Acta Paediatr. Int. J. Paediatr.* **2009**, *98*, 229–238. [CrossRef]
20. Sakata, H.; Yoshioka, H.; Fujita, K. Development of the intestinal flora in very low birth weight infants compared to normal full-term newborns. *Eur. J. Pediatr.* **1985**, *144*, 186–190. [CrossRef]

21. Vael, C.; Desager, K. The importance of the development of the intestinal microbiota in infancy. *Curr. Opin. Pediatr.* **2009**, *21*, 794–800. [CrossRef] [PubMed]
22. Yatsunenko, T.; Rey, F.E.; Manary, M.J.; Trehan, I.; Dominguez-Bello, M.G.; Contreras, M.; Magris, M.; Hidalgo, G.; Baldassano, R.N.; Anokhin, A.P.; et al. Human gut microbiome viewed across age and geography. *Nature* **2012**, *486*, 222–227. [CrossRef] [PubMed]
23. Wang, M.; Li, M.; Wu, S.; Lebrilla, C.B.; Chapkin, R.S.; Ivanov, I.; Donovan, S.M. Fecal Microbiota Composition of Breast-fed Infants is Correlated with Human Milk Oligosaccharides Consumed. *J. Pediatr. Gastroenterol. Nutr.* **2016**, *60*, 825–833. [CrossRef] [PubMed]
24. Jin, Y.; Chen, K.; Wang, Z.; Wang, Y.; Liu, J.; Lin, L.; Shao, Y.; Gao, L.; Yin, H.; Cui, C.; et al. DNA in serum extracellular vesicles is stable under different storage conditions. *BMC Cancer* **2016**, *16*, 1–9. [CrossRef]
25. Wu, G.D.; Chen, J.; Hoffmann, C.; Bittinger, K.; Chen, Y.Y.; Keilbaugh, S.A.; Bewtra, M.; Knights, D.; Walters, W.A.; Knight, R.; et al. Life long term dietary patterns with gut microbial enterotypes. *Science* **2011**, *334*, 105–108. [CrossRef]
26. Ruengsomwong, S.; Korenori, Y.; Sakamoto, N.; Wannissorn, B.; Nakayama, J.; Nitisinprasert, S. Senior thai fecal microbiota comparison between vegetarians and non-vegetarians using PCR-DGGE and real-time PCR. *J. Microbiol. Biotechnol.* **2014**, *24*, 1026–1033. [CrossRef]
27. Glick-Bauer, M.; Yeh, M.C. The health advantage of a vegan diet: Exploring the gut microbiota connection. *Nutrients* **2014**, *6*, 4822–4838. [CrossRef]
28. Cordain, L.; Eaton, S.B.; Sebastian, A.; Mann, N.; Lindeberg, S.; Watkins, B.; Okeefe, J.; Brand-Miller, J. Origins and Evolution of the Western Diet: Health Implications for the 21st Century. *Am. J. Clin. Nutr.* **2005**, *81*, 341–354. [CrossRef]
29. Cordain, L.; Miller, J.B.; Eaton, S.B.; Mann, N.; Holt, S.H.A.; Speth, J.D. Plant-animal subsistence ratios and macronutrient energy estimations in worldwide hunter-gatherer diets. *Am. J. Clin. Nutr.* **2000**, *71*, 682–692. [CrossRef]
30. Milton, K. Hunter-gatherer diets—A different perspective. *Am. J. Clin. Nutr.* **2000**, *71*, 665–667. [CrossRef]
31. De Filippo, C.; Cavalieri, D.; Di Paola, M.; Ramazzotti, M.; Poullet, J.B.; Massart, S.; Collini, S.; Pieraccini, G.; Lionetti, P. Impact of diet in shaping gut microbiota revealed by a comparative study in children from Europe and rural Africa. *Proc. Natl. Acad. Sci. USA* **2010**, *107*, 14691–14696. [CrossRef] [PubMed]
32. Clemente, J.C.; Pehrsson, E.; Blaser, M.J.; Sandhu, K.; Gao, Z.; Wang, B.; Magris, M.; Hidalgo, G.; Contreras, M.; Noya-Alarcón, O.; et al. The microbiome of uncontacted Amerindians. *Sci. Adv.* **2015**, *1*, e1500183. [CrossRef] [PubMed]
33. Schnorr, S.L.; Candela, M.; Rampelli, S.; Centanni, M.; Consolandi, C.; Basaglia, G.; Turroni, S.; Biagi, E.; Peano, C.; Severgnini, M.; et al. Gut microbiome of the Hadza hunter-gatherers. *Nat. Commun.* **2014**, *15*, 3654. [CrossRef] [PubMed]
34. Rampelli, S.; Schnorr, S.L.; Consolandi, C.; Turroni, S.; Severgnini, M.; Peano, C.; Brigidi, P.; Crittenden, A.N.; Henry, A.; Candela, M. Metagenome Sequencing of the Hadza Hunter-Gatherer Gut Microbiota. *Curr. Biol.* **2015**, *25*, 1682–1693. [CrossRef]
35. Morton, E.R.; Lynch, J.; Froment, A.; Lafosse, S.; Heyer, E.; Przeworski, M.; Blekhman, R.; Segurel, L. Variation in Rural African Gut Microbiota Is Strongly Correlated with Colonization by *Entamoeba* and Subsistence. *PLoS Genet.* **2015**, *11*, e1005658. [CrossRef]
36. Gomez, A.; Petrzelkova, K.J.; Burns, M.B.; Yeoman, C.J.; Amato, K.R.; Vlčková, K.; Modry, D.; Todd, A.; Robinson, C.A.J.; Remis, M.J.; et al. Gut Microbiome of Coexisting BaAka Pygmies and Bantu Reflects Gradients of Traditional Subsistence Patterns. *Cell Rep.* **2016**, *14*, 2142–2153. [CrossRef]
37. Ou, J.; Carbonero, F.; Zoetendal, E.G.; Delany, J.P.; Wang, M.; Newton, K.; Gaskins, H.R.; O'Keefe, S.J.D. Diet, microbiota, and microbial metabolites in colon cancer risk in rural Africans and African Americans. *Am. J. Clin. Nutr.* **2013**, *98*, 111–120. [CrossRef]
38. Skolnik, H. History, Evolution, and Status of Agriculture and Food Science and Technology. *J. Chem. Doc.* **1968**, *8*, 95–98. [CrossRef]
39. Carmody, R.N.; Wrangham, R.W. The energetic significance of cooking. *J. Hum. Evol.* **2009**, *57*, 379–391. [CrossRef]
40. Luca, F.; Perry, G.; Di Rienzo, A. Evolutinary Adaptations to Dietary Changes. *Annu. Rev. Nutr.* **2010**, *30*, 291–314. [CrossRef]

41. Arnoni, Y.; Berry, E.M. On the Origins and Evolution of the Mediterranean Diet. The Mediterranean Diet: An Evidence-Based Approach. *Mediterr. Diet* **2014**, 3–11. [CrossRef]
42. Altomare, R.; Cacciabaudo, F.; Damiano, G.; Palumbo, V.D.; Gioviale, M.C.; Bellavia, M.; Tomasello, G.; Lo Monte, A.I. The mediterranean diet: A history of health. *Iran. J. Public Health* **2013**, *42*, 449–457. [PubMed]
43. Underwood, B.A.; Galal, O. Human Nutrition: An Overview. In *Encyclopedia of Life Support Systems*; Eolss Publishers Co. UK: Oxford, UK, 2000.
44. Kearney, J. Food consumption trends and drivers. *Philos. Trans. R. Soc. B Boil. Sci.* **2010**, *365*, 2793–2807. [CrossRef]
45. Fanzo, J. The role of farming and rural development as central to our diets. *Physiol. Behav.* **2018**, *193*, 291–297. [CrossRef] [PubMed]
46. Monteiro, C.A.; Moubarac, J.C.; Cannon, G.; Ng, S.W.; Popkin, B. Ultra-processed products are becoming dominant in the global food system. *Obes. Rev.* **2013**, *14 (Suppl. S2)* (Suppl. S2), 21–28. [CrossRef]
47. Statovci, D.; Aguilera, M.; MacSharry, J.; Melgar, S. The impact of western diet and nutrients on the microbiota and immune response at mucosal interfaces. *Front. Immunol.* **2017**, *8*, 838. [CrossRef] [PubMed]
48. Zinöcker, M.K.; Lindseth, I.A. The Western Diet-Microbiome-Host Interaction and Its Role in Metabolic Disease. *Nutrients* **2018**, *10*, 365. [CrossRef]
49. Ley, R.E.; Peterson, D.A.; Gordon, J.I. Ecological and evolutionary forces shaping microbial diversity in the human intestine. *Cell* **2006**, *124*, 837–848. [CrossRef]
50. Gonzalez, A.; Clemente, J.C.; Shade, A.; Metcalf, J.L.; Song, S.; Prithiviraj, B.; Palmer, B.E.; Knight, R. Our microbial selves: What ecology can teach us. *EMBO Rep.* **2011**, *12*, 755–784. [CrossRef]
51. Warinner, C.; Speller, C.; Collins, M.J.; Lewis, C.M., Jr. Ancient human microbiomes. *J. Hum. Evol.* **2015**, 125–136. [CrossRef]
52. Dethlefsen, L.; McFall-Ngai, M.; Relman, D.A. An ecological and evolutionary perspective on humang-microbe mutualism and disease. *Nature* **2007**, *449*, 811–818. [CrossRef] [PubMed]
53. El Aidy, S.; Dinan, T.G.; Cryan, J.F. Gut Microbiota: The Conductor in the Orchestra of Immune-Neuroendocrine Communication. *Clin. Ther.* **2015**, *37*, 954–967. [CrossRef]
54. Arumugam, M.; Raes, J.; Pelletier, E.; Le Paslier, D.; Yamada, T.; Mende, D.R.; Fernandes, G.D.R.; Tap, J.; Bruls, T.; Batto, J.M.; et al. Enterotypes of the human gut microbiome. *Nature* **2011**, *473*, 174–180. [CrossRef] [PubMed]
55. Edwards, C.A.; Parrett, A.M. Intestinal flora during the first months of life: New perspectives. *Br. J. Nutr.* **2002**, *88* (Suppl. S1), s11. [CrossRef] [PubMed]
56. Gill, S.R.; Pop, M.; DeBoy, R.T.; Eckburg, P.B.; Turnbaugh, P.J.; Samuel, B.S.; Gordon, J.I.; Relman, D.A.; Fraser, C.M.; Nelson, K.E. Metagenomic Analysis of the Human Distal Gut Microbiome. *Science.* **2006**, *312*, 1355–1359. [CrossRef] [PubMed]
57. Guarner, F.; Malagelada, J.R. Gut flora in health and disease. *Lancet* **2003**, *361*, 512–519. [CrossRef]
58. Mariat, D.; Firmesse, O.; Levenez, F.; Guimarães, V.D.; Sokol, H.; Doré, J.; Corthier, G.; Furet, J.-P. The firmicutes/bacteroidetes ratio of the human microbiota changes with age. *BMC Microbiol.* **2009**, *9*, 1–6. [CrossRef]
59. Bäckhed, F.; Fraser, C.M.; Ringel, Y.; Sanders, M.E.; Sartor, R.B.; Sherman, P.M.; Versalovic, J.; Young, V.; Finlay, B.B. Defining a healthy human gut microbiome: Current concepts, future directions, and clinical applications. *Cell Host Microbe* **2012**, *12*, 611–622. [CrossRef]
60. Turnbaugh, P.J.; Hamady, M.; Yatsunenko, T.; Cantarel, B.L.; Duncan, A.; Ley, R.E.; Sogin, M.L.; Jones, W.J.; Roe, B.A.; Affourtit, J.P.; et al. A core gut microbiom in obese and lean twins. *Nature* **2009**, *457*, 480–484. [CrossRef]
61. Le Chatelier, E.; MetaHIT Consortium; Nielsen, T.; Qin, J.; Prifti, E.; Hildebrand, F.; Falony, G.; Almeida, M.; Arumugam, M.; Batto, J.M.; et al. Richness of human gut microbiome correlates with metabolic markers. *Nature* **2013**, *500*, 541–546. [CrossRef]
62. Flint, H.J.; Duncan, S.H.; Scott, K.P.; Louis, P. Links between diet, gut microbiota composition and gut metabolism. *Proc. Nutr. Soc.* **2015**, *74*, 13–22. [CrossRef] [PubMed]
63. Ubaldi, M.; Luciani, S.; Marota, I.; Fornaciari, G.; Cano, R.J.; Rollo, F. Sequence analysis of bacterial DNA in the colon of an Andean mummy. *Am J Phys. Anthropol.* **1998**, *107*, 285–295. [CrossRef]

64. Cano, R.J.; Tiefenbrunner, F.; Ubaldi, M.; Del Cueto, C.; Luciani, S.; Cox, T.; Orkand, P.; Künzel, K.H.; Rollo, F. Sequence analysis of bacterial DNA in the colon and stomach of the Tyrolean Iceman. *Am. J. Phys. Anthropol.* **2000**, *112*, 297–309. [CrossRef]
65. Lugli, G.A.; Milani, C.; Mancabelli, L.; Turroni, F.; Ferrario, C.; Duranti, S.; Van Sinderen, U.; Ventura, M. Ancient bacteria of the Ötzi's microbiome: A genomic tale from the Copper Age. *Microbiome* **2017**, *5*, 5. [CrossRef] [PubMed]
66. Obregon-Tito, A.J.; Tito, R.Y.; Metcalf, J.; Sankaranarayanan, K.; Clemente, J.C.; Ursell, L.K.; Xu, Z.Z.; Van Treuren, W.; Knight, R.; Gaffney, P.M.; et al. Subsistence strategies in traditional societies distinguish gut microbiomes. *Nat. Commun.* **2015**, *6*, 6505. [CrossRef]
67. Contreras, M.; Costello, E.K.; Hidalgo, G.; Magris, M.; Knight, R.; Dominguez-Bello, M.G. The bacterial microbiota in the oral mucosa of rural Amerindians. *Microbiology* **2010**, *156*, 3282–3287. [CrossRef]
68. Broussard, J.L.; Devkota, S. The changing microbial landscape of Western society: Diet, dwellings and discordance. *Mol. Metab.* **2016**, *5*, 737–742. [CrossRef]
69. Muegge, B.D.; Kuczynski, J.; Knights, D.; Clemente, J.C.; González, A.; Fontana, L.; Henrissat, B.; Knight, R.; Gordon, J.I. Diet drives convergence in gut microbiome functions across mammalian phylogeny and within humans. *Science* **2011**, *332*, 970–974. [CrossRef]
70. David, L.A.; Maurice, C.F.; Carmody, R.N.; Gootenberg, D.; Button, J.E.; Wolfe, B.E.; Ling, A.V.; Devlin, A.S.; Varma, Y.; Fischbach, M.A.; et al. Diet rapidly and reproducibly alters the human gut microbiome. *Nature* **2014**, *505*, 559–563. [CrossRef]
71. Sonnenburg, E.D.; Smits, S.A.; Tikhonov, M.; Higginbottom, S.K.; Wingreen, N.S.; Sonnenburg, J.L. Diet-induced extinctions in the gut microbiota compound over generations. *Nature* **2016**, *529*, 212–215. [CrossRef]
72. Parkar, S.G.; Kalsbeek, A.; Cheeseman, J.F. Potential Role for the Gut Microbiota in Modulating Host Circadian Rhythms and Metabolic Health. *Microorganisms* **2019**, *7*, 41. [CrossRef] [PubMed]
73. Stenvers, D.J.; Jonkers, C.F.; Fliers, E.; Bisschop, P.H.; Kalsbeek, A. Nutrition and the circadian timing system. *Prog. Brain Res.* **2012**, *199*, 359–376. [PubMed]
74. Voigt, R.M.; Summa, K.C.; Forsyth, C.B.; Green, S.J.; Engen, P.; Naqib, A.; Vitaterna, M.H.; Turek, F.W.; Keshavarzian, A. The circadian clock mutation promotes intestinal dysbiosis. *Alcohol. Clin. Exp. Res.* **2016**, *40*, 335–347. [CrossRef] [PubMed]
75. Shankar, V.; Gouda, M.; Moncivaiz, J.; Gordon, A.; Reo, N.V.; Hussein, L.; Paliy, O. Differences in gut metabolites and microbial composition and functions between Egyptian and U.S. children are consistent with their diets. *mSystems* **2017**, *2*, e00169-16. [CrossRef] [PubMed]
76. Cammarota, G.; Ianiro, G.; Bibbò, S.; Gasbarrini, A. Gut microbiota modulation: Probiotics, antibiotics or fecal microbiota transplantation? *Intern. Emerg. Med.* **2014**, *9*, 365–373. [CrossRef] [PubMed]
77. Depner, C.M.; Stothard, E.R.; Wright, K.P., Jr. Metabolic consequences of sleep and circadian disorders. *Curr. Diabetes Rep.* **2014**, *14*, 507. [CrossRef]
78. Shi, S.Q.; Ansari, T.S.; McGuinness, O.P.; Wasserman, D.H.; Johnson, C.H. Circadian disruption leads to insulin resistance and obesity. *Curr. Biol.* **2013**, *23*, 372–381. [CrossRef]
79. Spreadbury, I. Comparison with ancestral diets suggests dense acellular carbohydrates promote an inflammatory microbiota, and may be the primary dietary cause of leptin resistance and obesity. *Diabetes Metab. Syndr. Obes. Targets Ther.* **2012**, *5*, 175–189. [CrossRef]
80. Warinner, C.; Rodrigues, J.F.M.; Vyas, R.; Trachsel, C.; Shved, N.; Grossmann, J.; Radini, A.; Hancock, Y.; Tito, R.Y.; Fiddyment, S.; et al. Pathogens and host immunity in the ancient human oral cavity. *Nat. Genet.* **2014**, *46*, 336–344. [CrossRef]
81. Bengmark, S. Gut microbiota, immune development and function. *Pharmacol. Res.* **2013**, *69*, 87–113. [CrossRef]
82. Oriach, C.S.; Robertson, R.C.; Stanton, C.; Cryan, J.F.; Dinan, T.G. Food for thought: The role of nutrition in the microbiota-gut-brain axis. *Clin. Nutr. Exp.* **2016**, *6*, 25–38. [CrossRef]
83. Rothschild, D.; Weissbrod, O.; Barkan, E.; Kurilshikov, A.; Korem, T.; Zeevi, D.; Costea, P.I.; Godneva, A.; Kalka, I.N.; Bar, N.; et al. Environment dominates over host genetics in shaping human gut microbiota. *Nature* **2018**, *555*, 210–215. [CrossRef] [PubMed]

84. Costea, P.I.; Hildebrand, F.; Arumugam, M.; Bäckhed, F.; Blaser, M.J.; Bushman, F.D.; De Vos, W.M.; Ehrlich, S.D.; Fraser, C.M.; Hattori, M.; et al. Enterotypes in the landscape of gut microbial community composition. *Nat. Microbiol.* **2017**, *3*, 8–16. [CrossRef] [PubMed]
85. Goodrich, J.K.; Davenport, E.; Beaumont, M.; Jackson, M.; Knight, R.; Ober, C.; Spector, T.D.; Bell, J.T.; Clark, A.G.; Ley, R.E. Genetic Determinants of the Gut Microbiome in UK Twins. *Cell Host Microbe* **2016**, *19*, 731–743. [CrossRef]
86. Antonopoulos, D.A.; Huse, S.M.; Morrison, H.G.; Schmidt, T.M.; Sogin, M.L.; Young, V.B. Reproducible community dynamics of the gastrointestinal microbiota following antibiotic perturbation. *Infect. Immun.* **2009**, *77*, 2367–2375. [CrossRef]
87. Raymond, F.; Deraspe, M.; Boissinot, M.; Bergeron, M.G.; Corbeil, J. Partial recovery of microbiomes after antibiotic treatment. *Gut Microbes* **2016**, *7*, 1–7. [CrossRef]
88. Laubitz, D.; Midura-Kiela, M.; Ghishan, F.K.; Kiela, P.R. Dynamics of Gut Microbiome Recovery after Broad-Spectrum Antibiotic Treatment in Young and Old Mice. *Gastroenterology* **2018**, *154*, S-643. [CrossRef]
89. Iizumi, T.; Battaglia, T.; Ruiz, V.; Perez, G.I. Gut Microbiome and Antibiotics. *Arch Med Res.* **2017**, *48*, 727–734. [CrossRef]
90. Francino, M.P. Antibiotics and the human gut microbiome: Dysbioses and accumulation of resistances. *Front. Microbiol.* **2016**, *6*, 1–11. [CrossRef]
91. Hill, C.; Guarner, F.; Reid, G.; Gibson, G.R.; Merenstein, D.J.; Pot, B.; Morelli, L.; Canani, R.B.; Flint, H.J.; Salminen, S.; et al. The International Scientific Association for Probiotics and Prebiotics consensus statement on the scope and appropriate use of the term probiotic. *Nat. Rev. Gastroenterol. Hepatol.* **2014**, *11*, 506–514. [CrossRef]
92. Gibson, G.R.; Hutkins, R.; Sanders, M.E.; Prescott, S.L.; Reimer, R.A.; Salminen, S.J.; Scott, K.; Stanton, C.; Swanson, K.S.; Cani, P.D.; et al. The International Scientific Association for Probiotics and Prebiotics (ISAPP) consensus statement on the definition and scope of prebiotics. *Nat. Rev. Gastroenterol. Hepatol.* **2017**, *14*, 491–502. [CrossRef] [PubMed]
93. Druart, C.; Alligier, M.; Salazar, N.; Neyrinck, A.M.; Delzenne, N.M. Modulation of the gut microbiota by nutrients with prebiotic and probiotic properties. *Adv. Nutr. Int. Rev. J.* **2014**, *5*, 624S–633S. [CrossRef] [PubMed]
94. Mensink, M.A.; Frijlink, H.W.; Maarschalk, K.V.D.V.; Hinrichs, W.L.J. Inulin, a flexible oligosaccharide I: Review of its physicochemical characteristics. *Carbohydr. Polym.* **2015**, *130*, 405–419. [CrossRef] [PubMed]
95. Liu, F.; Li, P.; Chen, M.; Luo, Y.; Prabhakar, M.; Zheng, H.; He, Y.; Qi, Q.; Long, H.; Zhang, Y.; et al. Fructooligosaccharide (FOS) and Galactooligosaccharide (GOS) Increase *Bifidobacterium* but Reduce Butyrate Producing Bacteria with Adverse Glycemic Metabolism in healthy young population. *Sci. Rep.* **2017**, *7*, 1–12. [CrossRef]
96. Tandon, D.; Haque, M.M.; Gote, M.; Jain, M.; Bhaduri, A.; Dubey, A.K.; Mande, S.S. A prospective randomized, doubleblind, placebo-controlled, doseresponse relationship study to investigate efcacy of fructooligosaccharides (FOS) on human gut microfora. *Sci. Rep.* **2019**, *9*, 5473. [CrossRef]
97. Walton, G.; Heuvel, E.V.D.; Kosters, M.H.W.; Rastall, R.A.; Tuohy, K.; Gibson, G.R. A Randomised Crossover Study Investigating the Effects of Galacto-Oligosaccharides on the Faecal Microbiota in Men and Women Over 50 Years of Age. *Br. J. Nutr.* **2012**, *107*, 1466–1475. [CrossRef]
98. Zaman, S.A.; Sarbini, S.R. The potential of resistant starch as a prebiotic. *Crit. Rev. Biotechnol.* **2016**, *36*, 578–584. [CrossRef]
99. Fuentes-Zaragoza, E.; Sánchez-Zapata, E.; Sendra, E.; Sayas, E.; Navarro, C.; Fernández-Lõpez, J.; Pérez-Alvarez, J.A. Resistant starch as prebiotic: A review. *Starch Stärke* **2011**, *63*, 406–415. [CrossRef]
100. EFSA. Scientific Opinion on the maintenance of the list of QPS biological agents intentionally added to food and feed (2011 update). *EFSA J.* **2011**, *9*, 2497. [CrossRef]
101. Rodríguez, J.M. Probióticos: Del laboratorio al consumidor. *Nutr. Hosp.* **2015**, *31*, 33–47.
102. Sanders, M.E.; Akkermans, L.M.; Haller, D.; Hammerman, C.; Heimbach, J.T.; Hörmannsperger, G.; Huys, G.; Levy, D.D.; Lutgendorff, F.; Mack, D.; et al. Safety assessment of probiotics for human use. *Gut Microbes* **2010**, *1*, 1–22. [CrossRef] [PubMed]
103. Walker, A.W.; Lawley, T.D. Therapeutic modulation of intestinal dysbiosis. *Pharmacol. Res.* **2013**, *69*, 75–86. [CrossRef] [PubMed]

104. Versalovic, J. The human microbiome and probiotics: Implications for pediatrics. *Ann. Nutr. Metab.* **2013**, *63* (Suppl. 2), 42–52. [CrossRef] [PubMed]
105. Islam, K.S.; Fukiya, S.; Hagio, M.; Fujii, N.; Ishizuka, S.; Ooka, T.; Ogura, Y.; Hayashi, T.; Yokota, A. Bile acid is a host factor that regulates the composition of the cecal microbiota in rats. *Gastroenterology* **2013**, *141*, 1773–1781. [CrossRef] [PubMed]
106. Ilhan, Z.E.; Marcus, A.K.; Kang, D.; Rittmann, B.E.; Krajmalnik-Brown, R. pH-Mediated Microbial and Metabolic Interactions in Fecal Enrichment Cultures. *Ecol. Evol. Sci.* **2017**, *2*, 1–12. [CrossRef] [PubMed]
107. García-Mantrana, I.; Selma-Royo, M.; Alcántara, C.; Collado, M.C. Shifts on gut microbiota associated to mediterranean diet adherence and specific dietary intakes on general adult population. *Front. Microbiol.* **2018**, *9*, 1–11. [CrossRef] [PubMed]
108. Chambers, E.S.; Preston, T.; Frost, G.; Morrison, D.J. Role of Gut Microbiota-Generated Short-Chain Fatty Acids in Metabolic and Cardiovascular Health. *Curr. Nutr. Rep.* **2018**, *7*, 198–206. [CrossRef]
109. De Filippis, F.; Pellegrini, N.; Vannini, L.; Jeffery, I.B.; La Storia, A.; Laghi, L.; Serrazanetti, D.I.; Di Cagno, R.; Ferrocino, I.; Lazzi, C.; et al. High-level adherence to a Mediterranean diet beneficially impacts the gut microbiota and associated metabolome. *Gut* **2016**, *65*, 1–10. [CrossRef]
110. Cotillard, A.; Kennedy, S.P.; Kong, L.C.; Prifti, E.; Pons, N.; Le Chatelier, E.; Almeida, M.; Quinquis, B.; Levenez, F.; Galleron, N.; et al. Dietary intervention impact on gut microbial gene richness. *Nature* **2013**, *500*, 585–588. [CrossRef]
111. Erny, D.; De Angelis, A.L.H.; Jaitin, D.; Wieghofer, P.; Staszewski, O.; David, E.; Keren-Shaul, H.; Mahlakoiv, T.; Jakobshagen, K.; Buch, T.; et al. Host microbiota constantly control maturation and function of microglia in the CNS. *Nat. Neurosci.* **2015**, *18*, 965–977. [CrossRef]
112. Hoban, A.E.; Stilling, R.M.; Ryan, F.J.; Shanahan, F.; Dinan, T.G.; Claesson, M.J.; Clarke, G.; Cryan, J.F. Regulation of prefrontal cortex myelination by the microbiota. *Transl. Psychiatry* **2016**, *6*, e774-9. [CrossRef] [PubMed]
113. Braniste, V.; Al-Asmakh, M.; Kowal, C.; Anuar, F.; Abbaspour, A.; Tóth, M.; Korecka, A.; Bakocevic, N.; Gascoigne, N.R.J.; Kundu, P.; et al. The gut microbiota influences blood brain barrier permeability in mice. *Sci. Transl. Med.* **2014**, *6*, 263ra158. [CrossRef] [PubMed]
114. Kootte, R.S.; Levin, E.; Salojärvi, J.; Smits, L.P.; Hartstra, A.V.; Udayappan, S.D.; Hermes, G.; Bouter, K.E.; Koopen, A.M.; Holst, J.J.; et al. Improvement of Insulin Sensitivity after Lean Donor Feces in Metabolic Syndrome Is Driven by Baseline Intestinal Microbiota Composition. *Cell Metab.* **2017**, *26*, 611–619.e6. [CrossRef] [PubMed]

*nutrients*

MDPI

*Review*

# A Revolutionizing Approach to Autism Spectrum Disorder Using the Microbiome

**Dinyadarshini Johnson [1], Vengadesh Letchumanan [1], Sivakumar Thurairajasingam [2] and Learn-Han Lee [1,*]**

[1] Novel Bacteria and Drug Discovery Research Group (NBDD), Microbiome and Bioresource Research Strength (MBRS), Jeffrey Cheah School of Medicine and Health Sciences, Monash University Malaysia, Bandar Sunway 47500, Malaysia; dinyadarshini.johnson@monash.edu (D.J.); vengadesh.letchumanan1@monash.edu (V.L.)

[2] Clinical School Johor Bahru, Jeffrey Cheah School of Medicine and Health Sciences, Monash University Malaysia, Johor Bahru 80100, Malaysia; sivakumar.thurairajasingam@monash.edu

* Correspondence: lee.learn.han@monash.edu or leelearnhan@yahoo.com

Received: 29 May 2020; Accepted: 30 June 2020; Published: 3 July 2020

**Abstract:** The study of human microbiota and health has emerged as one of the ubiquitous research pursuits in recent decades which certainly warrants the attention of both researchers and clinicians. Many health conditions have been linked to the gut microbiota which is the largest reservoir of microbes in the human body. Autism spectrum disorder (ASD) is one of the neurodevelopmental disorders which has been extensively explored in relation to gut microbiome. The utilization of microbial knowledge promises a more integrative perspective in understanding this disorder, albeit being an emerging field in research. More interestingly, oral and vaginal microbiomes, indicating possible maternal influence, have equally drawn the attention of researchers to study their potential roles in the etiopathology of ASD. Therefore, this review attempts to integrate the knowledge of microbiome and its significance in relation to ASD including the hypothetical aetiology of ASD and its commonly associated comorbidities. The microbiota-based interventions including diet, prebiotics, probiotics, antibiotics, and faecal microbial transplant (FMT) have also been explored in relation to ASD. Of these, diet and probiotics are seemingly promising breakthrough interventions in the context of ASD for lesser known side effects, feasibility and easier administration, although more studies are needed to ascertain the actual clinical efficacy of these interventions. The existing knowledge and research gaps call for a more expanded and resolute research efforts in establishing the relationship between autism and microbiomes.

**Keywords:** clinician; autism spectrum disorder; microbiome; aetiology; comorbidities; diet; prebiotics; probiotics; faecal microbial transplant

---

## 1. Introduction

Autism spectrum disorder (ASD) is identified with persistent deficit in social communication and phenotypic behaviours which are typically repetitive and restrictive [1]. This disorder affects more boys than girls in a ratio close to 3:1 [2]. There has been a significant increase in the prevalence of ASD over the decades, it is currently estimated to affect 1% of the general population [3]. The most recent Global Burden of Diseases, Injuries, and Risk Factors Study in 2016 estimates 62.2 million individuals live with ASD globally [4]. It was also demonstrated that the prevalence of ASD based on special education enrolment data within the United States (US) over 11 years, from 2000 (1.2 per 1000) to 2010 (5.2 per 1000) alone shows an increase of 331%. It was inferred that the diagnostic recategorization may be the possible explanation for this significant rise [5]. However, it will be an understatement if it is solely attributable to the changes introduced in the diagnostic criteria of autism and increased awareness which will be explored in this review. It has been speculated that the evolving environmental influence which contributes significantly to the occurrence of ASD could be partly responsible for the rise in the ASD prevalence globally [6]. A rising prevalence in ASD is a matter of concern and calls for a more effective management of this disorder.

The study of human microbiomes has become a prime hope in understanding this disorder and catering to the growth of a more clinical-based intervention more than to merely augment the widely advocated behavioural therapies. The understanding of microbial–human host relationship which was once thought to be purely commensal in nature, if not pathogenic, skewing to a one-sided relationship has now evolved into a complex interaction holding imperative roles in key physiological processes in the human body [7]. Determining the role of the microbiome in human health has become an intriguing quest in recent decades. A plethora of health conditions have been associated with microbiome across a wide range of populations identifying a distinctive spread of microbiome in a selected patient cohort compared to a healthy cohort pointing to its potential etiological role in the occurrence of a commonly identified disorder. It is not an understatement if this could potentially invent a breakthrough management approach in curbing a wide variety of health disorders and pandemics which include obesity and cardiometabolic diseases, infections, respiratory, allergic, gastrointestinal, neuropsychiatric disorders and even cancers [7–14]. The National Institutes of Health (NIH) Human Microbiome Project, launched in 2007, was one of the large-scale projects initiated to support and catalyst the growth of this emerging field of research. The microbiome in five major habitats in the human body, which include the gastrointestinal tract, airway, skin, oral cavity and vagina, were explored in this pursuit [15]. The gut microbiome has been the most widely studied area, accounting for four-fifths of total publications in microbiome over the last four decades [9]. Germ-free rodents and controls—conventionally colonized rodents with specific pathogen free (SPF) rodent models—have been utilized to ascertain the microbial influence in behavioural outcomes which have been grouped into four domains including, ASD-mimicking behaviours, stress and anxiety-related behaviours, learning and memory and motor controls. The germ-free rodents exhibited significant deficits in social interactions, cognitive and motor functions compared to SPF rodents suggesting the imperative role of microbiome in neurobehavioral outcome [16]. The gut microbial composition, if disturbed, has an impact on the various physiological activities regulated by these microbes principally through its metabolites and has a bidirectional communication with the brain involving autonomic nervous system. Neuronal, neuroendocrine and immunologic pathways have been described through which the microbes contribute to the bidirectional signalling between the gut and the brain [17,18]. The bidirectional transfer of information between the gut and the brain is principally controlled by the vagus nerve. The gut microbiota communicate to the brain via endocrine and neurocrine pathways while the brain impacts the microbial composition via immune and humoral systems mediated by autonomic nervous system, thus establishing the gut–brain–microbiota axis [18,19]. In the context of ASD, the exploration of other habitats of the microbiome, in the vagina and oral region, highlights the influence of maternal factors in the development of ASD. The vertical transmission of disrupted maternal vaginal microbes to the offspring at birth predisposes the offspring to the prenatal risk of developing ASD [20]. The direct relation of the oral microbiome and ASD has yet to be established; however, the resemblance of infants' oral microbiome to the maternal microbiome during the first six months of birth ascertains the significant contribution of maternal microbiome in early stage of oral microbial colonization [21].

Although the aetiology of ASD largely remains unanswered, the emerging microbial knowledge may be a key finding in explaining the etiopathogenesis of ASD; however, with more extensive work needed to understand its involvement at the molecular level. In an attempt to understand the molecular involvement in ASD, a study on post-mortem brain tissue and small intestines of ASD subjects revealed that blood–brain barrier (BBB) and gut barrier were disrupted with significant neuroinflammation evident by increased expression of genes and markers associated with brain inflammation. It was further inferred that the gut–brain axis disruption may be associated with non-self-antigens which triggers neuroinflammatory reaction by crossing the damaged gut barriers, thus leading to ASD in genetically susceptible subjects [22]. The BBB has a pivotal role in early phase of brain development and neuronal functions [23–25]. It was found that adult germ-free mice models and the foetuses of germ-free mice's mothers had more permeable BBB compared to mice with pathogen-free gut microbiota. The faecal transfer from a pathogen- to germ-free mice models and administration of short-chain fatty acid (SCFA) producing bacteria were able to restore the permeability of BBB, emphasizing the roles of intestinal microbes and SCFA in guarding the integrity of BBB [26]. The genetic component has primarily conditioned ASD as a highly heritable disorder through twin studies and large population-based studies which assess familial risk [27]. Nevertheless, the influential role of environment may potentially supersede the sole genetic involvement in the development of ASD, especially with the recent data pointing to the integrative approach of epigenetics in the study of pathogenesis of ASD [28]. In a study involving 192 twins, genetic factors were made accountable for only 38% of ASD risk, whereas the remaining 68% was attributed to environmental factors [29]. The microbial composition is regulated and influenced by many factors which could be broadly classified into extrinsic and intrinsic factors. The extrinsic factors consist of mainly environmental elements which include diet, lifestyle, microbial exposure in early developmental phase and infection, whereas intrinsic factors are naturally occurring elements within an individual which include genetic make-up, metabolites, immunologic and hormonal aspects [30]. The understanding of microbial involvement as part of an interplay between the intrinsic and extrinsic factors gives rise to an immense possibility of associating various factors in explaining its relation to ASD; however, the research progress in understanding the actual etiological mechanism involved in ASD development is still at an infant stage with a lack of defined explanations [18,31,32]. The co-existing health conditions in ASD is another major challenge in both research and clinical fronts. At least one comorbidity exists in about 70% of children with ASD, while 41% have two or more, reflecting the onerous disease burden in this population [33]. There are limited numbers of published studies on comorbidities alone in ASD due the complex and large spectrum of heterogeneity associated with this disorder. The widely reported and studied medical and psychiatric comorbidities in ASD include gastrointestinal (GI) disorders, epilepsy, depression and anxiety disorder [33–35]. These comorbidities can be individually linked to the microbiome which either overlaps with a similar ASD pattern of microbial dysbiosis, or that the related use of medication induces microbial dysbiosis which could contribute to ASD occurrence. This relation points towards the need to include the assessment of associated comorbidities to further understand the possible shared aetiology in ASD development, as many microbial studies using ASD subjects often do not take into account their existing comorbidities.

This review therefore focuses on autism spectrum disorder and microbiome in general while examining its relation to the hypothetical aetiology of ASD and its commonly associated comorbidities.

## 2. Evolving Conceptualization of Autism Spectrum Disorder

ASD is a behaviourally defined, neurodevelopmental disorder which lacks specific clinical biomarkers and has seen an evolving conceptualization over the last seven decades since it was first described [36,37]. The Centers for Disease Control and Prevention (CDC) reports that the earliest known median age of ASD diagnosis is at 52 months in the United States (US), whereas in the United Kingdom (UK), the median age of ASD diagnosis is reported at 55 months [38,39]. In general, ASD is diagnosed by three years of age in most children, although roughly 39% of children are not first evaluated until after four years of age [38]. A psychiatric diagnosis which has behaviour as its basis of definition heavily relies upon a meticulous observation and clinical expertise as it lacks standardised biomarkers [40]. Therefore, it is crucial to understand the core concept of a disorder to facilitate a more effective diagnostic process. The idea of autism emerged as early as in the 19th century when a Swiss psychiatrist, Eugen Blueler described the aloofness he observed in individuals with schizophrenia as a form of schizophrenic trait itself [41,42]. It was in the 1943, when two child psychiatrists, Leo Kanner and Hans Asperger, garnered the attention of scientific community through their publications, which eventually led to recognising autism as a distinct category of diagnosis in children years later [42]. They individually published clinical cases of children with distinctive behaviours, primarily reflecting social deficits and echoing the descriptive term, autism coined by Blueler [43,44]. However, autism was often identified as a form of schizophrenia until a more refined and stand-alone diagnostic classification was first introduced in 1980 by American Psychiatric Association in Statistical Manual of Mental Disorders (DSM-III) and subsequently by the World Health Organization in International Statistical Classification of Diseases and Related Health Problems (ICD-10) in 1990 [36]. These are the two broadly adopted sources of reference for both research and clinical purposes with revised versions to date. The most recent definition, based on the DSM-5 published in 2013, includes subtypes of autistic disorders, Asperger's syndrome and Pervasive developmental disorder not otherwise specified (PDD-NOS) under one umbrella term known as ASD. DSM-5 allows classification of ASD based on its severity and takes into account intellectual ability and other comorbidities [1]. Despite an evolving conceptualization from a disorder to a spectrum highly denoting its clinical diversity, the core definition of autism has remained central to deficits in social interaction and stereotypic behaviours. It took several decades to define ASD within a smaller framework of classification to facilitate diagnostic process. However, it still remains a debate to perfectly fit ASD in a confined framework of definition due to the heterogenous expression of the core symptoms which are further influenced by age and development factors. The progressive work to define ASD within a narrower framework of theoretical concept inadvertently reflects the complexity that revolves around this neurodevelopmental disorder, which makes it even more challenging to conclusively define other aspects pertaining to ASD.

## 3. Microbiome and Autism Spectrum Disorder

The terms microbiota and microbiome have a slight semantic difference, however, collectively refer to the entire microbial community residing on and in human body encompassing bacteria, eukaryotic viruses, fungi, protozoa, archaea and bacteriophage. The number of bacteria is overwhelmingly larger than the other taxa to an extent where microbes interchangeably could simply refer to bacteria alone. Our human body is home to trillions of microbial cells which encode 100-fold more genes than human genome, with the latest revision estimating a ratio of 1.3 bacterial cells for every human cell, showing a reduction from widely quoted 10:1 and 100:1 ratios, respectively [15,45,46]. The enormous spread of the microbiome in the human body has both therapeutic and pathogenic roles in health depending on the microbial composition [47]. In the context of ASD (Figure 1), the link between gut, vaginal and oral microbiomes and ASD have been studied thus far using animal models and human subjects.

Figure 1 illustrates the possible mechanism involved in the microbiome–brain interaction in the context of autism spectrum disorder (ASD). The neural, neuroendocrine, and immunologic and humoral pathways are the potential mediators in the bidirectional communication between the gut microbiome and the brain. The maternal contribution is significant in determining the early intestinal colonization in the offspring while in general, the environmental factors that significantly alter maternal microbiome during the prenatal and perinatal periods influence the microbial composition in the offspring. Ectopic transfer and dissemination of pathogenic oral bacteria mediated by the olfactory nerve via the blood, disrupted blood–brain barrier (BBB), perivascular space and circumventricular organs to the gut and brain, respectively, are plausible mechanisms resulting in neuroinflammation and metabolic disruption in the brain, thus indicating the influence of oral microbiota and dysbiosis in ASD occurrence. Another plausible exchange pathway in the gut–brain axis is thought to be mediated by the oropharynx which has a significant role in the pathology of ASD. In general, the interaction between microbiome and the brain in the context of ASD involves a complex mechanism and interplay between the genetic and various environmental factors, in which, some could be explained through epigenetic mechanisms involving short-chain fatty acids (SCFAs) and brain-derived neurotrophic factors (BDNFs).

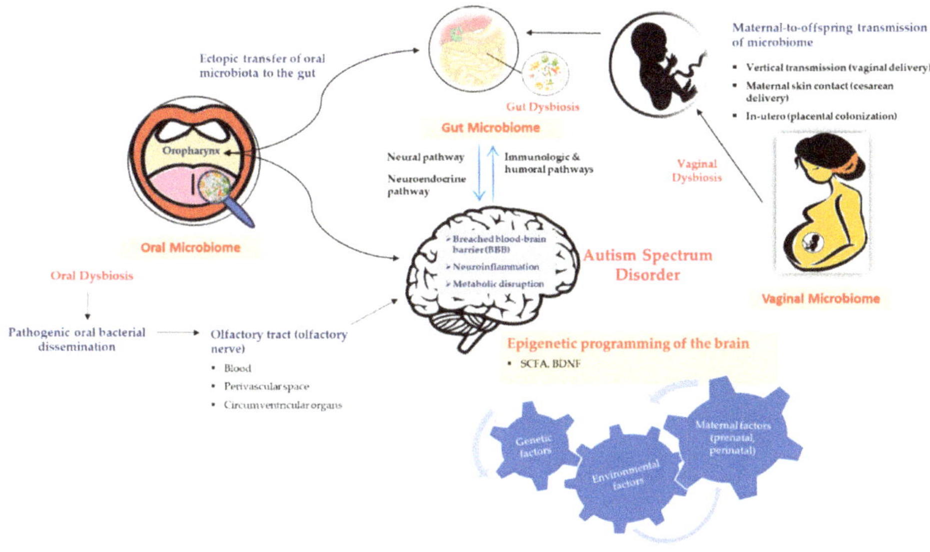


**Figure 1.** Illustration of autism spectrum disorder (ASD) and its association with microbiome.

### *3.1. Gut Microbiome*

Gut microbiome has been the most extensively explored area in relation to ASD compared to microbiome in other habitats of the body. The years between 2013 and 2017 saw the largest number of publications focusing on gut microbiota alone accounting for more than 80% of total publications on microbiota in the last four decades, implying the immense possibility of gut microbe in relation to human health [9]. The human gut is the largest reservoir of bacteria in human body which has imperative physiological roles in metabolism, digestion, immunity, endocrine and neurological activities [48]. The gut microbes interact closely with multiple human cells and any imbalance in the microbes, otherwise known as dysbiosis, will have an impact on host key physiological processes and has a potential aetiological role in many health conditions [47]. The immense possibility of gut microbes in human health has prompted researches to highlight the importance of recognizing it as an individual organ system in the human body [49].

In ASD, the relation between gut microbes and central nervous system resulting in manifestation of ASD behaviour has been established using rodent models and clinical subjects, although the latter is still largely inadequate to strongly affirm the causative relation between gut microbes and ASD symptomatology. *Bacteroidetes* (e.g., *Bacteriodes* and *Prevotella*), *Firmicutes* (e.g., *Lactobacillus*, *Clostridium*, *Ruminococcus*), *Proteobacteria* (e.g., *Enterobacter*) and *Actinobacteria* (e.g., *Bifidobacterium*) are four major phyla that are constituent of a healthy adult gut with *Bacteriodes* and *Firmicutes*, representing more than 90% of gut microbes [45,50]. It was found that, in the ASD population, the ratio of *Bacteroidetes* to *Firmicutes* phyla were increased compared to neurotypical cohort [51]. This pattern is also observed prominently in Western ASD children and has been inferred to be influenced by environment and dietary habits [52]. There is also another study which found increased *Firmicutes* compared to *Bacteroidetes* in the ASD group compared to neurotypical group where both groups have GI symptoms [53]. A large number of gram-negative bacteria (e.g., *Bacteroidetes*), exhibit pathogenic nature because of their cell wall which contains lipopolysaccharide (LPS) that has malefic effect on the immune system of the host. LPS confers the ability to breach the blood–brain barrier, increasing the mercury level in the cerebrum and decrease glutathione level which is a key antioxidant in the detoxification of heavy metals. Two species of bacteria with LPS in their cell walls, namely *Bacteriodes vulgatus* and *Desulfovibrio*, have been significantly raised in ASD children compared to neurotypical children. The pathogenic feature of these bacteria may possibly contribute to ASD development. Another noteworthy genus of gram-negative bacteria is *Prevotella*, a healthy-gut biomarker which is at a greater abundance in neurotypical group while almost absent in the autistic group. *Prevotella* is found in abundance in individuals whose diet is rich in plant-based carbohydrates and includes fish oil which is vital to normal brain development. It has a metabolic role in vitamin B1 production which is known to alleviate ASD symptoms. The lack of *Prevotella* suggests a distinctive dietary habit in autistic children which significantly alters the gut microbial composition and has influence on neurodevelopment which points to its therapeutic potential if restored [48,54]. Another concerning microbe is *Clostridium*, a gram-positive genus in the *Firmicutes* phyla, which has been found higher in ASD group. *Clostridum boltae, C. histolyticum*, and subgroups of *I* and *XI* are the associated species. These gram-positive microbes release enterotoxins which damage the intestinal tissue resulting in diarrhoea and may cater to the increased absorption of large molecules like casein and gluten. On the other hand, beneficial bacteria *Bifidobacterium* were found less abundant in the ASD group [48]. The increased abundance of potentially pathogenic bacteria and decreased beneficial bacteria affirms the existence of gut dysbiosis in the ASD population. In another perspective, the general lack of gut microbial richness and diversity in ASD group, predisposes them to a vulnerable gut environment which could lead to GI disturbances, infections and autistic behaviours [48]. In general, a significant alteration in gut microbial composition interferes with key physiological processes which have an influence on the neurobehavioral manifestation of ASD symptoms either through the absence of beneficial microbial metabolites, release of harmful microbial endotoxins, pathogenic invasion of the intestinal wall and/or through immune mediators catering to neuroimmune inflammation [6,47,48,50,55].

Many factors have been associated with the dysbiosis of gut microbes which include diet, medication and hygiene as well as numerous maternal factors which include maternal stress, infection and a high-fat diet during pregnancy [6,56–59]. The frequent use of oral antibiotics in ASD children during the first three years of life is another factor that has also been hypothesized to disturb the natural balance of the gut microbes while some antibiotics confer benefits [48,55,60]. For instance, the use of macrolides except penicillin within the first six months has been associated with decreased *Actinobacteria* and increased *Bacteroidetes* and *Proteobacteria* and this microbial alteration persisted for a year [61]. On the other hand, the administration of oral vancomycin in a small eight-week clinical trial, which targets gram-positive bacteria including *Clostridium*, temporarily but significantly relieves gastrointestinal and ASD symptoms in 8 out of 11 children with regressive onset autism [62].

### *3.2. Vaginal Microbiome*

The vaginal microbiome denotes the influence of maternal factors in the occurrence of ASD. As it has been understood that we are born germ-free, the first colonization of microbes in human gut is thought to begin at birth while passing through the vagina, although there are emerging data suggesting it occurs even earlier in utero via placental colonization [63,64]. In cases of caesarean delivery, the colonization occurs after in contact with the maternal skin and environmental microbes [65]. The passing of microbial community through vertical transmission determines the child's gut microbial composition, thus any significant disturbance to maternal vaginal microbe (e.g., reduced *Lactobacillus*) inadvertently interferes with the normal neurodevelopment in offspring due to high metabolic demand during the critical time of early brain development [20,66].

The vaginal tract houses more than fifty microbial species, dominantly populated by *Lactobacillus* in a healthy woman [67]. Maternal stress during early pregnancy has been found to exert a suppressive effect on the vaginal immune response and the abundance of *Lactobacillus*, thus resulting in gut microbial dysbiosis in offspring via vertical transmission [66]. Bacterial vaginosis, a common infection among women characterized by a marked reduction in *Lactobacillus*, predisposes a pregnant woman to the risk of preterm delivery [68]. It has been reported that infants born extremely preterm have 10-fold higher risk of developing ASD compared to infants born at term [69,70]. Further, numerous clinical studies and large epidemiological studies have ascertained that prenatal maternal infection and elevated level of pro-inflammatory cytokines increases the risk of ASD in offspring. The injection of antigens activating such maternal immune response in pregnant mice, rats or monkeys have resulted in neurobehavioral deficits mimicking ASD in their offspring [71,72].

### 3.3. Oral Microbiome

Oral cavity is home to more than 700 bacterial species or phyla of which more than 50% are yet to be cultivated and these microbes have an influence on individual oral health. Poor oral health is one of the concerning issues amongst ASD children and it is more prevalent in this group compared to neurotypical group [73]. In the context of ASD, only a handful of studies have explored the differences between oral microbiota in autism and neurotypical children. Notable differences in the distribution of oral microbes were detected in ASD children compared to neurotypical children [54,74,75]. In one of the largest cross-sectional studies which studied the oral microbiome profiles in ASD and typically developed children, eight oral taxa that could distinguish children with ASD from typically developed children were identified. Further, 28 taxa that distinguish ASD children with and without GI disturbances were also identified. It was inferred that the gut microbial disruption could potentially extend to the oropharynx. The analysis of oral microbiome to aid the clinical diagnosis of ASD was also suggested [74].

ASD children often have speech-related difficulties and are very selective with food choices where each of these have motor and sensory involvement, respectively. Oropharynx serves as an important bridge to the GI tract and is innervated by five cranial nerves of both motor and sensory origins. It is thought to play a significant role in the pathology of ASD and has a plausible exchange pathway in the gut–brain axis [74,76–78]. Other plausible pathways in which the oral bacteria could reach the brain have also been hypothesized resulting neuroinflammation and metabolic disruption in the central nervous system. It has been hypothesized that the olfactory nerve in the olfactory tract may act as a potential mediator in bacterial dissemination to the brain through blood, disrupted blood–brain barrier, perivascular spaces or circumventricular organs [79–81]. Further, similar distribution pattern of gut and oral microbiota were demonstrated in a study of oral microbiota in ASD children with significantly higher amount of *Proteobacteria* and *Firmicutes* and lower amount of *Bacteroidetes* and *Actinobacteria* suggesting a potential interaction between gut and oral microbiota leading to a shared pathway in the etiopathology of ASD [54,75,81]. It was also found that oral bacteria resemble 45% of the stool bacteria in the Human Microbiome Project, suggesting a possible interaction between the gut and the oral microbiome [82]. Another example is the increased oral *Bacilli* genus (*Firmicutes* phylum) in ASD children which is also found in abundance in the gut of ASD children and those with inflammatory bowel disease [75,83]. This intestinal colonization by oral bacteria has been hypothesized to occur through ectopic transfer of pathogenic oral bacteria (e.g., *Porphyromonas gingivalis* in chronic periodontitis) to the gut which could induce gut microbial dysbiosis and trigger systemic inflammation [81,84,85]. Poor oral health and hygiene, dental caries and lack of dental care have been found more prevalent in ASD children, implying that ectopic transfer of pathogenic bacteria to the gut associated with oral dysbiosis may possibly explain its relation in ASD occurrence [86]. A marked increase in potential pathogens in analysis of oral samples of ASD children which include *Streptococcus* and *Haemophilus* and reduced abundance of beneficial bacteria like *Prevotella* also point to a significant oral microbial dysbiosis in the ASD group. It is hypothesized that *Haemophilus parainfluenza*, a gram-negative bacteria associated with oral diseases, and its metabolites could potentially cross the blood–brain barrier and impose a detrimental effect on brain development which results in ASD [54,87].

There are specific bacterial species found in subjects with periodontal disease which were otherwise not detected in individuals with a healthy oral cavity [88]. Periodontal disease is associated with an increased risk of preterm birth by 2 to 7 times [89]. Microbial species which were detected in the oral cavity and not in the urogenital tract have been found as causative organisms in intrauterine infection which confers a high risk for preterm birth [90]. More interestingly, the microbiome of 48 term placentae were found to resemble oral microbiome more compared to vaginal microbiome with theory suggesting hematogenous dissemination, especially with underlying periodontal disease, and oral sex may be possible route for such colonization resulting in intrauterine infection [91–93]. The prenatal maternal infection which increases the risk of preterm birth inevitably pave the way for ASD occurrence risk in infants born to mothers with the associated conditions which implies a possible but indirect association with the maternal oral microbiome. The maternal influence was more significant during the infant stage and early childhood. During the first six months after birth, 85% of the oral microbiota of infants resembled their mother's [21]. The pregnancy term, mode of delivery and feeding method were other identified factors that influence the development of oral microbiota in early childhood [94].

## 4. Aetiology of Autism Spectrum Disorder

The aetiology of ASD has always remained a puzzle that has yet to be fully understood. The diverse expression of ASD symptomatology, its associated risk factors and comorbidities make it difficult to pinpoint the exact mechanism involved in the etiopathology of ASD. Nevertheless, the aetiology of ASD can be broadly classified into genetic and environmental causes. Although these are not direct causative factors, strong associations have been established and linked to the development of ASD. Individually, genetic and a myriad of environmental factors have been identified to contribute to the occurrence of ASD through various mechanisms. The microbes and microbiome have both environmental and genetical origins in the manifestation of ASD. More interestingly, a complex interaction between gene and environment through an epigenetic mechanism has also been implicated in the pathogenesis of ASD [95–97].

### *4.1. Genetic Factors*

Heritability confers a large accountability in cases of ASD, with an estimate of 83% of familial risk in a meta-analysis of twin and family studies. This reanalysis sees a drop in the percentage risk compared to previously published meta-analysis of twin studies which estimates heritability ranging between 64% and 90% with minimal influence of shared environmental factors between the twins [27,98]. The diverse clinical phenotype in ASD is due to the genetic heterogeneity in the ASD population, particularly when comorbid conditions exist. Both common and rare genetic variations which are either heritable or occurs newly as de novo mutation (DNM) contribute to the occurrence of ASD [99,100]. Rare genetic variants have a larger effect compared to the smaller effect of common variants in ASD phenotypes and can combine to create an ASD risk [101]. DNM is a rare variant and newly occurs during gamete formation or at the early phase of embryonic development which are not inherited from either of the parents and unique to the child, resulting in the sporadic occurrence of ASD. DNM is more frequently associated with ASD subjects with co-occurring intellectual disability or developmental delay [102,103]. DNM and common genetic variants provoke the idea of environmental influences which could potentially surpass the etiological weightage of genetic factors and heritability in ASD.

Despite the high accountability to genetic origins, there is no individual genetic marker that has been identified to date. However, continuous attempts are being made to identify novel genes which are significantly attributable to ASD. In a meta-analysis which was attempted to identify the genetic risk of ASD, two novel ASD risk genes, namely, YBX3 and HSPA1A, were found to be associated with the pathogenesis of ASD through an indirect regulation of neuronal pathways involved in behavioural manifestation of ASD [104]. These genes confer protective mechanism against the development of ASD, but it was later found to be of a weak association [105]. This further ascertains the lack of homogeneity inevitably challenges the genetic evaluation in ASD populations.

### *4.2. Environmental Factors*

A vast number of environmental factors have been linked to the development of ASD. The role of environment has been speculated to begin since pre-conception and extends to post-natal period in a child. Pre-conceptionally, advanced paternal age and maternal age at more than 50 and 40 years old, respectively, confers a significant risk for ASD development, apart from low level of education status in parents [106,107]. The greatest number of environmental factors have been linked during the prenatal period, particularly during the first and second trimester of pregnancy which include maternal infections, comorbid cardiometabolic conditions, certain antidepressant and antiepileptic medications, toxic exposures, diet and lifestyle. Perinatal factors which include mode of delivery, obstetrics complications, prematurity, hypoxia, low birth weight and low Apgar score have been associated with ASD risk [108,109]. Postnatally, congenital infection, the use of steroid therapy in very low birth weight infants, birth asphyxia and neonatal jaundice have all been found more prevalent in children with ASD [110,111]. The environmental factors may be explained by direct biological impact on neuronal activities of developing brain of the foetus or foetal activation of neuroinflammatory responses and gene dysregulation which are hypothesized to result from maternal immune activation which crosstalk through the placenta [109,112]. Largely, these environmental factors have been studied retrospectively in cohorts of ASD children and mothers of ASD children, while only limited number of studies done using rodent models. There are studies which deem environmental factors to play a more prominent role in the liability and genetical variance of ASD, thus paving way for the eminent role of epigenetics in ASD [96,113,114].

### 4.3. Epigenetic Factors

While traditional methods of identifying genes and environmental factors independent of each other has seen a lack of integrative approach in studying the aetiology of ASD, epigenetics has become the prime pursuits in recent decades incorporating the role of environmental influence on genes. Epigenetic refers to non-genetical influences which changes the phenotypic expression of a gene without actual mutation or alteration occurring in the original DNA sequence and these changes are heritable [97]. It is an epitomic description of a gene–environment interaction in ASD which can be explained by integrating various environmental factors at a cellular level in the occurrence of ASD, particularly through gut microbiota-derived metabolites. At the molecular level, the epigenetic modification occurs via two broadly studied mechanisms, which include DNA-methylation and histone modification besides Ribonucleic acid (RNA) interference [28,115–117]. Epigenetic programming of the brain explained through stress-regulating pathways and reduced expression of brain-derived neurotrophic factor (BDNF) by environmental factors during pre-natal and postnatal periods have been thought to exert a long-lasting effect on the neural functioning and behavioural outcome [118,119]. BDNF has a crucial role in the regulation of neurodevelopment, neuronal functions and neuroplasticity and it is frequently associated with depressive disorder and neuroinflammation [120,121]. However, recent findings on new-borns later diagnosed with ASD have demonstrated a significantly reduced level of BDNF in the blood samples [122]. In mice models, decreased levels of BDNF transcript variants were observed in germ-free mice models and antibiotic-treated SPF mice models in the amygdala regions [123–125]. These associations of BDNF with the epigenetic mechanism, ASD subjects as well as microbiome, warrant the need to further study and explore the role of BDNF in relation to ASD as currently there are no available studies to affirm this relationship.

The gut microbiome is thought to modulate gene expression through an epigenetic mechanism which may either institute the etiopathogenesis of ASD and/or aggravate ASD symptomatology primarily [126]. It has also been hypothesized that the epigenetic regulation is diet-dependent where the role of microbial metabolite, SCFAs come into play. SCFAs are the product of the ASD-associated bacterial (such as *Clostridia, Bacteriodetes, Desulfovibrio*) fermentation of dietary carbohydrates and are regulated by the gut microbial composition [115,127,128]. However, an overproduction of SCFAs (e.g., propionic acid, butyrate, acetate) have been indicated in ASD based on analysis of faecal samples of ASD children and it was inferred that this possibility could be due impaired fermentation process and utilization of its by-products without eliminating the possibility of increased faecal SCFA could also be due to an overall increased SCFA production in ASD children, which is indeed regarded beneficial [129,130]. Therefore, the microbial dysbiosis (reduced SCFA-producing, ASD-associated bacteria) manipulates the level of nutrients and metabolites like SCFA which regulate the DNA methylation and histone modification resulting in immune activation that have a potential role in ASD development. These metabolites either directly inhibit the enzymes which catalyse these epigenetic processes or alter the substrate availability required for the enzymatic process, thus highlighting the possible epigenetic mechanism in the occurrence of ASD [127].

## 5. Comorbidities in Autism Spectrum Disorder

Gastrointestinal comorbidities have been found to be significantly higher and more common in the ASD group compared to neurotypical children, with 46.8% of ASD children exhibited at least one GI symptom [131–133]. GI disorders have been associated with gut dysbiosis, although whether underlying GI disturbances result in dysbiosis or the dysbiosis results in GI disturbance largely remains a chicken-or-the-egg conundrum. Gastrointestinal symptoms (e.g., constipation, diarrhoea, bloating) have been correlated to gut dysbiosis and severity of autistic symptoms [6,65,134,135]. It was hypothesized that the gut dysbiosis gives rise to the pathogenic invasion of intestinal wall, thus resulting in the breach of the gut barrier which ultimately results in neuroinflammation responsible for ASD's behavioural manifestation [9,48]. However, some of the recent findings demonstrated reduced diversity and altered microbial pattern in autistic group compared to a neurotypical group; however, no significant correlation was found between GI symptoms and severity of ASD symptoms [31,48,136,137]. It was further inferred that GI disturbances alone could possibly trigger ASD symptoms possibly due to heightened sensory perception and experience in ASD children [134]. An interesting correlation between GI symptoms, intestinal mucosal dysbiosis, gene expression and ASD was highlighted in a study which looked at the molecular mechanism involved in explaining the GI disturbances in ASD children by analysing the intestinal biopsies obtained from 15 children with ASD and GI disease, and seven children with GI disease alone. The messenger ribonucleic acid (mRNA) deficiencies in genes encoding disaccharidases and hexose transporters responsible for carbohydrate digestion and transport across the intestinal epithelium was linked to the intestinal dysbiosis (decreased *Bacteroidetes*, increased *Firmicutes* to *Bacteroidetes* ratio, and increased *Betaproteobacteria*) in the ASD group. The pre-existing impaired carbohydrate digestion and absorption manifests as GI disturbances with dietary intake of carbohydrates which leads to fermentation, increased gas production and osmotic load in the gut of ASD children. Although no dietary evaluation was done in the children involved in this study, its finding provides a significant insight regarding the GI disturbances in ASD children, especially on how gluten-free/low-carbohydrate diets may benefit some of the ASD children [53]. In another study, the restoration of microbial imbalance with significant and lasting improvement in terms of GI symptoms and autistic behaviours in children diagnosed with ASD was made possible through microbiota transfer therapy (MTT) [138]. This highlights the influence of microbiome in both GI symptoms and autistic behaviours; however, no distinctive microbial patterns specific to ASD subjects, with or without the presence of GI disturbances, have been elicited so far.

Epilepsy and ASD have often been associated with one another; however, the possible pathophysiological link has yet to be fully understood. It has been reported that ASD subjects possess a seven-fold increased risk of developing epilepsy compared to general population, with an estimated prevalence of epilepsy in ASD up to 50% [139,140]. The prenatal use of anti-epileptics, mainly valproate, has a teratogenic effect on the neurodevelopment of the offspring and predisposes them with a risk of developing ASD [141,142]. The valproate has been demonstrated to cause microbial disruption in mouse models of ASD in the context of in utero usage [143]. Interestingly, ketogenic diet in ASD cases with refractory epilepsy has shown significant improvement in both ASD symptoms and seizure episodes through gut microbial modulation [144]. *Akkermansia* and *Parabacteroides* are two ketogenic diet associated microbes which confer the anti-seizure effect [145].

Depression and anxiety are part of neuropsychiatric disorders which have been elucidated through the gut–brain–microbiota axis [146]. These disorders are associated with the alteration of gut microbial composition, thus impacting the neurobehavioral outcome as narrated in the context of ASD [147,148].

Similar to gut microbial observation in ASD subjects, the ratio of Bacteroidetes to Firmicutes phyla were markedly increased in both animal models and human subjects with depression [149,150]. Further, a recent review on psychotropics and the microbiome has highlighted gut dysbiosis and anti-microbial exertions of antipsychotic, antidepressant and antianxiety drugs which are commonly prescribed in ASD subjects with associated psychiatric comorbidities [151,152].

## 6. Microbiota-Based Interventions and ASD

### 6.1. Dietary and Supplementary Interventions

Diet has an influential role in determining the intestinal microbial composition and in rodent models; it has been demonstrated that significant dietary change has the potential to alter the microbiome as rapidly as over one day [153]. The microbiota evaluation of ASD children differed across countries, suggesting the influence of geographical location in microbiota profile, and diet has been thought to be an important factor to explain such differences [32,154]. SCFA, the colonic by-product of bacterial fermentation of dietary fibres and resistant starch, is gaining attention as one of the important mediators of the gut microbiota–brain communication, although the exact mechanism on how this metabolite influences the brain physiology and neurobehavioral outcome is yet to be fully understood [155–159]. An intervention using a diet that promotes increased SCFA levels (Mediterranean diet, plant-based proteins, dietary fibre) may potentially exert a positive impact on ASD behavioural outcomes, although no studies have been done so far to support this relation in human subjects [158]. Dietary intervention which has been commonly advocated amongst ASD children include gluten- and casein-free diets which have been reported to confer a beneficial outcome in ASD symptoms, while more recent study shows no significant changes in ASD symptoms after a six-month trial [160–162]. This intervention is recommended only when there is a clinically diagnosed intolerance to gluten and casein [162]. In terms of maternal aspect, rodent models have significantly demonstrated that offspring born to mothers fed with high-fat diet eight weeks before mating had impaired social interaction and repetitive behaviours mimicking ASD, with altered gut microbial composition and notable nine-fold decrease in *Lactobacillus reuteri* [163]. The Western diet, also referred to as the high-fat diet, has been associated with reduced microbial diversity and richness [164]. Dietary intervention should be exercised with caution to avoid malnourishment and nutritional deficiency. Vitamin D deficiency has been commonly reported in ASD children and low prenatal vitamin D has been associated with an increased risk of ASD occurrence [165–167]. It was also found that GI problems are more evident in ASD subjects with vitamin D deficiency than those without this deficiency [166]. The supplementation of vitamin $D_3$ shows a remarkable improvement in ASD core symptoms [168].

### 6.2. Prebiotics, Probiotics, Synbiotics and Antibiotics

Prebiotics are mainly fibres consisting of non-digestible diet components which benefit the host's health by selectively promoting microbial proliferation in the colon, generally *Lactobacilli* and *Bifidobacteria*. The probiotics are live, non-pathogenic microorganisms that confer health benefits to the host when adequately administered, whereas synbiotics are formulations consisting of both prebiotic and probiotic which were meant to improve the efficiency of probiotics [169]. These interventions have been recommended as adjuvant therapies given their promising benefits in terms of improving ASD behavioural symptoms and ameliorating GI disturbances in ASD children [170,171]. The administration of probiotics consisting of three strains (*Lactobacillus acidophilus*, *Lactobacillus rhamnosus* and *Bifidobacteria longum*) for a duration of three months significantly improved autistic behaviours and GI symptoms in ASD children with increased count of the beneficial bacteria *Bifidobacteria* and *Lactobacilli* [170]. In a pre-clinical study, *Bacteroides fragilis* has also been demonstrated to restore the microbial dysbiosis, improve gut permeability and autism behaviours in the offspring of maternal immune activation (MIA) mouse models which displayed ASD behavioural features [172]. In ASD children, probiotics—namely, *Lactobacillus reuteri*, *Lactobacillus plantarum*, *Lactobacillus acidophilus*, *Lactobacillus rhamnosus* and *Bifidobacterium longum*—have been utilized in clinical trials and have significantly improved autistic and GI outcomes. These periods of trial ranged from three weeks to six months [170,173–175]. In a randomized control trial, which was carried out on 75 infants, where *Lactobacillus rhamnosus* was administered in 40 infants and a placebo in the remaining infants in control group for the first six months, the outcome revealed that the early administration of probiotic may potentially reduce the risk of developing ASD. These infants were followed over 13 years and 6 out 35 infants in the control group were diagnosed with either Asperger's syndrome (now part of ASD) or attention deficit hyperactivity disorder (ADHD), while none were diagnosed in the probiotic group [175]. It is also worth mentioning that the a causal link between maternal diet, altered gut microbiome and social behaviour was demonstrated in a breakthrough study where the administration of *Lactobacillus reuteri* over four weeks in rodent offspring born to mothers fed with a high-fat diet with reduced gut microbial diversity, particularly *Lactobacillus reuteri*, was able to significantly improve social behaviour [163]. However, one of the recent systematic reviews has pinpointed the lack of evidence in supporting the beneficial roles of probiotics and prebiotics in ASD and ascertained that more studies are needed to validate the benefits of these interventions in ASD subjects [176].

In terms of antibiotics, vancomycin and metronidazole have been used in treating ASD symptoms. However, metronidazole is not a preferred choice due to its possible risk of causing systemic adverse effects [177]. There is one case report on the administration of amoxicillin over a 10-day course which deemed to improve autism symptoms in a child as reported by his parents [178]. However, in rodent models, the maternal use of oral antibiotics (non-absorbable sulfonamide, neomycin, bacitracin, pimaricin) either during preconception or early gestation period yielded offspring with anxiety-like behaviours and impaired social interactions [179,180]. Further analysis of the faecal samples of the offspring exposed to antibiotics also showed a 50% reduction in the abundance of the *Lactobacillales* and increased *Clostridium*, thus implying that early exposure of antibiotics have a negative impact on the behavioural outcome in offspring [179].

### *6.3. Fecal Microbial Transplant (FMT) and Microbiota Transfer Therapy (MTT)*

In a study using mice models, it was demonstrated that the colonization of germ-free mothers with microbiota obtained from SPF mice 30 days prior to mating was able to "normalize" the behavioural outcome in the germ-free mice while the same produced no changes in adult germ-free mice. The elevated stress hormones level was also reversed through colonization of the germ-free mice with the microbiota of SPF mice before 6 weeks of age [124]. This does not only indicate the importance of maternal microbiome at the time of conception but also the need for an earlier intervention for an improved behavioural outcome, which may be challenging in human subjects as the ASD-like behaviour may not be apparent during the perinatal period. In an open-label study involving 18 children diagnosed with ASD, the MTT involving antibiotic treatment for an initial two weeks, followed by extended faecal microbiota transplant (FMT) over a period of 7–8 weeks, significantly improved GI disturbances (constipation, diarrhoea, abdominal pain, indigestion) as well as ASD's behavioural symptoms. Further, improved bacterial diversity with significant increase in *Bifidobacterium*, *Desulfovibrio* and *Prevotella* were observed and all these changes persisted for eight weeks after the cessation of treatment [138]. Despite its promising benefits, the faecal transplantation may possess the risk of transmitting norovirus and some autoimmune conditions (rheumatoid arthritis, Sjogren's syndrome), aspiration and even inducing obesity in recipients [181,182]. However, these are theoretical speculations which should not be simply disregarded.

## 7. Discussion

While the relation between microbiome and ASD is extensive, the immense possibility of the microbial role in ASD is indisputable, inclusive of whether viewed from an aetiological point of view or its association with various comorbidities affecting the ASD population. However, research efforts in understanding the exact mechanism involved in the context of ASD are still largely inadequate; relationships are yet to be established as causal and are rather associative at this stage, mainly due to the heterogeneous nature of this disorder. Although germ-free and animal models have been utilized to a great extent to demonstrate the clear associations of microbes in host physiology, coupled with the fact that human genome resembles 85% of mouse genome, it is near impossible to create germ-free human samples and eliminate all the confounding factors in establishing the same outcome [183]. The genetic makeup, age, sex, life exposure, environmental influences, medication and comorbidities are all possible confounding factors which need to be carefully defined in human samples [16]. Further, a larger sample size is required to affirm the clinical outcomes especially those involving microbiota-based interventions such as dietary, probiotics and faecal experiments. The influence of geographical locations and diet in ASD and its associated microbiota profile are becoming more relevant; therefore, studies fostering larger cohorts from diverse geographical locations with standardized specifications (e.g., age, gender, comorbidities) are needed to understand if the findings are consistent across the different locations and dietary habits.

Clinically, ASD patients are managed by a multi-disciplinary healthcare team primarily by a child psychiatrist, paediatrician, clinical psychologists and other physicians as per the associated comorbidities, although a regular follow-up and monitoring remains largely questionable. The role of parents and caregivers are of utmost importance in ASD. Although many maternal factors and few paternal factors have been associated with the risk of ASD, a proper parental counselling should be provided to ensure a rational understanding and awareness about ASD development. A prenatal counselling to avoid potentially teratogenic factors especially in mothers with underlying medical or psychiatric comorbidities and those with existing ASD children should not be neglected. However, it has be acknowledged that the maternal influence has been mainly described using animal models and no defined relationship between maternal factors and ASD occurrence have been established. This gap in the existing studies need to be understood to ensure the proper channelling of information to the parents, for this may possibly trigger a detrimental emotional response (anxiety, paranoia), especially in mothers who are either planning to conceive or have given birth to children with ASD. The knowledge about the availability of genetic testing should also be given to parents who wish to conceive or those with a known familial risk of neurodevelopmental disorders.

The incorporation of microbial knowledge in clinical context of ASD management provides a promising insight into defining the neurodevelopmental disorder using potential clinical biomarkers, although no definite classification of biomarkers have been established and widely adapted to date [40,75]. Clinical trials revolving around the microbial field are inadequate with a smaller sample size, lack of large population-based studies, and this is certainly attributable to the highly heterogenous and complex nature of ASD itself. The potential clinical intervention targeting gut microbiome which include prebiotics, probiotics, antibiotics, dietary and supplementary interventions, faecal microbial transplant (FMT) and a modified protocol of FMT, namely microbiota transfer therapy (MTT), are important breakthrough interventions which need to be carefully evaluated and validated using double-blinded, randomized, controlled trials involving larger sample size and standardized treatment regimens and durations. FMT and MTT seem the most promising treatments in restoring gut dysbiosis in ASD, but the safety and tolerability in a long run is still questionable [138]. The various clinical interventions have to be carefully evaluated and catered based on individual suitability and requirements. In the management of highly heterogenous disorder like ASD, personalized treatment based on appropriate clinical judgement will be more beneficial.

Although ASD has been largely identified as a childhood disorder, it should not be forgotten that the adult population is not spared from the gigantic wave of ASD. Epidemiological studies on adult autism is very limited and still remains a poorly explored topic. A survey on psychiatric morbidity in adults carried out in England revealed the prevalence rate of autism is about 1% in adults, affecting more males than females, consistent with the findings involving the cohort of children [184]. Adults with autism has also been reported to be at greater risk of socially deprived and isolated, lack of financial security and poor access to specialist care other than having poor physical and mental health outcomes [185]. The children with ASD who will eventually grow into adults should be examined and followed-up closely and consistently.

## 8. Conclusions

ASD has a major health and economic burden to the affected population and care providers. It is becoming an alarming global epidemic which calls for a greater attention and effort. The wide range of serious health comorbidities associated with ASD is often a huge concern when it comes to clinical interventions. The possible explanation and relation of microbes in terms of aetiology and ASD comorbidities point to the need for an integrative approach in treating ASD subjects. Newly emerging fields of research, such as epigenetics, provide more integrative insights and approaches in directing future research, incorporating knowledge of the microbiome. The study of epigenetics incorporating the microbiome reflects the highly dynamic process which is involved in shaping and reprogramming the growth and development of an individual. The systematic identification of environmental factors which interfere with gene regulation could possibly create new avenues in the clinical management of ASD. The treatment approach in ASD should be more individualized to ensure the best clinical outcomes. Diet and probiotics are important and promising breakthrough microbiota-based interventions in the context of ASD. These interventions have better feasibility and an easier method of administration with less known side effects compared to other interventions. Research efforts should be expanded and focused in this direction to ascertain the clinical efficacy of these interventions.

**Author Contributions:** D.J. performed the literature search, critical data analysis as well as the manuscript writing. Technical supports and proofreading were contributed by V.L., S.T., and L.-H.L. L.-H.L. set up the research project. All authors have read and agreed to the published version of the manuscript.

**Funding:** This work was supported by PVC Award Grant (Project No. PVC-ECR-2016), External Industry Grant (Biotek Abadi—Vote No. GBA-808138 and GBA-808813) awarded to L.-H.L.

**Acknowledgments:** Mohamed Shajahan Yasin, Professor and Head of School, Jeffrey Cheah School of Medicine and Health Sciences, Monash University Malaysia.

**Conflicts of Interest:** The authors declare no conflict of interest.

## Glossary

| | |
|---|---|
| Microbiome | Collectively refers to the genetic material of the entire microbial community residing on and in human body encompassing bacteria, eukaryotic viruses, fungi, protozoa, archaea and bacteriophage. |
| Microbiota | Refers to the entire microbial community residing on and in human body encompassing bacteria, eukaryotic viruses, fungi, protozoa, archaea and bacteriophage. |

## References

1. Association, A.P. *Diagnostic and Statistical Manual of Mental Disorders (DSM-5®)*; American Psychiatric Pub: Washington, DC, USA, 2013.
2. Loomes, R.; Hull, L.; Mandy, W.P.L. What is the male-to-female ratio in autism spectrum disorder? A systematic review and meta-analysis. *J. Am. Acad. Child Adolesc. Psychiatry* **2017**, *56*, 466–474. [CrossRef] [PubMed]
3. Catalá-López, F.; Ridao, M.; Hurtado, I.; Núñez-Beltrán, A.; Gènova-Maleras, R.; Alonso-Arroyo, A.; Tobías, A.; Aleixandre-Benavent, R.; Catalá, M.A.; Tabarés-Seisdedos, R. Prevalence and comorbidity of autism spectrum disorder in Spain: Study protocol for a systematic review and meta-analysis of observational studies. *Syst. Rev.* **2019**, *8*, 141. [CrossRef]
4. Vos, T.; Abajobir, A.A.; Abate, K.H.; Abbafati, C.; Abbas, K.M.; Abd-Allah, F.; Abdulkader, R.S.; Abdulle, A.M.; Abebo, T.A.; Abera, S.F. Global, regional, and national incidence, prevalence, and years lived with disability for 328 diseases and injuries for 195 countries, 1990–2016: A systematic analysis for the Global Burden of Disease Study 2016. *Lancet* **2017**, *390*, 1211–1259. [CrossRef]
5. Polyak, A.; Kubina, R.M.; Girirajan, S. Comorbidity of intellectual disability confounds ascertainment of autism: Implications for genetic diagnosis. *Am. J. Med. Genet. Part B Neuropsychiatr. Genet.* **2015**, *168*, 600–608. [CrossRef]

6. Mayer, E.A.; Padua, D.; Tillisch, K. Altered brain-gut axis in autism: Comorbidity or causative mechanisms? *Bioessays* **2014**, *36*, 933–939. [CrossRef] [PubMed]
7. Young, V.B. The role of the microbiome in human health and disease: An introduction for clinicians. *BMJ* **2017**, *356*, j831. [CrossRef]
8. Khanna, S.; Tosh, P.K. A clinician's primer on the role of the microbiome in human health and disease. *Mayo Clin. Proc.* **2014**, *89*, 107–114. [CrossRef]
9. Cani, P.D. Human gut microbiome: Hopes, threats and promises. *Gut* **2018**, *67*, 1716–1725. [CrossRef]
10. Lee, L.-H.; Ser, H.-L.; Khan, T.M.; Gan, K.-G.; Goh, B.-H.; Ab Mutalib, N.-S. Relationship between autism and gut microbiome: Current status and update. *Gut* **2019**, *68*. [CrossRef]
11. Lee, L.-H.; Letchumanan, V.; Khan, T.M.; Long, M.; Chan, K.-G.; Goh, B.-H.; Ab Mutalib, N.-S. Role of human microbiota in skin dermatitis and eczema: A systematic review. *Gut* **2018**, *67*, A1–A118. [CrossRef]
12. Selvaraj, S.M.; Wong, S.H.; Ser, H.-L.; Lee, L.-H. Role of low FODMAP diet and probiotics on gut microbiome in irritable bowel syndrome (IBS). *Prog. Microbes Mol. Biol.* **2020**, *3*. [CrossRef]
13. Durganaudu, H.; Kunasegaran, T.; Ramadas, A. dietary glycaemic index and Type 2 diabetes mellitus: Potential modulation of gut microbiota. *Prog. Microbes Mol. Biol.* **2020**, *3*. [CrossRef]
14. Lee, L.-H.; Letchumanan, V.; Khan, T.M.; Chan, K.-G.; Goh, B.-H.; Ab Mutalib, N.-S. IDDF2019-ABS-0322 Dissecting the gut and skin: Budding association between gut microbiome in the development to psoriasis? *Gut* **2019**, *68*. [CrossRef]
15. Proctor, L.M. The national institutes of health human microbiome project. *Semin. Fetal Neonatal. Med.* **2016**, *21*, 368–372. [CrossRef] [PubMed]
16. Vuong, H.E.; Yano, J.M.; Fung, T.C.; Hsiao, E.Y. The microbiome and host behavior. *Annu. Rev. Neurosci.* **2017**, *40*, 21–49. [CrossRef]
17. Collins, S.M.; Surette, M.; Bercik, P. The interplay between the intestinal microbiota and the brain. *Nat. Rev. Microbiol.* **2012**, *10*, 735–742. [CrossRef] [PubMed]
18. Sivamaruthi, B.S.; Suganthy, N.; Kesika, P.; Chaiyasut, C. The Role of Microbiome, Dietary Supplements, and Probiotics in Autism Spectrum Disorder. *Int. J. Environ. Res. Public Health* **2020**, *17*, 2647. [CrossRef] [PubMed]
19. Bonaz, B.; Bazin, T.; Pellissier, S. The vagus nerve at the interface of the microbiota-gut-brain axis. *Front. Neurosci.* **2018**, *12*, 49. [CrossRef] [PubMed]
20. Jašarević, E.; Rodgers, A.B.; Bale, T.L. A novel role for maternal stress and microbial transmission in early life programming and neurodevelopment. *Neurobiol. Stress* **2015**, *1*, 81–88. [CrossRef]
21. Mason, M.R.; Chambers, S.; Dabdoub, S.M.; Thikkurissy, S.; Kumar, P.S. Characterizing oral microbial communities across dentition states and colonization niches. *Microbiome* **2018**, *6*, 67. [CrossRef]
22. Fiorentino, M.; Sapone, A.; Senger, S.; Camhi, S.S.; Kadzielski, S.M.; Buie, T.M.; Kelly, D.L.; Cascella, N.; Fasano, A. Blood–brain barrier and intestinal epithelial barrier alterations in autism spectrum disorders. *Mol. Autism* **2016**, *7*, 49. [CrossRef] [PubMed]
23. Engelhardt, B. Development of the blood-brain barrier. *Cell Tissue Res.* **2003**, *314*, 119–129. [CrossRef]
24. Hensch, T.K. Critical period regulation. *Annu. Rev. Neurosci.* **2004**, *27*, 549–579. [CrossRef] [PubMed]
25. Knudsen, E.I. Sensitive periods in the development of the brain and behavior. *J. Cogn. Neurosci.* **2004**, *16*, 1412–1425. [CrossRef]
26. Braniste, V.; Al-Asmakh, M.; Kowal, C.; Anuar, F.; Abbaspour, A.; Tóth, M.; Korecka, A.; Bakocevic, N.; Ng, L.G.; Kundu, P. The gut microbiota influences blood-brain barrier permeability in mice. *Sci. Transl. Med.* **2014**, *6*, 263ra158. [CrossRef] [PubMed]
27. Sandin, S.; Lichtenstein, P.; Kuja-Halkola, R.; Hultman, C.; Larsson, H.; Reichenberg, A. The heritability of autism spectrum disorder. *JAMA* **2017**, *318*, 1182–1184. [CrossRef]
28. Siu, M.T.; Weksberg, R. Epigenetics of autism spectrum disorder. In *Neuroepigenomics in Aging and Disease*; Springer: New York, NY, USA, 2017; pp. 63–90.
29. Hallmayer, J.; Cleveland, S.; Torres, A.; Phillips, J.; Cohen, B.; Torigoe, T.; Miller, J.; Fedele, A.; Collins, J.; Smith, K. Genetic heritability and shared environmental factors among twin pairs with autism. *Arch. Gen. Psychiatry* **2011**, *68*, 1095–1102. [CrossRef]
30. Zhu, S.; Jiang, Y.; Xu, K.; Cui, M.; Ye, W.; Zhao, G.; Jin, L.; Chen, X. The progress of gut microbiome research related to brain disorders. *J. Neuroinflamm.* **2020**, *17*, 25. [CrossRef]

31. Ahmed, S.A.; Elhefnawy, A.M.; Azouz, H.G.; Roshdy, Y.S.; Ashry, M.H.; Ibrahim, A.E.; Meheissen, M.A. Study of the gut Microbiome Profile in Children with Autism Spectrum Disorder: A Single Tertiary Hospital Experience. *J. Mol. Neurosci.* **2020**, *70*, 887–896. [CrossRef]
32. Saurman, V.; Margolis, K.G.; Luna, R.A. Autism Spectrum Disorder as a Brain-Gut-Microbiome Axis Disorder. *Dig. Dis. Sci.* **2020**, *65*, 818–828. [CrossRef]
33. Mannion, A.; Leader, G. Comorbidity in autism spectrum disorder: A literature review. *Res. Autism Spectr. Disord.* **2013**, *7*, 1595–1616. [CrossRef]
34. Buckley, A.W.; Holmes, G.L. Epilepsy and autism. *Cold Spring Harb. Perspect. Med.* **2016**, *6*, a022749. [CrossRef] [PubMed]
35. Davignon, M.N.; Qian, Y.; Massolo, M.; Croen, L.A. Psychiatric and medical conditions in transition-aged individuals with ASD. *Pediatrics* **2018**, *141*, S335–S345. [CrossRef]
36. Volkmar, F.R.; McPartland, J.C. From Kanner to DSM-5: Autism as an evolving diagnostic concept. *Annu. Rev. Clin. Psychol.* **2014**, *10*, 193–212. [CrossRef] [PubMed]
37. Wiggins, L.D.; Rice, C.E.; Barger, B.; Soke, G.N.; Lee, L.-C.; Moody, E.; Edmondson-Pretzel, R.; Levy, S.E. DSM-5 criteria for autism spectrum disorder maximizes diagnostic sensitivity and specificity in preschool children. *Soc. Psychiatry Psychiatr. Epidemiol.* **2019**, *54*, 693–701. [CrossRef] [PubMed]
38. Baio, J.; Wiggins, L.; Christensen, D.L.; Maenner, M.J.; Daniels, J.; Warren, Z.; Kurzius-Spencer, M.; Zahorodny, W.; Rosenberg, C.R.; White, T. Prevalence of autism spectrum disorder among children aged 8 years—Autism and developmental disabilities monitoring network, 11 sites, United States, 2014. *MMWR Surveill. Summ.* **2018**, *67*, 1. [CrossRef]
39. Brett, D.; Warnell, F.; McConachie, H.; Parr, J.R. Factors affecting age at ASD diagnosis in UK: No evidence that diagnosis age has decreased between 2004 and 2014. *J. Autism Dev. Disord.* **2016**, *46*, 1974–1984. [CrossRef] [PubMed]
40. Venigalla, H.; Mekala, H.M.; Hassan, M.; Ahmed, R.; Zain, H.; Dar, S.; Veliz, S. An update on biomarkers in psychiatric disorders–are we aware, do we use in our clinical practice. *Ment. Health Fam. Med.* **2017**, *13*, 471–479.
41. Crespi, B.J. Revisiting Bleuler: Relationship between autism and schizophrenia. *Br. J. Psychiatry* **2010**, *196*, 495. [CrossRef]
42. Cook, K.A.; Willmerdinger, A.N. The History of Autism. 2015. Available online: https://scholarexchange.furman.edu/schopler-about/1 (accessed on 29 May 2020).
43. Kanner, L. Autistic disturbances of affective contact. *Nerv. Child* **1943**, *2*, 217–250.
44. Asperger, H. Die "Autistischen psychopathen" im kindesalter. *Arch. Psychiatr. Nervenkr.* **1944**, *117*, 76–136. [CrossRef]
45. Qin, J.; Li, R.; Raes, J.; Arumugam, M.; Burgdorf, K.S.; Manichanh, C.; Nielsen, T.; Pons, N.; Levenez, F.; Yamada, T. A human gut microbial gene catalogue established by metagenomic sequencing. *Nature* **2010**, *464*, 59–65. [CrossRef] [PubMed]
46. Sender, R.; Fuchs, S.; Milo, R. Revised estimates for the number of human and bacteria cells in the body. *PLoS Biol.* **2016**, *14*, e1002533. [CrossRef] [PubMed]
47. Umbrello, G.; Esposito, S. Microbiota and neurologic diseases: Potential effects of probiotics. *J. Transl. Med.* **2016**, *14*, 298. [CrossRef] [PubMed]
48. Krajmalnik-Brown, R.; Kang, D.-W.; Park, J.G.; Labaer, J.; Ilhan, Z. Microbiome Markers and Therapies for Autism Spectrum Disorders. U.S. Patent No. 16/118,061, 16 May 2019.
49. Anwar, H.; Irfan, S.; Hussain, G.; Faisal, M.N.; Muzaffar, H.; Mustafa, I.; Mukhtar, I.; Malik, S.; Ullah, M.I. Gut Microbiome: A New Organ System in Body. In *Eukaryotic Microbiology*; IntechOpen: London, UK, 2019. [CrossRef]
50. Principi, N.; Esposito, S. Gut microbiota and central nervous system development. *J. Infect.* **2016**, *73*, 536–546. [CrossRef] [PubMed]
51. Finegold, S.M.; Dowd, S.E.; Gontcharova, V.; Liu, C.; Henley, K.E.; Wolcott, R.D.; Youn, E.; Summanen, P.H.; Granpeesheh, D.; Dixon, D. Pyrosequencing study of fecal microflora of autistic and control children. *Anaerobe* **2010**, *16*, 444–453. [CrossRef]
52. Zhang, M.; Ma, W.; Zhang, J.; He, Y.; Wang, J. Analysis of gut microbiota profiles and microbe-disease associations in children with autism spectrum disorders in China. *Sci. Rep.* **2018**, *8*, 13981. [CrossRef]

53. Williams, B.L.; Hornig, M.; Buie, T.; Bauman, M.L.; Cho Paik, M.; Wick, I.; Bennett, A.; Jabado, O.; Hirschberg, D.L.; Lipkin, W.I. Impaired carbohydrate digestion and transport and mucosal dysbiosis in the intestines of children with autism and gastrointestinal disturbances. *PLoS ONE* **2011**, *6*, e24585. [CrossRef] [PubMed]
54. Qiao, Y.; Wu, M.; Feng, Y.; Zhou, Z.; Chen, L.; Chen, F. Alterations of oral microbiota distinguish children with autism spectrum disorders from healthy controls. *Sci. Rep.* **2018**, *8*, 1597. [CrossRef]
55. Van Ameringen, M.; Turna, J.; Patterson, B.; Pipe, A.; Mao, R.Q.; Anglin, R.; Surette, M.G. The gut microbiome in psychiatry: A primer for clinicians. *Depress. Anxiety* **2019**, *36*, 1004–1025. [CrossRef]
56. Bronson, S.L.; Bale, T.L. Prenatal stress-induced increases in placental inflammation and offspring hyperactivity are male-specific and ameliorated by maternal antiinflammatory treatment. *Endocrinology* **2014**, *155*, 2635–2646. [CrossRef]
57. Estes, M.L.; McAllister, A.K. Maternal immune activation: Implications for neuropsychiatric disorders. *Science* **2016**, *353*, 772–777. [CrossRef] [PubMed]
58. Connolly, N.; Anixt, J.; Manning, P.; Ping-I Lin, D.; Marsolo, K.A.; Bowers, K. Maternal metabolic risk factors for autism spectrum disorder—An analysis of electronic medical records and linked birth data. *Autism Res.* **2016**, *9*, 829–837. [CrossRef]
59. Wang, Y.; Kasper, L.H. The role of microbiome in central nervous system disorders. *Brain. Behav. Immun.* **2014**, *38*, 1–12. [CrossRef] [PubMed]
60. Yassour, M.; Vatanen, T.; Siljander, H.; Hämäläinen, A.-M.; Härkönen, T.; Ryhänen, S.J.; Franzosa, E.A.; Vlamakis, H.; Huttenhower, C.; Gevers, D. Natural history of the infant gut microbiome and impact of antibiotic treatment on bacterial strain diversity and stability. *Sci. Transl. Med.* **2016**, *8*, 343ra381. [CrossRef]
61. Korpela, K.; Salonen, A.; Virta, L.J.; Kekkonen, R.A.; Forslund, K.; Bork, P.; De Vos, W.M. Intestinal microbiome is related to lifetime antibiotic use in Finnish pre-school children. *Nat. Commun.* **2016**, *7*, 10410. [CrossRef]
62. Sandler, R.H.; Finegold, S.M.; Bolte, E.R.; Buchanan, C.P.; Maxwell, A.P.; Väisänen, M.-L.; Nelson, M.N.; Wexler, H.M. Short-term benefit from oral vancomycin treatment of regressive-onset autism. *J. Child Neurol.* **2000**, *15*, 429–435. [CrossRef] [PubMed]
63. Collado, M.C.; Rautava, S.; Aakko, J.; Isolauri, E.; Salminen, S. Human gut colonisation may be initiated in utero by distinct microbial communities in the placenta and amniotic fluid. *Sci. Rep.* **2016**, *6*, 23129. [CrossRef]
64. Jiménez, E.; Marín, M.L.; Martín, R.; Odriozola, J.M.; Olivares, M.; Xaus, J.; Fernández, L.; Rodríguez, J.M. Is meconium from healthy newborns actually sterile? *Res. Microbiol.* **2008**, *159*, 187–193. [CrossRef]
65. Fattorusso, A.; Di Genova, L.; Dell'Isola, G.B.; Mencaroni, E.; Esposito, S. Autism spectrum disorders and the gut microbiota. *Nutrients* **2019**, *11*, 521. [CrossRef]
66. Jašarević, E.; Howerton, C.L.; Howard, C.D.; Bale, T.L. Alterations in the vaginal microbiome by maternal stress are associated with metabolic reprogramming of the offspring gut and brain. *Endocrinology* **2015**, *156*, 3265–3276. [CrossRef] [PubMed]
67. Cribby, S.; Taylor, M.; Reid, G. Vaginal microbiota and the use of probiotics. *Interdiscip. Perspect. Infect. Dis.* **2008**, *2008*. [CrossRef]
68. Saunders, S.; Bocking, A.; Challis, J.; Reid, G. Effect of Lactobacillus challenge on Gardnerella vaginalis biofilms. *Colloids Surf. B. Biointerfaces* **2007**, *55*, 138–142. [CrossRef] [PubMed]
69. Bokobza, C.; Van Steenwinckel, J.; Mani, S.; Mezger, V.; Fleiss, B.; Gressens, P. Neuroinflammation in preterm babies and autism spectrum disorders. *Pediatr. Res.* **2019**, *85*, 155–165. [CrossRef]
70. Joseph, R.M.; O'Shea, T.M.; Allred, E.N.; Heeren, T.; Hirtz, D.; Paneth, N.; Leviton, A.; Kuban, K.C. Prevalence and associated features of autism spectrum disorder in extremely low gestational age newborns at age 10 years. *Autism Res.* **2017**, *10*, 224–232. [CrossRef] [PubMed]
71. Careaga, M.; Murai, T.; Bauman, M.D. Maternal immune activation and autism spectrum disorder: From rodents to nonhuman and human primates. *Biol. Psychiatry* **2017**, *81*, 391–401. [CrossRef]
72. Brown, A.S. Epidemiologic studies of exposure to prenatal infection and risk of schizophrenia and autism. *Dev. Neurobiol.* **2012**, *72*, 1272–1276. [CrossRef]
73. Lai, B.; Milano, M.; Roberts, M.W.; Hooper, S.R. Unmet dental needs and barriers to dental care among children with autism spectrum disorders. *J. Autism Dev. Disord.* **2012**, *42*, 1294–1303. [CrossRef]

74. Hicks, S.D.; Uhlig, R.; Afshari, P.; Williams, J.; Chroneos, M.; Tierney-Aves, C.; Wagner, K.; Middleton, F.A. Oral microbiome activity in children with autism spectrum disorder. *Autism Res.* **2018**, *11*, 1286–1299. [CrossRef]
75. Kong, X.; Liu, J.; Cetinbas, M.; Sadreyev, R.; Koh, M.; Huang, H.; Adeseye, A.; He, P.; Zhu, J.; Russell, H. New and Preliminary Evidence on Altered Oral and Gut Microbiota in Individuals with Autism Spectrum Disorder (ASD): Implications for ASD Diagnosis and Subtyping Based on Microbial Biomarkers. *Nutrients* **2019**, *11*, 2128. [CrossRef]
76. Bercik, P.; Park, A.; Sinclair, D.; Khoshdel, A.; Lu, J.; Huang, X.; Deng, Y.; Blennerhassett, P.; Fahnestock, M.; Moine, D. The anxiolytic effect of Bifidobacterium longum NCC3001 involves vagal pathways for gut–brain communication. *Neurogastroenterol. Motil.* **2011**, *23*, 1132–1139. [CrossRef] [PubMed]
77. Cermak, S.A.; Curtin, C.; Bandini, L.G. Food selectivity and sensory sensitivity in children with autism spectrum disorders. *J. Am. Diet. Assoc.* **2010**, *110*, 238–246. [CrossRef] [PubMed]
78. Tierney, C.; Mayes, S.; Lohs, S.R.; Black, A.; Gisin, E.; Veglia, M. How valid is the checklist for autism spectrum disorder when a child has apraxia of speech? *J. Dev. Behav. Pediatr.* **2015**, *36*, 569–574. [CrossRef] [PubMed]
79. Olsen, I.; Singhrao, S.K. Can oral infection be a risk factor for Alzheimer's disease? *J. Oral Microbiol.* **2015**, *7*, 29143. [CrossRef] [PubMed]
80. Ranjan, R.; Abhinay, A.; Mishra, M. Can oral microbial infections be a risk factor for neurodegeneration? A review of the literature. *Neurol. India* **2018**, *66*, 344.
81. Olsen, I.; Hicks, S.D. Oral microbiota and autism spectrum disorder (ASD). *J. Oral Microbiol.* **2020**, *12*, 1702806. [CrossRef]
82. Segata, N.; Haake, S.K.; Mannon, P.; Lemon, K.P.; Waldron, L.; Gevers, D.; Huttenhower, C.; Izard, J. Composition of the adult digestive tract bacterial microbiome based on seven mouth surfaces, tonsils, throat and stool samples. *Genome Biol.* **2012**, *13*, R42. [CrossRef]
83. Winter, S.E.; Lopez, C.A.; Bäumler, A.J. The dynamics of gut-associated microbial communities during inflammation. *EMBO Rep.* **2013**, *14*, 319–327. [CrossRef]
84. Hajishengallis, G.; Darveau, R.P.; Curtis, M.A. The keystone-pathogen hypothesis. *Nat. Rev. Microbiol.* **2012**, *10*, 717–725. [CrossRef]
85. Darveau, R.; Hajishengallis, G.; Curtis, M. Porphyromonas gingivalis as a potential community activist for disease. *J. Dent. Res.* **2012**, *91*, 816–820. [CrossRef]
86. Jaber, M.A. Dental caries experience, oral health status and treatment needs of dental patients with autism. *J. Appl. Oral Sci.* **2011**, *19*, 212–217. [CrossRef]
87. Granulicatella, C. Changes in the salivary microbiota of oral leukoplakia and oral cancer. *Oral Oncol.* **2016**, *56*, e6–e8.
88. Aas, J.A.; Paster, B.J.; Stokes, L.N.; Olsen, I.; Dewhirst, F.E. Defining the normal bacterial flora of the oral cavity. *J. Clin. Microbiol.* **2005**, *43*, 5721–5732. [CrossRef] [PubMed]
89. Jeffcoat, M.K.; Hauth, J.C.; Geurs, N.C.; Reddy, M.S.; Cliver, S.P.; Hodgkins, P.M.; Goldenberg, R.L. Periodontal disease and preterm birth: Results of a pilot intervention study. *J. Periodontol.* **2003**, *74*, 1214–1218. [CrossRef] [PubMed]
90. Han, Y.W.; Ikegami, A.; Bissada, N.F.; Herbst, M.; Redline, R.W.; Ashmead, G.G. Transmission of an uncultivated Bergeyella strain from the oral cavity to amniotic fluid in a case of preterm birth. *J. Clin. Microbiol.* **2006**, *44*, 1475–1483. [CrossRef]
91. Aagaard, K.; Ganu, R.; Ma, J.; Racusin, D.; Arndt, M.; Riehle, K.; Petrosino, J.; Versalovic, J. 8: Whole metagenomic shotgun sequencing reveals a vibrant placental microbiome harboring metabolic function. *Am. J. Obstet. Gynecol.* **2013**, *208*, S5. [CrossRef]
92. Bearfield, C.; Davenport, E.S.; Sivapathasundaram, V.; Allaker, R.P. Possible association between amniotic fluid micro-organism infection and microflora in the mouth. *BJOG* **2002**, *109*, 527–533. [CrossRef]
93. Alanen, A.; Laurikainen, E. Second-trimester abortion caused by Capnocytophaga sputigena: Case report. *Am. J. Perinatol.* **1999**, *16*, 181–183. [CrossRef]
94. Xiao, J.; Fiscella, K.A.; Gill, S.R. Oral microbiome: Possible harbinger for children's health. *Int. J. Oral Sci.* **2020**, *12*, 1–13. [CrossRef]
95. Xiong, J.; Chen, S.; Pang, N.; Deng, X.; Yang, L.; He, F.; Wu, L.; Chen, C.; Yin, F.; Peng, J. Neurological diseases with autism spectrum disorder: Role of ASD risk genes. *Front. Neurosci.* **2019**, *13*, 349. [CrossRef]

96. Mazina, V.; Gerdts, J.; Trinh, S.; Ankenman, K.; Ward, T.; Dennis, M.Y.; Girirajan, S.; Eichler, E.E.; Bernier, R. Epigenetics of autism-related impairment: Copy number variation and maternal infection. *J. Dev. Behav. Pediatr.* **2015**, *36*, 61–67. [CrossRef] [PubMed]
97. Landgrave-Gómez, J.; Mercado-Gómez, O.; Guevara-Guzmán, R. Epigenetic mechanisms in neurological and neurodegenerative diseases. *Front. Cell. Neurosci.* **2015**, *9*, 58. [CrossRef] [PubMed]
98. Tick, B.; Bolton, P.; Happé, F.; Rutter, M.; Rijsdijk, F. Heritability of autism spectrum disorders: A meta-analysis of twin studies. *J. Child Psychol. Psychiatry* **2016**, *57*, 585–595. [CrossRef]
99. Kember, R.; Ji, X.; Zhang, J.; Brown, C.; Rader, D.; Almasy, L.; Bucan, M. Spectrum of common and rare mutations contributing to autism risk in families. *Eur. Neuropsychopharmacol.* **2019**, *29*, S962–S963. [CrossRef]
100. Leblond, C.S.; Cliquet, F.; Carton, C.; Huguet, G.; Mathieu, A.; Kergrohen, T.; Buratti, J.; Lemière, N.; Cuisset, L.; Bienvenu, T. Both rare and common genetic variants contribute to autism in the Faroe Islands. *NPJ Genom. Med.* **2019**, *4*, 1. [CrossRef]
101. Banerjee-Basu, S.; Packer, A. *SFARI Gene: An Evolving Database for the Autism Research Community*; The Company of Biologists Ltd.: Cambridge, UK, 2010.
102. Kosmicki, J.; He, L.; Samocha, K.; Robinson, E.; Barrett, J.; Daly, M. Meta-analysis of 9246 neurodevelopmental disorder probands identifies 8 novel genes and finds de novo mutations in prior associated autism spectrum disorder genes are more often observed in probands without ASD. *Eur. Neuropsychopharmacol.* **2019**, *29*, S785–S786. [CrossRef]
103. Wilfert, A.B.; Sulovari, A.; Turner, T.N.; Coe, B.P.; Eichler, E.E. Recurrent de novo mutations in neurodevelopmental disorders: Properties and clinical implications. *Genome Med.* **2017**, *9*, 101. [CrossRef]
104. Liu, D.; Wang, Z. Identification and Validation Novel Risk Genes for Autism Spectrum Disorder—A Meta-Analysis. *J. Psychiatry Brain Sci.* **2017**, *2*. [CrossRef]
105. Xu, C.; Zhang, F.; Amey, R.; Yao, Y. Weak Association between Autism Spectrum Disorder and Two Genes YBX3 and HSPA1A–A Meta-Analysis. *J. Psychiatr. Brain Sci.* **2017**, *2*, 1–9. [CrossRef]
106. De Kluiver, H.; Buizer-Voskamp, J.E.; Dolan, C.V.; Boomsma, D.I. Paternal age and psychiatric disorders: A review. *Am. J. Med. Genet.* **2017**, *174*, 202–213. [CrossRef]
107. Chiang, T.-L.; Lin, S.-J.; Lee, M.-C.; Shu, B.-C. Advanced maternal age and maternal education disparity in children with autism spectrum disorder. *Matern. Child Health J.* **2018**, *22*, 941–949. [CrossRef]
108. Alibek, K.; Farmer, S.; Tskhay, A.; Moldakozhayev, A.; Isakov, T. Prevalence of Prenatal, Neonatal and Postnatal Complications among Healthy Children and Children Diagnosed with ASD in Central Asia and Eastern Europe. *J. Gynaecol. Neonatal* **2019**, *2*, 103.
109. Bölte, S.; Girdler, S.; Marschik, P.B. The contribution of environmental exposure to the etiology of autism spectrum disorder. *Cell. Mol. Life Sci.* **2019**, *76*, 1275–1297. [CrossRef] [PubMed]
110. Sahana, K.; Bhat, S.S.; Kakunje, A. Study of prenatal, natal, and neonatal risk factors associated with autism. *Indian J. Child Health* **2018**. [CrossRef]
111. Davidovitch, M.; Kuint, J.; Lerner-Geva, L.; Zaslavsky-Paltiel, I.; Rotem, R.S.; Chodick, G.; Shalev, V.; Reichman, B. Postnatal steroid therapy is associated with autism spectrum disorder in children and adolescents of very low birth weight infants. *Pediatr. Res.* **2019**, *87*, 1045–1051. [CrossRef]
112. Abdallah, M.W.; Larsen, N.; Grove, J.; Nørgaard-Pedersen, B.; Thorsen, P.; Mortensen, E.L.; Hougaard, D.M. Amniotic fluid inflammatory cytokines: Potential markers of immunologic dysfunction in autism spectrum disorders. *World J. Biol. Psychiatry* **2013**, *14*, 528–538. [CrossRef]
113. Kim, D.; Volk, H.; Girirajan, S.; Pendergrass, S.; Hall, M.A.; Verma, S.S.; Schmidt, R.J.; Hansen, R.L.; Ghosh, D.; Ludena-Rodriguez, Y. The joint effect of air pollution exposure and copy number variation on risk for autism. *Autism Res.* **2017**, *10*, 1470–1480. [CrossRef]
114. Reed, Z.; Larsson, H.; Haworth, C.; Thomas, R.; Boyd, A.; Smith, G.D.; Plomin, R.; Lichtenstein, P.; Davis, O. Geographical gene-environment interaction in ASD and ADHD traits. *Behav. Genet.* **2019**, *49*, 519.
115. Bhat, M.I.; Kapila, R. Dietary metabolites derived from gut microbiota: Critical modulators of epigenetic changes in mammals. *Nutr. Rev.* **2017**, *75*, 374–389. [CrossRef]
116. Loke, Y.J.; Hannan, A.J.; Craig, J.M. The role of epigenetic change in autism spectrum disorders. *Front. Neurol.* **2015**, *6*, 107. [CrossRef]
117. Stilling, R.M.; Dinan, T.G.; Cryan, J.F. Microbial genes, brain & behaviour–epigenetic regulation of the gut–brain axis. *Genes Brain Behav.* **2014**, *13*, 69–86. [CrossRef]

118. Forssberg, H. Microbiome programming of brain development: Implications for neurodevelopmental disorders. *Dev. Med. Child Neurol.* **2019**, *61*, 744–749. [CrossRef] [PubMed]
119. Murgatroyd, C.; Spengler, D. Epigenetics of early child development. *Front. Psychiatry* **2011**, *2*, 16. [CrossRef] [PubMed]
120. Butler, M.I.; Cryan, J.F.; Dinan, T.G. Man and the microbiome: A new theory of everything? *Annu. Rev. Clin. Psychol.* **2019**, *15*, 371–398. [CrossRef]
121. Giacobbo, B.L.; Doorduin, J.; Klein, H.C.; Dierckx, R.A.; Bromberg, E.; de Vries, E.F. Brain-derived neurotrophic factor in brain disorders: Focus on neuroinflammation. *Mol. Neurobiol.* **2019**, *56*, 3295–3312. [CrossRef] [PubMed]
122. Skogstrand, K.; Hagen, C.M.; Borbye-Lorenzen, N.; Christiansen, M.; Bybjerg-Grauholm, J.; Bækvad-Hansen, M.; Werge, T.; Børglum, A.; Mors, O.; Nordentoft, M. Reduced neonatal brain-derived neurotrophic factor is associated with autism spectrum disorders. *Transl. Psychiatry* **2019**, *9*, 1–9. [CrossRef]
123. Arentsen, T.; Raith, H.; Qian, Y.; Forssberg, H.; Heijtz, R.D. Host microbiota modulates development of social preference in mice. *Microb. Ecol. Health Dis.* **2015**, *26*, 29719. [CrossRef]
124. Heijtz, R.D.; Wang, S.; Anuar, F.; Qian, Y.; Björkholm, B.; Samuelsson, A.; Hibberd, M.L.; Forssberg, H.; Pettersson, S. Normal gut microbiota modulates brain development and behavior. *Proc. Natl. Acad. Sci. USA* **2011**, *108*, 3047–3052. [CrossRef]
125. Bercik, P.; Denou, E.; Collins, J.; Jackson, W.; Lu, J.; Jury, J.; Deng, Y.; Blennerhassett, P.; Macri, J.; McCoy, K.D. The intestinal microbiota affect central levels of brain-derived neurotropic factor and behavior in mice. *Gastroenterology* **2011**, *141*, 599–609.e3. [CrossRef]
126. Frye, R.E.; Slattery, J.; MacFabe, D.F.; Allen-Vercoe, E.; Parker, W.; Rodakis, J.; Adams, J.B.; Krajmalnik-Brown, R.; Bolte, E.; Kahler, S. Approaches to studying and manipulating the enteric microbiome to improve autism symptoms. *Microb. Ecol. Health Dis.* **2015**, *26*, 26878. [CrossRef]
127. Miro-Blanch, J.; Yanes, O. Epigenetic Regulation at the Interplay Between Gut Microbiota and Host Metabolism. *Front. Genet.* **2019**, *10*. [CrossRef] [PubMed]
128. MacFabe, D.F. Short-chain fatty acid fermentation products of the gut microbiome: Implications in autism spectrum disorders. *Microb. Ecol. Health Dis.* **2012**, *23*, 19260. [CrossRef] [PubMed]
129. Wang, L.; Christophersen, C.T.; Sorich, M.J.; Gerber, J.P.; Angley, M.T.; Conlon, M.A. Elevated fecal short chain fatty acid and ammonia concentrations in children with autism spectrum disorder. *Dig. Dis. Sci.* **2012**, *57*, 2096–2102. [CrossRef] [PubMed]
130. Bird, A.; Conlon, M.; Christophersen, C.; Topping, D. Resistant starch, large bowel fermentation and a broader perspective of prebiotics and probiotics. *Benef. Microbes* **2010**, *1*, 423–431. [CrossRef]
131. McElhanon, B.O.; McCracken, C.; Karpen, S.; Sharp, W.G. Gastrointestinal symptoms in autism spectrum disorder: A meta-analysis. *Pediatrics* **2014**, *133*, 872–883. [CrossRef] [PubMed]
132. Holingue, C.; Newill, C.; Lee, L.C.; Pasricha, P.J.; Daniele Fallin, M. Gastrointestinal symptoms in autism spectrum disorder: A review of the literature on ascertainment and prevalence. *Autism Res.* **2018**, *11*, 24–36. [CrossRef]
133. Fulceri, F.; Morelli, M.; Santocchi, E.; Cena, H.; Del Bianco, T.; Narzisi, A.; Calderoni, S.; Muratori, F. Gastrointestinal symptoms and behavioral problems in preschoolers with Autism Spectrum Disorder. *Dig. Liver Dis.* **2016**, *48*, 248–254. [CrossRef]
134. Neuhaus, E.; Bernier, R.A.; Tham, S.W.; Webb, S.J. Gastrointestinal and psychiatric symptoms among children and adolescents with autism spectrum disorder. *Front. Psychiatry* **2018**, *9*, 515. [CrossRef]
135. Adams, J.B.; Johansen, L.J.; Powell, L.D.; Quig, D.; Rubin, R.A. Gastrointestinal flora and gastrointestinal status in children with autism–comparisons to typical children and correlation with autism severity. *BMC Gastroenterol.* **2011**, *11*, 22. [CrossRef]
136. Kang, D.-W.; Park, J.G.; Ilhan, Z.E.; Wallstrom, G.; LaBaer, J.; Adams, J.B.; Krajmalnik-Brown, R. Reduced incidence of Prevotella and other fermenters in intestinal microflora of autistic children. *PLoS ONE* **2013**, *8*, e68322. [CrossRef]
137. Mazefsky, C.A.; Schreiber, D.R.; Olino, T.M.; Minshew, N.J. The association between emotional and behavioral problems and gastrointestinal symptoms among children with high-functioning autism. *Autism* **2014**, *18*, 493–501. [CrossRef] [PubMed]

138. Kang, D.-W.; Adams, J.B.; Gregory, A.C.; Borody, T.; Chittick, L.; Fasano, A.; Khoruts, A.; Geis, E.; Maldonado, J.; McDonough-Means, S.; et al. Microbiota Transfer Therapy alters gut ecosystem and improves gastrointestinal and autism symptoms: An open-label study. *Microbiome* **2017**, *5*, 10. [CrossRef] [PubMed]
139. Spence, S.J.; Schneider, M.T. The role of epilepsy and epileptiform EEGs in autism spectrum disorders. *Pediatr. Res.* **2009**, *65*, 599–606. [CrossRef] [PubMed]
140. Thomas, S.; Hovinga, M.E.; Rai, D.; Lee, B.K. Brief report: Prevalence of co-occurring epilepsy and autism spectrum disorder: The US National Survey of Children's Health 2011–2012. *J. Autism Dev. Disord.* **2017**, *47*, 224–229. [CrossRef] [PubMed]
141. Gerard, E.E.; Meador, K.J. An update on maternal use of antiepileptic medications in pregnancy and neurodevelopment outcomes. *J. Pediat. Genet.* **2015**, *4*, 94–110. [CrossRef]
142. Tartaglione, A.M.; Schiavi, S.; Calamandrei, G.; Trezza, V. Prenatal valproate in rodents as a tool to understand the neural underpinnings of social dysfunctions in autism spectrum disorder. *Neuropharmacology* **2019**, *159*, 107477. [CrossRef]
143. De Theije, C.G.; Wopereis, H.; Ramadan, M.; van Eijndthoven, T.; Lambert, J.; Knol, J.; Garssen, J.; Kraneveld, A.D.; Oozeer, R. Altered gut microbiota and activity in a murine model of autism spectrum disorders. *Brain. Behav. Immun.* **2014**, *37*, 197–206. [CrossRef]
144. El-Rashidy, O.; El-Baz, F.; El-Gendy, Y.; Khalaf, R.; Reda, D.; Saad, K. Ketogenic diet versus gluten free casein free diet in autistic children: A case-control study. *Metab. Brain Dis.* **2017**, *32*, 1935–1941. [CrossRef]
145. Olson, C.A.; Vuong, H.E.; Yano, J.M.; Liang, Q.Y.; Nusbaum, D.J.; Hsiao, E.Y. The gut microbiota mediates the anti-seizure effects of the ketogenic diet. *Cell* **2018**, *173*, 1728–1741.e13. [CrossRef]
146. Lach, G.; Schellekens, H.; Dinan, T.G.; Cryan, J.F. Anxiety, depression, and the microbiome: A role for gut peptides. *Neurotherapeutics* **2018**, *15*, 36–59. [CrossRef]
147. Park, A.; Collins, J.; Blennerhassett, P.; Ghia, J.; Verdu, E.; Bercik, P.; Collins, S. Altered colonic function and microbiota profile in a mouse model of chronic depression. *Neurogastroenterol. Motil.* **2013**, *25*, 733-e575. [CrossRef] [PubMed]
148. O'Malley, D.; Julio-Pieper, M.; Gibney, S.M.; Dinan, T.G.; Cryan, J.F. Distinct alterations in colonic morphology and physiology in two rat models of enhanced stress-induced anxiety and depression-like behaviour. *Stress* **2010**, *13*, 114–122. [CrossRef] [PubMed]
149. Jiang, H.; Ling, Z.; Zhang, Y.; Mao, H.; Ma, Z.; Yin, Y.; Wang, W.; Tang, W.; Tan, Z.; Shi, J. Altered fecal microbiota composition in patients with major depressive disorder. *Brain. Behav. Immun.* **2015**, *48*, 186–194. [CrossRef] [PubMed]
150. Yu, M.; Jia, H.; Zhou, C.; Yang, Y.; Zhao, Y.; Yang, M.; Zou, Z. Variations in gut microbiota and fecal metabolic phenotype associated with depression by 16S rRNA gene sequencing and LC/MS-based metabolomics. *J. Pharm. Biomed. Anal.* **2017**, *138*, 231–239. [CrossRef]
151. Alfageh, B.H.; Man, K.K.; Besag, F.M.; Alhawassi, T.M.; Wong, I.C.; Brauer, R. Psychotropic Medication Prescribing for Neuropsychiatric Comorbidities in Individuals Diagnosed with Autism Spectrum Disorder (ASD) in the UK. *J. Autism Dev. Disord.* **2019**, *50*, 625–633. [CrossRef]
152. Houghton, R.; Ong, R.C.; Bolognani, F. Psychiatric comorbidities and use of psychotropic medications in people with autism spectrum disorder in the United States. *Autism Res.* **2017**, *10*, 2037–2047. [CrossRef]
153. Turnbaugh, P.J.; Ridaura, V.K.; Faith, J.J.; Rey, F.E.; Knight, R.; Gordon, J.I. The effect of diet on the human gut microbiome: A metagenomic analysis in humanized gnotobiotic mice. *Sci. Transl. Med.* **2009**, *1*, 6ra14. [CrossRef]
154. Krajmalnik-Brown, R.; Kang, D.-W.; Park, J.G.; Labaer, J.; Ilhan, Z. Microbiome Markers and Therapies for Autism Spectrum Disorders. U.S. Patent US9719144B2, 1 August 2017.
155. Silva, Y.P.; Bernardi, A.; Frozza, R.L. The role of short-chain fatty acids from gut microbiota in gut-brain communication. *Front. Endocrinol. (Lausanne)* **2020**, *11*, 25. [CrossRef]
156. Dalile, B.; Van Oudenhove, L.; Vervliet, B.; Verbeke, K. The role of short-chain fatty acids in microbiota–gut–brain communication. *Nat. Rev. Gastroenterol. Hepatol.* **2019**, *16*, 461–478. [CrossRef]
157. Salvucci, E. The human-microbiome superorganism and its modulation to restore health. *Int. J. Food Sci. Nutr.* **2019**, *70*, 781–795. [CrossRef]
158. Dong, T.S.; Gupta, A. Influence of early life, diet, and the environment on the microbiome. *Clin. Gastroenterol. Hepatol.* **2019**, *17*, 231–242. [CrossRef] [PubMed]

159. Pascale, A.; Marchesi, N.; Marelli, C.; Coppola, A.; Luzi, L.; Govoni, S.; Giustina, A.; Gazzaruso, C. Microbiota and metabolic diseases. *Endocrine* **2018**, *61*, 357–371. [CrossRef] [PubMed]
160. Adams, J.B.; Audhya, T.; Geis, E.; Gehn, E.; Fimbres, V.; Pollard, E.L.; Mitchell, J.; Ingram, J.; Hellmers, R.; Laake, D. Comprehensive nutritional and dietary intervention for autism spectrum disorder—A randomized, controlled 12-month trial. *Nutrients* **2018**, *10*, 369. [CrossRef]
161. Lange, K.W.; Hauser, J.; Reissmann, A. Gluten-free and casein-free diets in the therapy of autism. *Curr. Opin. Clin. Nutr. Metab. Care* **2015**, *18*, 572–575. [CrossRef] [PubMed]
162. González-Domenech, P.J.; Atienza, F.D.; Pablos, C.G.; Soto, M.L.F.; Martínez-Ortega, J.M.; Gutiérrez-Rojas, L. Influence of a Combined Gluten-Free and Casein-Free Diet on Behavior Disorders in Children and Adolescents Diagnosed with Autism Spectrum Disorder: A 12-Month Follow-Up Clinical Trial. *J. Autism Dev. Disord.* **2019**, *50*, 935–948. [CrossRef]
163. Buffington, S.A.; Di Prisco, G.V.; Auchtung, T.A.; Ajami, N.J.; Petrosino, J.F.; Costa-Mattioli, M. Microbial reconstitution reverses maternal diet-induced social and synaptic deficits in offspring. *Cell* **2016**, *165*, 1762–1775. [CrossRef]
164. Gupta, V.K.; Paul, S.; Dutta, C. Geography, ethnicity or subsistence-specific variations in human microbiome composition and diversity. *Front. Microbiol.* **2017**, *8*, 1162. [CrossRef]
165. Fernell, E.; Bejerot, S.; Westerlund, J.; Miniscalco, C.; Simila, H.; Eyles, D.; Gillberg, C.; Humble, M.B. Autism spectrum disorder and low vitamin D at birth: A sibling control study. *Mol. Autism* **2015**, *6*, 3. [CrossRef]
166. Alzghoul, L.; AL-Eitan, L.N.; Aladawi, M.; Odeh, M.; Hantash, O.A. The Association between Serum Vitamin D3 Levels and Autism among Jordanian Boys. *J. Autism Dev. Disord.* **2019**, 1–6. [CrossRef]
167. Khamoushi, A.; Aalipanah, E.; Sohrabi, Z.; Akbarzadeh, M. Vitamin D and Autism Spectrum Disorder: A Review. *Int. J. Nutr. Sci.* **2019**, *4*, 9–13. [CrossRef]
168. Jia, F.; Wang, B.; Shan, L.; Xu, Z.; Staal, W.G.; Du, L. Core symptoms of autism improved after vitamin D supplementation. *Pediatrics* **2015**, *135*, e196–e198. [CrossRef]
169. Pandey, K.R.; Naik, S.R.; Vakil, B.V. Probiotics, prebiotics and synbiotics-a review. *J. Food Sci. Technol.* **2015**, *52*, 7577–7587. [CrossRef] [PubMed]
170. Shaaban, S.Y.; El Gendy, Y.G.; Mehanna, N.S.; El-Senousy, W.M.; El-Feki, H.S.; Saad, K.; El-Asheer, O.M. The role of probiotics in children with autism spectrum disorder: A prospective, open-label study. *Nutr. Neurosci.* **2018**, *21*, 676–681. [CrossRef] [PubMed]
171. Grimaldi, R.; Gibson, G.R.; Vulevic, J.; Giallourou, N.; Castro-Mejía, J.L.; Hansen, L.H.; Gibson, E.L.; Nielsen, D.S.; Costabile, A. A prebiotic intervention study in children with autism spectrum disorders (ASDs). *Microbiome* **2018**, *6*, 133. [CrossRef] [PubMed]
172. Hsiao, E.Y.; McBride, S.W.; Hsien, S.; Sharon, G.; Hyde, E.R.; McCue, T.; Codelli, J.A.; Chow, J.; Reisman, S.E.; Petrosino, J.F. Microbiota modulate behavioral and physiological abnormalities associated with neurodevelopmental disorders. *Cell* **2013**, *155*, 1451–1463. [CrossRef] [PubMed]
173. Urbańska, M.; Gieruszczak-Białek, D.; Szajewska, H. Systematic review with meta-analysis: *Lactobacillus reuteri* DSM 17938 for diarrhoeal diseases in children. *Aliment. Pharmacol. Ther.* **2016**, *43*, 1025–1034. [CrossRef]
174. Golnik, A.E.; Ireland, M. Complementary alternative medicine for children with autism: A physician survey. *J. Autism Dev. Disord.* **2009**, *39*, 996–1005. [CrossRef]
175. Pärtty, A.; Kalliomäki, M.; Wacklin, P.; Salminen, S.; Isolauri, E. A possible link between early probiotic intervention and the risk of neuropsychiatric disorders later in childhood: A randomized trial. *Pediatr. Res.* **2015**, *77*, 823–828. [CrossRef]
176. Ng, Q.X.; Loke, W.; Venkatanarayanan, N.; Lim, D.Y.; Soh, A.Y.S.; Yeo, W.S. A systematic review of the role of prebiotics and probiotics in autism spectrum disorders. *Medicina* **2019**, *55*, 129. [CrossRef]
177. Mellon, A.; Deshpande, S.; Mathers, J.; Bartlett, K. Effect of oral antibiotics on intestinal production of propionic acid. *Arch. Dis. Child.* **2000**, *82*, 169–172. [CrossRef]
178. Rodakis, J. An n = 1 case report of a child with autism improving on antibiotics and a father's quest to understand what it may mean. *Microb. Ecol. Health Dis.* **2015**, *26*, 26382. [CrossRef] [PubMed]
179. Tochitani, S.; Ikeno, T.; Ito, T.; Sakurai, A.; Yamauchi, T.; Matsuzaki, H. Administration of non-absorbable antibiotics to pregnant mice to perturb the maternal gut microbiota is associated with alterations in offspring behavior. *PLoS ONE* **2016**, *11*, e0138293. [CrossRef] [PubMed]

180. Degroote, S.; Hunting, D.J.; Baccarelli, A.A.; Takser, L. Maternal gut and fetal brain connection: Increased anxiety and reduced social interactions in Wistar rat offspring following peri-conceptional antibiotic exposure. *Prog. Neuropsychopharmacol. Biol. Psychiatry* **2016**, *71*, 76–82. [CrossRef] [PubMed]
181. Bowman, K.A.; Broussard, E.K.; Surawicz, C.M. Fecal microbiota transplantation: Current clinical efficacy and future prospects. *Clin. Exp. Gastroenterol.* **2015**, *8*, 285. [PubMed]
182. Kelly, C.R.; Kahn, S.; Kashyap, P.; Laine, L.; Rubin, D.; Atreja, A.; Moore, T.; Wu, G. Update on fecal microbiota transplantation 2015: Indications, methodologies, mechanisms, and outlook. *Gastroenterology* **2015**, *149*, 223–237. [CrossRef]
183. Hugenholtz, F.; de Vos, W.M. Mouse models for human intestinal microbiota research: A critical evaluation. *Cell. Mol. Life Sci.* **2018**, *75*, 149–160. [CrossRef] [PubMed]
184. Brugha, T.S.; McManus, S.; Bankart, J.; Scott, F.; Purdon, S.; Smith, J.; Bebbington, P.; Jenkins, R.; Meltzer, H. Epidemiology of autism spectrum disorders in adults in the community in England. *Arch. Gen. Psychiatry* **2011**, *68*, 459–465. [CrossRef]
185. Howlin, P.; Moss, P. Adults with autism spectrum disorders. *Can. J. Psychiatry* **2012**, *57*, 275–283. [CrossRef]

nutrients

MDPI

Article

# Gut Health-Promoting Benefits of a Dietary Supplement of Vitamins with Inulin and Acacia Fibers in Rats

**Malén Massot-Cladera [1,2], Ignasi Azagra-Boronat [1,2], Àngels Franch [1,2], Margarida Castell [1,2], Maria J. Rodríguez-Lagunas [1,2] and Francisco J. Pérez-Cano [1,2,*]**

1 Physiology Section, Department of Biochemistry and Physiology, Faculty of Pharmacy and Food Science, University of Barcelona (UB), 08028 Barcelona, Spain; malen.massot@ub.edu (M.M.-C.); ignasiazagra@ub.edu (I.A.-B.); angelsfranch@ub.edu (À.F.); margaridacastell@ub.edu (M.C.); mjrodriguez@ub.edu (M.J.R.-L.)

2 Nutrition and Food Safety Research Institute (INSA-UB), 08921 Santa Coloma de Gramenet, Spain

* Correspondence: franciscoperez@ub.edu; Tel.: +34-93-402-45-05

Received: 7 June 2020; Accepted: 20 July 2020; Published: 23 July 2020

**Abstract:** The study's objective was to ascertain whether a nutritional multivitamin and mineral supplement enriched with two different dietary fibers influences microbiota composition, mineral absorption, and some immune and metabolic biomarkers in adult rats. Nine-week-old Wistar rats were randomly assigned into four groups: the reference group; the group receiving a daily supplement based on a food matrix with proteins, vitamins, and minerals; and two other groups receiving this supplement enriched with inulin (V + I) or acacia (V + A) fiber for four weeks. Microbiota composition was determined in cecal content and mineral content in fecal, blood, and femur samples. Intestinal IgA concentration, hematological, and biochemical variables were evaluated. Both V + I and V + A supplementations increased *Firmicutes* and *Actinobacteria* phyla, which were associated with a higher presence of *Lactobacillus* and *Bifidobacterium* spp. V + A supplementation increased calcium, magnesium, phosphorus, and zinc concentrations in femur. V + I supplementation increased the fecal IgA content and reduced plasma total cholesterol and uric acid concentration. Both fiber-enriched supplements tested herein seem to be beneficial to gut-health, although differently.

**Keywords:** inulin fiber; acacia fiber; immune system; microbiota; mineral absorption; IgA

## 1. Introduction

The intake in an appropriate dose (20–35 g/day for healthy adults) of dietary fiber (DF) has long been linked to reduction of metabolic diseases incidence, including diabetes, cardiovascular disease and obesity, among others, due to its capacity to lower blood cholesterol and C Reactive Protein (CRP), to attenuate glucose absorption and to improve insulin response [1–3]. Moreover, when non-digestible fiber reaches the colon unaltered and is selectively metabolized by microbiota, it induces specific changes, both in the composition and/or functionality of one or a limited number of bacteria potentially associated with health and well-being [4–6]. Meeting all these criteria, non-digestible fiber is considered a prebiotic as defined by Gibson and Roberfroid [1,4–6].

Prebiotic consumption is also believed to improve the immune system in both humans and animals [7]. The most examined mechanism involved in this effect is the indirect modulation of the immune response by changing the microbiota composition [8,9] and, therefore, its crosstalk with the immune system. Moreover, the enrichment of beneficial bacteria induced by prebiotic intake can result in the modulated release of pro- and anti-inflammatory cytokines [7] as well as in increasing the intestinal and fecal immunoglobulin (Ig) A content [10]. In addition, although little information is available, direct effects of prebiotics on the immune system such as changes of the intestinal gene expression, such as toll-like receptors (TLRs), have also been reported [11,12].

The favorable shift in the gut microbiota composition after prebiotic fiber consumption [1] is proposed as a potential mechanism by which prebiotics improve mineral absorption [13,14]. In this regard, the most widely accepted theory supporting this effect is associated with the microbial fermentation of the prebiotic fiber into short-chain fatty acids (SCFAs) in the colon. This metabolic process acidifies the intestinal compartment; thereby, preventing the formation of complexes between minerals and negatively charged metabolites. Consequently, it increases the extent of mineral absorption [15]. Alternatively, prebiotic consumption may also influence tissue morphology by increasing cell density, intestinal crypt depth, and blood flow in the large intestine, a mechanism that is believed to increase the intestinal surface area and lead to a higher mineral absorption [16,17].

There are several well-documented prebiotic fibers, such as the inulin-type fructans [18,19]. Inulin is part of everyday human diet. It can be found naturally among others in a range of plants such as chicory, garlic, tomato, and banana [20]. Its bifidogenic effects have been widely described in vitro, in vivo, and in clinical studies [19,21]. In the last few years, a new prebiotic fiber has emerged: acacia gum. It is a soluble DF obtained from the stems and branches of *Acacia senegal* and *Acacia seyal* and it is mainly composed of complex polysaccharides [22]. It resists digestion in the upper gastrointestinal tract; thus, reaching the large intestine and it can induce an increase in *Bifidobacterium* spp. in vitro [23,24] and in human [25] studies. However, unlike the inulin, little is known about the impact of acacia gum on health benefits.

On the other hand, the maintaining of a healthy diet, defined as an appropriate balance of energy, macro- and micronutrients and water, is important for adults but it is particularly relevant for members of the elderly population who are more vulnerable to malnutrition. Moreover, the efficiency of nutrient absorption may be impaired in this population; thus, involving different nutritional requirements. Moreover, the presence of oral problems, for example with dentition, together with a decrease in smell and taste perception induce a change in dietary patterns. Additionally, there is a concomitant decline in the normal function of the immune system (immunosenescence) that may contribute to an increase in the risk of infection and frailty [26]. On this basis, the hypothesis of the present study is that the intake of a nutritional supplement containing proteins, vitamins, minerals, and fiber is beneficial for the adult population and if so, its impact should be important to take into account for the elderly. Therefore, the aim of the current study was to ascertain whether a nutritional multivitamin and mineral supplement enriched with two different DFs influences microbiota composition, mineral absorption, and some immune and metabolic biomarkers in adult rats.

## 2. Materials and Methods

### 2.1. Animals and Supplements

Nine-week-old female and male Wistar rats, purchased from Janvier Labs (Saint Berthevin Cedex, France), were housed individually in polycarbonate cages with large fibrous-particle bedding and tissue papers as enrichment, in a controlled environment of temperature and humidity and in a 12/12 h light/dark cycle at the Faculty of Pharmacy and Food Science animal facility. All rats were fed a commercial diet corresponding to the American Institute of Nutrition 93 M formulation [27] (Teklad Global 14% Protein Rodent Maintenance Diet, Envigo, Indianapolis, IN, USA), which contains 5% of cellulose, and water ad libitum throughout the study.

After the acclimation period, animals were randomly assigned into four experimental groups ($n$ = 10/each, 5 females and 5 males). One constituted the reference (REF) group which did not receive any supplement; another group received a daily supplement based on a food matrix with proteins, vitamins and minerals (V) (Table S1); and two other groups received this supplement enriched with inulin (V + I) or acacia fiber (V + A), containing 4.5 g of fiber/100 g of product each (La Piara S.A, Manlleu, Barcelona). The chow consumption was measured every other day to adjust the dose of the supplements that were administered during 4 weeks in daily small portions. With this aim, the weights of each serving in V + I and V + A were readjusted periodically for each animal in order to receive an extra 20% of fiber daily. Accordingly, the same amount of supplement was used for the V group.

All experimental procedures were approved by the Ethical Committee for Animal Experimentation of the University of Barcelona and the Government of Catalonia (CEEA UB ref. 351/17 and CG 9735, respectively), in accordance with the EU Directive 2010/63/EU.

With regard to sample size estimation ($n$ = 10/group), the Appraising Project Office's program from the Universidad Miguel Hernández de Elche (Alicante, Spain) was used to calculate the minimum number of animals providing statistically significant differences among groups, assuming that there is no dropout rate, a beta risk of 0.2 (80% power) and a type I error of 0.05 (two-sided). We used the IgA-coating bacteria percentage data from a previous study [11] with similar design: mean values in the REF group were 25.3%, the estimated common standard deviation was 13 and the minimum expected difference was 12. In addition, the sample size was adjusted to the minimum needed to follow the University Ethical Committee guidelines.

### 2.2. Monitoring, Sample Collection and Processing

Body weight and food and water intake were monitored three times per week throughout the study. Fecal samples were collected weekly in order to determine changes in the fecal wet weight, humidity, and pH. Additionally, fecal samples collected at the end of the study allowed the concentration of IgA and the proportion of IgA-coated bacteria to be quantified, as previously described [28], as well as the mineral elimination.

At the end of the nutritional intervention, animals were anesthetized intramuscularly with ketamine (90 mg/kg) (Merial Laboratorios, S.A. Barcelona, Spain) and xylazine (10 mg/kg) (Bayer A. G., Leverkusen, Germany) in order to obtain tissue samples. The body weight and naso–anal length were measured to calculate the body mass index (BMI) as body weight/length$^2$ (g/cm$^2$). Urine samples for mineral quantification were obtained by direct puncture of the bladder. The weight of stomach, duodenum, jejunum, ileum, cecum, colon and rectum, spleen, liver, thymus, kidneys, heart, submandibular gland, and the length of the small and large intestines were recorded. Blood samples were collected in heparin-treated tubes to determine the hematological and biochemical variables, and plasma mineral content. For IgA quantification, the gut wash (GW) from the distal part of the small intestine was obtained as previously described [11]. Finally, the central part of the left femur was excised for mineral quantification.

### 2.3. Fecal Variables

Fresh feces collected weekly were used to determine fecal pH using a surface electrode (Crison Instruments, S.A., Barcelona, Spain). Afterwards, fecal samples were dried for 24 h at 60 °C. Fecal humidity of each sample was calculated considering the weight difference between before and after the drying process.

### 2.4. Mineral Analysis in Biological Samples

Firstly, blood, feces, femur, and urine samples underwent a chemical cleavage process to obtain aqueous solutions without precipitation or colloids. For that, the same amount of each type of sample was introduced into a tared high-pressure vessel made of polytetrafluoroethylene. Then, 2 mL of concentrated nitric acid ($HNO_3$) and 2 mL of concentrated hydrogen peroxide ($H_2O_2$) were also added in order to ensure a better oxidation of the organic matrix. All the high-pressure vessels were incubated overnight at 90 °C. After digestion, the samples were diluted with 16 mL of ultra-pure water, and then the vessels were weighed again in order to determine the weight of the aqueous sample solutions. Additionally, for each digestion cycle, triplicates of digestion blanks containing only $HNO_3$, $H_2O_2$ and ultra-pure water were also prepared. Finally, the digested solutions were transferred into the test tubes. The concentrations of calcium (Ca), magnesium (Mg), phosphorous (P), and zinc (Zn) were determined using an inductively coupled plasma-optical emission spectrometer (ICP-OES, Optima 3200 RL, Perkin-Elmer, Massachusetts, USA), whereas iron (Fe) and zinc (Zn) concentrations were measured by an inductively coupled plasma-mass spectrometer (ICP-MS, Nexlon 350 D, Perkin-Elmer, Massachusetts, MA, USA) using standard conditions. The analysis was carried out at the Unit of Metal Analysis of the Scientific and Technological Centers of the University of Barcelona (CCiT-UB). Results are expressed as mg/g of sample.

### 2.5. Hematological and Biochemical Analysis

Heparin-treated blood was immediately used to count platelets and white and red blood cells and related variables using an automated hematology analyzer (Spincell3, MonLab, Barcelona), following the manufacturer's instructions.

Plasma samples were used to quantify total cholesterol (TC) by cholesterol oxidase (CHOD)-peroxidase (POD) method; high-density lipoprotein cholesterol (HDL-C) by colorimetric method; triglycerides (TG) by glycerol phosphate oxidase (GPO) method; glucose by glucose oxidase (GOD)-POD method and uric acid by Uricasa-POD method using kits provided by Química Clínica Aplicada, S.A. (Química Clínica Aplicada, S.A., Tarragona, Spain) and following the manufacturer's instructions. Low-density lipoprotein cholesterol (LDL-C) was assessed according to the formula by Friedewald et al. in 1972, in which cLDL = TC − (cHDL + TG/5).

### 2.6. Immunoglobulin A and IgA-Coated Bacteria Quantification

The concentration of IgA in GW, fecal homogenates, and plasma was quantified at the end of the nutritional intervention by ELISA as previously described [11,29]. Moreover, the proportion of bacteria coated with IgA in feces was determined and analyzed by flow cytometry, as previously established [28]. The results concerning IgA are expressed as ng/g of tissue, ng/mg of fecal sample, and ng/mL of plasma, whereas those related to the IgA-coated bacteria are expressed as percentage.

### 2.7. Analysis of Cecal Microbiota Composition by 16S rRNA Sequencing

Cecal samples from all the animals at the end of the study ($n$ = 10 animals/group) ranging from 500–1000 mg collected were used to extract genomic DNA using QIAmp DNA Stool Mini Kit (Qiagen) with some previous modification as has been previously reported [30]. Briefly, extra purification and concentration were performed following the cleaning protocol from QIAmp Micro Kit (Qiagen, Madrid, Spain). Then, for massive sequencing, the hypervariable region V3-V4 of the bacterial 16s rRNA gene was amplified using key-tagged eubacterial primers (forward: S-D-Bact-0341-b-S-17, 5′-CCTACGGGNGGCWGCAG-3′ and reverse S-D-Bact-0785-a-A-21, 5′-GACTACHVGGGTATCTAATCC-3′) [31] and sequenced with a MiSeq Illumina Platform (Illumina Inc., San Diego, CA, USA) by the Life Sequencing facilities (ADM Life Sequencing, Valencia, Spain), following the Illumina recommendation for Library preparation and sequencing for metagenomics studies.

The software Paired-End read merger (PEAR v 0.9.6, Exelixis Lab, Heidelberg, Germany) was used to merge raw sequences forward and reverse. Using this approach, the ends of the obtained sequences were overlapped in order to get complete sequences. The amplification primers from the sequences obtained in the sequencing step were trimmed with Cutadapt v1.8.1 [32], using parameters by default, in order to reduce the bias in the annotation step. Once the primers were removed, sequences lower than 200 nucleotides were excluded from the analysis because short sequences have a higher chance of generating wrong taxonomical group associations. After obtaining the clean complete sequences, a quality filter was applied in order to delete sequences of poor quality. The resulting sequences were inspected for PCR chimera constructs (Uchime, USEARCH) [33], that may occur during the different experimental process, which were removed from further analysis. Later, each group of sequences was compared to a database of rRNA using an alignment BLAST strategy to associate taxonomic groups.

The relative proportions of phyla, families, and genera were calculated. Moreover, to estimate the specific genus biodiversity, the Shannon–Wiener and CHAO1 indexes were calculated.

Results of the qualitative analyses relative to the most abundant phyla, families, and genera are represented with stacked bars separated by gender. The category "others" represented in each graph includes those phyla whose presence was lower than 0.05% in the REF group; and those families and genera whose presence was lower than 0.8% in the same group.

To estimate the presence or absence of certain bacterial genera in the experimental groups, it was agreed that all bacterial genera present in all animals belonging to the same group with a proportion higher than 0.01% were computed as "present". Otherwise, they were computed as "absent" in such groups. Based on that, the Venn diagrams were created for all groups together allowing the way the genera were distributed among the groups to be seen numerically, in order to compare their coincidences and differences.

### 2.8. Principal Components Analysis

The principal components analysis (PCA) was performed to evaluate the dimensionality of microbiota with regard to the supplements. The model was done using Simca v14.1 (Umetrics, Umeå, Sweden) as previously reported [30].

### 2.9. Statistical Analysis

The Statistical Package for the Social Sciences (SPSS v22.0) (IBM, Chicago, IL, USA) and the R software (R-3.6.3) were used for statistical analysis. Data were tested for homogeneity of variance and normality distribution by the Levene's and Shapiro–Wilk tests, respectively. When data were homogeneous and normally distributed, a two-way ANOVA test was applied. When no differences between genders were observed, the data were analyzed together using a conventional-one-way ANOVA. Otherwise, they were analyzed separately. When significant differences among groups were detected, Bonferroni's post hoc test was performed. Kruskal–Wallis test was used when results were neither equally nor normally distributed, followed by Nemenyi post hoc test in the case of significant difference among groups. To compare variables along the study, a repeated-measures ANOVA or Friedman test were applied followed by Student's t-test or Nemenyi post hoc test, respectively. Significant differences were considered when $p < 0.05$, except regarding repeated comparisons, when $p$ value was corrected, dividing it by the number of applied tests.

## 3. Results

### 3.1. Effects of Supplements on Morphometry and Food and Water Intake

Throughout the study period, male rats from all groups had a higher body weight (447.51 ± 5.80 g), chow intake (28.09 ± 0.68 g/day/rat), and water consumption (29.23 ± 0.56 mL/day/rat) than female rats (255.25 ± 2.5 g/day/rat, 17.73 ± 0.33 g/day/rat, and 21.66 ± 0.61 mL/day/rat, respectively) ($p < 0.05$), and none of the dietary supplementations modified these variables (Figure S1a,b).

Regarding the body mass index (BMI) at the end of the nutritional intervention, only sex-associated significant differences were observed within all experimental groups. In particular, BMI in female rats was 0.64 ± 0.01 whereas it was 0.85 ± 2.5 in male rats, considering all animals independently of the experimental group, and no effects due to supplementation were detected (Figure S1c).

Moreover, whereas male rats showed significantly lower relative weight in most of the organs analyzed than female rats in the same group ($p < 0.05$) (Table S2), supplementation did not result in changes. No differences between groups were found when the small intestine and large intestine were measured, their mean length being 80.69 ± 1.43 cm and 16.76 ± 0.34 cm, respectively, for all experimental groups considered together at the end of the study.

### 3.2. Effects of Supplements on Fecal Variables

No sex-associated differences were detected in all the fecal variables studied; thus, these results were analyzed considering both female and male rats together (Figure S2).

Fecal weight registered was similar throughout the study and this was around 0.25 ± 0.01 g/day for all experimental groups, without being influenced by the dietary supplementations (Figure S2a). Similar results were observed when humidity of the feces was measured. All samples had around 53.24 ± 0.48% of water independently of the experimental group (Figure S2b).

When pH was measured in fecal samples, it was similar during the first two weeks of supplementation, but higher pH was observed at the end of the study. This was not associated with any of the nutritional interventions given that the pH was similar (5.94 ± 0.04) in all experimental groups (Figure S2c).

### 3.3. Effects of Supplements on Mineral Concentration

Mineral content measured in blood, feces, femur, and urine samples at the end of the nutritional intervention for all experimental groups is summarized in Table 1.

**Table 1.** Mineral concentration (mg/g) in blood, feces, femur, and urine samples at the end of the study for all experimental groups considering both sexes together.

| | | [Ca] (mg/g) | [Fe] (mg/g) | [Mg] (mg/g) | [P] (mg/g) | [Zn] (mg/g) |
|---|---|---|---|---|---|---|
| Blood | REF | 60.11 ± 2.51 | 0.45 ± 0.01 | 34.36 ± 1.10 | 0.37 ± 0.01 | 0.005 ± 0.000 |
| | V | 63.45 ± 2.20 | 0.44 ± 0.01 | 32.28 ± 0.76 | 0.35 ± 0.01 | 0.005 ± 0.000 |
| | V + I | 57.21 ± 3.16 $^{\beta}$ | 0.44 ± 0.00 | 31.80 ± 0.68 | 0.34 ± 0.01 | 0.005 ± 0.000 |
| | V + A | 59.53 ± 4.24 $^{\beta}$ | 0.44 ± 0.01 | 31.73 ± 0.36 $^{*\beta}$ | 0.34 ± 0.01 | 0.005 ± 0.000 |
| Feces | REF | 19.35 ± 1.43 | 0.41 ± 0.03 | 3.70 ± 0.14 | 12.39 ± 0.61 | 0.19 ± 0.01 |
| | V | 18.55 ± 1.21 | 0.39 ± 0.02 | 3.67 ± 0.15 | 12.02 ± 0.51 | 0.19 ± 0.01 |
| | V + I | 18.16 ± 1.36 | 0.43 ± 0.04 | 3.57 ± 0.04 | 12.13 ± 0.59 | 0.19 ± 0.01 |
| | V + A | 19.96 ± 1.04 | 0.39 ± 0.02 | 3.87 ± 0.13 | 13.20 ± 0.50 | 0.21 ± 0.01 |
| Femur | REF | 140.60 ± 6.33 | 0.05 ± 0.00 | 3.00 ± 0.13 | 66.95 ± 2.93 | 0.15 ± 0.01 |
| | V | 126.03 ± 8.05 | 0.06 ± 0.01 | 2.71 ± 0.15 | 61.61 ± 3.40 | 0.14 ± 0.00 |
| | V + I | 120.05 ± 6.73 | 0.05 ± 0.00 | 2.60 ± 0.16 | 58.93 ± 2.90 | 0.14 ± 0.00 |
| | V + A | 157.90 ± 4.77 $^{\beta\varepsilon}$ | 0.05 ± 0.00 | 3.45 ± 0.10 $^{\beta\varepsilon}$ | 76.87 ± 2.33 $^{\beta\varepsilon}$ | 0.16 ± 0.00 $^{\beta\varepsilon}$ |
| Urine | REF | 0.08 ± 0.03 | 0.002 ± 0.001 | 0.33 ± 0.12 | 0.04 ± 0.02 | 0.001 ± 0.000 |
| | V | 0.11 ± 0.04 | 0.001 ± 0.000 | 0.30 ± 0.08 | 0.13 ± 0.11 | 0.003 ± 0.003 |
| | V + I | 0.08 ± 0.02 | 0.000 ± 0.000 | 0.23 ± 0.05 | 0.04 ± 0.02 | 0.002 ± 0.001 |
| | V + A | 0.17 ± 0.07 | 0.001 ± 0.000 | 0.33 ± 0.10 | 0.02 ± 0.00 | 0.001 ± 0.001 |

Results are expressed as mean ± SEM ($n$ = 10/group). Calcium and magnesium blood concentrations are expressed as the mean ± SEM of mg of each mineral × $10^{-3}$/g of sample. REEF: animals no receiving supplement; V: animals receiving a daily supplement based on a food matrix with proteins, vitamins and minerals; V + I: inulin-enriched supplement-fed animals, and V + A: acacia-enriched supplement-fed animals.* $p < 0.05$ vs. REF group; $^{\beta}$ $p < 0.05$ vs. V group; $^{\varepsilon}$ $p < 0.05$ vs. V + I group. Ca: calcium; Fe: iron; Mg: magnesium; P: phosphorus; Zn: zinc.

In blood samples the most abundant mineral studied was iron, followed by phosphorus, calcium, and magnesium, whereas zinc was the one detected in the lowest concentration. Regarding the nutritional intervention, V + I and V + A-supplemented animals had lower calcium concentration compared to the group receiving non-fiber-enriched supplement (V group) ($p < 0.05$). Moreover, the supplement containing acacia fiber resulted in a lower magnesium blood concentration compared to the REF and V groups ($p < 0.05$).

The most abundant mineral detected in feces was calcium, followed by potassium, magnesium, and iron. No effects due to dietary intervention were identified in this compartment.

When mineral content in femur was analyzed, the most abundant was calcium, followed by phosphorus, magnesium, and zinc. Iron was detected in a very low proportion. Interestingly, only acacia-enriched supplement significantly increased the concentration of calcium, magnesium, phosphorous, and zinc in comparison with both the V and V + I groups ($p < 0.05$).

In urine samples the mineral found in most abundant concentration was magnesium, followed by calcium and potassium. Both zinc and iron were detected in very low concentrations. No effects due to dietary intervention were observed in this compartment.

### 3.4. *Effects of Supplements on Hematological and Biochemical Variables*

From all the parameters related to the leucocytes, erythrocytes, and platelets, only punctual differences in the erythrocyte parameters were seen (Table S3). The inulin-enriched supplement-fed animals (V + I) showed a slightly lower mean corpuscular hemoglobin (MCH) and a reduction in its concentration (MCHC) when compared to that of the REF and V groups ($p < 0.05$).

The inulin-enriched supplement (V + I) intake resulted in a significantly lower plasma concentration of total cholesterol and uric acid in comparison to that of the REF group and to the acacia-supplemented animals ($p < 0.05$) (Figure 1a,f). Moreover, the acacia-enriched supplement intake reduced the glucose concentration compared to the supplement without fiber ($p < 0.05$) (Figure 1e).

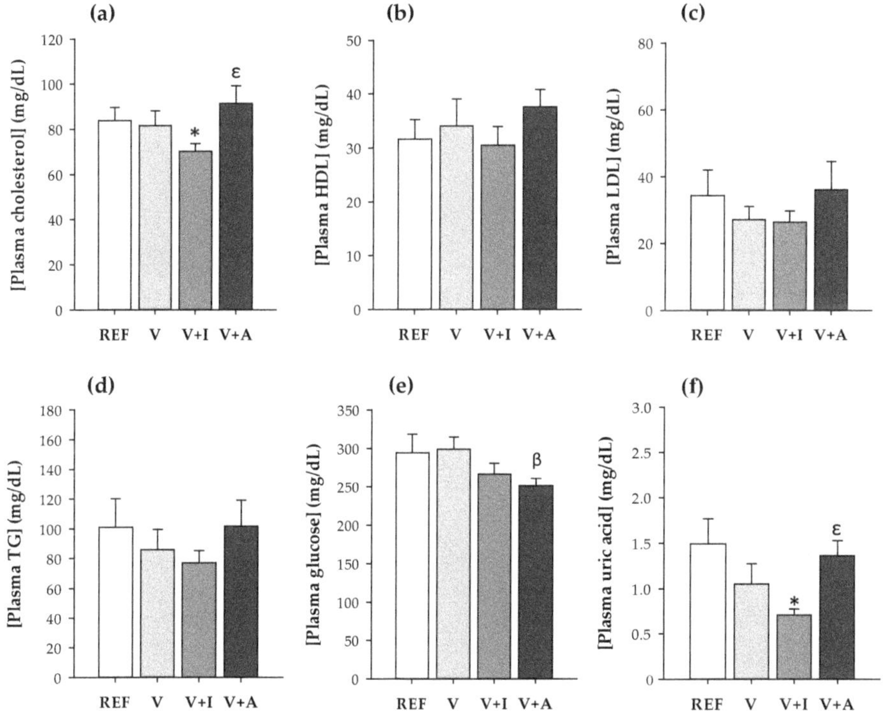


**Figure 1.** (**a**) Total cholesterol; (**b**) high-density lipoprotein cholesterol (HDL-C); (**c**) low-density lipoprotein cholesterol (LDL-C); (**d**) triglycerides (TG); (**e**) glucose; and (**f**) uric acid concentration in blood samples at the end of the study for all experimental groups considering both sexes together. Results are expressed as mean ± SEM ($n$ = 10/group). Statistical significance: * $p < 0.05$ vs. REF group; $^{\beta}$ $p < 0.05$ vs. V group; $^{\varepsilon}$ $p < 0.05$ vs. V + I group.

No effects either on HDL-C, LDL-C, or triglycerides were found due to supplementation with the DF.

### 3.5. *Effects of Supplements on IgA Concentration*

The mean fecal IgA content was 1.5–2-fold higher in animals receiving both the inulin- (V + I) and acacia-enriched (V + A) supplements, compared to the REF and V groups (Figure 2). However, only the increase caused by inulin was statistically significant compared to both the REF and V groups ($p < 0.05$).

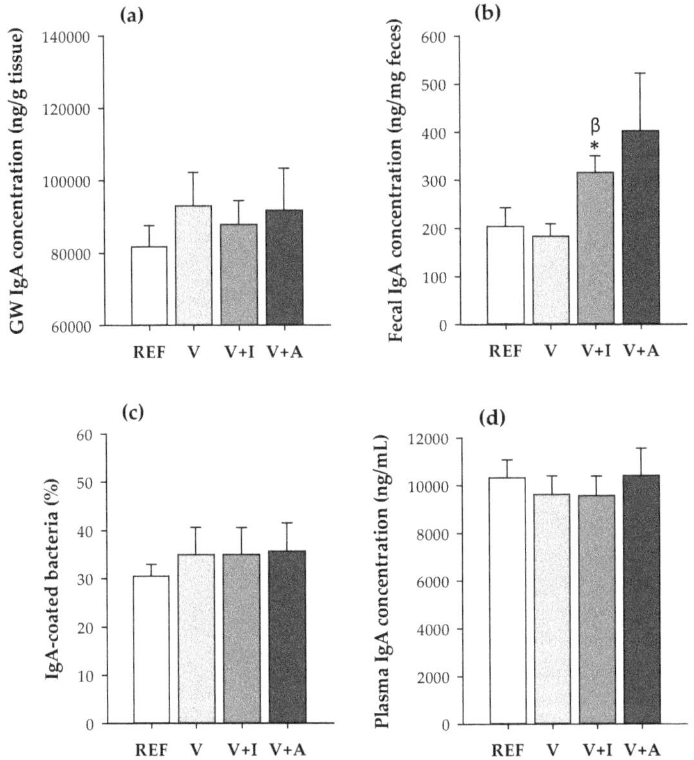


**Figure 2.** (**a**) IgA concentration in gut wash (GW) and (**b**) fecal samples, (**c**) proportion of fecal IgA-coating bacteria and (**d**) plasma IgA concentration quantified at the end of the study for all experimental groups considering both sexes together. Results are expressed as mean ± SEM ($n$ = 10/group). Statistical significance: * $p < 0.05$ vs. REF group; $^{\beta}$ $p < 0.05$ vs. V group.

No changes due to dietary intervention were detected either in the gut wash and plasma IgA concentrations or in the proportion of fecal IgA-coated bacteria.

### 3.6. *Effects of Supplements on Cecal Microbiota Composition*

#### 3.6.1. Diversity and Taxonomic Analysis

No changes on the Shannon–Wiener Index (3.52 ± 0.06) and CHAO1 (448.13 ± 10.69), as indicators of the diversity and richness of the microbial community, respectively, were observed after any dietary supplementation for all experimental groups at the end of the study.

Although it is well established that *Firmicutes* and *Bacteroidetes* are the most abundant phyla in the cecal microbiota, male rats had a lower proportion of *Bacteroidetes* in favor of that of Firmicutes. This increase in *Firmicutes* in males was also associated with a higher proportion of the family *Lactobacillaceae* spp. and in particular, the genus *Lactobacillus* (Figure 3).

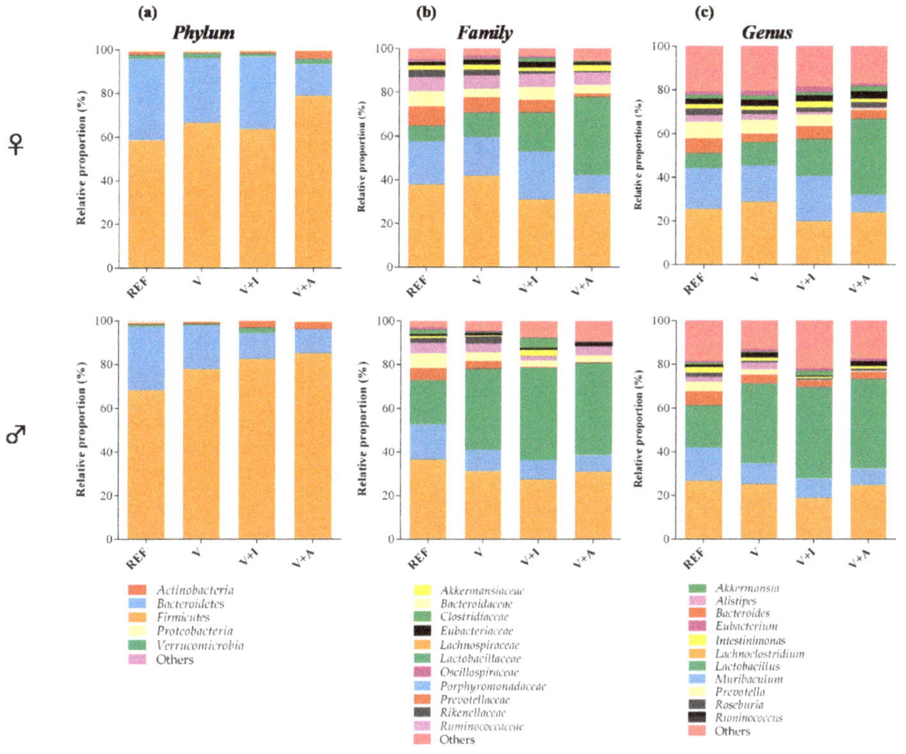


**Figure 3.** Main taxonomic ranks showing the proportion of bacterial populations in the cecal content at the end of the study in males and females. The relative proportion of the bacteria was calculated in each taxonomic rank: (**a**) phylum, (**b**) family, and (**c**) genus. Results are expressed as mean ($n = 5$ female or male/group). Significant differences not shown.

Regarding the nutritional intervention, the acacia-enriched supplement (V + A) increased the proportion of the *Firmicutes* (up to 81.76%) and decreased *Bacteroidetes* (up to 12.92%), compared to the REF group whose proportions were 63.28% and 33.29%, respectively ($p < 0.05$) (Figure 3a). These changes were more evident in female than in male rats. These changes after acacia fiber-enriched supplement (V + A) intake were associated with an increase of *Lactobacillaceae* family, this effect being stronger in female than in male rats (Figure 3b), whose *Lactobacillaceae* proportion was already increased at the baseline. Moreover, acacia supplementation significantly increased genera belonging to the *Firmicutes* and *Actinobacteria* phyla (Figure 3c). In particular, a significant increase in *Bifidobacterium* spp. (up to 0.07% and 0.06% in females and males, respectively) and *Lactobacillus* spp. (up to 28% and 33% in females and males, respectively) was observed in V + A-fed animals in comparison with those in the REF (< 0.015% and < 15% for *Bifidobacterium* spp. and *Lactobacillus* spp., respectively, in both females and males) and V (< 0.03% and < 30% for *Bifidobacterium* spp. and *Lactobacillus* spp., respectively, in both females and males) groups ($p < 0.05$) (Figure 4a,b).

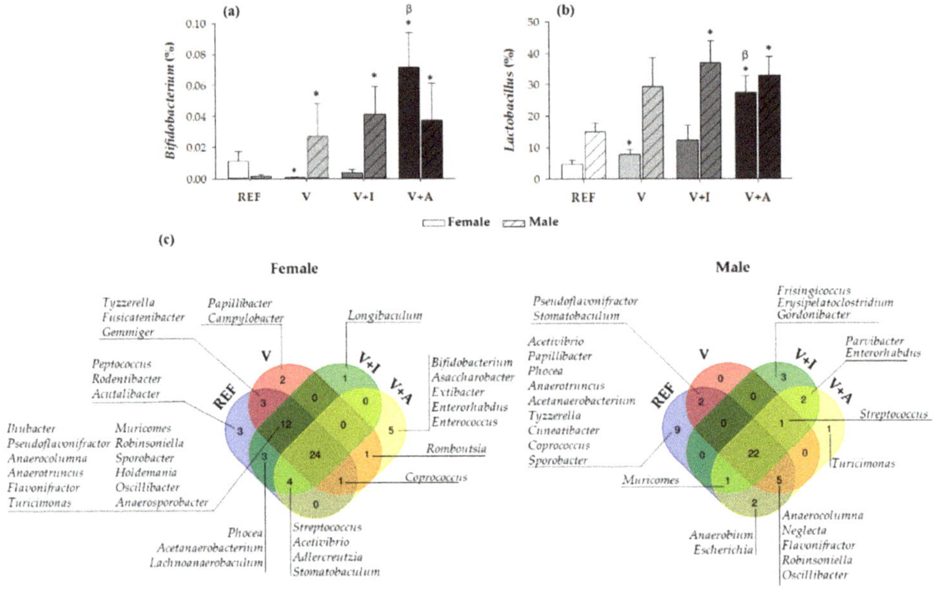


**Figure 4.** The relative proportion of the (**a**) *Bifidobacterium* and (**b**) *Lactobacillus* genera in cecal content at the end of the study differentiating between sexes. Results are expressed as mean ± SEM ($n$ = 10/group). Statistical significance: * $p < 0.05$ vs. REF group; $^{\beta}$ $p < 0.05$ vs. V group. (**c**) A representation of Venn diagrams showing the diversity in all genera differentiating between sexes. Results derived from $n$ = 10/group.

### 3.6.2. Venn Diagrams and Principal Components Analysis: Genera

The analysis of the genera distribution in Venn diagrams revealed that there was a core of 24 and 22 genera, in female and male rats, respectively, that persisted in all four experimental groups when considering both sexes separately (Figure 4c). Moreover, when comparing the microbiota depending on the supplementation, it could be observed that both the inulin (V + I) and acacia-enriched supplements (V + A) were able to exclusively promote the colonization of new genera in both female and male rats. On the one hand, inulin was able to promote the colonization of one new genus (*Longibaculum*) in female rats and five new genera in male rats (*Frisingicoccus, Erysipelatoclostridium, Gordonibacter, Parvibacter,* and *Enterorhabdus*), two of which also appeared with acacia supplementation. On the other hand, acacia supplementation resulted in the colonization of five new genera (*Bifidobacterium, Asaccharobacter, Extibacter, Enterorhabdus,* and *Enterococcus)* in female rats and three new genera in male rats (*Streptococcus*, *Parvibacter*, and *Enterorhabdus*), two of which also appeared in the inulin. In addition, some particular genera present in both the REF and V groups were absent in the V + I and V + A groups, this being the case in eight for females and eleven for males.

The PCA score plot revealed that the microbiota of the acacia-fiber supplemented animals (V + A) clustered differentially compared to those of both the REF and V groups in the genera analysis (Figure 5a). Moreover, the loading plot revealed that the *Bifidobacterium* spp. and *Lactobacillus* spp. were variables involved in the clustering of the V + A group (Figure 5b).

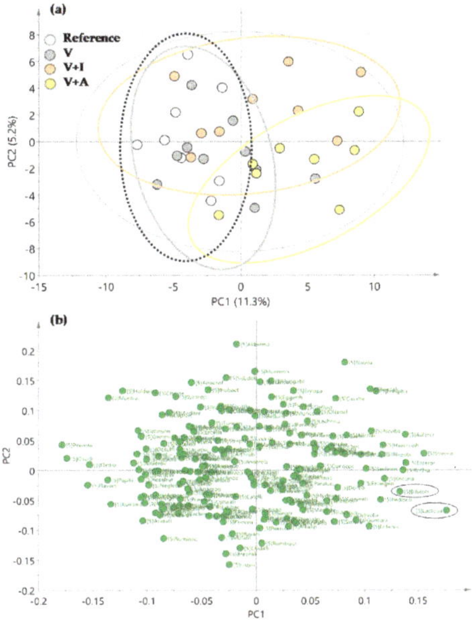


**Figure 5.** (**a**) Representation of Principal Components Analysis (PCA) for all experimental groups in a score plot and (**b**) a loading plot. Results derived from $n = 10$/group.

## 4. Discussion

Changes in dietary pattern, implying a lower intake of vitamins and minerals, may lead also to changes in microbiota composition. This situation, besides having a poor nutrient absorption and reduced bacterial diversity, is even more evident in the elderly [34]. In the current study, adult rats have been used to mimic the feasible impact of the fiber-enriched nutritional supplements tested herein on the immunological, hematological, and biochemical variables. Moreover, due to the disparity existing in the immune response, microbiota composition and the susceptibility to disease between gender [35–37], both female and male rats have been included in the present interventional study.

Herein, we demonstrate that the supplementations with inulin, a well-known prebiotic, and acacia gum fibers modify the adult rat microbiota composition with different intensity. Rats aged nine weeks that received acacia gum supplementation daily for four weeks showed an increased proportion in *Firmicutes* and *Actinobacteria*. In this regard, the acacia supplementation resulted in a significantly higher presence of *Lactobacillus* and the appearance of *Bifidobacterium* in both genders. In fact, this is not the first time that acacia gum's potential as a prebiotic agent has been described in both in vitro [38] and clinical studies [22,25]. Indeed, an interventional study carried out in human volunteers demonstrated that consumption of 10 and 15 g/day of acacia gum for 10 days increased the counts of both lactic acid-producing bacteria and *Bifidobacterium* in feces [22]. Moreover, its ability to selectively prevent the overgrowth of unwanted bacteria such as *Clostridium difficile* or *Clostridium histolyticum*, has also been studied in vitro [38,39], although some controversy exists.

Moreover, acacia fiber supplementation produced the appearance of the genus *Asaccharobacter* (which belongs to the *Actinobacteria* phylum) in female rats, a single species of which has been reported to be a powerful equol producer [40]. Therefore, older people receiving the acacia fiber-enriched supplement may benefit from equol's health-promoting benefits, as has been reported, for example on osteoporosis, prostate cancer, and cardiovascular diseases [41,42].

In this study, the appearance of the genus *Enterorhabdus* (*Actinobacteria* phylum) has been associated with the consumption of acacia supplementation in both female and male rats and also with the inulin supplement in male rats. Although little is known about the possible role of this *Actinobacteria* genus, its higher relative abundance has been negatively correlated to serum TC, TG and LDL-C and hepatic TC, TG, bile acids, and non-esterified fatty acids in *Grifola frondosa* polysaccharide-chromium III-treated type 2 diabetes mellitus (T2DM) mice [43]. Although further, more in-depth studies should be carried out into this association, it seems that *Enterorhabdus* genus could exert some kind of hypoglycemic and hypolipidemic activities in T2DM, one of the multifactorial chronic metabolic disorders affecting mainly adults worldwide. Therefore, this result suggests that the inclusion of acacia or inulin fiber in the diet of adult people would be beneficial for them.

One of the objectives of the present study was to compare the prebiotic activity of both inulin and acacia fibers in adult rats. Surprisingly, the effects observed on microbiota composition after acacia fiber intake in female rats were significantly stronger than those exerted by inulin in the same gender. These differential results agree with those reported by Calame et al. who evidenced that acacia gum was able to produce a higher increase in both bifidobacteria and lactobacilli than an equal dose of inulin in healthy men [25].

The shift in microbiota composition has been proposed as a potential mechanism by which prebiotics improve mineral absorption [13,14]. In the current study, increased calcium, magnesium, phosphorus, and zinc concentrations were observed in femur from both female and male acacia-supplemented animals; thus, suggesting that acacia-enriched supplement could be beneficial for bone mineralization. Similar findings to those described herein have been reported after galactooligosaccharides (GOS), fructooligosaccharides (FOS), a mixture of GOS/FOS, and inulin supplementation in vitro, in vivo, and in human studies [15,44]. However, the fact that in our study no effects were observed in the inulin-supplemented group could be due to an insufficient dose, the type of inulin used, or the duration of the treatment. On the other hand, either the promotion of the lactic-acid bacteria or the production of SCFA may result in an acidification of the colon compartment; thus, preventing the formation of complexes between mineral and negatively charged metabolites, and therefore improving the bioavailability of minerals [15]. In the present study we found no changes in either fecal or cecal pH after the dietary supplementations that would explain this mechanism. The lack of intestinal acidification after inulin intake has already been reported in younger rats fed a diet with inulin for three weeks [28]. Further studies should be carried out in order to elucidate the impact of both fiber supplementations on SCFA production and their relationship with the mineral absorption in rats.

Most research has only been focused on the indirect effects of prebiotics over a considerable period of time. However, it has recently been evidenced that prebiotics may also cause direct effects, such as immunomodulation in the gastrointestinal tract [45]. In this regard, it is well known that prebiotic administration, such as inulin, generally results in increased fecal IgA concentration [18]. This fact is in line with those results obtained here because the intake of inulin-enriched supplement for four weeks increased fecal IgA content; thus, enhancing the intestinal immune system, with no difference between genders. With regard to the acacia-enriched supplement-fed animals, although a relevant tendency to increase fecal IgA levels was observed, it did not reach statistical significance. None of the dietary supplementations tested herein modified the proportion of IgA-coated bacteria, contrary to what was observed in the youngest rats whose proportion increased after three weeks of inulin intake [28]. Further studies should confirm these results and should be aimed at understanding them.

On the other hand, fiber and prebiotic intake in appropriate doses is associated with less incidence of metabolic diseases due to its indirect capacity to modulate the blood lipid profile and other metabolic variables [46]. In the current study, the intake of the inulin-enriched supplement for four weeks exerted lipid-lowering effects by significantly reducing the cholesterol and the uric acid in plasma. These results are partially in line with those reported in animal models [47] and in hypercholesterolemic [48] and healthy [49,50] subjects receiving an inulin supplementation for 3–16 weeks. Nevertheless, the modulation of biochemical variables after acacia fiber intake is quite controversial. Whereas some authors did not observe significant effects either in hypercholesterolemic [51] or in healthy [52] subjects, others have attributed to acacia fiber significant benefits for metabolic disorders [53,54]. In particular, the intake of 30 g of acacia for three months resulted in a significant reduction of blood triglyceride and fasting plasma glucose concentrations in type 2 diabetic patients [53]. However, the conditions (dose and length) tested within the study evidenced a tendency to reduce the glucose and uric acid concentrations. This lack of effect after the acacia-enriched supplement on biochemical variables agrees with that reported in hypercholesterolemic [51] or in healthy [52] subjects. Further studies are required to clarify the protective effects of acacia gum on cardiometabolic diseases.

## 5. Conclusions

Overall, both fiber-enriched supplements tested in the present study show the potential to be beneficial to gut-health, although differently. Whereas inulin-enriched supplement shows intestinal immune enhancement, acacia fiber supplement has stronger prebiotic activity, which may lead to increasing mineral absorption.

**Supplementary Materials:** The following are available online at http://www.mdpi.com/2072-6643/12/8/2196/s1, Figure S1: (**a**) Body weight and (**b**) food intake registered throughout the study and (**c**) body mass index (BMI) measured at the end of the study differentiating between sexes. Results are expressed as mean ± SEM ($n$ = 5/group). No significant differences were observed. Figure S2: (**a**) Fecal wet weight, (**b**) humidity, and (**c**) pH registered throughout the study considering both sexes together. Results are expressed as mean ± SEM ($n$ = 10/group). No significant differences were observed. Table S1: Composition and content of macronutrients and micronutrients (fiber, vitamins, and minerals) of the experimental supplements. Table S2: Relative weight of organs expressed as percentage (%) with respect to the body weight at the end of the study for all experimental groups differentiating between sexes. Results are expressed as mean ± SEM ($n$ = 5 female and male/group). Statistical significance: $^{\alpha}$ $p < 0.05$ vs. REF. Table S3: Hematological parameters in blood samples at the end of the study for all experimental groups considering both sexes together. Results are expressed as mean ± SEM ($n$ = 10/group). Statistical significance: * $p < 0.05$ vs. REF group; $^{\beta}$ $p < 0.05$ vs. V group; $^{\varepsilon}$ $p < 0.05$ vs. V + I group.

**Author Contributions:** Conceptualization, M.M.-C., À.F., M.C., and F.J.P.-C.; methodology, M.M.-C., I.A.-B., and M.J.R.-L.; formal analysis, M.M.-C. and F.J.P.-C.; writing—original draft preparation, M.M.-C. and F.J.P.-C.; writing—review and editing, I.A.-B., À.F., M.C., and M.J.R.-L.; supervision, F.J.P.-C.; and funding acquisition, F.J.P.-C. All authors have read and agreed to the published version of the manuscript.

**Funding:** This research was funded through "Identificació de Nous Ingredients Moduladors de la Microbiota Humana i Alimentació Fent Ús de la Biotecnologia Industrial, les Tecnologies Òmiques i les Tecnologies de Big Data–MICROBIOTA (COMRDI-15-1-0029)" project included in the FEDER funding (Fondo Europeo de Desarrollo Regional 2014-2020).

**Acknowledgments:** The authors gratefully acknowledge the Units of Metal Analysis and Cytometry of the "Centres Científics i Tecnològics" of the university of Barcelona (CCiT-UB) for their help and advice with the ICP-MS and ICP-OES and flow cytometry analysis, respectively. In addition, we also would like to thank the Faculty of Pharmacy and Food Science animal facility's workers for their technical assistance and the students involved in the laboratory work (Alex Llorca, Mara Carmona, and Sergi Casanova).

**Conflicts of Interest:** The authors declare no conflict of interest. The funders had no role in the design of the study; in the collection, analyses, or interpretation of data; in the writing of the manuscript, or in the decision to publish the results.

## References

1. Slavin, J. Fiber and prebiotics: Mechanisms and health benefits. *Nutrients* **2013**, *5*, 1417–1435. [CrossRef]
2. Ma, Y.; Griffith, J.A.; Chasan-Taber, L.; Olendzki, B.C.; Jackson, E.; Stanek, E.J.; Li, W.; Pagoto, S.L.; Hafner, A.R.; Ockene, I.S. Association between dietary fiber and serum C-reactive protein. *Am. J.* **2006**, *83*, 760–766. [CrossRef]
3. Keogh, G.F.; Cooper, G.J.; Mulvey, T.B.; Mcardle, B.H.; Coles, G.D.; Monro, J.A.; Poppitt, S.D. Randomized controlled crossover study of the effect of a highly-glucan-enriched barley on cardiovascular disease risk factors in mildly hypercholesterolemic men. *Am. J. Clin. Nutr.* **2003**, *78*, 711–719. [CrossRef]
4. Gibson, G.R.; Roberfroid, M.B. Dietary modulation of the human colonic microbiota: Introducing the concept of prebiotics. *J. Nutr.* **1995**, *125*, 1401–1412. [CrossRef]

5. Gibson, G.R.; Scott, K.P.; Rastall, R.A.; Tuohy, K.M.; Hotchkiss, A.; Dubert-Ferrandon, A.; Gareau, M.; Murphy, E.F.; Saulnier, D.; Loh, G.; et al. Dietary prebiotics: Current status and new definition. *Food Sci. Technol. Bull. Funct. Foods* **2010**, *7*, 1–19. [CrossRef]
6. Gibson, G.R.; Probert, H.M.; Van Loo, J.; Rastall, R.A.; Roberfroid, M.B. Dietary modulation of the human colonic microbiota: Updating the concept of prebiotics. *Nutr. Res. Rev.* **2004**, *17*, 259. [CrossRef]
7. Shokryazdan, P.; Faseleh Jahromi, M.; Navidshad, B.; Liang, J.B. Effects of prebiotics on immune system and cytokine expression. *Med. Microbiol. Immunol.* **2017**, *206*, 1–9. [CrossRef]
8. Looijer-Van Langen, M.A.; Dieleman, L.A. Prebiotics in chronic intestinal inflammation. *Inflamm. Bowel Dis.* **2009**, *15*, 454–462. [CrossRef] [PubMed]
9. Steed, H.; Macfarlane, S. Mechanisms of prebiotic impact on health. In *Prebiotics and Probiotics Science and Technology*; Charalampopoulos, D., Rastall, R.A., Eds.; Springer: New York, NY, USA, 2009; pp. 135–161.
10. Ito, H.; Takemura, N.; Sonoyama, K.; Kawagishi, H.; Topping, D.L.; Conlon, M.A.; Morita, T. Degree of polymerization of inulin-type fructans differentially affects number of lactic acid bacteria, intestinal immune functions, and immunoglobulin a secretion in the rat cecum. *J. Agric. Food Chem.* **2011**, *59*, 5771–5778. [CrossRef] [PubMed]
11. Massot-Cladera, M.; Franch, À.; Pérez-Cano, F.J.; Castell, M. Cocoa and cocoa fibre differentially modulate IgA and IgM production at mucosal sites. *Br. J. Nutr.* **2016**, *115*, 1539–1546. [CrossRef] [PubMed]
12. Massot-Cladera, M.; Franch, À.; Castell, M.; Pérez-Cano, F.J. Cocoa polyphenols and fiber modify colonic gene expression in rats. *Eur. J. Nutr.* **2017**, *56*. [CrossRef] [PubMed]
13. Mccabe, L.; Britton, R.A.; Parameswaran, N. Prebiotic and probiotic regulation of bone health: Role of the intestine and its microbiome. *Curr. Osteoporos. Rep.* **2015**, *13*, 363–371. [CrossRef] [PubMed]
14. Weaver, C.M. Diet, gut microbiome, and bone health. *Curr. Osteoporos. Rep.* **2015**, *13*, 125–130. [CrossRef]
15. Whisner, C.M.; Castillo, L.F. Prebiotics, bone and mineral metabolism. *Calcif. Tissue Int.* **2018**, *102*, 443–479. [CrossRef]
16. Scholz-Ahrens, K.E.; Schrezenmeir, J. Inulin, oligofructose and mineral metabolism—Experimental data and mechanism. *Br. J. Nutr.* **2002**, *87*, S179–S186. [CrossRef] [PubMed]
17. Raschka, L.; Daniel, H. Mechanisms underlying the effects of inulin-type fructans on calcium absorption in the large intestine of rats. *Bone* **2005**, *37*, 728–735. [CrossRef]
18. Vogt, L.; Meyer, D.; Pullens, G.; Faas, M.; Smelt, M.; Venema, K.; Ramasamy, U.; Schols, H.A.; De Vos, P. Immunological properties of inulin-type fructans. *Crit. Rev. Food Sci. Nutr.* **2015**, *55*, 414–436. [CrossRef]
19. Shoaib, M.; Shehzad, A.; Omar, M.; Rakha, A.; Raza, H.; Sharif, H.R.; Shakeel, A.; Ansari, A.; Niazi, S. Inulin: Properties, health benefits and food applications. *Carbohydr. Polym.* **2016**, *147*, 444–454. [CrossRef]
20. Kolida, S.; Meyer, D.; Gibson, G.R. A double-blind placebo-controlled study to establish the bifidogenic dose of inulin in healthy humans. *Eur. J. Clin. Nutr.* **2007**, *61*, 1189–1195. [CrossRef]
21. Meyer, D.; Stasse-Wolthuis, M. The bifidogenic effect of inulin and oligofructose and its consequences for gut health. *Eur. J. Clin. Nutr.* **2009**, *63*, 1277–1289. [CrossRef]
22. Cherbut, C.; Michel, C.; Raison, V.; Kravtchenko, T.; Severine, M. Acacia gum is a bifidogenic dietary fibre with high digestive tolerance in healthy humans. *Microb. Ecol. Health Dis.* **2003**, *15*, 43–50.
23. Terpend, K.; Possemiers, S.; Daguet, D.; Marzorati, M. Arabinogalactan and fructo-oligosaccharides have a different fermentation profile in the Simulator of the Human Intestinal Microbial Ecosystem (SHIME®). *Environ. Microbiol. Rep.* **2013**, *5*, 595–603. [CrossRef] [PubMed]
24. Marzorati, M.; Qin, B.; Hildebrand, F.; Klosterbuer, A.; Roughead, Z.; Roessle, C.; Rochat, F.; Raes, J.; Possemiers, S. Addition of acacia gum to a FOS/inulin blend improves its fermentation profile in the Simulator of the Human Intestinal Microbial Ecosystem (SHIME®). *J. Funct. Foods* **2015**, *16*, 211–222. [CrossRef]
25. Calame, W.; Weseler, A.R.; Viebke, C.; Flynn, C.; Siemensma, A.D. Gum arabic establishes prebiotic functionality in healthy human volunteers in a dose-dependent manner. *Br. J. Nutr.* **2008**, *100*, 1269–1275. [CrossRef]
26. Salazar, N.; Valdés-Varela, L.; González, S.; Gueimonde, M.; de los Reyes-Gavilán, C.G. Nutrition and the gut microbiome in the elderly. *Gut Microbes* **2017**, *8*, 82–97. [CrossRef] [PubMed]
27. Reeves, P.G.; Nielsen, F.H.; Fahey, G.C. AIN-93 Purified Diets for Laboratory Rodents: Final Report of the American Institute of Nutrition Ad Hoc Writing Committee on the Reformulation of the AIN-76A Rodent Diet. *J. Nutr.* **1993**, *123*, 1939–1951. [CrossRef] [PubMed]

28. Massot-Cladera, M.; Costabile, A.; Childs, C.E.; Yaqoob, P.; Franch, À.; Castell, M.; Pérez-Cano, F.J. Prebiotic effects of cocoa fibre on rats. *J. Funct. Foods* **2015**, *19*, 341–352. [CrossRef]
29. Pérez-Berezo, T.; Franch, À.; Ramos-Romero, S.; Castellote, C.; Pérez-Cano, F.J.; Castell, M. Cocoa-enriched diets modulate intestinal and systemic humoral immune response in young adult rats. *Mol. Nutr. Food Res.* **2011**, *55* (Suppl. 1), S56–S66. [CrossRef]
30. Azagra-Boronat, I.; Massot-Cladera, M.; Knipping, K.; Van't Land, B.; Tims, S.; Stahl, B.; Knol, J.; Garssen, J.; Franch, À.; Castell, M.; et al. Oligosaccharides modulate rotavirus-associated dysbiosis and TLR gene expression in neonatal rats. *Cells* **2019**, *8*, 876. [CrossRef]
31. Klindworth, A.; Pruesse, E.; Schweer, T.; Peplies, J.; Quast, C.; Horn, M.; Glöckner, F.O. Evaluation of general 16S ribosomal RNA gene PCR primers for classical and next-generation sequencing-based diversity studies. *Nucleic Acids Res.* **2013**, *41*, e1. [CrossRef]
32. Martin, M. Cutadapt removes adapter sequences from high-throughput sequencing reads. *EMBnet.journal* **2011**, *17*, 10–12. [CrossRef]
33. Edgar, R.C.; Haas, B.J.; Clemente, J.C.; Quince, C.; Knight, R. UCHIME improves sensitivity and speed of chimera detection. *Bioinformatics* **2011**, *27*, 2194–2200. [CrossRef] [PubMed]
34. Salazar, N.; Arboleya, S.; Valdés, L.; Stanton, C.; Ross, P.; Ruiz, L.; Gueimonde, M.; de los Reyes-Gavilán, C.G. The human intestinal microbiome at extreme ages of life. Dietary intervention as a way to counteract alterations. *Front. Genet.* **2014**, *5*, 1–9.
35. Christoforidou, Z.; Mora Ortiz, M.; Poveda, C.; Abbas, M.; Walton, G.; Bailey, M.; Lewis, M.C. Sexual dimorphism in immune development and in response to nutritional intervention in neonatal piglets. *Front. Immunol.* **2019**, *10*, 1–17. [CrossRef]
36. Vom Steeg, L.G.; Klein, S.L. SeXX matters in infectious disease pathogenesis. *PLoS Pathog.* **2016**, *12*, 1–6. [CrossRef]
37. Gubbels Bupp, M.R.; Potluri, T.; Fink, A.L.; Klein, S.L. The confluence of sex hormones and aging on immunity. *Front. Immunol.* **2018**, *9*, 1269. [CrossRef]
38. Alarifi, S.; Bell, A.; G, W. In vitro fermentation of gum acacia - impact on the faecal microbiota. *J. Food Sci. Nutr.* **2018**, *69*, 696–704. [CrossRef]
39. Michel, C.; Kravtchenko, T.P.; David, A.; Gueneau, S.; Kozlowski, F.; Cherbut, C. In vitro prebiotic effects of Acacia gums onto the human intestinal microbiota depends on both botanical origin and environmental pH. *Anaerobe* **1998**, *4*, 257–266. [CrossRef]
40. Minamida, K.; Ota, K.; Nishimukai, M.; Tanaka, M.; Abe, A.; Sone, T.; Tomita, F.; Hara, H.; Asano, K. Asaccharobacter celatus gen. nov., sp. nov., isolated from rat caecum. *Int. J. Syst. Evol. Microbiol.* **2008**, *58*, 1238–1240. [CrossRef]
41. Jackson, R.L.; Greiwe, J.S.; Schwen, R.J. Emerging evidence of the health benefits of S-equol, an estrogen receptor β agonist. *Nutr. Rev.* **2011**, *69*, 432–448. [CrossRef]
42. Sekikawa, A.; Ihara, M.; Lopez, O.; Kakuta, C.; Lopresti, B.; Higashiyama, A.; Aizenstein, H.; Chang, Y.-F.; Mathis, C.; Miyamoto, Y.; et al. Effect of S-equol and soy isoflavones on heart and brain. *Curr. Cardiol. Rev.* **2018**, *15*, 114–135. [CrossRef] [PubMed]
43. Guo, W.L.; Chen, M.; Pan, W.L.; Zhang, Q.; Xu, J.X.; Lin, Y.C.; Li, L.; Liu, B.; Bai, W.D.; Zhang, Y.Y.; et al. Hypoglycemic and hypolipidemic mechanism of organic chromium derived from chelation of Grifola frondosa polysaccharide-chromium (III) and its modulation of intestinal microflora in high fat-diet and STZ-induced diabetic mice. *Int. J. Biol. Macromol.* **2020**, *145*, 1208–1218. [CrossRef]
44. Bryk, G.; Coronel, M.Z.; Pellegrini, G.; Mandalunis, P.; Rio, M.E.; de Portela, M.L.P.M.; Zeni, S.N. Effect of a combination GOS/FOS® prebiotic mixture and interaction with calcium intake on mineral absorption and bone parameters in growing rats. *Eur. J. Nutr.* **2015**, *54*, 913–923. [CrossRef] [PubMed]
45. Azagra-Boronat, I.; Rodríguez-Lagunas, M.J.; Castell, M.; Pérez-Cano, F.J. Prebiotics for gastrointestinal infections and acute diarrhea. In *Dietary Interventions in Gastrointestinal Diseases*; Watson, J.R., Preedy, V.R., Eds.; Elsevier: Amsterdam, The Netherlands, 2019; pp. 179–191.
46. Ooi, L.G.; Liong, M.T. Cholesterol-lowering effects of probiotics and prebiotics: A review of in vivo and in vitro findings. *Int. J. Mol. Sci.* **2010**, *11*, 2499–2522. [CrossRef] [PubMed]
47. Mortensen, A.; Poulsen, M.; Frandsen, H. Effect of a long-chained fructan Raftiline HP on blood lipids and spontaneous atherosclerosis in low density receptor knockout mice. *Nutr. Res.* **2002**, *22*, 473–480. [CrossRef]

48. Causey, J.L.; Feirtag, J.M.; Gallaher, D.D.; Tungland, B.C.; Slavin, J.L. Effects of dietary inulin on serum lipids, blood glucose and the gastrointestinal environment in hypercholesterolemic men. *Nutr. Res.* **2000**, *20*, 191–201. [CrossRef]
49. Letexier, D.; Diraison, F.; Beylot, M. Addition of inulin to a moderately high-carbohydrate diet reduces hepatic lipogenesis and plasma triacylglycerol concentrations in humans. *Am. J. Clin. Nutr.* **2003**, *77*, 559–564. [CrossRef]
50. Brighenti, F.; Casiraghi, M.C.; Canzi, E.; Ferrari, A. Effect of consumption of a ready-to-eat breakfast cereal containing inulin on the intestinal milieu and blood lipids in healthy male volunteers. *Eur. J. Clin. Nutr.* **1999**, *53*, 726–733. [CrossRef]
51. Jensen, C.D.; Spiller, G.A.; Gates, J.E.; Miller, A.F.; Whittam, J.H. The effect of acacia gum and a water-soluble dietary fiber mixture on blood lipids in humans. *J. Am. Coll. Nutr.* **1993**, *12*, 147–154. [CrossRef]
52. Haskell, W.L.; Spiller, G.A.; Jensen, C.D.; Ellis, B.K.; Gates, J.E. Role of water-soluble dietary fiber in the management of elevated plasma cholesterol in healthy subjects. *Am. J. Cardiol.* **1992**, *69*, 433–439. [CrossRef]
53. Babiker, R.; Elmusharaf, K.; Keogh, M.B.; Banaga, A.S.I.; Saeed, A.M. Metabolic effect of gum Arabic (Acacia Senegal) in patients with type 2 diabetes mellitus (T2DM): Randomized, placebo controlled double blind trial. *Funct. Foods Health Dis.* **2018**, *7*, 222. [CrossRef]
54. Ross, A.H.; Eastwood, M.A.; Brydon, W.G.; Anderson, J.R.; Anderson, D.M. A study of the effects of dietary gum arabic in humans. *Am. J. Clin. Nutr.* **1983**, *37*, 368–375. [CrossRef] [PubMed]

*nutrients*

MDPI

*Article*

# Differences in Gut Microbiome Composition between Senior Orienteering Athletes and Community-Dwelling Older Adults

**Frida Fart [1,†], Sukithar Kochappi Rajan [1,†], Rebecca Wall [1], Ignacio Rangel [1], John Peter Ganda-Mall [1,2], Lina Tingö [1], Robert J. Brummer [1], Dirk Repsilber [1,‡], Ida Schoultz [1,‡] and Carl Mårten Lindqvist [1,*,‡]**

1 School of Medical Sciences, Faculty of Medicine and Health, Örebro University, 702 81 Örebro, Sweden; frida.fart@oru.se (F.F.); sukithar.rajan@oru.se (S.K.R.); rebecca.wall@oru.se (R.W.); ignacio.rangel@oru.se (I.R.); john-peter.ganda-mall@oru.se (J.P.G.-M.); lina.tingo@oru.se (L.T.); robert.brummer@oru.se (R.J.B.); dirk.repsilber@oru.se (D.R.); ida.schoultz@oru.se (I.S.)

2 Laboratory of Translational Mucosal Immunology, Digestive Diseases Research Unit, Vall d'Hebron Institut de Recerca, Hospital Universitari Vall d'Hebron, 08035 Barcelona, Spain

* Correspondence: marten.lindqvist@oru.se

† Authors contributed equally to this work.

‡ Shared senior authorship.

Received: 18 July 2020; Accepted: 18 August 2020; Published: 27 August 2020

**Abstract:** Background: Gastrointestinal (GI) health is an important aspect of general health. Gastrointestinal symptoms are of specific importance for the elderly, an increasing group globally. Hence, promoting the elderly's health and especially gastrointestinal health is important. Gut microbiota can influence gastrointestinal health by modulation of the immune system and the gut–brain axis. Diverse gut microbiota have been shown to be beneficial; however, for the elderly, the gut microbiota is often less diverse. Nutrition and physical activity, in particular, are two components that have been suggested to influence composition or diversity. Materials and Methods: In this study, we compared gut microbiota between two groups of elderly individuals: community-dwelling older adults and physically active senior orienteering athletes, where the latter group has less gastrointestinal symptoms and a reported better well-being. With this approach, we explored if certain gut microbiota were related to healthy ageing. The participant data and faecal samples were collected from these two groups and the microbiota was whole-genome sequenced and taxonomically classified with MetaPhlAn. Results: The physically active senior orienteers had a more homogeneous microbiota within the group and a higher abundance of *Faecalibacterium prausnitzii* compared to the community-dwelling older adults. *Faecalibacterium prausnitzii* has previously shown to have beneficial properties. Senior orienteers also had a lower abundance of *Parasutterella excrementihominis* and *Bilophila* unclassified, which have been associated with impaired GI health. We could not observe any difference between the groups in terms of Shannon diversity index. Interestingly, a subgroup of community-dwelling older adults showed an atypical microbiota profile as well as the parameters for gastrointestinal symptoms and well-being closer to senior orienteers. Conclusions: Our results suggest specific composition characteristics of healthy microbiota in the elderly, and show that certain components of nutrition as well as psychological distress are not as tightly connected with composition or diversity variation in faecal microbiota samples.

**Keywords:** gut microbiota; metagenomics; aged; *Faecalibacterium prausnitzii*; orienteering

## 1. Introduction

During the last decade, longevity has increased among the elderly population, resulting in a global ageing phenomenon that is having a major impact on healthcare systems worldwide. This has led to an increased awareness of the importance of promoting healthy ageing and quality of life throughout an individual's lifespan. To promote and initiate healthy ageing, it is important to understand and reveal the underlying mechanisms.

The gastrointestinal (GI) tract is an essential part of the human body and physiological system through which health and well-being might be promoted [1]. A well-functioning GI tract has previously been identified as crucial for subjective health and well-being among older adults [2]. GI symptoms are common among community-dwelling older adults (i.e., older adults residing in their own household) in Sweden, and as many as 65% experience one or several gut symptoms that correlate with increased psychological distress, including anxiety and depression [3]. On the contrary, physically active seniors engaged in orienteering (a sport involving finding specific locations using a map and a compass) have previously been identified as a potential model of healthy ageing [4], as they display the three main components of successful ageing—physical endurance, cognitive skills, and social interaction [5]. Indeed, our previous data show fewer GI symptoms among senior orienteers and a better overall health compared to community-dwelling older adults [4,6]. This indicates that gut health may reveal important factors of well-being in the elderly, especially its association with various factors that are known to influence gut microbiota during the entire lifespan. The microbial composition of an individual depends on factors such as age, diet, geography, environmental exposure, and many others, as shown in Figure 1 [7–10].

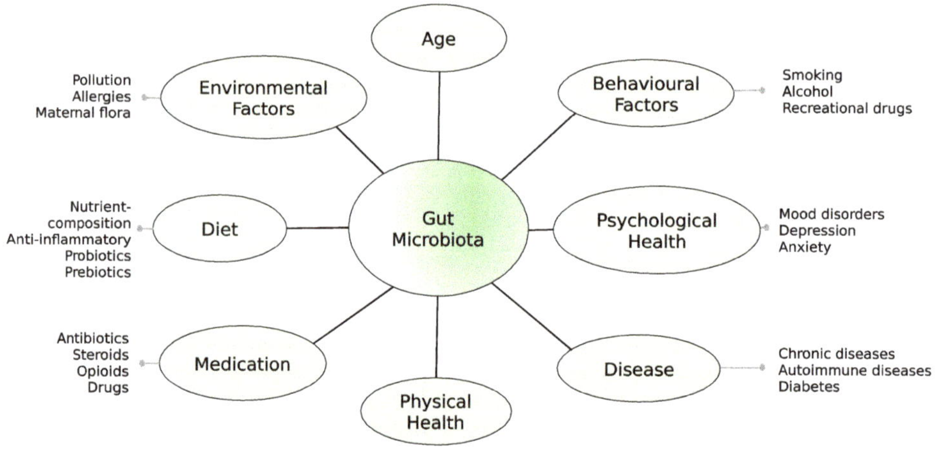


**Figure 1.** Factors affecting the composition of gut microbiota.

The human GI tract is a complex ecosystem where the gut microbiome interplays with host cells and dietary-derived components, both of which have been implicated in playing a major role in health and disease [1]. A diverse gut microbiome has been related to several essential mechanisms for both a well-functioning GI tract as well as well-being, including modulating the immune system, maintaining an intact intestinal barrier, and being a part of the regulation of the gut-brain axis [11,12], where a decreased diversity has been linked to both GI and psychiatric disorders [13]. Ageing has been associated with a loss of diversity of the gut microbiome; specifically, bacteria belonging to the phyla *Firmicutes* and *Actinobacteria* decrease, whereas *Proteobacteria* increase in abundance [14]. These changes could be due to nutritional deficiencies such as lower intakes of specific nutrients, e.g., dietary fibres and proteins, that are important for maintaining the immune and GI functions [15]. Recent evidence further indicates that physical activity, independent of diet, could induce positive alterations of the gut microbiome composition [16,17]. However, the relationship between physical activity and gut microbiota across the life course has not been entirely elucidated. It is also less known to what extent microbiome composition and diversity are influenced by certain factors when other factors change at the same time, especially in a diverse population such as the elderly. For example, it is still not clear which specific influence could be attributed to nutrition components or psychological factors such as distress or anxiety.

In the present study, we investigated the gut microbiota profile in senior orienteering athletes, as a proposed model of healthy ageing, in relation to GI symptoms and macronutrient intake and compared it to the gut microbiota composition of community-dwelling older adults, representing the general older adult population, to identify possible patterns specifically related to healthy ageing.

## 2. Materials and Methods

### 2.1. Study Participants, Data Collection, and Ethics

Samples were available from two previously established cohorts: community-dwelling older adults (hereafter referred to as older adults), representing a cross-section of the general older adult population [3,18] ($n$ = 70) and physically active senior orienteers (hereafter referred to as senior orienteers) as a model of healthy ageing [6] ($n$ = 28). All participants were ≥65 years of age; the inclusion and exclusion criteria are presented in Table 1. The study received approval from the Regional Ethics Board in Uppsala, Sweden (dnr: 2012/309, 2013/037, 2015/357) and was conducted in accordance with the Declaration of Helsinki.

**Table 1.** Inclusion and exclusion criteria.

| Older Adults | Senior Orienteering Athletes |
|---|---|
| Inclusion criteria | |
| Informed consent signed by the study participant<br>Age ≥ 65 years<br>Mentally and physically fit to complete questionnaires during the study period | Informed consent signed by the study participant<br>Age ≥ 65 years<br>Mentally and physically fit to complete questionnaires during the study period<br>Actively performing and competing in orienteering |
| Exclusion criteria | |
| Any known gastrointestinal disease, malignancies, and ischemia<br>Inflammatory bowel disease<br>Participation in another clinical trial in the past three months | Any known gastrointestinal disease, malignancies, and ischemia<br>Inflammatory bowel disease<br>Participation in another clinical trial in the past three months |

### 2.2. Gastrointestinal Symptoms, Psychological Distress, and Physical Activity

Data regarding GI symptoms, psychological distress, and physical activity were available from the two previously established cohorts for all orienteers and a subset of older adults ($n$ = 54) [3,6]. GI symptoms and psychological distress were assessed through the following validated questionnaires: the Gastrointestinal Symptom Rating Scale (GSRS) [19] and the Hospital Anxiety and Depression Scale (HADS) [20]. Briefly, the GSRS comprises 15 questions assessing five GI symptoms (i.e., reflux, abdominal pain, dyspepsia, diarrhoea, and constipation) that are scored from 1 to 7 depending on their severity. A total score is then calculated as the average from the five symptom scores. The HADS includes 14 questions and is divided into two subscales assessing anxiety and depression (7 questions/scale) together giving an estimation of psychological distress. The Frändin–Grimby Activity Scale (FGAS) [21], a 6-point scale with fixed response alternatives, was used to assess the level of physical activity.

### 2.3. Macronutrient Intake

The nutrient intake was estimated by a validated semi-quantitative Food Frequency Questionnaire (FFQ) [22] asking for dietary intake during the past year. The questionnaire has previously been described and used in an elderly population [3]. Raw data were available from the previously established cohorts [3,18] and were further analysed according to a standard procedure to assess the following macronutrients: fibre, protein, saturated fat, unsaturated fat, and carbohydrates as well as estimated added sugar. Briefly, participants estimated their intake of 66 food items from 0–8 (0 = never, 8 = 4 or more times a day). To facilitate inter-individual comparisons, the intake per day was expressed as energy percentage (E%) and the intake of fibre was expressed as gram per megajoule (MJ) energy intake.

### 2.4. Medications

Medications were self-reported and grouped according to the Anatomical Therapeutic Chemical (ATC) classification system, controlled by the WHO's Collaborating Centre for Drug Statistics Methodology, by a physician (author F.F.), using a national tool [23].

### 2.5. Next-Generation Sequencing for Determination of the Microbiota Composition

Stool samples were collected according to standard operating procedures [18] and were analysed using next-generation sequencing (NGS) for assessment of the faecal microbial composition [24]. Total DNA was extracted from faecal samples using a QIAmp DNA stool mini kit according to the manufacturer's instructions (Qiagen, Hilden, Germany), coupled with an initial bead-beating step. The total microbial content was further assessed through whole-genome sequencing (WGS) at SciLifeLab, (Stockholm, Sweden) using an Illumina HiSeq 2500 device (Illumina, San Diego, CA, USA) with four samples per lane, yielding approximately 50 million read pairs per sample. Whole-genome sequences were taxonomically classified using MetaPhlAn v2.0 (Huttenhower Lab, Boston, MA, USA) [25] at default settings. Relative abundances for the taxomic levels of genera and species were extracted from the output of MetaPhlAn and further analysed in R (3.6.1, R Core Team, New Zealand) [26].

### 2.6. Data Analysis

Continuous demographic data were analysed using the Mann–Whitney U test, and categorical demographic data were analysed using the chi-square test. Relative abundances for microbiota at genus and species level were calculated and considered for further analysis. Welch's two-sample *t*-tests followed by the Benjamini–Hochberg procedure for multiple testing correction were used to assess the difference in bacterial abundance between the two groups [27]. Top genera, differentially occurring in orienteers and general elderly, were selected based on a false discovery rate (FDR) less than 5%. Given our abundance data and group sizes (orienteers, older adults), we estimated to be able to detect a 20% difference in abundance, with 80% power and 5% significance level. Species representing the top predicted genera were considered for further downstream statistical analysis. To estimate if a difference was consistent after the effect of the covariate was taken into account, we fitted a zero-inflated negative binomial (ZINB) regression model with each covariate as an explanatory variable [28–30]. The resulting residuals were considered as corrected bacterial abundances with the effect of the covariate removed. The differences of these corrected bacterial abundances between groups were tested using a ZINB model and ANOVA type III sums of squares test for the bacterial abundances. The relative importance of all covariates was assessed by performing a model comprising all covariates using likelihood-ratio chi-square statistics [31].

All plots were produced in R (version 3.6.1, R Core Team, New Zealand) [32] using either the base graphics package or ggplot2 version 3.2.1 [33]. Boxplots were produced with the graphics package using the notch option, where box encapsulates the first to third quantiles and whiskers are the minimum of 1.5 interquartile range (IQR) from the box or the min/max value. Boxplot notches visualise a non-parametric estimation of the 95% confidence interval of the median calculated as +/-1.58 IQR/√n [34]. ANOVA with type III sums of squares analysis was performed as implemented in the car package v 3.0-3. Zero-inflated negative binomial regression was performed using the function zeroinfl within the pscl package v 1.5.5 [35]. FDR values were estimated using the package multtest, following the approach adopted by Benjamini and Hochberg [36]. Bray-Curtis distances and Shannon diversity index were calculated from species abundance profiles using the vegan (v2.5-6) package [37]. PCoA analyses were performed with the R package labdsv [38]. Participants outside the 95% confidence area formed the subset atypical older adults. Student's $t$-test was performed to compare average values between senior orienteers, typical older adults, and the subset of atypical older adults for each covariate. Taxonomy prediction and statistical analysis were automated using in-house scripts written in Bash and R (Figure S1) [26].

## 3. Results

### 3.1. Demographic Data

All demographic data are presented in Table 2. The degree of anxiety ($p = 0.006$) and depression ($p = 0.002$) were significantly higher among older adults compared to senior orienteers, whereas physical activity was higher among senior orienteers ($p < 0.001$).

**Table 2.** Participant characteristics.

| Parameter | Community-Dwelling Older Adults $n = 70$ | Senior Orienteering Athletes $n = 28$ | $p$-Value |
|---|---|---|---|
| Sex<br>Median $n$ (%)<br>Female<br>Male | <br><br>33 (47%)<br>37 (53%) | <br><br>12 (43%)<br>16 (57%) | 0.701 |
| Age<br>Median (IQR) | 72 (69–76) | 68.5 (67–72) | 0.034 |
| Smoking<br>$n$ (%) | 1 (1%) | 0 (0%) | 0.537 |
| Physical activity<br>Median (IQR) | 3.5 (3–4) | 4 (4–5) | <0.001 * |
| Polypharmacy<br>$n$ (%) | 8 (12%) | 2 (7%) | 0.487 |
| Number of medications<br>Median (IQR) | 2 (1–4) | 1 (0–2) | 0.016 |
| GI symptoms<br>Median (IQR)<br>Indigestion<br>Constipation<br>Abdominal pain<br>Diarrhoea<br>Reflux | <br><br>2.0 (1.3–3.1)<br>1.3 (1.0–3.3)<br>1.3 (1.0–2.0)<br>1.0 (1.0–3.3)<br>1.0 (1.0–1.5) | <br><br>1.5 (1.3–1.9)<br>1.3 (1.0–1.6)<br>1.0 (1.0–1.7)<br>1.3 (1.0–1.7)<br>1.0 (1.0–1.0) | <br><br>0.011<br>0.569<br>0.009<br>0.497<br>0.043 |
| Total GI symptoms | 1.8 (1.1–2.5) | 1.3 (1.1–1.5) | 0.021 |
| Depression<br>Median (IQR) | 2 (1–4) | 0 (0–1) | 0.002 * |
| Anxiety<br>Median (IQR) | 2 (0.5–5.5) | 0.5 (0–2.8) | 0.006 * |

* Retained significant difference after multiple testing corrections. Physical activity, GI symptoms, and psychological distress (depression and anxiety) are all measured with questionnaires, see the Materials and Methods section for a more detailed description of each questionnaire. Interquartile range (IQR) is presented within parentheses where applicable. GI = gastrointestinal.

### 3.2. Microbiota Composition

Faecal microbiota profiles of the two established cohorts of older individuals (senior orienteers and older adults) were analysed on both genera and species levels from shotgun metagenomic sequences. *Faecalibacterium* was on average the most prominent genus and a total of 111 genera were found in at least one sample (Figure 2, Supplementary Table S1). Three of these genera showed significantly different proportions between senior orienteers and older adults (Figure 3A). These three genera are represented by four species that were used for further analysis. Of these four species, *Faecalibacterium prausnitzii* and *Bilophila* unclassified were the most abundant (Figure 3B).

To investigate whether differences of microbiota composition were due to confounding factors, twelve covariates were included in the analyses, i.e., five macronutrients (carbohydrates, protein, unsaturated fat, saturated fat, and fibre), two parameters assessing psychological distress (anxiety and depression), three parameters assessing medicines associated with dynamic changes in the microbiota (antibiotics during the previous six months, acetylsalicylic acid, and any medicine affecting the GI tract), sex, and age (Figures 4–6). Several covariates were significantly different between the groups. The senior orienteers had a significantly higher intake of carbohydrates and a lower intake of saturated fat in their diet compared to older adults ($p = 0.006$ and $p = 0.038$, nominal $p$-values). Older adults reported a higher level of depression and anxiety (Table 1). One species was significantly increased in senior orienteers after correcting for all covariates, namely *Faecalibacterium prausnitzii*. *Bilophila* unclassified was significantly different for 8/15 covariates or combinations of covariates, and *Bilophila wadsworthia* as significantly different for 5/15 covariates or combinations of covariates (more abundant in older adults; see Figure 5).

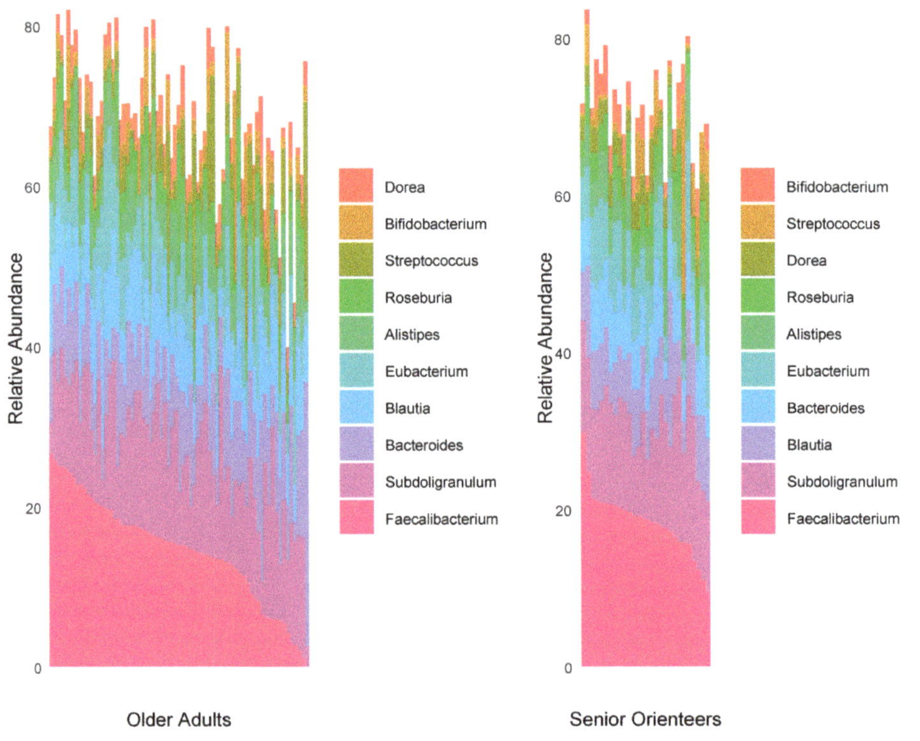


**Figure 2.** Relative abundance of the 10 most abundant genera across 98 samples.

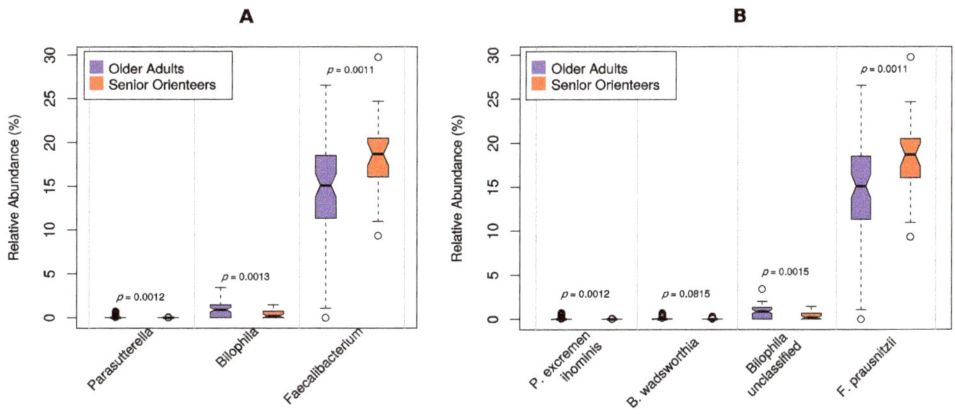


**Figure 3.** Relative abundance of significantly different genera and selected species stratified for group (senior orienteers compared to older adults). Cut-off for significance was set at false discovery rate (FDR) <5%. Descriptive *p*-values for each comparison are shown. (**A**) Genera; (**B**) Species.

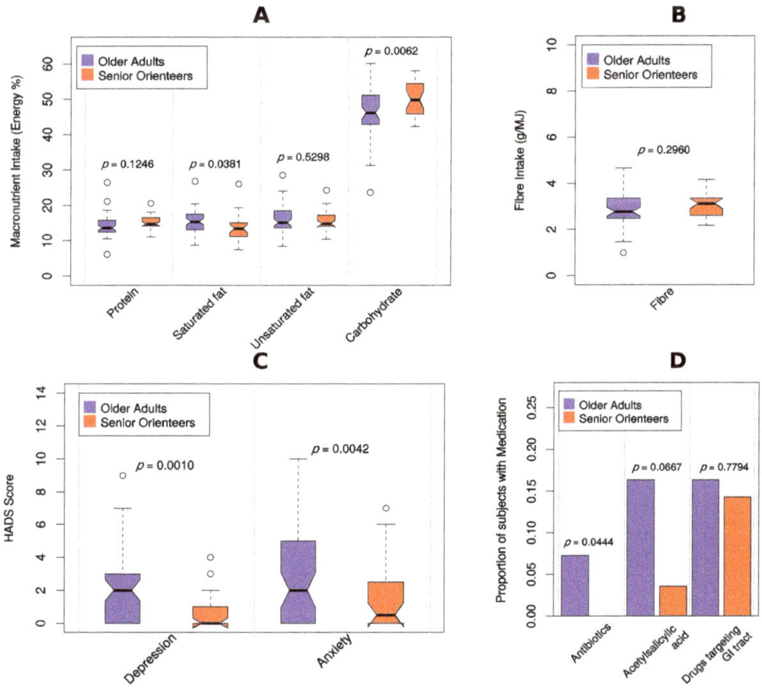


**Figure 4.** Comparison of covariates. Boxplots of covariates stratified for older adults and senior orienteers, including descriptive p-values from Welch's *t*-test. (**A**) Macronutrients measured by energy percentage (E%). (**B**) Fibre measured by grams per megajoule (MJ). (**C**) Hospital Anxiety and Depression Scale (HADS) score. (**D**) Bar plot for medication covariates for older adults and senior orienteers, including descriptive *p*-values from chi-square test.

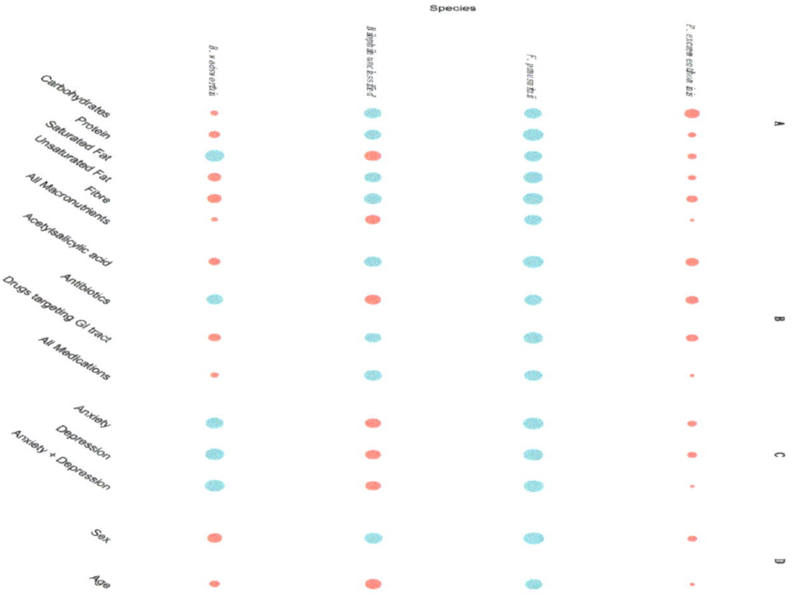


**Figure 5.** Significance of difference between older adults and senior orienteers after correction for macronutrients, psychological distress, and medication variables. Corrected bacterial composition values were compared between groups for each species and false discovery rates (FDRs) calculated. The dots represent negative log10 *p*-values belonging to respective species, where blue denotes significance and red denotes non-significance, with a significance threshold at FDR <5%. A Results for models with a single macronutrient variable and with all macronutrient variables in a multi-variable model. B Results for models with single medication variables and with all variables in a multi-variable model. C Results for models regarding anxiety and depression separately with single Hospital Anxiety and Depression Scale (HADS) variables and with both HADS variables in a multi-variable model. D Results for models with sex and age.

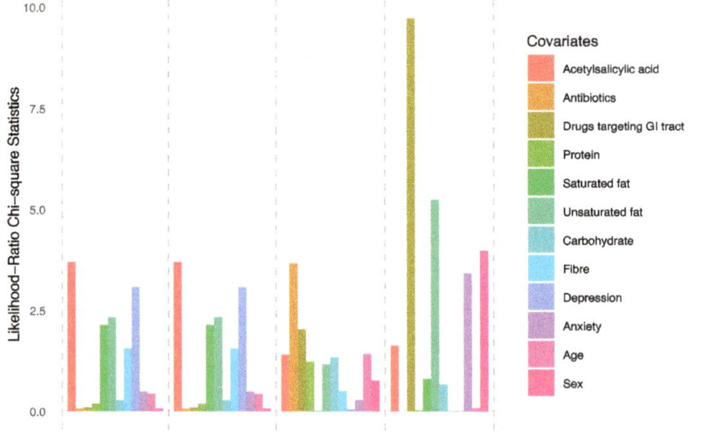


**Figure 6.** Assessment of relative importance of all covariates. A complete model comprising all covariates for assessing variable importance. The relative importance of each covariate was measured as likelihood-ratio chi-square statistics.

### 3.3. Ecological Diversity and Homogeneity

No difference in alpha diversity in terms of Shannon index was observed between the groups (Supplementary Figure S2). To estimate the beta diversity, principal coordinates analysis (PCoA) with a Bray–Curtis dissimilarity score was used. In the PCoA, the microbiota profiles of senior orienteers appear more homogenous than the profiles of older adults (95% confidence ellipse area, 0.2013 for older adults and 0.1094 for senior orienteers; see Figure 7A). When analysing only the four species that are significantly different between the groups, the homogeneity difference became even larger (0.1861 for older adults and 0.0179 for senior orienteers; see Figure 7B). Based on the PCoA with the four selected species, there appeared to be a subset of individuals, all from the older adult group, that have an atypical microbiota profile. This atypical participant group (atypical older adults, $n = 12$) was compared with orienteers ($n = 28$) and the rest of the older adults (typical older adults, $n = 42$) regarding covariates (Figure 8). Significant differences were observed only between senior orienteers and the majority group of typical older adults. Protein, saturated fat, carbohydrates, depression, anxiety, and GSRS variables showed significant differences between these two groups. Interestingly, the atypical group of older adults seems to be closer to the senior orienteering group than the typical older adults for these covariates. Confidence intervals of correlation values between *F. prausnitzii* and fibre showed a trend towards a weak correlation. A trend for a positive correlation was found between *F. prausnitzii* and fibre in the orienteering group although not significant (cor = 0.33, 95% CI = [−0.046, 0.63]). For the older adults, no such correlation could be seen (cor = −0.062, 95% CI = [−0.36, 0.25] for typical adults and cor = 0.14, 95% CI = [−0.53, 0.57] for atypical adults) (Figure 9).

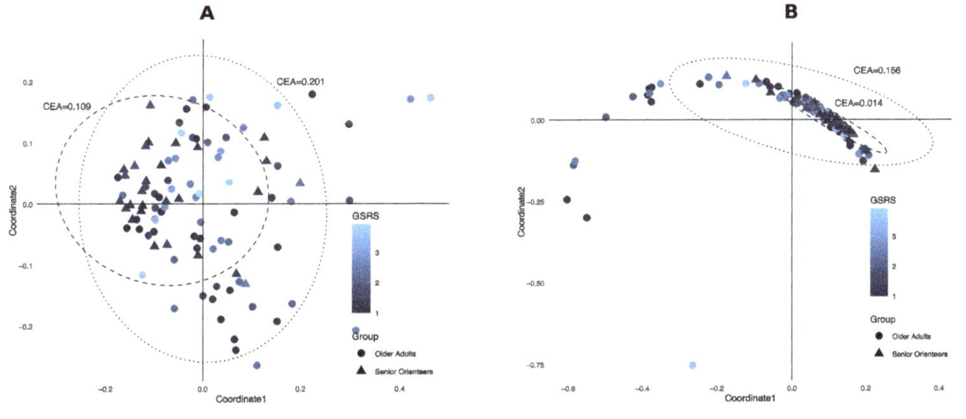


**Figure 7.** Principal coordinates analysis (PCoA) plots. Principal coordinates were estimated using Bray–Curtis distance on the predicted species. Each dot represents an individual sample, shape depicts groups, and blue scale codes for the gastrointestinal symptom scores measured with Gastrointestinal Symptom Rating Scale (GSRS) values. Dotted ellipse indicates 95% confidence region of older adults and dashed ellipse indicates 95% confidence region of senior orienteers. CEA = 95% confidence ellipse area. (**A**) PCoA using all predicted species; (**B**) PCoA using four selected species that were significantly different between older adults and senior orienteers.

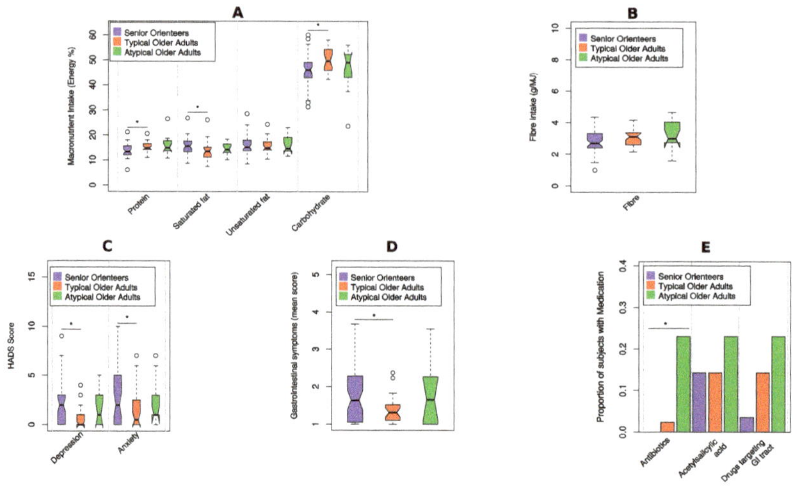


**Figure 8.** Comparison of covariates when older adults are stratified for typical and atypical. Atypical older adults are defined as samples outside of the confidence ellipse area in Figure 7. Statistically significant differences are marked with an asterisk. (**A**) Macronutrient intake measured by energy percentage (E%). (**B**) Fibre measured by grams per megajoule. (**C**) Anxiety and depression scores. (**D**) Mean score of gastrointestinal symptoms. (**E**) Representation of proportion of subjects with medications.

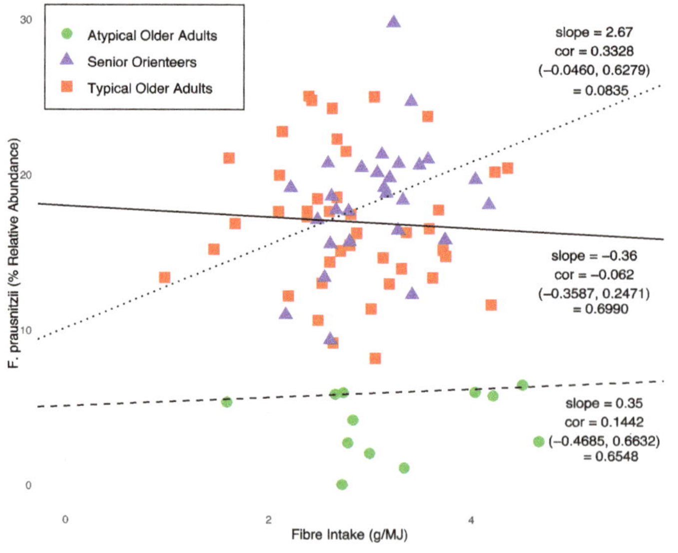


**Figure 9.** Correlation between *Faecalibacterium prausnitzii* and fibre intake. Shape depicts different groups. Dotted line, solid line, and dashed line represent regression lines for senior orienteers, typical older adults, and atypical older adults, respectively. Confidence interval (95%) values are given in brackets for respective observed correlations.

## 4. Discussion

The present study focused particularly on identifying gut microbiota profiles related to healthy ageing. Collectively, the novel data of the study show that senior orienteers, used as a model of healthy ageing, display a significantly different composition of the gut microbiota, with higher levels of *F. prausnitzii*, compared to older adults. Notably, these changes were found to be persistent even after correcting for macronutrient intake, psychological distress, and medical regimen affecting the GI tract. As a higher abundance of *F. prausnitzii* is associated with good gastrointestinal health [39], this result is coherent with our previous studies of the cohorts [4,6]. In addition, we observed more homogeneous overall compositions of gut microbiota in the senior orienteer cohort.

Senior orienteering athletes have previously been identified as a potential model of healthy ageing [4], where we have previously shown that signs of depression, anxiety, and gastrointestinal discomfort are lower in this group compared to older adults [6]. In the present study, assessment of the macronutrient intake further showed that senior orienteers had a lower intake of saturated fats and a higher intake of carbohydrates compared to older adults. This result further supports our previous findings that senior orienteers display several factors associated with health. The dietary intake has previously been shown to be a major factor influencing the composition of the gut microbiota and, subsequently, the metabolic output and function of the gut microbiome [40–42].

While a Western-style diet, rich in saturated fat and low in fibre, gives rise to a less diverse gut microbiota with a metabolic profile likely to be detrimental to health [42,43], the addition of dietary fibres, fruits, and vegetables is able to shift the composition to a more diverse composition associated with an increase in bacterial species, including *F. prausnitzii* [44–46]. *F. prausnitzii*, recognized as a marker of a healthy gut [39], is a non-spore forming and strict anaerobe, placed taxonomically within *Clostridium* cluster IV [47], which is a member of the *Clostridium leptum* group [48]. It is also one of the most important members among the butyrate-producing bacteria in the human colon [49,50]. The function of *F. prausnitzii* in the gut has been associated with its high capacity to contribute to the production of the short-chain fatty acid butyrate, the main nutrient for colonocytes known to display anti-inflammatory properties [51]. A diet high in fibre has previously been associated with an increased abundance of *F. prausnitzii* [52,53]. Within the senior orienteering group, we identified a trend towards a positive correlation between intake of fibres, including dietary fibres, and relative abundance of *F. prausnitzii*. As the trend is not visible in community-dwelling older adults, this result could suggest that fibre intake is linked to higher *F. prausnitzii* abundance only in a group with a lower degree of GI problems. However, it is important to note that a limitation of the study is the assessment of macronutrient intake via an FFQ estimating intake over a year. Therefore, the result may be affected by recall bias and a dietary diary would have been an excellent complement to estimate the intake during the days of stool sampling.

Our findings further show that *F. prausnitzii* accounts for approximately 18% of the total faecal gut microbiota in senior orienteers compared to 15% among community-dwelling older adults. This is in accordance with two previous independent studies showing that 5–15% of the microbiota consists of *F. prausnitzii* [39,54]. This observation may indicate that senior orienteers have a higher production of butyrate. However, butyrate was not assessed in the present study as the level of butyrate in the luminal content does not reveal whether the elevated levels are due to the gut microbiota composition or a disturbed uptake of butyrate in the intestinal mucosa. Hence, further studies are needed to elucidate how the abundance of *F. prausnitzii* correlates to butyrate production in older adults. Moreover, the high relative abundance of *F. prausnitzii* in the present study may be due to geographical location as both elderly and adult individuals in Sweden have been found to have a high abundance of this particular species compared to microbiota profiles found in other European countries [55].

Even though a higher relative abundance of *F. prausnitzii* was observed, we did not observe an enhanced microbial diversity among senior orienteers. This is in contrast to previous findings where regular exercise and sustained levels of increased physical activity have been shown to enhance microbial diversity independent of diet [16,17,56]. Interestingly, a recent report shows that, even though regular exercise among older adults is important to maintain a stable gut microbiota, the α-diversity was not significantly different between older adults performing regular exercise compared to those who did not [57]. On a family level, a change in relative abundance of several bacterial families was observed, but not in the Ruminococcaceae family, to which *F. prausnitzii* belongs. A recent systematic review further summarizes the field and shows that higher levels of physical activity and cardiorespiratory fitness are associated with higher faecal concentration of short-chain fatty acids in adults [58]. However, it was not possible to distinguish whether short-term or medium-/long-term exercise had a more positive effect on the gut microbiota composition. It is therefore possible that orienteering among elderly may only have moderate effects on the gut microbiota. It is further important to note that the level of physical activity is self-reported and does not give an exact indication of how hard the participants exercised. In addition, the present study is limited by the low number of senior orienteers, and the absence of significant differences may reflect low statistical power rather than true negative findings. Hence, more in-depth future studies are needed to thoroughly elucidate the relationship between physical activity and gut microbiota composition in the elderly.

Moreover, a physically inactive lifestyle together with a diet high in refined carbohydrate and low in dietary fibre is associated with a depleted microbiome and the elevated risk to develop chronic diseases [59]. In our study, the bacterial species *Parasutterella excrementihominis* and *Bilophila wadsworthia* were found in a higher relative abundance in community-dwelling older adults. Although little is known regarding their function, it is intriguing to note that both species have been associated with decreased intestinal health. *Parasutterella excrementihominis* belongs to the class *Betaproteobacteria* (one of eight classes of *Proteobacteria*). The relative abundance of *Parasutterella excrementihominis* has previously been associated with different host health outcomes such as inflammatory bowel disease, irritable bowel syndrome, obesity, diabetes, and fatty liver disease [60–63]. *Bilophila* is a member of the hydrogen sulphide ($H_2S$)-producing family *Desulfovibrionaceae*. *Bilophila* metabolizes sulphated compounds and produces $H_2S$ that can trigger inflammation, exert genotoxic and cytotoxic effects on epithelial cells, and impair intestinal barrier function [64]. Correlations of sulfidogenic bacteria to the aetiology of chronic metabolic diseases have recently been shown [65,66]. However, little is known about the genus *Bilophila*. *Bilophila wadsworthia* has been associated with a variety of human and animal infections [67–70].

Another possible environmental factor that can influence gut microbiota is medications [71]. Common drugs, including antibiotics, have been found to alter the gut microbiota composition [72]. Repeated courses of antibiotic treatment may result in the loss of microbial species that may not be restored [73]. Prescribed medication from medical records would have provided appropriate data to investigate an accurate list of medications since our data did not include dosage or common usage. However, the prescribed medications do not include over-the-counter medications, which comprise several agents affecting the gastrointestinal canal directly (such as proton-pump inhibitors, laxatives, etc.). As the differences between senior orienteers and older adults are still significant after diet and medications are taken into account, the distinctive features of the former group are further accentuated as important for the differences in microbiota.

Moreover, it is important to acknowledge that senior orienteers have a lifestyle represented not only by a high level of physical activity, but also by an active social life. Loneliness and lack of contact often increase the risk of depression and anxiety among older adults [74]. In accordance with previous data, we show that depression and anxiety are significantly lower among senior orienteering athletes compared to community-dwelling older adults. Depression and anxiety are known to be associated with an altered gut microbiota composition that is most likely due to changes in the microbiota–gut–brain axis, the bidirectional relationship between the gut microbiota and brain [75]. One of the major factors influencing this pathway is diet and, among other factors, a change in eating habits due to increased psychological distress has been proposed to contribute to the alterations of the gut microbiota associated with depression and anxiety [76]. However, the relationship between diet and depression and anxiety needs to be further investigated as the results from dietary intervention studies are contradictory and the directionality and mechanisms are currently unclear as reviewed by Bear et al. [75]. In accordance with these observations, a recent systematic review of the field shows that so far there is no consensus within human studies regarding the question about which bacterial taxa would be most relevant to depression [77].

This study focused only on a few factors that could possibly have an impact on the gut microbiome. However, there are many other environmental, behavioural, socio-economic, and health-related variables that contribute to the gut microbial composition (Figure 1). The scope of this study was to investigate the difference between senior orienteers and older adults after correcting for a variety of factors. Many of the factors that were used for correcting the microbiota composition are not independent, but are different between the two studied groups and, therefore, confounded with each other. An elaborate analysis of predicted function profiles of proteins, pathways, and metabolite levels will provide more insight into the functional aspects of healthy ageing, but remains outside the scope of this study.

In healthy adults, the gut microbiome is a very stable community of microbes composed of highly adapted microbial species [78,79]. The composition of the gut microbiome has been shown to be shaped more by environment than by host genetics [80]. Our analyses showed that senior orienteers as a group had a more homogenous microbiota, which makes individually stable microbiota profiles also more likely. This is not the case in the older adults' samples. Individual stability over time has been observed as a feature that distinguishes the microbiota of healthy individuals compared with individuals with gastrointestinal disease [81]. Nevertheless, future studies need to validate our findings in a longitudinal study to verify that the gut microbiota is homogenous and stable among senior orienteers.

## 5. Conclusions

In conclusion, our data show that senior orienteers can be seen as a model of healthy ageing also from the perspective of the microbiota. Their faecal microbiota shows a higher abundance of Faecalibacterium prausnitzii that has been previously associated with positive health benefits, as well as an active lifestyle. In contrast, the senior orienteers have a lower abundance of Parasutterella excrementihominis and Bilophila wadsworthia, two species that previously have been associated with decreased intestinal health. Furthermore, our observation of senior orienteer faecal microbiota being more homogenous suggests this group of older adults as a model of healthy ageing.

**Supplementary Materials:** The following are available online at http://www.mdpi.com/2072-6643/12/9/2610/s1, Figure S1: Work flow, Figure S2: Shannon_diversity, Table S1: Relative abundances: genera, species

**Author Contributions:** Conceptualization, F.F., S.K.R., J.P.G.-M., L.T., R.J.B., D.R., I.S., C.-M.L.; methodology, F.F., S.K.R., R.W., I.R., J.P.G.-M., L.T., R.J.B., D.R., I.S., C.M.L. formal analysis, F.F., S.K.R., R.W., I.R., J.P.G.-M., L.T., D.R., I.S. and C.-M.L., data curation, F.F., S.K.R., D.R., I.S. and C.-M.L.; writing—original draft preparation, F.F., S.K.R., R.W., I.R., D.R., I.S., and C.-M.L.; writing—review and editing, F.F., S.K.R., J.-P.M., L.T., R.J.B., D.R., I.S. C.-M.L. visualization, F.F., S.K.R., D.R. and C.-M.L.; funding acquisition, I.S. and R.J.B. All authors have read and agreed to the published version of the manuscript.

**Funding:** This research was funded by Bo Rydin foundation grant number: F0514 and the knowledge foundation, grant number: 20110225.

**Conflicts of Interest:** There is no conflict of interest to declare.

## References

1. Dinan, T.G.; Cryan, J.F. The Microbiome-Gut-Brain Axis in Health and Disease. *Gastroenterol. Clin. N. Am.* **2017**, *46*, 77–89. [CrossRef]
2. Algilani, S.; Östlund-Lagerström, L.; Kihlgren, A.; Blomberg, K.; Brummer, R.J.; Schoultz, I. Exploring the concept of optimal functionality in old age. *J. Multidiscip. Healthc.* **2014**, *7*, 69–79. [CrossRef] [PubMed]
3. Fart, F.; Tingoe, L.; Engelheart, S.; Lindqvist, C.-M.; Brummer, R.J.; Kihlgren, A.; Schoultz, I. Gut Health, nutrient intake and well-being in community-dwelling older adults. *medRxiv* **2019**. [CrossRef]

4. Östlund-Lagerström, L.; Blomberg, K.; Algilani, S.; Schoultz, M.; Kihlgren, A.; Brummer, R.J.; Schoultz, I. Senior orienteering athletes as a model of healthy aging: A mixed-method approach. *BMC Geriatr.* **2015**, *15*, 76. [CrossRef] [PubMed]
5. Rowe, J.W.; Kahn, R.L. Human aging: Usual and successful. *Science* **1987**, *237*, 143–149. [CrossRef]
6. Ganda Mall, J.-P.; Östlund-Lagerström, L.; Lindqvist, C.M.; Algilani, S.; Rasoal, D.; Repsilber, D.; Brummer, R.J.; Keita, Å.V.; Schoultz, I. Are self-reported gastrointestinal symptoms among older adults associated with increased intestinal permeability and psychological distress? *BMC Geriatr.* **2018**, *18*, 75. [CrossRef]
7. Mahnic, A.; Rupnik, M. Different host factors are associated with patterns in bacterial and fungal gut microbiota in Slovenian healthy cohort. *PLoS ONE* **2018**, *13*, e0209209. [CrossRef]
8. Valdes, A.M.; Walter, J.; Segal, E.; Spector, T.D. Role of the gut microbiota in nutrition and health. *BMJ* **2018**, *361*, k2179. [CrossRef]
9. Hasan, N.; Yang, H. Factors affecting the composition of the gut microbiota, and its modulation. *PeerJ* **2019**, *7*, e7502. [CrossRef]
10. Kho, Z.Y.; Lal, S.K. The Human Gut Microbiome—A Potential Controller of Wellness and Disease. *Front. Microbiol.* **2018**, *9*, 1835. [CrossRef]
11. Cryan, J.F.; O'Riordan, K.J.; Cowan, C.S.M.; Sandhu, K.V.; Bastiaanssen, T.F.S.; Boehme, M.; Codagnone, M.G.; Cussotto, S.; Fulling, C.; Golubeva, A.V.; et al. The Microbiota-Gut-Brain Axis. *Physiol. Rev.* **2019**, *99*, 1877–2013. [CrossRef] [PubMed]
12. Takiishi, T.; Fenero, C.I.M.; Câmara, N.O.S. Intestinal barrier and gut microbiota: Shaping our immune responses throughout life. *Tissue Barriers* **2017**, *5*, e1373208. [CrossRef] [PubMed]
13. Rea, K.; Dinan, T.G.; Cryan, J.F. Gut Microbiota: A Perspective for Psychiatrists. *Neuropsychobiology* **2020**, *79*, 50–62. [CrossRef] [PubMed]
14. Claesson, M.J.; Jeffery, I.B.; Conde, S.; Power, S.E.; O'Connor, E.M.; Cusack, S.; Harris, H.M.B.; Coakley, M.; Lakshminarayanan, B.; O'Sullivan, O.; et al. Gut microbiota composition correlates with diet and health in the elderly. *Nature* **2012**, *488*, 178–184. [CrossRef] [PubMed]
15. Salazar, N.; Valdés-Varela, L.; González, S.; Gueimonde, M.; de Los Reyes-Gavilán, C.G. Nutrition and the gut microbiome in the elderly. *Gut Microbes* **2017**, *8*, 82–97. [CrossRef] [PubMed]
16. Mitchell, C.M.; Davy, B.M.; Hulver, M.W.; Neilson, A.P.; Bennett, B.J.; Davy, K.P. Does Exercise Alter Gut Microbial Composition? A Systematic Review. *Med. Sci. Sports Exerc.* **2019**, *51*, 160–167. [CrossRef]
17. O'Donovan, C.M.; Madigan, S.M.; Garcia-Perez, I.; Rankin, A.; O' Sullivan, O.; Cotter, P.D. Distinct microbiome composition and metabolome exists across subgroups of elite Irish athletes. *J. Sci. Med. Sport* **2020**, *23*, 63–68. [CrossRef]
18. Ganda Mall, J.P.; Löfvendahl, L.; Lindqvist, C.M.; Brummer, R.J.; Keita, Å.V.; Schoultz, I. Differential effects of dietary fibres on colonic barrier function in elderly individuals with gastrointestinal symptoms. *Sci. Rep.* **2018**, *8*, 13404. [CrossRef]
19. Svedlund, J.; Sjödin, I.; Dotevall, G. GSRS—A clinical rating scale for gastrointestinal symptoms in patients with irritable bowel syndrome and peptic ulcer disease. *Dig. Dis. Sci.* **1988**, *33*, 129–134. [CrossRef]
20. Roberts, M.H.; Fletcher, R.B.; Merrick, P.L. The validity and clinical utility of the hospital anxiety and depression scale (HADS) with older adult New Zealanders. *Int. Psychogeriatr.* **2014**, *26*, 325–333. [CrossRef]
21. Frändin, K.; Mellström, D.; Sundh, V.; Grimby, G. A life span perspective on patterns of physical activity and functional performance at the age of 76. *Gerontology* **1995**, *41*, 109–120. [CrossRef] [PubMed]
22. Johansson, I.; Hallmans, G.; Wikman, A.; Biessy, C.; Riboli, E.; Kaaks, R. Validation and calibration of food-frequency questionnaire measurements in the Northern Sweden Health and Disease cohort. *Public Health Nutr.* **2002**, *5*, 487–496. [CrossRef] [PubMed]
23. FASS.se. Available online: https://www.fass.se/LIF/startpage (accessed on 18 January 2020).
24. Brumfield, K.D.; Huq, A.; Colwell, R.R.; Olds, J.L.; Leddy, M.B. Microbial resolution of whole genome shotgun and 16S amplicon metagenomic sequencing using publicly available NEON data. *PLoS ONE* **2020**, *15*, e0228899. [CrossRef] [PubMed]
25. Truong, D.T.; Franzosa, E.A.; Tickle, T.L.; Scholz, M.; Weingart, G.; Pasolli, E.; Tett, A.; Huttenhower, C.; Segata, N. MetaPhlAn2 for enhanced metagenomic taxonomic profiling. *Nat. Methods* **2015**, *12*, 902–903. [CrossRef] [PubMed]
26. Bioinfopublic/Differences in Gut Microbiome Composition. Available online: https://git.oru.se/bioinfoPublic/Paper-differences-in-gut-microbiome (accessed on 19 February 2020).

27. Welch, B.L. The generalisation of student's problems when several different population variances are involved. *Biometrika* **1947**, *34*, 28–35. [CrossRef] [PubMed]
28. Mwalili, S.M.; Lesaffre, E.; Declerck, D. The zero-inflated negative binomial regression model with correction for misclassification: An example in caries research. *Stat. Methods Med. Res.* **2008**, *17*, 123–139. [CrossRef] [PubMed]
29. Calle, M.L. Statistical Analysis of Metagenomics Data. *Genom. Inform.* **2019**, *17*, e6. [CrossRef]
30. Xia, Y.; Sun, J. Hypothesis Testing and Statistical Analysis of Microbiome. *Genes Dis.* **2017**, *4*, 138–148. [CrossRef]
31. Boulesteix, A.-L. Maximally selected chi-square statistics for ordinal variables. *Biom. J.* **2006**, *48*, 451–462. [CrossRef]
32. R Core Team. *R: A Language and Environment for Statistical Computing*; R Foundation for Statistical Computing: Vienna, Austria, 2018.
33. Wickham, H. *ggplot2: Elegant Graphics for Data Analysis*; Springer: New York, NY, USA, 2016; ISBN 978-3-319-24277-4.
34. McGill, R.; Tukey, J.W.; Larsen, W.A. Variations of Box Plots. *Am. Stat.* **1978**, *32*, 12. [CrossRef]
35. Zeileis, A.; Kleiber, C.; Jackman, S. Regression Models for Count Data in R. *J. Stat. Soft.* **2008**, *27*. [CrossRef]
36. Benjamini, Y.; Hochberg, Y. Controlling the False Discovery Rate: A Practical and Powerful Approach to Multiple Testing. *J. R. Stat. Soc. Ser. B* **1995**, *57*, 289–300. [CrossRef]
37. Oksanen, J.; Blanchet, F.G.; Friendly, M.; Kindt, R.; Legendre, P.; McGlinn, D.; Minchin, P.R.; O'Hara, R.B.; Simpson, G.L.; Solymos, P.; et al. *Vegan: Community Ecology Package*; 2019.
38. Roberts, D.W. Labdsv: Ordination and Multivariate Analysis for Ecology. 2019.
39. Miquel, S.; Martín, R.; Rossi, O.; Bermúdez-Humarán, L.G.; Chatel, J.M.; Sokol, H.; Thomas, M.; Wells, J.M.; Langella, P. Faecalibacterium prausnitzii and human intestinal health. *Curr. Opin. Microbiol.* **2013**, *16*, 255–261. [CrossRef] [PubMed]
40. Saxelin, M.; Lassig, A.; Karjalainen, H.; Tynkkynen, S.; Surakka, A.; Vapaatalo, H.; Järvenpää, S.; Korpela, R.; Mutanen, M.; Hatakka, K. Persistence of probiotic strains in the gastrointestinal tract when administered as capsules, yoghurt, or cheese. *Int. J. Food Microbiol.* **2010**, *144*, 293–300. [CrossRef] [PubMed]
41. De Filippis, F.; Pellegrini, N.; Vannini, L.; Jeffery, I.B.; La Storia, A.; Laghi, L.; Serrazanetti, D.I.; Di Cagno, R.; Ferrocino, I.; Lazzi, C.; et al. High-level adherence to a Mediterranean diet beneficially impacts the gut microbiota and associated metabolome. *Gut* **2016**, *65*, 1812–1821. [CrossRef]
42. Wan, Y.; Wang, F.; Yuan, J.; Li, J.; Jiang, D.; Zhang, J.; Li, H.; Wang, R.; Tang, J.; Huang, T.; et al. Effects of dietary fat on gut microbiota and faecal metabolites, and their relationship with cardiometabolic risk factors: A 6-month randomised controlled-feeding trial. *Gut* **2019**, *68*, 1417–1429. [CrossRef]
43. Russell, W.R.; Gratz, S.W.; Duncan, S.H.; Holtrop, G.; Ince, J.; Scobbie, L.; Duncan, G.; Johnstone, A.M.; Lobley, G.E.; Wallace, R.J.; et al. High-protein, reduced-carbohydrate weight-loss diets promote metabolite profiles likely to be detrimental to colonic health. *Am. J. Clin. Nutr.* **2011**, *93*, 1062–1072. [CrossRef]
44. David, L.A.; Maurice, C.F.; Carmody, R.N.; Gootenberg, D.B.; Button, J.E.; Wolfe, B.E.; Ling, A.V.; Devlin, A.S.; Varma, Y.; Fischbach, M.A.; et al. Diet rapidly and reproducibly alters the human gut microbiome. *Nature* **2014**, *505*, 559–563. [CrossRef]
45. Venkataraman, A.; Sieber, J.R.; Schmidt, A.W.; Waldron, C.; Theis, K.R.; Schmidt, T.M. Variable responses of human microbiomes to dietary supplementation with resistant starch. *Microbiome* **2016**, *4*, 33. [CrossRef]
46. Martínez, I.; Kim, J.; Duffy, P.R.; Schlegel, V.L.; Walter, J. Resistant starches types 2 and 4 have differential effects on the composition of the fecal microbiota in human subjects. *PLoS ONE* **2010**, *5*, e15046. [CrossRef]
47. Duncan, S.H. Growth requirements and fermentation products of Fusobacterium prausnitzii, and a proposal to reclassify it as Faecalibacterium prausnitzii gen. nov., comb. nov. *Int. J. Syst. Evol. Microbiol.* **2002**, *52*, 2141–2146. [CrossRef] [PubMed]
48. Collins, M.D.; Lawson, P.A.; Willems, A.; Cordoba, J.J.; Fernandez-Garayzabal, J.; Garcia, P.; Cai, J.; Hippe, H.; Farrow, J.A.E. The Phylogeny of the Genus Clostridium: Proposal of Five New Genera and Eleven New Species Combinations. *Int. J. Syst. Bacteriol.* **1994**, *44*, 812–826. [CrossRef] [PubMed]
49. Louis, P.; Flint, H.J. Diversity, metabolism and microbial ecology of butyrate-producing bacteria from the human large intestine. *FEMS Microbiol. Lett.* **2009**, *294*, 1–8. [CrossRef] [PubMed]
50. Ferreira-Halder, C.V.; de Sousa Faria, A.V.; Andrade, S.S. Action and function of Faecalibacterium prausnitzii in health and disease. *Best Pract. Res. Clin. Gastroenterol.* **2017**, *31*, 643–648. [CrossRef] [PubMed]

51. Plöger, S.; Stumpff, F.; Penner, G.B.; Schulzke, J.-D.; Gäbel, G.; Martens, H.; Shen, Z.; Günzel, D.; Aschenbach, J.R. Microbial butyrate and its role for barrier function in the gastrointestinal tract: Butyrate and the gastrointestinal barrier. *Ann. N. Y. Acad. Sci.* **2012**, *1258*, 52–59. [CrossRef]
52. Gutiérrez-Díaz, I.; Fernández-Navarro, T.; Salazar, N.; Bartolomé, B.; Moreno-Arribas, M.V.; de Andres-Galiana, E.J.; Fernández-Martínez, J.L.; de los Reyes-Gavilán, C.G.; Gueimonde, M.; González, S. Adherence to a Mediterranean Diet Influences the Fecal Metabolic Profile of Microbial-Derived Phenolics in a Spanish Cohort of Middle-Age and Older People. *J. Agric. Food Chem.* **2017**, *65*, 586–595. [CrossRef]
53. Lin, D.; Peters, B.A.; Friedlander, C.; Freiman, H.J.; Goedert, J.J.; Sinha, R.; Miller, G.; Bernstein, M.A.; Hayes, R.B.; Ahn, J. Association of dietary fibre intake and gut microbiota in adults. *Br. J. Nutr.* **2018**, *120*, 1014–1022. [CrossRef]
54. Arumugam, M.; Raes, J.; Pelletier, E.; Le Paslier, D.; Yamada, T.; Mende, D.R.; Fernandes, G.R.; Tap, J.; Bruls, T.; Batto, J.-M.; et al. Enterotypes of the human gut microbiome. *Nature* **2011**, *473*, 174–180. [CrossRef]
55. Mueller, S.; Saunier, K.; Hanisch, C.; Norin, E.; Alm, L.; Midtvedt, T.; Cresci, A.; Silvi, S.; Orpianesi, C.; Verdenelli, M.C.; et al. Differences in fecal microbiota in different European study populations in relation to age, gender, and country: A cross-sectional study. *Appl. Environ. Microbiol.* **2006**, *72*, 1027–1033. [CrossRef]
56. Allen, J.M.; Mailing, L.J.; Niemiro, G.M.; Moore, R.; Cook, M.D.; White, B.A.; Holscher, H.D.; Woods, J.A. Exercise Alters Gut Microbiota Composition and Function in Lean and Obese Humans. *Med. Sci. Sports Exerc.* **2018**, *50*, 747–757. [CrossRef]
57. Zhu, Q.; Jiang, S.; Du, G. Effects of exercise frequency on the gut microbiota in elderly individuals. *Microbiologyopen* **2020**, e1053. [CrossRef]
58. Ortiz-Alvarez, L.; Xu, H.; Martinez-Tellez, B. Influence of Exercise on the Human Gut Microbiota of Healthy Adults: A Systematic Review. *Clin. Transl. Gastroenterol.* **2020**, *11*, e00126. [CrossRef] [PubMed]
59. Fragiadakis, G.K.; Smits, S.A.; Sonnenburg, E.D.; Van Treuren, W.; Reid, G.; Knight, R.; Manjurano, A.; Changalucha, J.; Dominguez-Bello, M.G.; Leach, J.; et al. Links between environment, diet, and the hunter-gatherer microbiome. *Gut Microbes* **2019**, *10*, 216–227. [CrossRef]
60. Blasco-Baque, V.; Coupé, B.; Fabre, A.; Handgraaf, S.; Gourdy, P.; Arnal, J.-F.; Courtney, M.; Schuster-Klein, C.; Guardiola, B.; Tercé, F.; et al. Associations between hepatic miRNA expression, liver triacylglycerols and gut microbiota during metabolic adaptation to high-fat diet in mice. *Diabetologia* **2017**, *60*, 690–700. [CrossRef] [PubMed]
61. Chen, Y.-J.; Wu, H.; Wu, S.-D.; Lu, N.; Wang, Y.-T.; Liu, H.-N.; Dong, L.; Liu, T.-T.; Shen, X.-Z. Parasutterella, in association with irritable bowel syndrome and intestinal chronic inflammation. *J. Gastroenterol. Hepatol.* **2018**, *33*, 1844–1852. [CrossRef] [PubMed]
62. Morotomi, M. The Family Sutterellaceae. In *The Prokaryotes: Alphaproteobacteria and Betaproteobacteria*; Rosenberg, E., DeLong, E.F., Lory, S., Stackebrandt, E., Thompson, F., Eds.; Springer: Berlin/Heidelberg, Germany, 2014; pp. 1005–1012. ISBN 978-3-642-30197-1.
63. Shin, N.-R.; Whon, T.W.; Bae, J.-W. Proteobacteria: Microbial signature of dysbiosis in gut microbiota. *Trends Biotechnol.* **2015**, *33*, 496–503. [CrossRef]
64. Scanlan, P.D.; Shanahan, F.; Marchesi, J.R. Culture-independent analysis of desulfovibrios in the human distal colon of healthy, colorectal cancer and polypectomized individuals. *FEMS Microbiol. Ecol.* **2009**, *69*, 213–221. [CrossRef]
65. Xiao, S.; Fei, N.; Pang, X.; Shen, J.; Wang, L.; Zhang, B.; Zhang, M.; Zhang, X.; Zhang, C.; Li, M.; et al. A gut microbiota-targeted dietary intervention for amelioration of chronic inflammation underlying metabolic syndrome. *FEMS Microbiol. Ecol.* **2014**, *87*, 357–367. [CrossRef]
66. Yazici, C.; Wolf, P.G.; Kim, H.; Cross, T.-W.L.; Vermillion, K.; Carroll, T.; Augustus, G.J.; Mutlu, E.; Tussing-Humphreys, L.; Braunschweig, C.; et al. Race-dependent association of sulfidogenic bacteria with colorectal cancer. *Gut* **2017**, *66*, 1983–1994. [CrossRef]
67. Baron, E.J. Bilophila wadsworthia: A unique Gram-negative anaerobic rod. *Anaerobe* **1997**, *3*, 83–86. [CrossRef]
68. Baron, E.J.; Curren, M.; Henderson, G.; Jousimies-Somer, H.; Lee, K.; Lechowitz, K.; Strong, C.A.; Summanen, P.; Tunér, K.; Finegold, S.M. Bilophila wadsworthia isolates from clinical specimens. *J. Clin. Microbiol.* **1992**, *30*, 1882–1884. [CrossRef] [PubMed]

69. Feng, Z.; Long, W.; Hao, B.; Ding, D.; Ma, X.; Zhao, L.; Pang, X. A human stool-derived Bilophila wadsworthia strain caused systemic inflammation in specific-pathogen-free mice. *Gut Pathog.* **2017**, *9*, 59. [CrossRef] [PubMed]
70. McOrist, A.L.; Warhurst, M.; McOrist, S.; Bird, A.R. Colonic infection by Bilophila wadsworthia in pigs. *J. Clin. Microbiol.* **2001**, *39*, 1577–1579. [CrossRef] [PubMed]
71. Jackson, M.A.; Goodrich, J.K.; Maxan, M.-E.; Freedberg, D.E.; Abrams, J.A.; Poole, A.C.; Sutter, J.L.; Welter, D.; Ley, R.E.; Bell, J.T.; et al. Proton pump inhibitors alter the composition of the gut microbiota. *Gut* **2016**, *65*, 749–756. [CrossRef]
72. Mu, C.; Zhu, W. Antibiotic effects on gut microbiota, metabolism, and beyond. *Appl. Microbiol. Biotechnol.* **2019**, *103*, 9277–9285. [CrossRef]
73. Livanos, A.E.; Greiner, T.U.; Vangay, P.; Pathmasiri, W.; Stewart, D.; McRitchie, S.; Li, H.; Chung, J.; Sohn, J.; Kim, S.; et al. Antibiotic-mediated gut microbiome perturbation accelerates development of type 1 diabetes in mice. *Nat Microbiol* **2016**, *1*, 16140. [CrossRef]
74. Santini, Z.I.; Jose, P.E.; York Cornwell, E.; Koyanagi, A.; Nielsen, L.; Hinrichsen, C.; Meilstrup, C.; Madsen, K.R.; Koushede, V. Social disconnectedness, perceived isolation, and symptoms of depression and anxiety among older Americans (NSHAP): A longitudinal mediation analysis. *Lancet Public Health* **2020**, *5*, e62–e70. [CrossRef]
75. Bear, T.L.K.; Dalziel, J.E.; Coad, J.; Roy, N.C.; Butts, C.A.; Gopal, P.K. The Role of the Gut Microbiota in Dietary Interventions for Depression and Anxiety. *Adv. Nutr.* **2020**, *11*, 890–907. [CrossRef]
76. Bastiaanssen, T.F.S.; Cussotto, S.; Claesson, M.J.; Clarke, G.; Dinan, T.G.; Cryan, J.F. Gutted! Unraveling the Role of the Microbiome in Major Depressive Disorder. *Harv. Rev. Psychiatry* **2020**, *28*, 26–39. [CrossRef] [PubMed]
77. Cheung, S.G.; Goldenthal, A.R.; Uhlemann, A.-C.; Mann, J.J.; Miller, J.M.; Sublette, M.E. Systematic Review of Gut Microbiota and Major Depression. *Front. Psychiatry* **2019**, *10*, 34. [CrossRef] [PubMed]
78. Caporaso, J.G.; Lauber, C.L.; Costello, E.K.; Berg-Lyons, D.; Gonzalez, A.; Stombaugh, J.; Knights, D.; Gajer, P.; Ravel, J.; Fierer, N.; et al. Moving pictures of the human microbiome. *Genome Biol.* **2011**, *12*, R50. [CrossRef] [PubMed]
79. McBurney, M.I.; Davis, C.; Fraser, C.M.; Schneeman, B.O.; Huttenhower, C.; Verbeke, K.; Walter, J.; Latulippe, M.E. Establishing What Constitutes a Healthy Human Gut Microbiome: State of the Science, Regulatory Considerations, and Future Directions. *J. Nutr.* **2019**, *149*, 1882–1895. [CrossRef] [PubMed]
80. Rothschild, D.; Weissbrod, O.; Barkan, E.; Kurilshikov, A.; Korem, T.; Zeevi, D.; Costea, P.I.; Godneva, A.; Kalka, I.N.; Bar, N.; et al. Environment dominates over host genetics in shaping human gut microbiota. *Nature* **2018**, *555*, 210–215. [CrossRef] [PubMed]
81. Halfvarson, J.; Brislawn, C.J.; Lamendella, R.; Vázquez-Baeza, Y.; Walters, W.A.; Bramer, L.M.; D'Amato, M.; Bonfiglio, F.; McDonald, D.; Gonzalez, A.; et al. Dynamics of the human gut microbiome in inflammatory bowel disease. *Nature Microbiol.* **2017**, *2*, 17004. [CrossRef]

nutrients

MDPI

*Article*

# Whole Blueberry and Isolated Polyphenol-Rich Fractions Modulate Specific Gut Microbes in an In Vitro Colon Model and in a Pilot Study in Human Consumers

**Alexandra Ntemiri [1,2], Tarini S. Ghosh [1,2], Molly E. Gheller [4], Tam T. T. Tran [1,2], Jamie E. Blum [4], Paola Pellanda [1,2], Klara Vlckova [1,2], Marta C. Neto [1,2], Amy Howell [3], Anna Thalacker-Mercer [4,5] and Paul W. O'Toole [1,2,*]**

1 School of Microbiology, University College Cork, T12 K8AF Cork, Ireland; alexandra.ntemiri@ucc.ie (A.N.); tarini.ghosh@ucc.ie (T.S.G.); tam.tran@ucc.ie (T.T.T.T.); paola.pellanda@ucc.ie (P.P.); klara.vlckova@ucc.ie (K.V.); marta.neto@ucc.ie (M.C.N.)

2 APC Microbiome Ireland, University College Cork, T12 K8AF Cork, Ireland

3 Marucci Center for Blueberry Cranberry Research, Rutgers University, Chatsworth, NJ 08019, USA; ahowell@njaes.rutgers.edu

4 Division of Nutritional Sciences, Cornell University, Ithaca, NY 14853, USA; meg369@cornell.edu (M.E.G.); jeb462@cornell.edu (J.E.B.); aet74@cornell.edu (A.T.-M.)

5 Department of Cell, Developmental and Integrative Biology, University of Alabama at Birmingham, Birmingham, Alabama, AL 35294, USA

* Correspondence: pwotoole@ucc.ie

Received: 26 June 2020; Accepted: 8 September 2020; Published: 12 September 2020

**Abstract:** Blueberry (BB) consumption is linked to improved health. The bioconversion of the polyphenolic content of BB by fermentative bacteria in the large intestine may be a necessary step for the health benefits attributed to BB consumption. The identification of specific gut microbiota taxa that respond to BB consumption and that mediate the bioconversion of consumed polyphenolic compounds into bioactive forms is required to improve our understanding of how polyphenols impact human health. We tested the ability of polyphenol-rich fractions purified from whole BB—namely, anthocyanins/flavonol glycosides (ANTH/FLAV), proanthocyanidins (PACs), the sugar/acid fraction (S/A), and total polyphenols (TPP)—to modulate the fecal microbiota composition of healthy adults in an *in vitro* colon system. In a parallel pilot study, we tested the effect of consuming 38 g of freeze-dried BB powder per day for 6 weeks on the fecal microbiota of 17 women in two age groups (i.e., young and older). The BB ingredients had a distinct effect on the fecal microbiota composition in the artificial colon model. The ANTH/FLAV and PAC fractions were more effective in promoting microbiome alpha diversity compared to S/A and TPP, and these effects were attributed to differentially responsive taxa. Dietary enrichment with BB resulted in a moderate increase in the diversity of the microbiota of the older subjects but not in younger subjects, and certain health-relevant taxa were significantly associated with BB consumption. Alterations in the abundance of some gut bacteria correlated not only with BB consumption but also with increased antioxidant activity in blood. Collectively, these pilot data support the notion that BB consumption is associated with gut microbiota changes and health benefits.

**Keywords:** polyphenols; blueberries; gut microbiota; in vitro; human study; oxidative stress

## 1. Introduction

The gut microbiota is a recognized modulator of human health and is shaped by host genetics, environment, lifestyle, and diet [1,2]. Most studies investigated cohorts representing western populations and lifestyle and compared healthy subjects and individuals with diverse conditions. Despite significant inter-individual variations in the gut microbiota composition, a general description of the healthy adult gut microbiota has emerged [3], but the variation range of phylum proportional abundances in healthy subjects is still large. Factors working throughout the lifespan such as repeated antibiotic use, significant changes in dietary habits, and infections may lead to perturbations and reductions in the composition and phylogenetic diversity of the gut microbiota that are associated with disease [4].

Immuno-senescence, hospitalization, and changes in dietary habits may collectively contribute to the age-related gut microbiota alterations observed in older individuals and that are linked in turn to the increase in the inflammatory state of older adults, a known risk factor for mortality in humans and animal models [5,6]. In conditions characterized by an altered microbiome at any point in life, gut microbiota manipulation can be a target for prevention, improvement, or even therapy [7,8]. This could be achieved with the use of probiotics, fecal microbiota transplants (FMT), live biotherapeutics, or prebiotics [9]. Accumulating evidence suggests that polyphenols are a dietary component with potential prebiotic activity [10,11].

Polyphenols are plant-derived dietary components that can be grouped into non-flavonoids and flavonoids [12]. Non-flavonoids include compounds such as tannins, phenolic acids, and lignans [12]. Flavonoids include isoflavones; neoflavonoids; and others such as chalcones, flavones, and flavonols, which are the building blocks of proanthocyanidins and anthocyanins [13]. Anthocyanins are present in plants as glycosylated anthocyanidins conjugated with sugars including glucose, galactose, arabinose, rhamnose, and xylose [14]. The dietary intake of these compounds has been associated with health benefits based on in vitro and in vivo experimental models and human studies [10,15].

The levels of flavonoid consumption reported for adult populations vary significantly, likely due at least in part to differences in the analytical methods and associated reference standards used to assess the flavonoid content in food products, but also because of widely varying dietary habits [14,16,17]. Adults in the US, Europe, and the UK have a daily consumption of flavonoids that ranges from 177 mg/d up to 428 mg/d, and a consumption of anthocyanidins that ranges from 4.2 mg/d up to 19 mg/d [16,18,19].

Unabsorbed phenolic compounds reach the colon, where they may serve as substrates for fecal microbiota fermentation [20]. Several *in vitro* [21,22], *in vivo* [23–27], and human studies [28–30] indicate health benefits and the potential for polyphenols to modulate the gut microbiota. There is also scientific interest in the combined effect of dietary polyphenols and fiber on the gut microbiota [31]. In this context, blueberry (BB) consumption may provide adequate amounts of dietary polyphenols with potential health benefits [32].

To further elucidate the interactions between BB ingredients and the human gut microbiome, we profiled the compositional changes that occurred in batch fermentations inoculated with the fecal microbiota of healthy young adults and supplemented with isolated BB polyphenol-rich fractions. These fractions are individually enriched for different major classes of presumptive bioactive BB ingredients, and have been tested for their activity in multiple previous publications [33–37]. To detect microbiota taxa that are responsive to human BB consumption, 17 healthy female volunteers in two age groups consumed freeze-dried BB for 6 weeks, and their gut microbiota composition was analyzed before and after the dietary intervention. The data indicate that BB ingredients or whole BB fruit can affect the microbiota both in *in vitro* colon model systems and in human consumers, but the effects on microbiota diversity are greater in older consumers.

## 2. Materials and Methods

### 2.1. Faecal Inocula and the In Vitro Colon Model

A pool of 5 fecal microbiota samples was used to inoculate the fermenter vessels comprising an artificial colon model. The five fecal samples were collected from healthy young donors (coded as follows: HYD3 32 years old (yrs), HA4 26 yrs, HA6 29 yrs, HA7 29 yrs, HA8 35 yrs) under a procedure approved by the local clinical research ethics committee. All the subjects gave their informed consent for inclusion before they participated in the study. The study was conducted in accordance with the Declaration of Helsinki, and the protocol was approved by the Ethics Committee/Institutional Review Board at Cornell University (Protocol ID#: 1706007263).

Fecal samples were collected and transferred to an anaerobic cabinet no later than one hour after passing. Each fecal sample was homogenized in a reduced sterile solution of PBS containing 20% glycerol and stored at −80 °C. Before each fermentation run, an aliquot of each of the 5 fecal samples (i.e., microbiota samples) was thoroughly thawed in an anaerobic cabinet and mixed in equal volumes for the inoculum.

Batch fermentations were used to simulate the colonic bacterial fermentation of the selected substrates [38]. One percent (*w/v*) fecal inoculum was prepared to inoculate each of three parallel single vessels with a 150 mL working volume in each vessel. A continuous flow of $NO_2$ was used to maintain anaerobic conditions during the 24 h pH (6.8) and temperature (37 °C)-controlled fermentation runs, with continuous stirring and atmosphere monitoring. Adaptation to in vitro media results in reduced microbiota diversity, as observed in previous studies [39]. To reduce this loss of microbiota diversity, the basal fermentation medium was supplemented with a mix of prebiotic fibres (referred to as MIX) (xylan 2 g/L; pectin 2 g/L; arabinogalactan 2 g/L; soluble starch 4 g/L) plus amylopectin (1 g/L), beta glucan (0.5 g/L), and glucose (2 g/L) [40,41].

The MIX medium was supplemented separately with each of 4 different BB polyphenol fractions: i. anthocyanins/flavonol glycosides (200 mg/L); ii. proanthocyanidins (200 mg/L); iii. sugar/acid fraction; and iv. total polyphenolics (333 mg/L). These were prepared as described in Section 2.4. A fermentation run without any supplementation was performed as a control. Samples from the fermentation culture were retrieved at 0, 16, and 24 h and centrifuged immediately, and the pellet and supernatants were kept at −20 °C for further analysis.

### 2.2. Bacterial DNA Extraction for In Vitro and Human Study

A 200 mg quantity of fecal pellet was weighed as instructed in the QIamp Fast DNA Stool (Qiagen, Manchester, UK) extraction kit protocol. The samples were homogenized mechanically in sterile tubes containing InhibitEX solution and zirconia glass beads of three sizes—0.1, 0.5, and 1.0 mm (Thistle Scientific Ltd., Glasgow, UK)—using a Mini-Beadbeader (Biospec Products, Inc., Bartlesville, OK, USA). The subsequent steps of gDNA extraction were performed as previously described by our laboratory [39].

### 2.3. Microbiome Profiling of In Vitro and Human Study Samples

The primers S-D-Bact-0341-b-S-17 (5′-CCTACGGGNGGCWGCAG-3′)/S-D-Bact-0785-a-A-21 (5′GACTACHVGGGTATC TAATC C-3′) 5′ [42] were used to amplify the V3/V4 variable region of the 16S rRNA gene for the profiling of the bacterial fecal microbiota using the Phusion High-Fidelity PCR Master Mix (ThermoFisher Scientific, Waltham, MA, USA). After PCR product purification, the Illumina MiSeq system protocol was used for library preparation. Indexing PCR was performed to amply the dual-index barcodes to the amplicon (Nextera XT V.2 Index Kits; Illumina, San Diego, CA, USA). The purification of the barcoded amplicons was performed with the Agencourt AMPure XP-PCR Purification system (Beckman Coutler, Inc., Brea, CA, USA). The Qubit dsDNA HS Assay Kit (Thermo Fischer Scientific, Waltham, MA, USA) was used to quantify the products. Equal concentrations of all the purified amplicons were pooled into a library that was sequenced (2 × 300 bp) on a MiSeq Illumina platform in the Teagasc Food Research Centre sequencing facility (Teagasc Moorepark, Fermoy, Ireland). ENA accession number: PRJEB39031.

### 2.4. Isolation of Enriched Fractions for Bioassay

Four major ingredient fractions of BB were prepared essentially as previously described in several previous publications [32–37]. In detail, whole frozen BB, cv. "Coville", were extracted, and fractions of total polyphenolics (TPP), proanthocyanidins (PACs), anthocyanin/flavonol glycosides (ANTH/FLAV), and sugars/acids (S/A) were isolated using solid-phase chromatography according to Howell et al., 2005 [43]. Briefly, BBs were homogenized with water in a blender and applied to a C18 column (Waters Corp., Milford, MA, USA) preconditioned with MeOH followed by $dH_2O$. The S/A fraction was collected as the column was washed with $dH_2O$ then $dH_2O$:MeOH (85:15) (*v*/*v*), followed by elution with acidified aqueous methanol. Solvents were removed from the S/A fraction under reduced pressure. The TPP fraction containing anthocyanins, flavonol glycosides, and PACs (confirmed using reverse-phase HPLC with diode array detection) was eluted with 1% HOAc in MeOH (*v*/*v*). All the fractions were dried under reduced pressure to remove the solvent. The TPP fraction was then suspended in 50% EtOH, and applied to a Sephadex™ LH-20 (Sigma Chemical Co., St. Louis, MO) column that was pre-equilibrated overnight in EtOH:$dH_2O$ (50:50) (*v*/*v*). The ANTH/FLAV fraction was eluted with 50% EtOH and dried to remove the solvent. The PAC fraction was eluted from the LH-20 column with 70% aqueous acetone, and monitored for purity using diode array detection at 280 nm. The absence of absorption at 360 nm and 450 nm confirmed that flavonol glycosides and anthocyanins, respectively, were removed. Acetone was evaporated under reduced pressure, and the resulting purified PAC fraction was dried. Analytical tools, including mass spectrometry and NMR spectrometry, have been routinely utilized to confirm the composition of these BB fractions using the method of Schmidt et al. (2004) and others [44–46].

### 2.5. *Human Study Design*

This study was approved by the Cornell University institutional review board and complies with the Helsinki Declaration. This trial is registered at clinicaltrials.gov (NCT04262258). All the participants provided written informed consent prior to participation in the study.

Seventeen healthy young (aged 21–39 yrs, $n = 11$) and old (aged 65–77 yrs, $n = 6$) women participated in the study. Potential participants were screened using an online survey to assess eligibility. After initial eligibility was established, the participants came to the Human Metabolic Research Unit at Cornell University to complete a health history questionnaire and provide information on current and recent medications. Inclusion criteria were females between the ages of 21 and 40 yrs y and 60 and 79 yrs. Participants were excluded if they had a musculoskeletal disease (e.g., rheumatoid arthritis) or other disorder that would impact skeletal muscle function (e.g., diabetes or cancer), were taking immunosuppressive medication, were pregnant or breastfeeding, had a high alcohol intake (>11 drink per week), had an allergy or intolerance to blueberries, and had antibiotic use within the past 6 months.

After the participants were enrolled in the study, they began a 2-week washout period in which they were asked to avoid foods rich in polyphenols and anthocyanins. Following the 2-week washout period, the participants began the blueberry enriched diet (BED); the participants were instructed to consume 38 g (two packages of 19 g) of freeze-dried BBs (*Vaccinium virgatum* (ashei)/*Vaccinium corymbosum*) with water daily for 6 weeks. Compliance was monitored through a supplement compliance log and empty BB packets returned by the participants to the study personnel.

Fasting stool and blood samples were obtained at four time points (washout, week 2, week 4, and week 6) throughout the BED study. The participants were given pre-labelled stool sample collection kits, and samples were collected outside of the lab. The participants transported samples to the lab with an insulated bag that contained an ice pack. The samples were immediately placed in −80 °C freezer until they were processed. To obtain plasma, venous blood samples (~10 mL) were collected in ethylenediaminetetraacetic acid (EDTA) tubes (Becton Dickinson Vacutainer system; Becton Dickinson, Franklin Lakes, NJ, USA) and then immediately centrifuged (4 °C at 1200× *g* for 10 min) to obtain plasma. Plasma was transferred to a new tube (volume ~500 uL) and stored in a −80 °C freezer until all the participants had completed the study and the samples were ready for analysis.

The participants BED_001, 002, 003, 006, 007, 008, 009, 011, 013, 015, and 016 belonged to the young age group and the participants BED_004, 005, 010, 014, 018, and 020 belonged to the older age group (Table 1).

**Table 1.** Human study participant demographics.

| | Young ($n$ = 10 [a]) | Old ($n$ = 6) |
|---|---|---|
| Age (yrs) | 28 ± 2 | 69 ± 2 |
| Weight (kg) | 64.31 ± 2.33 | 62.44 ± 3.84 |
| Height (cm) | 166.3 ± 1.5 | 161.3 ± 2.9 |
| BMI (kg/m$^2$) | 23.3 ± 0.9 | 24.2 ± 2.0 |
| Glucose (mg/dL, Range: 74–106) | 89.5 ± 2.44 | 98.0 ± 1.7 |
| CRP (mg/L, Range: <1.1) | 1.21 ± 0.33 | 0.76 ± 0.28 |

All values are presented as means ± standard error. BMI: body mass index; CRP: C-reactive protein. [a] One outlier value excluded.

### 2.6. Plasma FRAP Assay

The ferric-reducing antioxidant power (FRAP) was determined in plasma from the BED donors following established methods [47]. Briefly, a solution of sodium acetate (EMDMillipore, Burlington, MA, USA), 2,4,6 tripyridyl-S-triazine (TPTZ, ACROS Organics, Geel, Belgium), and ferric chloride (Fisher Scientific, Waltham, MA, USA) was incubated with plasma samples or ferrous sulfate (assay standard, Fisher Scientific, Waltham, MA, USA) for 4 min at 37 °C. The absorbance was measured at 593 nm and standardized to the absorbance of the ferrous sulfate standard to derive the FRAP value (μmol/L). Four technical replicates were measured. The absorbance of the blank was subtracted from each measurement, and the FRAP value of each sample was determined using the following formula: (sample absorbance)/(assay standard absorbance) * (assay standard concentration (1000 μmol/L)).

### 2.7. Plasma Glucose and CRP Measurements

Routine panels were conducted on stored fasting plasma samples (glucose and C-reactive protein) at Cornell University's Human Nutritional Chemistry Service Laboratory. Glucose was analyzed on a Dimension Xpand chemistry analyzer (Siemens Healthineers, Malvern, PA, USA), and CRP was measured on an Immulite 2000 immunoassay system (Siemens Healthineers, Malvern, PA, USA).

### 2.8. Microbiota Composition and Statistical Analysis

The pipeline for the microbiota composition analysis was described before [48] and comprised the following steps. Paired-end reads were joined with FLASH [49] and quality filtering was performed in Qiime (v.1.9.1) using the split_libraries_fastq.py script [50]. The forward and reverse primers were removed using cutadapt [51] and the script truncate_reverse_primer.py, respectively. For additional quality filtering and de novo operational taxonomic unit (OTU) clustering, USEARCH was used. Filtering by length and size was performed before single unique sequences were excluded. Clustering in OTUs was performed using 97% identity for the sequences after the various filtering steps. Chimeras were removed based on the use of UCHIME with the GOLD reference database. The OTUs were used to map sequences initially filtered for quality (97% identity). The mothur suite of tools (v1.36.1) and the RDP (trainset 14) were used for the OTU classification (classify.seqs) with a 80% confidence threshold [52]. The OTUs were classified down to the species level with SPINGO [53].

The PyNast tool [54] in Qiime (along with the diversity function of the vegan package version 2.4.3 of the R programming interface v 3.5.4) was used to align the sequences and calculate the alpha (α) diversity indices—i.e., Shannon, Simpson, Chao1, Phylogenetic Diversity (PD), Observed Species, and beta (β) diversity indices (i.e., Weighted Unifrac and Unweighted Unifrac). Weighted Unifrac and Spearman distances were used for principal coordinates analysis (PCoA) (ade4 package) using the R programming interface (v 3.5.4). PERMANOVA analysis was performed using the adonis function implemented in the vegan package (version 2.4.3) of the R programming interface. The reads assigned to taxa at various levels (OTU, species, and genus) were cumulated and divided by the total number of reads per sample.

For the artificial colon reactor analysis, the differentially abundant taxa in the different supplementations were identified by Kruskal–Wallis H-test followed by Dunns' test, and the $p$ values were adjusted for multiple testing by Benjamini Hochberg (BH) ($p_{adj}$). Significant results were indicated with * ($p_{adj} < 0.05$), and marginal differences with # ($p_{adj} < 0.1$). The dunn.test package of the R programming interface was used for this purpose (run with the method = "bh" argument to specify adjustment procedure to Benjamini Hochberg-BH). For this analysis, supplementation-specific abundances of the various taxa at 16 h and 24 h were combined. For each of these taxa, we also checked whether their abundances exhibited significant variations in their abundances at 16 h and 24 h using Wilcoxon Signed Rank Tests. The same approach (as described above) was also adopted for testing differences in the $\alpha$ diversity measures. The microbiota profiles resulting from the supplementation regimes were grouped into two groups (G1 and G2), as described in the results. The "Within G1" and "Within G2" distances were obtained as follows. For each microbiota resulting from a given supplementation, the median of the Weighted Unifrac distances of the microbiota with all the other microbiotas belonging to the same supplementation (that is, all the other microbiotas belonging to either G1 or G2 at 16 h and 24 h) was obtained. These median distances represented the microbiota variations within that given supplementation group. For the "Across G1 and G2" variations, for each sample belonging to a given supplementation, the median of the Weighted Unifrac distances of the microbiota with all the other microbiotas belonging to the other supplementations (that is, all the other microbiotas belonging to either G1 or G2 at 16 h and 24 h) was obtained. The wilcox.test function of the R programming interface v 3.5.4 was then used for comparing the within and across-group median distances.

For the human study, the OTU co-abundance groups (CAGs) demonstrating similar mean abundance pattern trends across time points were identified. The identification of taxa CAGs and their distinct taxonomic/temporal abundance profiles was conducted as follows. The mean abundance of each OTU was obtained for W0, W2, W4, and W6, providing a mean temporal trend of each OTU across time points. Subsequently, the Kendall correlations (taus) across all pairs of OTUs were obtained and then converted to Kendall distances. The Kendall distance between any two pairs of OTUs was calculated as 1—(Kendall tau)/2. The OTUs were clustered into CAGs based on their mutual abundance pattern (Ward-D2 method). The heatmap.2 function of R v 3.5.4 was used to visualize the clustering of the CAGs with the colors assigned using the RColorBrewer function. The Kruskal–Wallis H test followed by Dunns' test was used for a CAG abundance comparison across time points (using the dunn.test function, as described earlier for the differentially abundant taxa analysis in colonic reactor models). The OTUs were classified using SPINGO (0.65 threshold). For each CAG, OTUs with defined species classifications were obtained, and the frequency of each species in a given CAG was computed. The representation of the different species in the various CAGs were depicted using word clouds using the "wordcloud" module of the R programming interface v 3.5.4. A PERMANOVA analysis (Spearman distances at OTUs and CAG abundances level) was performed for the association between the taxa abundances (OTU and CAG level) and CRP, glucose, and FRAP.

For the FRAP values, the associations were further validated using Random Forest models (using the combination of the rfcv and randomForest functions of the randomForest module of R v 3.5.4). A total of 100 iterations of the Random Forest models were performed, each time taking 50% of the samples for training and testing on the other 50%. For each sample, the mean predicted FRAP values were then correlated with the actual FRAP values to ascertain the strength of the association. A key advantage of the Random Forest models is that they can not only be used to predict a given trait (either quantitative or categorical; in this case, the quantitative FRAP values) from a dataset of multiple predictor features (in this case, the OTUs), but they can provide the relative importance of each of the features to predict the given trait. This enables the identification of the most optimal set of associated features for predicting a given trait. Using these scores, the core list of FRAP-positive and FRAP-negative marker OTUs was identified, as described in detail in the Results section. The enrichment of the different CAGs in the two FRAP-associated marker OTUs was observed using Fishers' exact test. Spearman correlations (and the associated $p$ values) were computed using the corr.test function of the psych package of the R programming interface (run with the adjust = "fdr" argument for p value correction-false discovery rate FDR). The volcano plots showing the positive and negative associations of the various taxa with the clinical metadata were using the ggplot and ggrepel modules of R v 3.5.4.

## 3. Results

### *3.1. A Prebiotic MIX, PACs, or ANTH/FLAV Have a Similar Effect on Microbiota Structure in the In Vitro Colon Model*

The fecal microbiota α diversity (Shannon and Observed Species) was reduced over the 24 h of fermentation (observed at the 16 h and 24 h time points) compared to the baseline 0 h (Supplementary Materials Figure S1), which is a typical feature of in vitro colon models [43]. Supplementation with ANTH/FLAV, PACs, and prebiotic MIX resulted in microbiota communities displaying similar Shannon and Observed Species diversity index values that were noticeably higher than in the fermentations with either S/A and TPP supplementation (significant for PACs/MIX versus S/A with $p_{adj} < 0.5$) (Figure 1a; Figure S2). Thus, in the in vitro colon model used in this study, isolated BB components such as the polyphenol-rich fractions ANTH/FLAV and PACs were more efficient in promoting microbiota diversity than the other polyphenol-rich fractions used in this study—i.e., TPP or the S/A fraction.

To visualize the global effect of supplementing the fecal fermentation with BB polyphenol-rich fractions, we performed Principal Coordinate Analysis (PCoA) based on weighted and unweighted Unifrac distance measures (Figure 1b; Figure S3). Unweighted Unifrac analysis, where the taxa presence/absence is taken into account, showed that TPP supplementation led to a separation of the microbiota at 16 h that was not sustained through the 24 h (Figure S3). Weighted Unifrac analysis, in which the abundance of dominant taxa is more impactful on β diversity measurement, showed that TPP and S/A supplementation led to a microbiota profile separate from that of the other supplementation regimes at both 16 h (marginal but not significant variation; PERMANOVA $p < 0.06$) and 24 h (significant; PERMANOVA $p < 0.03$) (Figure 1b). Supplementation with the ANTH/FLAV and PACs fractions resulted in a microbiota β diversity close to that promoted by MIX (Figure 1b). Thus, the BB polyphenols tested had distinct effects on the fecal microbiota. The PCoA analysis showed that the supplementations could be grouped into G1, consisting of ANTH/FLAV, MIX, and PACs; and G2, consisting of S/A and TPP. The relatedness of samples within G1 and G2 and across G1 and G2 are shown in Figure 1c.

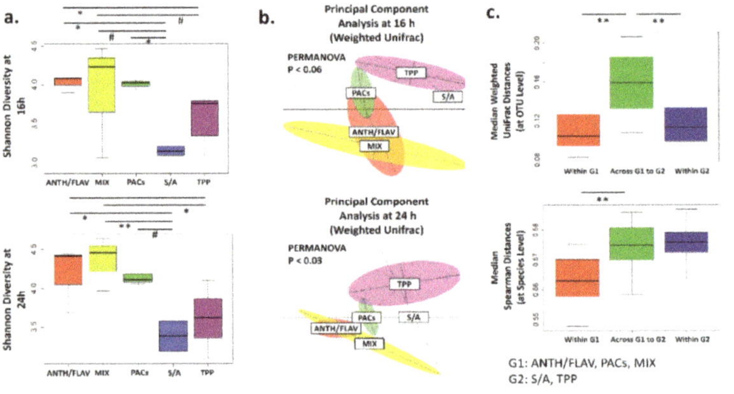


**Figure 1.** Differential fecal microbiota structure patterns due to the blueberry (BB) polyphenol-rich fraction supplementation in the *in vitro* colon model. (**a**). Boxplots showing the Shannon $\alpha$ diversity index in different supplementation regimes at 16 h and 24 h. (**b**). Principal Components Analysis (PCoA) based on the weighted Unifrac distances of the fecal microbiota for 16 h and 24 h *in vitro* fermentations. PERMANOVA $p$ values for each time point are indicated. (**c**). Boxplots showing the microbiota variation within in each group "Within G1" and "Within G2" (16 h and 24 h microbiotas combined) and "Across G1 to G2". Horizontal bar plots highlight the significant differences across the supplementation regimes: * $p_{adj} < 0.05$; ** $p_{adj} < 0.01$. Marginal differences are also noted: # $p_{adj} < 0.10$. ANTH/FLAV: anthocyanin/flavonols glycoside supplementation; MIX: prebiotic fibers mix supplementation; PACs: proanthocyanidins supplementation; S/A: sugar/acid fraction supplementation; TPP: total BB polyphenols.

The differences in $\alpha$ diversity observed between ANTH/FLAV, PACS, and MIX and TPP and S/A supplementation could be partially attributed to certain taxa dominating in relative abundance in the microbiota at 16 h and 24 h; non-significant microbiota differences were observed between the two time points per supplementation (Table S1; Figure S4). The supplementation-specific differences in various taxa were investigated by first performing a descriptive analysis and comparison of the supplementation-specific taxa at the family level (Figure S4), followed by a statistical comparison of the taxa abundances (genus and family level) (across supplementation combining the 16 h and 24 h time points) (Figure 2; Figure S5). The decrease in $\alpha$ diversity observed in the fecal microbiota fermented with TPP and S/A supplementation could be explained by the comparatively higher abundance of *Enterobacteriaceae* (49.77% and 47.49% average relative abundance, respectively) observed by compositional analysis of the fecal microbiota after 16 h and 24 h of fermentation (Figure S4). The lowest *Enterobacteriaceae* abundance was observed upon ANTH/FLAV supplementation and MIX (average relative abundance of 31.46% and 29.11%, respectively) (Figure S4). ANTH/FLAV supplementation resulted in a significantly lower *Escherichia/Shigella* (*Enterobacteriaceae*) relative abundance compared to the prebiotic MIX ($p_{adj} < 0.05$), S/A fraction ($p_{adj} < 0.05$), and TPP ($p_{adj} < 0.05$) supplementation (Figure 2). *Lachnospiraceae* was a major microbiota family that was reduced in abundance compared to baseline (34.4% average relative abundance) across fermentation regimes, potentially due to the in vitro conditions (Figure S4). Supplementation with ANTH/FLAV, PACs, or prebiotic MIX sustained the highest *Lachnospiraceae* (average relative abundance of 14.52%) in the microbiota (Figure S4). Similarly, the *Bacteroidaceae* relative abundance was higher upon ANTH/FLAV, PACs, and prebiotic MIX supplementation (16.61%, 15.1%, and 12.95% average relative abundance, respectively), and overall increased in abundance from baseline (average relative abundance of 6.89%) across supplementations (Figure S4).

The abundance of the health-relevant genus *Bifidobacterium* spp. (*Bifidobacteriaceae*; 2.54% average abundance at 0 h (Figure 2; Figure S4) was significantly increased upon TPP supplementation

(1.95% average relative abundance) compared to MIX (0.85% average relative abundance; $p_{adj} < 0.05$) and ANTH/FLAV supplementation (0.54% average relative abundance; $p_{adj} < 0.05$) (Figure 2). The health-relevant taxon *Faecalibacterium* was present at an average relative abundance of 6.04% (median abundance of 4%) upon MIX supplementation, which was marginally higher compared to PACs ($p_{adj} < 0.1$), and significantly higher compared to the TPP ($p_{adj} < 0.05$) and S/A ($p_{adj} < 0.05$) fraction supplementation (Figure 2).

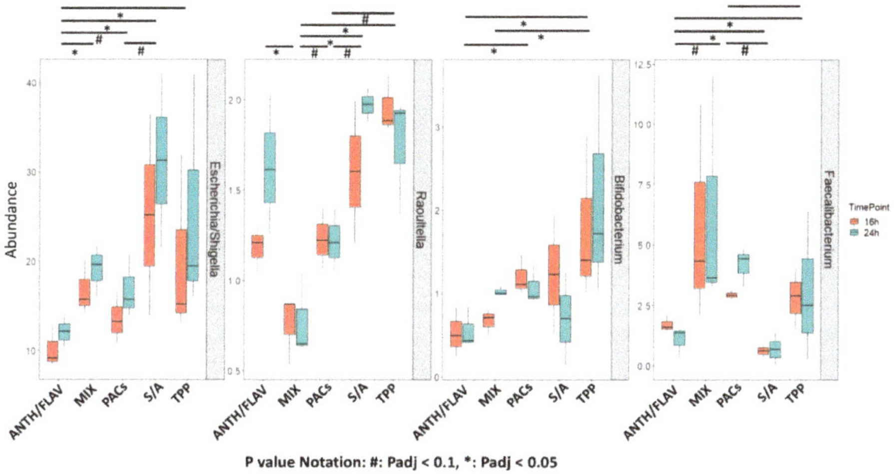


**Figure 2.** Differentially abundant taxa (genus level) in the fecal microbiota after *in vitro* supplementation with blueberry (BB) polyphenol-rich fractions. The results of relative abundance (>1%) from 16 h and 24 h fermentations are shown in pink and green color, respectively. Significant differences for comparisons combining the 16 h and 24 h data (post-hoc Dunn's test with Benjamini-Hochberg (BH) $p_{adj}$) between the corresponding pairs per supplementation across supplementations are indicated in horizontal bas plots: * $p_{adj} < 0.05$. Marginal differences are also indicated: # $p_{adj} < 0.10$. ANTH/FLAV: anthocyanin/flavonol glycoside supplementation; MIX: prebiotic fibers mix supplementation; PACs: proanthocyanidin supplementation; S/A: sugar/acid fraction supplementation; TPP: total BB polyphenols.

Apart from the aforementioned supplementation effects on the dominant taxa, the low abundance taxa (<1% average relative abundance) were also differentially abundant in the microbiota depending on the supplementation of the fermentation medium (Figure S5). Supplementation with ANTH/FLAV resulted in an increased relative abundance of *Phascolarctobacterium* compared to MIX (significant; $p_{adj} < 0.05$), S/A (marginal; $p_{adj} < 0.1$) and TPP (significant; $p_{adj} < 0.05$) supplementation, and of *Gemmiger* compared to MIX and PACs ($p_{adj} < 0.05$ for both supplementations). The *Clostridium* cluster XIVb relative abundance was increased with PACs supplementation ($p_{adj} < 0.05$ compared to ANTH/FLAV, S/A, and TPP). The *Sutterella* relative abundance was increased upon ANTH/FLAV supplementation ($p_{adj} < 0.05$ compared to S/A and TPP). The *Oscillibacter* and *Flavonifractor* relative abundance was significantly increased with MIX and PACs supplementation, whereas the *Burkholderiales* relative abundance was highest after the PAC supplementation (significantly higher: $p_{adj} < 0.05$ compared to TPP; marginally higher: $p_{adj} < 0.1$ as compared to ANTH/FLAV) (Figure S5). The unclassified *Erysipelotrichaceae* had the lowest relative abundance after S/A supplementation, whereas *Parasuterella* had the lowest relative abundance after the ANTH/FLAV and TPP supplementation (Figure S5).

### 3.2. A Trend Towards Increased Microbiota α Diversity in Older Women Consuming BB

The α diversity of the fecal microbiota in the human trial subjects did not show significant difference across time points for any of the α diversity indices measured (Shannon, Simpson, Chao1, PD, and Observed Species) (Figure S6). The lower sample size, especially of the group of older women, reduced the statistical power of the comparisons. However, investigating the time point-specific distributions of the diversity measures separately for the young and the older women indicated that, for the older group, for most of the measures (with the exception of Shannon) the α diversities at time points W4 and W6 were observed to be similar and higher than that at the pre-intervention W0 time point, indicating an increasing albeit non-significant trend for the elderly (Figure S6). We then investigated this further to check if any differences in the microbiota α diversity were observed by comparing the pre (W0) and post (W4 and W6) intervention time points on a per-individual basis (separately for each age group) (Figure S7a). In the fecal microbiota of five out of six older subjects, the Shannon diversity was increased during the intervention (mean of W4 and W6 aggregated) from the baseline W1 (Figure S7a). Similar results were not observed for the young subjects (Figure S7b).

### 3.3. Distinct CAGs Represented by Health-Promoting Taxa Were Associated with BB Consumption at Each Time Point

Based on β diversity, the fecal microbiota of the older subjects formed a distinct cluster at W4, albeit with high intra-sample variation (Figure S8a). Given the lower sample size and high intra-sample variability, the trends were not significant (PERMANOVA $R^2$ = 0.03). The fecal microbiota of the younger women showed no β diversity shifts throughout the intervention (PERMANOVA $R^2$ = 0.002) (Figure S8b).

Despite the lack of clear β diversity in the BB consumption-associated signatures, a significant lower intra-sample microbiota variability at intervention time points was observed for both sub-groups (at W4 versus W2 and W6, and at W6 compared to W2 and W4, respectively) (Figure S8c). This observation could indicate that, in spite of the high inter-individual variability, specific taxa groups may have changed across time points concurrently (enriched or depleted), resulting in a significantly lower inter-individual variability [55].

Microbiome configuration analysis offers a more refined approach to monitor microbiota changes compared to individual taxa analysis, because OTUs that co-occur at similar proportions may have trophic and functional interactions relevant for gut ecology [55]. In an analysis of the aggregated microbiota data from all study participants and based on the aforementioned variability trends, six CAGs of OTUs were identified (C1 to C6) (Figure 3a). Interestingly, while the abundance of each of these CAGs exhibited significant differences at W2, W4, and W6 (Figure 3b; Figure S9) when investigating for variability trends in old and young women, no significant differences in the OTUs' (of the six CAGs) cumulated abundance variation were observed. This indicated that, while the individual constituents may show a high inter-individual variability, the CAGs as a whole exhibit significant time point-specific trends (irrespective of the age group of the participants), even at the individual level (thereby indicating their reliability).

CAG-level PERMANOVA analysis at the different time points revealed significant differences (Figure 3c). The analysis revealed a distinct gut microbiome composition at W4, while W2 and W6 clustered closer; this was observed for both the old and young sub-groups. No significant differences in the cumulated abundances variation in the OTUs belonging to the six CAGs separately within the old and the young were observed (Figure S9).

The C1, C3, and C6 CAGs' cumulated abundances increased significantly with BB consumption (at W4 and W6). C1 increased from W0 to W4 and decreased at W6, whereas C3 and C6 progressively increased from W0 to W6 (Figure 3b). Several health-relevant species were abundant in these CAGs—e.g., *Faecalibacterium prausnitzii*, *Barnesiella intestinihominis*, *Eubacterium halii*, *Anaerostipes hadrus*, and *Ruminococcus bromii* (Figure 3d). These results indicate the putative beneficial effect of BB consumption on the gut microbiota.

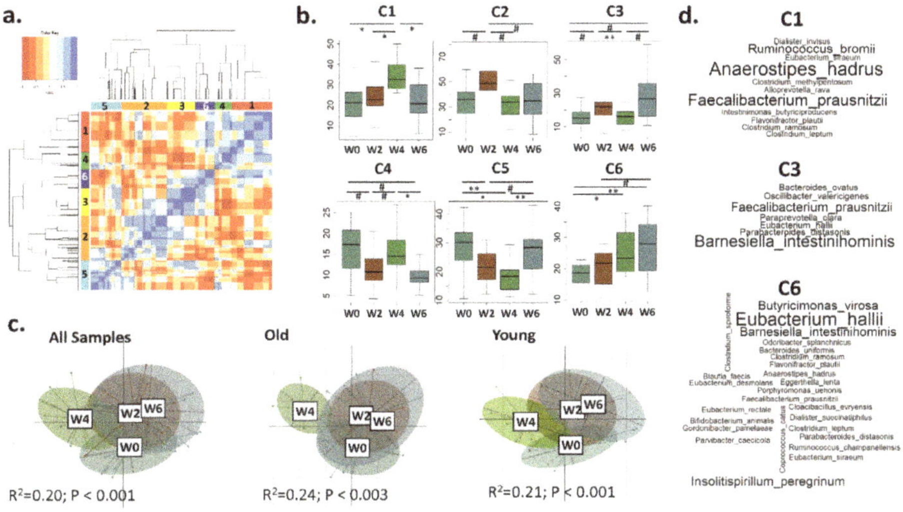

**Figure 3.** Associations of blueberry (BB) consumption with the enrichment of specific co-abundance taxonomic groups (CAGs). (**a**). Heatmap showing the Kendall tau between the different operational taxonomic units (OTUs) (that is, the OTU to OTU correlations) obtained based on their mean abundances across the different time points. Based on their association patterns, the OTUs were categorized into 6 co-abundance groups or CAGs. The 6 CAGs (C1 to C6) are indicated in colors on the left and top panels. (**b**). Boxplots showing the variation in the OTUs' cumulated relative abundances (y axis) belonging to the 6 CAGs across the four time points (x axis). $p_{adj}$ values showing the significant differences in the CAG abundances (Dunn's post-hoc test) across time points are indicated: *: $p_{adj} < 0.05$; **: $p_{adj} < 0.01$. Marginal differences are also noted: #: $p_{adj} < 0.1$. (**c**). Principal Coordinates Analysis (PCoA) showing gut microbiota grouping based on the abundances of the 6 different CAGs. The PCoA plots are shown for all the microbiotas aggregated and separated for the old and young sub-groups. The PERMANOVA $R^2$ and $p$ values are indicated in each plot. (**d**). Word clouds showing the species' enrichment in the CAGs C1, C3, and C6 dominant at either W4 or W6 or both. The species name is proportional to the frequency of that species.

### *3.4. Antioxidant Activity (FRAP) Is Significantly Associated with the Faecal Microbiota*

The levels of plasma CRP, glucose, and FRAP assay measures were collected at W0 and W6 for 10 young and 5 old women. Using PERMANOVA analysis, the association of the gut microbiota at both OTU and CAG level with CRP, glucose, and FRAP was investigated (Table 2). No association between the gut microbiota composition with either the CRP or glucose levels was observed. However, the FRAP assay measures showed significant association with the gut microbiota at both the OTU and CAG level (Table 2).

**Table 2.** Gut microbiota composition was significantly associated with ferric-reducing antioxidant power (FRAP). $R^2$ and $p$ values of the PERMANOVA analysis associating the clinical parameters with the gut microbiota at the operational taxonomic unit (OTU) and co-abundance group (CAG) level are shown in the table.

| Clinical Indicator | OTU-Level Microbiota | | CAG-Level Microbiota | |
|---|---|---|---|---|
| | $R^2$ | $p$ Value | $R^2$ | $p$ Value |
| CRP | 0.03 | 0.95 | 0.02 | 0.81 |
| Glucose | 0.03 | 0.54 | 0.02 | 0.79 |
| FRAP | 0.04 | 0.01 | 0.06 | 0.05 |

Random Forest models were built to predict the FRAP assay measure of an individual (at a given time point) based on their gut microbiota composition (at the OTU level) (as described in Materials and Methods). The validation of these models using an iterative leave-one-out-strategy (i.e., excluding from the training model the sample to be predicted) indicated a marginally positive Spearman correlation of 0.32 ($p < 0.07$), further indicating an association of the microbiome with plasma antioxidant activity (Figure 4a). Thus, in the current study the OTUs were initially ranked in increasing order of their feature importance scores, and subsequently the variation in these feature importance scores across them was investigated.

An exponential increase in scores for the last 150 taxa (or OTUs) as compared to the rest was observed (Figure S10a). The list of the 150 taxa was filtered by selecting only those OTUs that showed significant association with FRAP measures with BH-corrected FDR < 0.1 (Figure S10b). This provided the 30 top predictors of FRAP measures at the OTU level (Figure 4b). While 25 of these top markers were positively associated with FRAP (FRAP-positive markers), five were negatively associated with FRAP (FRAP-negative markers).

The efficacy of these top 30 markers was further evaluated using two variants of iterative Random Forest models, one using only these top 30 and the other using the remaining OTUs (Table S1). A comparison of the performances of the two variants indicated that models created using only these 30 top markers could still predict FRAP measures with a median Spearman Rho of 0.76 ($p < 1 \times 10^{-5}$), which was significantly higher than those created using the remaining 983 non-marker OTUs (median Rho = 0.02) ($p < 2.2 \times 10^{-16}$) (Figure 4c).

Distinct changes in the markers of FRAP assay measures during the intervention time points were as follows. The FRAP assay measures increased for 9 of the 15 subjects (4 out of 5 old, 5 out of 10 young) (Figure S10c). An overlap between some of the species that were positively associated with FRAP measures and those enriched during BB consumption was observed (Figure 4b). These included gut bacterial species such as *F. prausnitzii*, *E. halii*, *E. siraeum*, *C. catus*, and *A. hadrus*.

The representation of the different previously identified CAGs in the subset of FRAP-positive markers was explored. Seventeen out of the 25 FRAP positive markers belonged to either the C1 or the C6 CAGs that were significantly enriched in W4 and W6 time points, respectively (Figure 3b; Figure 4b) Thus, a subset of taxa that were identified as belonging to reportedly beneficial microbial groups enriched in the later stages of BB consumption also showed positive associations with the antioxidant activity. This indicates that BB consumption is associated with microbiome changes that are positively associated with antioxidant activity.

Interestingly, the across-time point changes in FRAP assay measures showed negative associations with the corresponding changes in the plasma glucose levels (Spearman Rho = −0.46; $p < 0.05$) (Figure S10d). A similar negative association was also observed between the FRAP positive-markers and the plasma glucose levels. Thus, these results overall seem to suggest a step-wise association between BB consumption and plasma glucose levels, wherein the consumption of BBs is associated with the enrichment of specific taxonomic groups, and a subset is positively associated with circulating antioxidant activity, which in turn is negatively associated with the plasma glucose levels. The associations of these OTUs with FRAP were both sample-specific and distinct for time points.

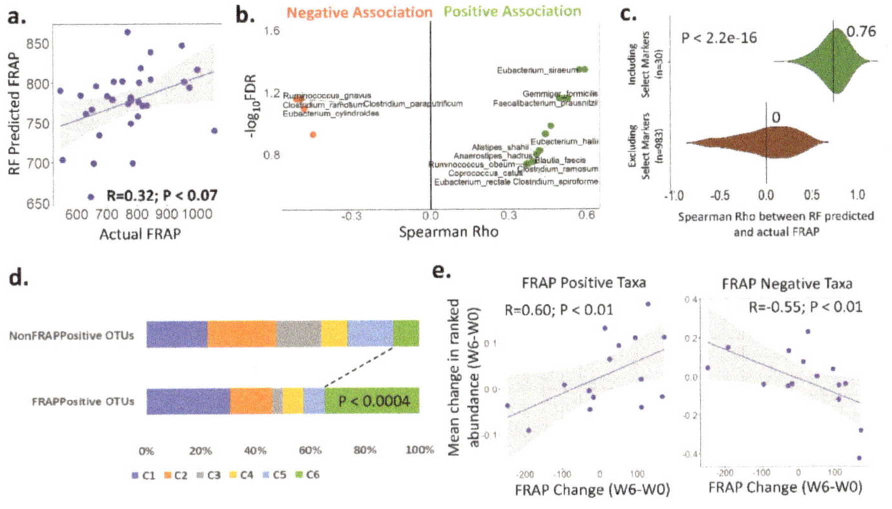


**Figure 4.** Fecal microbiota components were associated with increased ferric-reducing antioxidant power (FRAP) measures and with the co-abundance groups (CAGs) enriched upon blueberry (BB) consumption. (**a**). Scatter plot showing the correlation between the actual and the Random Forest-predicted FRAP values. (**b**). Violin plot showing the association of the top 30 operational taxonomic unit (OTU) markers with the FRAP assay measures. X axis: Spearman Rho between the OTU abundances and the FRAP assay measures. Y axis: log of the Benjamini-Hochberg (BH) false discovery rate (FDR) with base 10. OTUs on the left: negatively associated; OTUs on the right: positively associated. Green color: the top 30 OTU markers showing significant association with FRAP measures (with FDR < 0.2) (positively associated); red color: negatively associated markers. (**c**). Bean plots showing the Spearman Rho measures distribution obtained for the predicted and the actual FRAP across the 100 iterations of the two variants of Random Forest models. (**d**). Stacked bar plots showing the relative representation of the different CAGs in the FRAP positive OTUs and the other non-marker OTUs. Seventeen out of the 25 FRAP-positive OTUs belonged to either CAG C1 (8) or C6 (9). Fishers' exact test showed a significant association between C6 and the FRAP-positive OTUs (indicated in the Figure). (**e**). Scatter plots showing the correlation between the FRAP-positive and FRAP-negative OTUs' mean abundances change across time points with the corresponding changes in FRAP measures.

## 4. Discussion

We have previously reported the reduction in the fecal microbiota α diversity due to the loss of fastidious taxa while the microbiota is adapting to the in vitro conditions of the artificial colon model [39,56]. To retain much of the stool diversity throughout the fermentation period, the basal fermentation medium was supplemented with a mix of indigestible and prebiotic carbohydrates often used in continuous in vitro systems [57]. Importantly, the supplementation of the fermentation medium with the polyphenol-rich fractions ANTH/FLAV, PACs, or prebiotic MIX substrates resulted in a favorable (i.e., health-associated) microbiota profile. Although the content of these fractions was not yet investigated by chromatographic separation, the fractions tested were prepared in the same way as in multiple previous publications, and so they are directly comparable in terms of evaluating their bioactive properties. The α diversity and the abundance of the major microbiota families *Lachnospiraceae* and *Bacteroidaceae* were comparatively higher, whereas the Enterobacteriaceae relative abundance was lower compared to TPP and S/A supplementation. The sugar content of the S/A fractions and potential residual sugars in the TPP fraction, which would be expected to be absorbed in the small intestine in humans, may have resulted in the significantly increased relative abundance of these organisms that are potent utilizers of simple sugars [58]. Apart from this explanation, *Enterobacteriaceae* have been reported to be involved in the metabolism of polyphenols in the gut [27,59]. Importantly, not all *Enterobacteriaceae* are harmful, with some playing an important role in the "healthy" gut microbiota [9].

Bifidobacteria residing in the colon may be utilizing the polyphenolic sugar content that reaches distal gastrointestinal parts [60,61]. TTP followed by PACs (but not ANTH/FLAV) supplementation were the most efficient additives to maintain the *Bifidobacterium* abundance levels in the suboptimal in vitro environment. Previous in vitro fermentation studies have yielded conflicting results on the effect of polyphenol-rich fractions on *Bifidobacterium* spp. abundance in the microbiota [21,62]. Limitations of the in vitro systems and baseline microbiota variations may have contributed to these discrepancies. Human studies have confirmed some effect of polyphenols on bifidobacteria [63].

We observed a trend for microbiota α diversity increase in the group of older women consuming BB, and although the β diversity did not change throughout the intervention period, health-relevant taxa were significantly enriched with BB consumption in subjects of both age groups. We acknowledge the limitations of the small sample size in this study, due in part to the complexity of running a human dietary intervention trial in which the primary objective was to test the effect of BB consumption on the human muscle progenitor cell (hMPC) function [64]. Nevertheless, the current study serves adequately as a pilot study to investigate the potential of regular BB consumption to improve the microbiota diversity in older healthy people. Maintaining microbiota diversity is relevant throughout the lifespan. Risk factors for non-communicable disease are associated with a Western lifestyle [65–67]. Decreased gut microbiota diversity as in low species richness and low counts of bacterial genes may correlate to metabolic disease, and therefore global microbiota modulations can promote health in the general population [68,69].

The enrichment of the fecal microbiota in *Anaerostipes hadrus*, *F. prausnitzii*, and to a lesser extent *Ruminococcus bromii* (CAG C1)—all taxa of the "healthy" microbiota [70]—two weeks after BB intervention indicated a potential microbiota adaptation to BB consumption. Enrichment in the major fibrolytic taxon *R. bromii* [71,72] may represent an adaptation to the regular fibre derived from whole BB fruit. *Ruminococcus bromii* releases substrates from complex polysaccharides that other microbiota members such as *A. hadrus* and *F. prausnitzii* can metabolize [73]. *Anaerostipes hadrus* is a butyrate producer previously reported to be stimulated by prebiotic fibres [74,75]. The ecological context is important when evaluating the health benefit of taxa that are "prebiotically" stimulated. In the case of *A. hadrus*, it was reported that it exerted beneficial outcomes in "healthy" microbiota and adverse in dysbiotic microbiota in a mouse model [76], potentially involved in energy harvesting and blood glucose [77]. *Faecalibacterium prausnitzii* is a key butyrate producer with anti-inflammatory properties, and its reduced abundance in the microbiota has been associated with various gastrointestinal conditions [9,78]. A few studies in mice and humans have shown *Faecalibacterium* responsiveness to polyphenols, accompanied by metabolism improvement [79–81].

Taxa such as *E. hallii*, *B. intestinihominis*, and *Butyrisimonas virosa* (CAG C6) and *B. intestinihominis* and *F. prausnitzii* (CAG C3) showed a gradual increase in abundance from baseline towards later intervention time points. The increased abundance of *E. hallii*, a butyrate producer of the Lachnospiraceae family [70], identified here to be associated with BB consumption, may contribute to improved insulin sensitivity according to in vivo and human studies [82,83]. Other *Eubacterium* spp. taxa of the Lachnospiraceae family (e.g., *E. ramulus* and *E. rectale*) may be involved in the metabolism of polyphenolic compounds (e.g., flavonols, flavanols, and lignans) [84,85]. Similarly, intervention with a polyphenol-rich diet was associated with enrichment in *B. intestinihominis* and improved metabolism in mice and humans [86,87].

There was some moderate albeit significant correlation of OTUs and CAGs with the FRAP measurements that, in turn, positively correlated with BB consumption, especially in the older group. Conversely, FRAP measurements were negatively correlated with plasma glucose. Many studies on healthy adults have contributed evidence for the antioxidant and anti-inflammatory benefits of the regular consumption of polyphenol-rich foods, such as BB and other berry fruits [88]. There is evidence from cohort and clinical studies of reduced all-cause mortality, lower risk of CVD, improved insulin sensitivity, and lower type 2 diabetes (T2D) risk associated with BB and specifically anthocyanin intake [32]. Importantly, in older age groups anthocyanins appear to lower the risk of cognitive decline [32]. However, there is a lack of human studies investigating the consumption of polyphenol-rich berries, metabolic improvement, and the microbiota, with the majority of relevant data derived from animal studies [27,89–92]. Importantly, we recognize critics of FRAP assays and that, according to the literature, more oxidative damage markers should be added to allow robust conclusions to be drawn [88].

Here, we report taxa that not only were associated with BB consumption forming distinct CAGs (i.e., *E. hallii*, *B. intestinihominis*, *A. hadrus*, *F.prausnitzii*) as discussed, but that taxa that mostly belong to the significant CAGs were associated with improved FRAP measurements. In this part of the analysis, we found that *E. siraeum* (*Clostridium* cluster IV Ruminococcaceae taxon) and the phylogenetically close *F. prausnitzii* and *G. formicilis* [70] were positively associated with FRAP. Interestingly, in humans serum markers of insulin resistance were associated with reduced *E. siraeum* and *Butyrivibrio crossotus* abundance [93]. Conversely, *G. formicilis* and *C. catus*, identified in this study to be positively associated with FRAP, were associated with obesity [94–97].

## 5. Conclusions

In vitro conditions place constraints on microbiota responsiveness to supplementation tests [98]. Notwithstanding this, the investigational studies as presented here offer a straightforward experimental model to test the initial hypothesis and provide insight into informed in vivo study design [99]. In future studies, the potential of the in vitro-identified microbiota response to BB polyphenol-rich fractions can be extended to the development of next-generation symbiotics.

The human study contributed evidence for specific microbiota modulation due to BB consumption in correlation with antioxidant activity in healthy adults. The association of fibrolytic taxa with whole BB consumption may indicate that the BB can contribute to health by both its polyphenolic content and its fibre content that, in effect, may render the fibre-bound polyphenols more accessible to microbiota fermentation [100]. In the context of healthy ageing, BB consumption may increase colonic short chain fatty acid (SCFA) production through fiber contribution to the fibrolytic members of the microbiota and promote health [101]. Importantly, non-pathobionts in the "healthy" microbiota, such as the taxa *C. catus* or *A. hadrus* mentioned in our study, may play a variant role within a different health context and in response to external dietary stimuli [102,103]. Strain-level identification is important in order to explain why individuals may respond differently to microbiota modulation [101,104]. Interindividual variation can be of relevance in the way dietary polyphenols impact health, given the fact that their bioavailability largely depends on the gut microbiota enzymatic armor [105]. Future large-scale clinical studies including both women and men and examining the metabolic impact of BB consumption in correlation with microbiota changes, inflammatory markers, and gender will allow for a deeper understanding of the role of BB consumption in human health. At the same time, a detailed chromatographic analysis of the BB fractions described here is desirable to generate greater granularity and detail on what individual compounds are present in each starting fraction, and what they are metabolized into, in the context of the microbiome changes described already in this report.

**Supplementary Materials:** The following are available online at http://www.mdpi.com/2072-6643/12/9/2800/s1: Figure S1: Boxplots showing the supplementation-independent significant decrease in $\alpha$ diversity measures (a. Shannon and b. Observed Species) for 24 h fermentation. Figure S2: Observed Species $\alpha$ diversity of the fecal microbiota after different BB polyphenol-rich fractions supplementations at 16 h and 24 h. Figure S3: Principal Component Analysis (PCoA) based on Weighted Unifrac distances of microbiota after in vitro fermentation with BB polyphenols. Figure S4: Fecal microbiota compositional description at Family level during 24 h in vitro fermentation with BB polyphenol-rich fractions. Figure S5: Differentially abundant taxa (genus level) in the fecal microbiota after in vitro supplementation with BB polyphenol-rich fractions (average relative abundance < 1%). Figure S6: Alpha diversity of the fecal microbiota of a. young and b. older women of the human trial across all time points. Figure S7: Shannon diversity showing the fecal microbiota $\alpha$ diversity development pre (W0) and post (mean of W4 and W6) BB consumption intervention. Figure S8 Variation in within cohort beta diversity for the Young and Old women. Figure S9: Variation in the cumulated abundances of the OTUs belonging to the six CAGs specifying for old and the young women. Figure S10: Identification of FRAP responsive taxa, their variation across time points and their association with glucose levels. Table S1: $p_{adj}$ values of Wilcoxon Signed Rank tests comparing the abundances of the various taxa in fermenter samples belonging to the various supplementation groups. Table S2: Taxonomic classifications of (A) FRAP-positive and (B) FRAP-negative OUT markers obtained using SPINGO.

**Author Contributions:** A.N.: in vitro study co-design, fecal samples for fermenter experiments collection, performed the in vitro experiments, DNA extractions, 16S rRNA sequencing library preparation, 16S rRNA analysis pipeline, co-wrote and edited the manuscript. T.S.G.: performed bioinformatics and statistical analysis for the in vitro and human study, figure generation, manuscript editing, data deposition. T.T.T.T.: statistics for the in vitro study. M.C.N. collected fecal samples for fermenter inoculation. P.P. and K.V. performed DNA extractions. A.T.-M., J.E.B., and M.E.G. designed, conducted, and collected samples for the human intervention and analyzed clinical parameters and oxidative markers. A.H. performed chromatographic extractions on blueberries to obtain individual fractions. P.W.O. conceived and co-designed the in vitro fermenter analyses, co-analyzed the data, designed and co-analyzed the human microbiome studies and ensuing data, and co-wrote the manuscript. All authors have read and agreed to the published version of the manuscript.

**Funding:** This research was funded by the following: Work in PWOT's laboratory is supported by the Government of Ireland National Development Plan by way of a Department of Agriculture, Food, and the Marine (DAFM) under a Food Institutional Research Measure (FIRM) award (14F828), and by a centre award to APC Microbiome Ireland (12/RC/2273_P2) from Science Foundation Ireland. Work in ATM's laboratory is supported by the National Science Foundation Graduate Research Fellowship Program under Grant No. DGE-1650441 (to JEB). This research was also funded by a grant from the U.S. Highbush Blueberry Council to ATM.

**Acknowledgments:** We thank the US Highbush Blueberry Council for providing the standardized freeze-dried BB powder for the human dietary intervention. Any opinions, findings, and conclusions or recommendations expressed in this material are those of the author(s) and do not necessarily reflect the views of the National Science Foundation.

**Conflicts of Interest:** The human intervention study was funded by a grant from the U.S. Highbush Blueberry Council to ATM. The Highbush Blueberry Council was not involved in study design, analysis, or data interpretation, nor did they influence the decision to publish. The role of the Highbush Blueberry Council in the study implementation was limited to provision of the freeze-dried blueberries.

## References

1. Sonnenburg, J.L.; Bäckhed, F. Diet-microbiota interactions as moderators of human metabolism. *Nature* **2016**, *535*, 56–64. [CrossRef] [PubMed]
2. Messer, J.S.; Liechty, E.R.; Vogel, O.A.; Chang, E.B. Evolutionary and ecological forces that shape the bacterial communities of the human gut. *Mucosal Immunol.* **2017**, *10*, 567–579. [CrossRef] [PubMed]
3. Marchesi, J.R.; Adams, D.H.; Fava, F.; Hermes, G.D.; Hirschfield, G.M.; Hold, G.; Quraishi, M.N.; Kinross, J.; Smidt, H.; Tuohy, K.M.; et al. The gut microbiota and host health: A new clinical frontier. *Gut* **2016**, *65*, 330–339. [CrossRef] [PubMed]
4. Petersen, C.; Round, J.L. Defining dysbiosis and its influence on host immunity and disease. *Cell. Microbiol.* **2014**, *16*, 1024–1033. [CrossRef]
5. Jeffery, I.B.; Lynch, D.B.; O'Toole, P.W. Composition and temporal stability of the gut microbiota in older persons. *ISME J.* **2016**, *10*, 170–182. [CrossRef]
6. Thevaranjan, N.; Puchta, A.; Schulz, C.; Naidoo, A.; Szamosi, J.C.; Verschoor, C.P.; Loukov, D.; Schenck, L.P.; Jury, J.; Foley, K.P.; et al. Age-associated microbial dysbiosis promotes intestinal permeability, systemic inflammation, and macrophage dysfunction. *Cell Host Microbe* **2017**, *21*, 455–466.e4. [CrossRef]
7. McCarville, J.L.; Caminero, A.; Verdu, E.F. Novel perspectives on therapeutic modulation of the gut microbiota. *Therap. Adv. Gastroenterol.* **2016**, *9*, 580–593. [CrossRef]
8. Salazar, N.; Valdés-Varela, L.; González, S.; Gueimonde, M.; de Los Reyes-Gavilán, C.G. Nutrition and the gut microbiome in the elderly. *Gut Microbes* **2017**, *8*, 82–97. [CrossRef]
9. O'Toole, P.W.; Marchesi, J.R.; Hill, C. Next-generation probiotics: The spectrum from probiotics to live biotherapeutics. *Nat. Microbiol.* **2017**, *2*, 17057. [CrossRef]

10. Fraga, C.G.; Croft, K.D.; Kennedy, D.O.; Tomás-Barberán, F.A. The effects of polyphenols and other bioactives on human health. *Food Funct.* **2019**, *10*, 514–528. [CrossRef]
11. Van Hul, M.; Cani, P.D. Targeting carbohydrates and polyphenols for a healthy microbiome and healthy weight. *Curr. Nutr. Rep.* **2019**, *8*, 307–316. [CrossRef] [PubMed]
12. Durazzo, A.; Lucarini, M.; Souto, E.B.; Cicala, C.; Caiazzo, E.; Izzo, A.A.; Novellino, E.; Santini, A. Polyphenols: A concise overview on the chemistry, occurrence, and human health. *Phyther. Res.* **2019**, *33*, 2221–2243. [CrossRef] [PubMed]
13. Panche, A.N.; Diwan, A.D.; Chandra, S.R. Flavonoids: An overview. *J. Nutr. Sci.* **2016**, *5*, 1–15. [CrossRef] [PubMed]
14. Pojer, E.; Mattivi, F.; Johnson, D.; Stockley, C.S. The case for anthocyanin consumption to promote human health: A review. *Compr. Rev. Food Sci. Food Saf.* **2013**, *12*, 483–508. [CrossRef]
15. Li, D.; Wang, P.; Luo, Y.; Zhao, M.; Chen, F. Health benefits of anthocyanins and molecular mechanisms: Update from recent decade. *Crit. Rev. Food Sci. Nutr.* **2017**, *57*, 1729–1741. [CrossRef] [PubMed]
16. Chun, O.K.; Chung, S.J.; Song, W.O. Estimated dietary flavonoid intake and major food sources of U.S. adults. *J. Nutr.* **2007**, *137*, 1244–1252. [CrossRef] [PubMed]
17. Chun, O.K.; Lee, S.G.; Wang, Y.; Vance, T.; Song, W.O. Estimated flavonoid intake of the elderly in the United States and around the world. *J. Nutr. Gerontol. Geriatr.* **2012**, *31*, 190–205. [CrossRef] [PubMed]
18. Beking, K.; Vieira, A. An assessment of dietary flavonoid intake in the UK and Ireland. *Int. J. Food Sci. Nutr.* **2011**, *62*, 17–19. [CrossRef]
19. Vogiatzoglou, A.; Mulligan, A.A.; Lentjes, M.A.H.; Luben, R.N.; Spencer, J.P.E.; Schroeter, H.; Khaw, K.T.; Kuhnle, G.G.C. Flavonoid intake in European adults (18 to 64 Years). *PLoS ONE* **2015**, *10*, e0128132. [CrossRef] [PubMed]
20. Rowland, I.; Gibson, G.; Heinken, A.; Scott, K.; Swann, J.; Thiele, I.; Tuohy, K. Gut microbiota functions: Metabolism of nutrients and other food components. *Eur. J. Nutr.* **2018**, *57*, 1–24. [CrossRef]
21. Dueñas, M.; Muñoz-González, I.; Cueva, C.; Jiménez-Girón, A.; Sánchez-Patán, F.; Santos-Buelga, C.; Moreno-Arribas, M.V.; Bartolomé, B. A survey of modulation of gut microbiota by dietary polyphenols. *Biomed. Res. Int.* **2015**, *2015*, 15. [CrossRef] [PubMed]
22. Zhang, X.; Yang, Y.; Wu, Z.; Weng, P. The modulatory effect of anthocyanins from purple sweet potato on human intestinal microbiota in vitro. *J. Agric. Food Chem.* **2016**, *64*, 2582–2590. [CrossRef] [PubMed]
23. Lacombe, A.; Li, R.W.; Klimis-Zacas, D.; Kristo, A.S.; Tadepalli, S.; Krauss, E.; Young, R.; Wu, V.C.H. Lowbush wild blueberries have the potential to modify gut microbiota and xenobiotic metabolism in the rat colon. *PLoS ONE* **2013**, *8*, e67497. [CrossRef]
24. Anhê, F.F.; Roy, D.; Pilon, G.; Dudonné, S.; Matamoros, S.; Varin, T.V.; Garofalo, C.; Moine, Q.; Desjardins, Y.; Levy, E.; et al. A polyphenol-rich cranberry extract protects from diet-induced obesity, insulin resistance and intestinal inflammation in association with increased *Akkermansia* spp. population in the gut microbiota of mice. *Gut* **2015**, *64*, 872–883. [CrossRef]
25. Baldwin, J.; Collins, B.; Wolf, P.G.; Martinez, K.; Shen, W.; Chuang, C.C.; Zhong, W.; Cooney, P.; Cockrell, C.; Chang, E.; et al. Table grape consumption reduces adiposity and markers of hepatic lipogenesis and alters gut microbiota in butter fat-fed mice. *J. Nutr. Biochem.* **2016**, *27*, 123–135. [CrossRef]
26. Fotschki, B.; Juśkiewicz, J.; Jurgoński, A.; Kołodziejczyk, K.; Milala, J.; Kosmala, M.; Zduńczyk, Z. Anthocyanins in strawberry polyphenolic extract enhance the beneficial effects of diets with fructooligosaccharides in the rat cecal environment. *PLoS ONE* **2016**, *11*, 1–17. [CrossRef] [PubMed]
27. Lee, S.; Keirsey, K.I.; Kirkland, R.; Grunewald, Z.I.; Fischer, J.G.; de La Serre, C.B. Blueberry supplementation influences the gut microbiota, inflammation, and insulin resistance in high-fat-diet–fed rats. *J. Nutr.* **2018**, *148*, 209–219. [CrossRef]
28. Bekiares, N.; Krueger, C.G.; Meudt, J.J.; Shanmuganayagam, D.; Reed, J.D. Effect of sweetened dried cranberry consumption on urinary proteome and fecal microbiome in healthy human subjects. *Omi. A J. Integr. Biol.* **2018**, *22*, 145–153. [CrossRef]
29. Rodríguez-Morató, J.; Matthan, N.R.; Liu, J.; de la Torre, R.; Chen, C.-Y.O. Cranberries attenuate animal-based diet-induced changes in microbiota composition and functionality: A randomized crossover controlled feeding trial. *J. Nutr. Biochem.* **2018**, *62*, 76–86. [CrossRef]

30. Gómez-Juaristi, M.; Sarria, B.; Martínez-López, S.; Clemente, L.B.; Mateos, R. Flavanol bioavailability in two cocoa products with different phenolic content. A comparative study in humans. *Nutrients* **2019**, *11*, 1441. [CrossRef]
31. Fallani, M.; Amarri, S.; Uusijarvi, A.; Adam, R.; Khanna, S.; Aguilera, M.; Gil, A.; Vieites, J.M.; Norin, E.; Young, D.; et al. Determinants of the human infant intestinal microbiota after the introduction of first complementary foods in infant samples from five European centres. *Microbiology* **2011**, *157*, 1385–1392. [CrossRef] [PubMed]
32. Kalt, W.; Cassidy, A.; Howard, L.R.; Krikorian, R.; Stull, A.J.; Tremblay, F.; Zamora-ros, R. Recent research on the health benefits of blueberries and their anthocyanins. *Adv. Nutr.* **2020**, *11*, 224–236. [CrossRef] [PubMed]
33. Kalt, W.; McDonald, J.E.; Donner, H. Antioxidant capacity of processed lowbush blueberry products. *Food Chem. Toxicol.* **2000**, *65*, 390–393.
34. Naczk, M.; Amarowicz, R.; Zadernowski, R.; Pegg, R.; Shahidi, F. Antioxidant activity of crude phenolic extracts from wild blueberry leaves. *Pol. J. Food Nutr. Sci.* **2003**, *12*, 166–169.
35. Nackz, M.; Zadernowski, R.; Shahidi, F. *Antioxidant Capacity of Phenolic Extracts from Selected Food By-Products*; American Chemical Society: Washington, DC, USA, 2007; pp. 184–194. ISBN 0841274282.
36. Krikorian, R.; Shidler, M.D.; Nash, T.A.; Kalt, W.; Vinqvist-Tymchuk, M.R.; Shukitt-Hale, B.; Joseph, J.A. Blueberry supplementation improves memory in older adults. *J. Agric. Food Chem.* **2010**, *58*, 3996–4000. [CrossRef] [PubMed]
37. Zhou, T.; Wei, C.; Lan, W.; Zhao, Y.; Pan, Y.; Sun, X.; Wu, V.C.H. The effect of Chinese wild blueberry fractions on the growth and membrane integrity of various foodborne pathogens. *J. Food Sci.* **2020**, *85*, 1513–1522. [CrossRef] [PubMed]
38. Hidalgo, M.; Oruna-Concha, M.J.; Kolida, S.; Walton, G.E.; Kallithraka, S.; Spencer, J.P.E.; Gibson, G.R.; De Pascual-Teresa, S. Metabolism of anthocyanins by human gut microflora and their influence on gut bacterial growth. *J. Agric. Food Chem.* **2012**, *60*, 3882–3890. [CrossRef] [PubMed]
39. Ntemiri, A.; Chonchúir, F.N.; O'Callaghan, T.F.; Stanton, C.; Ross, R.P.; O'Toole, P.W. Glycomacropeptide sustains microbiota diversity and promotes specific taxa in an artificial colon model of elderly gut microbiota. *J. Agric. Food Chem.* **2017**, *65*, 1836–1846. [CrossRef]
40. Duncan, S.H.; Holtrop, G.; Lobley, G.E.; Calder, A.G.; Stewart, C.S.; Flint, H.J. Contribution of acetate to butyrate formation by human faecal bacteria. *Br. J. Nutr.* **2004**, *91*, 915–923. [CrossRef]
41. Tzounis, X. Flavanol monomer-induced changes to the human faecal microflora. *Br. J. Nutr.* **2008**, *99*, 782–792. [CrossRef]
42. Klindworth, A.; Pruesse, E.; Schweer, T.; Peplies, J.; Quast, C.; Horn, M.; Glöckner, F.O. Evaluation of general 16S ribosomal RNA gene PCR primers for classical and next-generation sequencing-based diversity studies. *Nucleic Acids Res.* **2012**, *41*, 1–11. [CrossRef] [PubMed]
43. Howell, A.B.; Reed, J.D.; Krueger, C.G.; Winterbottom, R.; Cunningham, D.G.; Leahy, M. A-type cranberry proanthocyanidins and uropathogenic bacterial anti-adhesion activity. *Phytochemistry* **2005**, *66*, 2281–2291. [CrossRef] [PubMed]
44. Youdim, K.A.; McDonald, J.; Kalt, W.; Joseph, J.A. Potential role of dietary flavonoids in reducing microvascular endothelium vulnerability to oxidative and inflammatory insults. *J. Nutr. Biochem.* **2002**, *13*, 282–288. [CrossRef]
45. Schmidt, B.M.; Howell, A.B.; McEniry, B.; Knight, C.T.; Seigler, D.; Erdman, J.W.; Lila, M.A. Effective separation of potent antiproliferation and antiadhesion components from wild blueberry (Vaccinium angustifolium Ait.) fruits. *J. Agric. Food Chem.* **2004**, *52*, 6433–6442. [CrossRef] [PubMed]
46. Yi, W.; Fischer, J.; Krewer, G.; Akoh, C.C. Phenolic compounds from blueberries can inhibit colon cancer cell proliferation and induce apoptosis. *J. Agric. Food Chem.* **2005**, *53*, 7320–7329. [CrossRef]
47. Benzie, I.F.F.; Devaki, M. The Ferric Reducing/Antioxidant Power (FRAP) assay for non-enzymatic antioxidant capacity: Concepts, procedures, limitations and applications. In *Measurement of Antioxidant Activity & Capacity Functional Food Science and Technology Series*, 1st ed.; Apak, R., Capanoglu, E., Shahidi, F., Eds.; John Wiley & Sons, Inc.: Hoboken, NJ, USA, 2018; ISBN 9781119135357.
48. Ntemiri, A.; Ribière, C.; Stanton, C.; Ross, R.P.; O'Connor, E.M.; O'Toole, P.W. Retention of microbiota diversity by lactose-free milk in a mouse model of elderly gut microbiota. *J. Agric. Food Chem.* **2019**, *67*, 2098–2112. [CrossRef]

49. Magoc, T.; Salzberg, S.L. FLASH: Fast length adjustment of short reads to improve genome assemblies. *Bioinformatics* **2011**, *27*, 2957–2963. [CrossRef]
50. Caporaso, J.G.; Kuczynski, J.; Stombaugh, J.; Bittinger, K.; Bushman, F.D.; Costello, E.K.; Fierer, N.; Gonzalez Peña, A.; Goodrich, J.K.; Gordon, J.I.; et al. QIIME allows analysis of high-throughput community sequencing data. *Nat. Methods* **2010**, *7*, 335–336. [CrossRef]
51. Martin, M. Cutadapt removes adapter sequences from high-throughput sequencing reads. *EMBnet. J.* **2011**, *17*, 10. [CrossRef]
52. Schloss, P.D.; Westcott, S.L.; Ryabin, T.; Hall, J.R.; Hartmann, M.; Hollister, E.B.; Lesniewski, R.A.; Oakley, B.B.; Parks, D.H.; Robinson, C.J.; et al. Introducing mothur: Open-source, platform-independent, community-supported software for describing and comparing microbial communities. *Appl. Environ. Microbiol.* **2009**, *75*, 7537–7541. [CrossRef]
53. Allard, G.; Ryan, F.J.; Jeffery, I.B.; Claesson, M.J. SPINGO: A rapid species-classifier for microbial amplicon sequences. *BMC Bioinform.* **2015**, *16*, 1–8. [CrossRef] [PubMed]
54. Caporaso, J.G.; Bittinger, K.; Bushman, F.D.; Desantis, T.Z.; Andersen, G.L.; Knight, R. PyNAST: A flexible tool for aligning sequences to a template alignment. *Bioinformatics* **2010**, *26*, 266–267. [CrossRef] [PubMed]
55. Ghosh, T.S.; Rampelli, S.; Jeffery, I.B.; Santoro, A.; Neto, M.; Capri, M.; Giampieri, E.; Jennings, A.; Candela, M.; Turroni, S.; et al. Mediterranean diet intervention alters the gut microbiome in older people reducing frailty and improving health status: The NU-AGE 1-year dietary intervention across five European countries. *Gut* **2020**, *69*, 1218–1228. [CrossRef] [PubMed]
56. Macfarlane, G.T.; Macfarlane, S. Models for intestinal fermentation: Association between food components, delivery systems, bioavailability and functional interactions in the gut. *Curr. Opin. Biotechnol.* **2007**, *18*, 156–162. [CrossRef] [PubMed]
57. Aguirre, M.; Eck, A.; Koenen, M.E.; Savelkoul, P.H.M.; Budding, A.E.; Venema, K. Evaluation of an optimal preparation of human standardized fecal inocula for in vitro fermentation studies. *J. Microbiol. Methods* **2015**, *117*, 78–84. [CrossRef]
58. Zoetendal, E.G.; Raes, J.; van den Bogert, B.; Arumugam, M.; Booijink, C.C.; Troost, F.J.; Bork, P.; Wels, M.; de Vos, W.M.; Kleerebezem, M. The human small intestinal microbiota is driven by rapid uptake and conversion of simple carbohydrates. *ISME J.* **2012**, *6*, 1415–1426. [CrossRef]
59. Shao, Y.; Ding, R.; Xu, B.; Hua, R.; Shen, Q.; He, K.; Yao, Q. Alterations of gut microbiota after Roux-en-Y gastric bypass and sleeve gastrectomy in sprague-dawley rats. *Obes. Surg.* **2017**, *27*, 295–302. [CrossRef]
60. Kawabata, K.; Yoshioka, Y.; Terao, J. Role of intestinal microbiota in the bioavailability and physiological functions of dietary polyphenols. *Molecules* **2019**, *24*, 370. [CrossRef]
61. Pokusaeva, K.; Fitzgerald, G.F.; van Sinderen, D. Carbohydrate metabolism in Bifidobacteria. *Genes Nutr.* **2011**, *6*, 285–306. [CrossRef]
62. Sánchez-Patán, F.; Barroso, E.; van de Wiele, T.; Jiménez-Girón, A.; Martín-Alvarez, P.J.; Moreno-Arribas, M.V.; Martínez-Cuesta, M.C.; Peláez, C.; Requena, T.; Bartolomé, B. Comparative in vitro fermentations of cranberry and grape seed polyphenols with colonic microbiota. *Food Chem.* **2015**, *183*, 273–282. [CrossRef]
63. Guglielmetti, S.; Fracassetti, D.; Taverniti, V.; Del Bo', C.; Vendrame, S.; Klimis-Zacas, D.; Arioli, S.; Riso, P.; Porrini, M. Differential modulation of human intestinal bifidobacterium populations after consumption of a wild blueberry (vaccinium angustifolium) drink. *J. Agric. Food Chem.* **2013**, *61*, 8134–8140. [CrossRef] [PubMed]
64. Blum, J.; Gheller, B.; Hwang, S.; Bender, E.; Gheller, M.; Thalacker-Mercer, A. Consumption of a blueberry-enriched diet by women for 6 weeks alters determinants of human muscle progenitor cell function. *J. Nutr.* **2020**, *1*, 2412–2418. [CrossRef] [PubMed]
65. Lusis, A.J.; Attie, A.D.; Reue, K. Metabolic syndrome: From epidemiology to systems biology. *Nat. Rev. Genet.* **2008**, *9*, 819–830. [CrossRef] [PubMed]
66. Catrysse, L.; van Loo, G. Inflammation and the metabolic syndrome: The tissue-specific functions of NF-κB. *Trends Cell Biol.* **2017**, *27*, 417–429. [CrossRef] [PubMed]
67. Christ, A.; Latz, E. The Western lifestyle has lasting effects on metaflammation. *Nat. Rev. Immunol.* **2019**, *19*, 267–268. [CrossRef] [PubMed]
68. Le Chatelier, E.; Nielsen, T.; Qin, J.; Prifti, E.; Hildebrand, F.; Falony, G.; Almeida, M.; Arumugam, M.; Batto, J.-M.; Kennedy, S.; et al. Richness of human gut microbiome correlates with metabolic markers. *Nature* **2013**, *500*, 541–546. [CrossRef]

69. Moeller, A.H. The shrinking human gut microbiome. *Curr. Opin. Microbiol.* **2017**, *38*, 30–35. [CrossRef]
70. Rajilić-Stojanović, M.; de Vos, W.M. The first 1000 cultured species of the human gastrointestinal microbiota. *FEMS Microbiol. Rev.* **2014**, *38*, 996–1047. [CrossRef]
71. Abell, G.C.J.; Cooke, C.M.; Bennett, C.N.; Conlon, M.A.; McOrist, A.L. Phylotypes related to Ruminococcus bromii are abundant in the large bowel of humans and increase in response to a diet high in resistant starch. *FEMS Microbiol. Ecol.* **2008**, *66*, 505–515. [CrossRef]
72. Ze, X.; Duncan, S.H.; Louis, P.; Flint, H.J. Ruminococcus bromii is a keystone species for the degradation of resistant starch in the human colon. *ISME J.* **2012**, *6*, 1535–1543. [CrossRef]
73. Cockburn, D.W.; Koropatkin, N.M. Polysaccharide degradation by the intestinal microbiota and its influence on human health and disease. *J. Mol. Biol.* **2016**, *428*, 3230–3252. [CrossRef] [PubMed]
74. Scott, K.P.; Martin, J.C.; Duncan, S.H.; Flint, H.J. Prebiotic stimulation of human colonic butyrate-producing bacteria and bifidobacteria, in vitro. *FEMS Microbiol. Ecol.* **2014**, *87*, 30–40. [CrossRef] [PubMed]
75. Vandeputte, D.; Falony, G.; Vieira-Silva, S.; Wang, J.; Sailer, M.; Theis, S.; Verbeke, K.; Raes, J. Prebiotic inulin-type fructans induce specific changes in the human gut microbiota. *Gut* **2017**, *66*, 1968–1974. [CrossRef] [PubMed]
76. Aron-Wisnewsky, J.; Prifti, E.; Belda, E.; Ichou, F.; Kayser, B.D.; Dao, M.C.; Verger, E.O.; Hedjazi, L.; Bouillot, J.-L.; Chevallier, J.-M.; et al. Major microbiota dysbiosis in severe obesity: Fate after bariatric surgery. *Gut* **2019**, *68*, 70–82. [CrossRef]
77. Zeevi, D.; Korem, T.; Zmora, N.; Israeli, D.; Rothschild, D.; Weinberger, A.; Ben-Yacov, O.; Lador, D.; Avnit-Sagi, T.; Lotan-Pompan, M.; et al. Personalized nutrition by prediction of glycemic responses. *Cell* **2015**, *163*, 1079–1095. [CrossRef] [PubMed]
78. Koh, A.; De Vadder, F.; Kovatcheva-Datchary, P.; Bäckhed, F. From dietary fiber to host physiology: Short-chain fatty acids as key bacterial metabolites. *Cell* **2016**, *165*, 1332–1345. [CrossRef] [PubMed]
79. Heyman-Lindén, L.; Kotowska, D.; Sand, E.; Bjursell, M.; Plaza, M.; Turner, C.; Holm, C.; Fåk, F.; Berger, K. Lingonberries alter the gut microbiota and prevent low-grade inflammation in high-fat diet fed mice. *Food Nutr. Res.* **2016**, *60*, 29993. [CrossRef] [PubMed]
80. Moreno-Indias, I.; Sánchez-Alcoholado, L.; Pérez-Martínez, P.; Andrés-Lacueva, C.; Cardona, F.; Tinahones, F.; Queipo-Ortuño, M.I. Red wine polyphenols modulate fecal microbiota and reduce markers of the metabolic syndrome in obese patients. *Food Funct.* **2016**, *7*, 1775–1787. [CrossRef]
81. González-Sarrías, A.; Romo-Vaquero, M.; García-Villalba, R.; Cortés-Martín, A.; Selma, M.V.; Espín, J.C. The endotoxemia marker lipopolysaccharide-binding protein is reduced in overweight-obese subjects consuming pomegranate extract by modulating the gut microbiota: A randomized clinical trial. *Mol. Nutr. Food Res.* **2018**, *62*, 1800160. [CrossRef]
82. Vrieze, A.; Van Nood, E.; Holleman, F.; Salojärvi, J.; Kootte, R.S.; Bartelsman, J.F.W.M.; Dallinga-Thie, G.M.; Ackermans, M.T.; Serlie, M.J.; Oozeer, R.; et al. Transfer of intestinal microbiota from lean donors increases insulin sensitivity in individuals with metabolic syndrome. *Gastroenterology* **2012**, *143*, 913–916.e7. [CrossRef]
83. Udayappan, S.; Manneras-Holm, L.; Chaplin-Scott, A.; Belzer, C.; Herrema, H.; Dallinga-Thie, G.M.; Duncan, S.H.; Stroes, E.S.G.; Groen, A.K.; Flint, H.J.; et al. Oral treatment with Eubacterium hallii improves insulin sensitivity in db/db mice. *NPJ Biofilms Microbiomes* **2016**, *2*, 16009. [CrossRef] [PubMed]
84. Marín, L.; Miguélez, E.M.; Villar, C.J.; Lombó, F. Bioavailability of dietary polyphenols and gut microbiota metabolism: Antimicrobial properties. *Biomed. Res. Int.* **2015**, *2015*, 1–18. [CrossRef] [PubMed]
85. Ozdal, T.; Sela, D.A.; Xiao, J.; Boyacioglu, D.; Chen, F.; Capanoglu, E. The reciprocal interactions between polyphenols and gut microbiota and effects on bioaccessibility. *Nutrients* **2016**, *8*, 78. [CrossRef] [PubMed]
86. Anhê, F.F.; Nachbar, R.T.; Varin, T.V.; Vilela, V.; Dudonné, S.; Pilon, G.; Fournier, M.; Lecours, M.-A.; Desjardins, Y.; Roy, D.; et al. A polyphenol-rich cranberry extract reverses insulin resistance and hepatic steatosis independently of body weight loss. *Mol. Metab.* **2017**, *6*, 1563–1573. [CrossRef] [PubMed]
87. Anhê, F.F.; Nachbar, R.T.; Varin, T.V.; Trottier, J.; Dudonné, S.; Le Barz, M.; Feutry, P.; Pilon, G.; Barbier, O.; Desjardins, Y.; et al. Treatment with camu camu ( Myrciaria dubia ) prevents obesity by altering the gut microbiota and increasing energy expenditure in diet-induced obese mice. *Gut* **2019**, *68*, 453–464. [CrossRef] [PubMed]
88. Del Bo, C.; Martini, D.; Porrini, M.; Klimis-Zacas, D.; Riso, P. Berries and oxidative stress markers: An overview of human intervention studies. *Food Funct.* **2015**, *6*, 2890–2917. [CrossRef] [PubMed]

89. Wankhade, U.D.; Zhong, Y.; Lazarenko, O.P.; Chintapalli, S.V.; Piccolo, B.D.; Chen, J.R.; Shankar, K. Sex-specific changes in gut microbiome composition following blueberry consumption in C57Bl/6J mice. *Nutrients* **2019**, *11*, 313. [CrossRef]
90. De Freitas Carvalho, M.M.; Lage, N.N.; de Souza Paulino, A.H.; Pereira, R.R.; de Almeida, L.T.; da Silva, T.F.; de Brito Magalhães, C.L.; de Lima, W.G.; Silva, M.E.; Pedrosa, M.L.; et al. Effects of açai on oxidative stress, ER stress, and inflammation-related parameters in mice with high fat diet-fed induced NAFLD. *Sci. Rep.* **2019**, *9*, 1–11. [CrossRef]
91. Frolinger, T.; Sims, S.; Smith, C.; Wang, J.; Cheng, H.; Faith, J.; Ho, L.; Hao, K.; Pasinetti, G.M. The gut microbiota composition affects dietary polyphenols-mediated cognitive resilience in mice by modulating the bioavailability of phenolic acids. *Sci. Rep.* **2019**, *9*, 1–10. [CrossRef]
92. Rodríguez-Daza, M.C.; Daoust, L.; Boutkrabt, L.; Pilon, G.; Varin, T.; Dudonné, S.; Levy, É.; Marette, A.; Roy, D.; Desjardins, Y. Wild blueberry proanthocyanidins shape distinct gut microbiota profile and influence glucose homeostasis and intestinal phenotypes in high-fat high-sucrose fed mice. *Sci. Rep.* **2020**, *10*, 1–16. [CrossRef]
93. Pedersen, H.K.; Gudmundsdottir, V.; Nielsen, H.B.; Hyotylainen, T.; Nielsen, T.; Jensen, B.A.H.; Forslund, K.; Hildebrand, F.; Prifti, E.; Falony, G.; et al. Human gut microbes impact host serum metabolome and insulin sensitivity. *Nature* **2016**, *535*, 376–381. [CrossRef] [PubMed]
94. Kasai, C.; Sugimoto, K.; Moritani, I.; Tanaka, J.; Oya, Y.; Inoue, H.; Tameda, M.; Shiraki, K.; Ito, M.; Takei, Y.; et al. Comparison of the gut microbiota composition between obese and non-obese individuals in a Japanese population, as analyzed by terminal restriction fragment length polymorphism and next-generation sequencing. *BMC Gastroenterol.* **2015**, *15*, 1–10. [CrossRef] [PubMed]
95. Lagkouvardos, I.; Kläring, K.; Heinzmann, S.S.; Platz, S.; Scholz, B.; Engel, K.-H.; Schmitt-Kopplin, P.; Haller, D.; Rohn, S.; Skurk, T.; et al. Gut metabolites and bacterial community networks during a pilot intervention study with flaxseeds in healthy adult men. *Mol. Nutr. Food Res.* **2015**, *59*, 1614–1628. [CrossRef]
96. Patrone, V.; Vajana, E.; Minuti, A.; Callegari, M.L.; Federico, A.; Loguercio, C.; Dallio, M.; Tolone, S.; Docimo, L.; Morelli, L. Postoperative changes in fecal bacterial communities and fermentation products in obese patients undergoing bilio-intestinal bypass. *Front. Microbiol.* **2016**, *7*, 1–13. [CrossRef] [PubMed]
97. Zeng, Q.; Li, D.; He, Y.; Li, Y.; Yang, Z.; Zhao, X.; Liu, Y.; Wang, Y.; Sun, J.; Feng, X.; et al. Discrepant gut microbiota markers for the classification of obesity-related metabolic abnormalities. *Sci. Rep.* **2019**, *9*, 1–10. [CrossRef] [PubMed]
98. Guerra, A.; Etienne-Mesmin, L.; Livrelli, V.; Denis, S.; Blanquet-Diot, S.; Alric, M. Relevance and challenges in modeling human gastric and small intestinal digestion. *Trends Biotechnol.* **2012**, *30*, 591–600. [CrossRef] [PubMed]
99. Schmidt, T.S.B.; Raes, J.; Bork, P. The human gut microbiome: From association to modulation. *Cell* **2018**, *172*, 1198–1215. [CrossRef] [PubMed]
100. Edwards, C.A.; Havlik, J.; Cong, W.; Mullen, W.; Preston, T.; Morrison, D.J.; Combet, E. Polyphenols and health: Interactions between fibre, plant polyphenols and the gut microbiota. *Nutr. Bull.* **2017**, *42*, 356–360. [CrossRef]
101. Duncan, S.H.; Flint, H.J. Probiotics and prebiotics and health in ageing populations. *Maturitas* **2013**, *75*, 44–50. [CrossRef]
102. Cani, P.D. Human gut microbiome: Hopes, threats and promises. *Gut* **2018**, *67*, 1716–1725. [CrossRef]
103. Cirstea, M.; Radisavljevic, N.; Finlay, B.B. Good bug, bad bug: Breaking through microbial stereotypes. *Cell Host Microbe* **2018**, *23*, 10–13. [CrossRef] [PubMed]
104. De Filippis, F.; Pellegrini, N.; Laghi, L.; Gobbetti, M.; Ercolini, D. Unusual sub-genus associations of faecal Prevotella and Bacteroides with specific dietary patterns. *Microbiome* **2016**, *4*, 1–6. [CrossRef] [PubMed]
105. Anhê, F.F.; Choi, B.S.Y.; Dyck, J.R.B.; Schertzer, J.D.; Marette, A. Host–microbe interplay in the cardiometabolic benefits of dietary polyphenols. *Trends Endocrinol. Metab.* **2019**, *30*, 384–395. [CrossRef] [PubMed]

*nutrients*

Article

# Putative Mechanisms Responsible for the Antihyperglycemic Action of *Lactobacillus paracasei* HII01 in Experimental Type 2 Diabetic Rats

**Parichart Toejing [1], Nuntawat Khat-Udomkiri [2], Jannarong Intakhad [1], Sasithorn Sirilun [2], Chaiyavat Chaiyasut [2] and Narissara Lailerd [1,2,*]**

[1] Department of Physiology, Faculty of Medicine, Chiang Mai University, Chiang Mai 50200, Thailand; lookplanoi_@hotmail.com (P.T.); Jannarong.7051@gmail.com (J.I.)

[2] Innovation Center for Holistic Health, Nutraceuticals and Cosmeceuticals, Department of Pharmaceutical Sciences, Faculty of Pharmacy, Chiang Mai University, Chiang Mai 50200, Thailand; mysan_t_u_s@hotmail.com (N.K.-U.); sasithorn.s@cmu.ac.th (S.S.); chaiyavat@gmail.com (C.C.)

* Correspondence: narissara.lailerd@cmu.ac.th; Tel.: +66-5393-5362-4

Received: 3 September 2020; Accepted: 30 September 2020; Published: 1 October 2020

**Abstract:** Despite the updated knowledge of the impact of gut dysbiosis on diabetes, investigations into the beneficial effects of individual bacteria are still required. This study evaluates the antihyperglycemic efficacy of *Lactobacillus paracasei* HII01 and its possible mechanisms in diabetic rats. Diabetic rats were assigned to receive vehicle, *L. paracasei* HII01 ($10^8$ CFU/day), metformin 30 (mg/kg) or a combination of *L. paracasei* HII01 and metformin. Normal rats given vehicle and *L. paracasei* HII01 were included. Metabolic parameters, including in vitro hemi-diaphragm glucose uptake, skeletal insulin-signaling proteins, plasma lipopolysaccharide (LPS), gut permeability, composition of gut microbiota and its metabolites, as well as short-chain fatty acids (SCFAs), were assessed after 12 weeks of experiment. The results clearly demonstrated that *L. paracasei* HII01 improved glycemic parameters, glucose uptake, insulin-signaling proteins including $pAkt^{Ser473}$, glucose transporter 4 (GLUT4) and phosphorylation of AMP-activated protein kinase ($pAMPK^{Thr172}$), tumor necrosis factor (TNF-α) and nuclear factor-κB (NF-kB) in diabetic rats. Modulation of gut microbiota was found together with improvement in leaky gut, endotoxemia and SCFAs in diabetic rats administered *L. paracasei* HII01. In conclusion, *L. paracasei* HII01 alleviated hyperglycemia in diabetic rats primarily by modulating gut microbiota along with lessening leaky gut, leading to improvement in endotoxemia and inflammation-disturbed insulin signaling, which was mediated partly by PI3K/Akt signaling and AMPK activation.

**Keywords:** type 2 diabetes mellitus; gut microbiota; *Lactobacillus paracasei*; antihyperglycemia

## 1. Introduction

Type 2 diabetes mellitus (T2DM), a multifactorial metabolic endocrine disorder, is characterized by persistent hyperglycemia, and it is basically a result of insulin resistance and impaired β-cell function. According to the International Diabetes Federation (IDF), the number of diabetic patients worldwide was 425 million in 2017 and will rise to 629 million by 2045 [1]. Although, several influences such as genetics, age, unhealthy lifestyle and obesity are accepted as risk factors of T2DM [2]. Nowadays, it is well accepted that gut microbiota is linked to the development of T2DM [3]. Changes in gut microbiota composition, known as gut dysbiosis, have been associated with disrupted gut barrier functions and increased gut permeability [4,5]. The enhancement of gut permeability might result in bacterial lipopolysaccharide (LPS) leak into blood circulation, followed by inflammatory activation through the LPS-Toll-like receptor 4-Nuclear factor-κB (LPS-TLR4-NF-kB) signaling pathway [6,7].

Moreover, tumor necrosis factor (TNF-α), a pro-inflammatory cytokine that is generated from the LPS-TLR4-NF-kB signaling pathway, induces the enhancement of the phosphorylation of insulin receptor substrate 1 (IRS-$1^{Ser307}$) [8]. The serine phosphorylation of IRS-1 blunts the activation of the phosphatidylinositol 3-kinase (PI3K)/protein kinase B (Akt) signaling pathway, resulting in a reduction in glucose transporter 4 (GLUT4) translocation and glucose uptake in the skeletal muscle, which causes insulin resistance and hyperglycemia [9]. Therefore, the modulation of gut microbiota is used as a strategy for prevention or adjuvant treatment of T2DM.

Probiotics are defined as "live microorganisms which when administered in adequate amounts confer a specific health benefit on the host" [7]. Conclusive evidence indicates that modulation of gut microbiota by probiotics provides beneficial health effects in both animal and clinical research of T2DM [10–12]. Among probiotics, Lactobacillus is one of the most popular strains that have been used for investigation [13]. The oral administration of *Lactobacillus reuteri* GMNL-263 decreased the plasma glucose level in high fructose-fed rats [14]. Lim et al., 2016, also revealed that gut tight junction, endotoxemia and inflammation were ameliorated after *Lactobacillus sakei* OK67 treatment in type 2 diabetic rat model [15]. Furthermore, a previous study demonstrated that the production of short-chain fatty acids (SCFAs) and gut microbial metabolites, including acetate, propionate and butyrate, seems to play an important role in the attenuation of T2DM [16].

Recently, a newly identified probiotic strain *Lactobacillus paracasei* spp. HII01, from the fermentation of northern Thai pickle, showed a significant improvement in gut dysbiosis and metabolic endotoxemia in obese rats [17]. In addition, *L. paracasei* HII01 restored kidney function by attenuating insulin resistance and hyperglycemia in obese rats [18]. However, no information is available on the antidiabetic potential of *L. paracasei* HII01. Therefore, this study was conducted to evaluate the antidiabetic effect of *L. paracasei* HII01 on experimental type 2 diabetic rats and explore the possible underlying mechanisms.

## 2. Materials and Methods

### 2.1. Animals and Ethical Approval

Adult male Wistar rats weighing approximately 180–200 g were used in this study. All rats were obtained from the National Laboratory Animal Center, Mahidol University, Thailand. The experimental protocol was approved by the Research Animal Care and Use Ethical Committee, Faculty of Pharmacy, Chiang Mai University, Thailand (Ethics approval no. 04/2015). All animals were housed under controlled temperature at 25 ± 2 °C with a 12 h light/dark cycle and were fed with a standard rodent chow diet and water ad libitum. The animals were given an acclimatization period of 1 week. The animals used in this study were cared for according to the principles and guidance of the "Guide for the Care and Use of Animals in compliance with the National Institute of Health Guideline for the Care and Treatment of Animals".

### 2.2. Stock and Cultivation of the Strain

Lactobacillus strain No. HII01 is a novel non-human origin-isolated strain of lactic acid-producing bacteria that has been approved by the Food and Drug Administration (FDA), Thailand. It was prepared at the Innovation Center for Holistic Health, Nutraceuticals and Cosmeceuticals, Faculty of Pharmacy, Chiang Mai University. The 16S rRNA gene sequence of the representative strain showed 99.0% similarity, 1511 bps, to *L. paracasei* accession number AP012541.1. The bacterial strain was revived in MRS (de Mann Rogosa Sharpe) (Difco Detroit, MI, USA) broth with pH of 6.5 + 0.2 at 25 °C. The stock culture of the HII01 was maintained at 20% (*v*/*v*) glycerol-MRS broth at −70 °C. The organism was activated 3 times in MRS broth using 1% (*v*/*v*) inoculum at 37 °C for 24 h until further use.

### 2.3. Bacterial Culture

The growth culture of the strain (1%) was inoculated into freshly prepared MRS. The bacterial cell of HII01 was prepared from the late exponential growth phase of cell growth. The inoculum of the strain in the culture medium was collected by centrifugation at 10,000× $g$, 4 °C for 10 min. The supernatant was discarded, and the cell pellet was washed 3 times with phosphate buffer saline (pH 7.0 ± 0.2). Then, the cell pellet was re-suspended, and a final concentration of approximately $10^8$ colony forming unit (CFU)/mL sterile distilled water was used in the experiment.

### 2.4. Induction of Experimental Diabetes

The establishment of a type 2 diabetic model was carried out as described by Srinivasan et al., 2005 [19]. The rats were assigned into two dietary regimens by feeding them with standard rodent chow diet (10.95% kcal energy from fat source) or high-fat diet (53.63% kcal energy from fat source) (Table S1) ad libitum. After 2 weeks of initial dietary period, diabetes mellitus was induced in overnight fasted rats with a single intraperitoneal injection of streptozotocin (STZ) (Sigma-Aldrich, St. Louis, MO, USA) dissolved in citrate buffer (pH 4.5) at a dose of 40 mg/kg. After 14 days of induction, diabetes mellitus was confirmed by the fasting plasma glucose levels. The rats with fasting plasma glucose level ≥250 mg/dL without hypoinsulinemia were considered to exhibit type 2 diabetes and were included in this study. A total of 60 male Wistar rats were randomly divided into six groups (n = 10 per group): normal diet control (NDC), normal rat supplemented with *L. paracasei* HII01 ($10^8$ CFU/day) (ND-L), diabetic rat control (DMC), diabetic rat supplemented with *L. paracasei* HII01 ($10^8$ CFU/day) (DM-L), diabetic rat treated with metformin (30 mg/kg) (DMM) as the positive control and diabetic rat supplemented with a combination of *L. paracasei* HII01 ($10^8$ CFU/day) and metformin (30 mg/kg) (DMM-L). After 12 weeks of supplementation, overnight fasted rats were sacrificed via an intraperitoneal injection of overdose Nembutal® (Liboume, France). Blood samples were collected in appropriate anticoagulant and then centrifuged at 13,000× $g$ for 1 min to obtain plasma. The soleus muscle, gastrocnemius muscle and liver were rapidly removed, frozen in liquid nitrogen and stored at −80 °C for further analysis.

### 2.5. Biochemical Analysis of Plasma

The plasma levels of glucose, triglyceride (TG), cholesterol, low-density lipoprotein cholesterol (LDL) and high-density lipoprotein cholesterol (HDL) were analyzed using a commercial kit (Biotech, Bangkok, Thailand). The plasma insulin, leptin and adiponectin levels were measured using a rat ELISA kit (LINCO Research, Charles, MO, USA) following the instructions of the manufacturer. The degree of insulin resistance was assessed by the homeostasis model assessment of insulin resistance (HOMA-IR), calculated from fasting plasma insulin and glucose concentrations [20]. The HOMA-IR index was calculated using the following formula:

$$\text{HOMA-IR} = [\text{fasting plasma insulin level (ng/dL)} \times \text{fasting plasma glucose level (mg/dL)}]/405.1$$

### 2.6. Oral Glucose Tolerance Test

Oral glucose tolerance test (OGTT) was performed on the 11th week. All rats were fasted overnight and the fasting plasma glucose was collected prior to glucose administration (time = 0) as the baseline value. Then, 2 g/kg of glucose solution was administered by oral gavage. The blood samples were collected at 15, 30, 60 and 120 min after glucose administration. The plasma glucose levels were determined, and the area under the curve (AUC) for glucose was calculated to assess glucose tolerance using the trapezoidal rule [21].

### 2.7. In Vitro Glucose Uptake by Isolated Rat Hemi-Diaphragm

Glucose uptake by isolated hemi-diaphragm was determined according to the methods described by Thabet et al., 2008, with some modifications [22]. The glucose uptake was divided into 2 experimental conditions, including without and with insulin (0.25 IU/mL) to determine the basal and insulin-stimulated glucose uptake, respectively. After overnight fasting, the rats were sacrificed with intraperitoneal injection of overdose Nembutal®. The diaphragm of the rat was rapidly removed with minimal trauma, divided into two halves and rinsed in cold balanced salt solution (BSS) to remove any blood clot. Each hemi-diaphragm was placed in a conical flask containing 3 mL of BSS and incubated with carbogen (95% $O_2$/5% $CO_2$) with shaking at 100 cycles/min for 90 min at 37 °C. At the end of the incubation period, the isolated hemi-diaphragm was removed, blotted with filter paper and weighed. An aliquot of the incubation medium was used for measurement of glucose concentration. Glucose uptake per gram of tissue was calculated as the difference between the initial and final glucose content in the incubated medium.

### 2.8. *In Vivo Intestinal Permeability Assay*

Gut permeability was assessed at the end of the experiment. This assay is an indirect measure of total intestinal permeability. The principle of this assessment is based on the intestinal leakage of 4000 Da Fluorescein isothiocyanate–dextran (FITC–dextran) into blood circulation. Briefly, rats were fasted overnight, and blood samples were collected as the negative control of the experiment to determine the background of rat plasma. Then, FITC–dextran (600 mg/kg) (Sigma-Aldrich, St. Louis, MO, USA) was administered to the rats by oral gavage, and blood samples were collected at 2.5 and 5 h later. The blood sample was immediately centrifuged at 6000 rpm for plasma separation, and the plasma was diluted with an equal volume of phosphate buffered saline (PBS) (pH 7.4). The plasma concentration of the FITC–dextran was determined using a Synergy™ H4 fluorescene microplate reader (BIOTEK® Instruments, Inc., Vermount, VT, USA) with an excitation wavelength of 485 nm and an emission wavelength of 535 nm compared with the standard curve of serially diluted FITC–dextran [23,24].

### 2.9. *Determination of Plasma Lipopolysaccharide (LPS)*

The plasma LPS level was determined using QCL-1000TM Endpoint Chromogenic Limulus Amebocyte Lysate (LAL) Assay Kit (Lonza, Verviers, Belgium) following the instructions of the manufacturer. Briefly, plasma was mixed with LAL reagent and incubated at 37 °C in a heating block for 10 min, followed by the addition of substrate solution and final incubation at 37 °C for 6 min. After that, the stop reagent was added. The presence of LPS in the plasma was inferred by the development of yellow color. The absorbance of the sample was quantified using spectrophotometry at 405–410 nm [25].

### 2.10. *Determination of Triglyceride Accumulation in Liver and Skeletal Muscle*

The liver and gastrocnemius muscle TG contents were measured according to the method of Frayn and Maycock, 1980, with slight modifications [26]. Briefly, a 0.05–0.2 g portion of the liver and muscle was minced and put into a glass tube containing 3 mL of chloroform-isopropanol 2:3 (*v*/*v*). The homogenate was pipetted into a glass tube and evaporated to dryness at 40 °C for 16 h. The dried residue was dissolved and mixed in 10% bovine serum albumin (BSA). The triglyceride contents were measured using a commercial colorimetric kit (Biotech, Bangkok, Thailand).

### 2.11. *DNA Extraction from Fecal Samples*

Bacterial DNA was collected from fecal samples (60–70 g) using NucleoSpin® DNA stool kit (Macherey-Nagel, Dueren, Germany). All procedures were performed according to the manufacturer's instructions. Qualitative analysis of bacterial DNA was evaluated by SPECTROstar Nano Absorbance microplate reader (BMG Labtech, Ortenberg, Germany). The ratio of the absorbance at 260 and 280 nm (A260/280) was used to identify the purity of nucleic acid specimen. An A260/280 value greater than 1.8 indicated a pure DNA sample. Bacterial DNA contents were evaluated using the relationship that 50 µg/mL of pure DNA sample represented an A260 of 1.

### 2.12. q-PCR Assay Conditions and Cycle Threshold

Quantitative PCR (qPCR) analyses were carried out in 96-well optical plates on the Quantstudio TM6 Flex Real Time PCR System (Applied Biosciences, Warrington, U.K.). The amplification reaction was performed in a total of 20 μL containing 10 μL of SYBR™ master mix, 2 μL of fecal bacterial DNA sample, 1 μL of reverse primer, 1 μL of forward primer and 6 μL of deionized water. The group-specific primers of bacterial targets based on 16S rDNA sequences are listed in Table S2. qPCR was conducted as follows: Uracil-DNA Glycosylase (UDG) activation step at 50 °C for 2 min followed by initial denaturation at 95 °C for 2 min and 40 cycles of denaturation step at 95 °C for 20 s and the annealing/extension step at 60 °C for 20 s. Melt curve analysis was then performed after each run to check the non-specific amplification of the primers. The cycle threshold (Ct) of bacterial DNA was calculated by absolute quantification strategy using the standard curve of the target bacterial strain. The result was expressed as log CFU/mL.

### 2.13. Measurement of Organic Acid Contents in Cecal Samples

The amounts of organic acids (acetic, propionic, butyric and lactic acids) in cecal content and fecal samples were measured by high-performance liquid chromatography (HPLC), as described previously [27]. Briefly, the sample was homogenized in 0.15 mM sulfuric acid and centrifuged at 10,000× *g* at 4 °C for 10 min. The supernatant was collected and filtered through 0.22 μm nylon syringe filter. The samples were analyzed by a Shimadzu HPLC system using Shodex SUGAR SH1011 (SHOWA DENKO K.K., Tokyo, Japan). The detection was carried out using a UV detector at 210 nm, and the column temperature was maintained at 75 °C. The samples were isocratically eluted with 5 mM sulfuric acid at 0.6 mL/min. The concentration of organic acids was quantified by comparison with the standard curve, and the results were expressed as μmol/g sample.

### 2.14. Western Blot Analysis

The soleus muscle of the rat was obtained after sacrifice. The homogenates were centrifuged at 4 °C for 10 min at 10,000× *g*, and the supernatants were used for Western blot. Total protein concentration was determined using a Bradford protein assay kit (Bio-Rad Laboratories, Hercules, CA, USA). Then, 30–50 ug of proteins was loaded in 10% sodium dodecyl sulphate–polyacrylamide gel electrophoresis (SDS-PAGE) for protein separation. The proteins were transferred to nitrocellulose membrane and blocked with blocking buffer for 1 h at room temperature with gentle shaking, followed by incubation overnight at 4 °C with specific primary antibody, Akt (Millipore Corporation, Burlington, MA, USA), phosphorylation of $pAkt^{Ser473}$ (Millipore Corporation, Burlington, MA, USA), AMP-activated protein kinase-α (AMPK-α) (Millipore Corporation, Burlington, MA, USA), $pAMPK\alpha^{Thr172}$ (Millipore Corporation, Burlington, MA, USA), GLUT4 (Chemicon International, Temecula, USA), TNF-α (Millipore Corporation, Burlington, MA, USA) and NF-kB (Santa Cruz biotechnology, Dallas, TX, USA). After incubation with the primary antibody, the membrane was washed and incubated with horseradish peroxidase-conjugated secondary antibody for 1 h at room temperature and rewashed again. The protein bands in the membranes were identified by enhanced chemiluminescence (ECL) detection reagent (GE Healthcare, Piscataway, NJ, USA). The concentration of protein was expressed by comparison with the mean value in the NDC group, which was arbitrarily set as 100.

### 2.15. Statistical Analysis

The results are presented as the mean value ± standard error of the mean (SEM). To detect the effects of treatment on the blood and fecal parameters among the six experimental groups, one-way analysis of variance (ANOVA) followed by Least-Significant Different (LSD) post-hoc analysis was used to determine significant differences between groups. The SPSS Advanced Statistics software (version17 SPSS Inc., Chicago, IL, USA) was used for statistical analysis. In all cases, a *p*-value less than 0.05 was used and considered to be statistically significant.

## 3. Results

### 3.1. *Effects of* L. paracasei *HII01 on Body Weight (BW), Visceral Fat (VF) Weight and Visceral Fat/100 g BW*

The BW, VF weight and VF/100 g BW of all experimental groups are represented in Table 1. The initial body weight was very similar in all experimental groups (394.5 ± 7.47 g, 389.00 ± 7.02 g, 389.5 ± 5.98 g, 386.5 ± 6.41 g, 383.75 ± 7.78 g and 393.50 ± 8.56 g, respectively). Following 12 weeks of oral administration of L. paracasei HII01, the BW, VF weight and VF/100 g BW did not differ among the normal rats. However, the DMC group had a significant increase in BW, VF weight and VF/100 g BW compared with the NDC group ($p < 0.05$), which indicates visceral obesity. Interestingly, the BW, VF weight and VF/100 g BW of the DM-L group were significantly decreased compared with the DMC group ($p < 0.05$). Likewise, the DMM and DMM-L groups had significantly lower values of the mentioned variables compared with the DMC group ($p < 0.05$). The above findings were observed in the absence of significant alterations in the food intake among the diabetic groups (93.86 ± 4.88 g/day, 90.00 ± 4.59 g/day, 89.43 ± 3.74 g/day and 97.00 ± 3.95 g/day, respectively).

**Table 1.** Effects of *L. paracasei* HII01 on BW, VF weight and VF/100 g BW in experimental groups.

| Parameters | NDC | ND-L | DMC | DM-L | DMM | DMM-L |
|---|---|---|---|---|---|---|
| BW (g) | 553.33 ± 18.60 $^{b}$ | 568.33 ± 26.13 $^{b}$ | 682.50 ± 35.25 $^{a}$ | 575.00 ± 13.78 $^{b}$ | 586.25 ± 11.43 $^{b}$ | 589.00 ± 4.00 $^{b}$ |
| VF (g) | 45.17 ± 2.36 $^{b}$ | 41.00 ± 3.20 $^{b}$ | 86.33 ± 7.84 $^{a}$ | 62.8 ± 2.99 $^{c}$ | 61.75 ± 2.46 $^{c}$ | 60.40 ± 5.62 $^{c}$ |
| VF/100g BW | 8.13 ± 0.22 $^{b}$ | 7.23 ± 0.53 $^{b}$ | 12.62 ± 0.51 $^{a}$ | 10.89 ± 0.27 $^{c}$ | 10.54 ± 0.41 $^{c}$ | 10.27 ± 1.00 $^{c}$ |

NDC, normal control rats; ND-L, normal control rats supplemented with *L. paracasei* HII01 ($10^8$ CFU/day); DMC, diabetic rats control; DM-L, diabetic rats supplemented with *L. paracasei* HII01 ($10^8$ CFU/day); DMM, diabetic rats treated with metformin 30 mg/kg; DMM-L, diabetic rats supplemented with combination of *L. paracasei* HII01 ($10^8$ CFU/day) and metformin 30 mg/kg; BW, body weight; VF, visceral fat. All data are expressed as mean ± SEM. Different lowercase letters indicate significant differences among different groups ($p < 0.05$).

### 3.2. *Effects of* L. paracasei *HII01 on Glycemic Control and Plasma Adipokine Hormones*

To explore the anti-hyperglycemic effect of *L. paracasei* HII01, the plasma biochemical parameters involved in glycemic control were measured at the end of the study. As shown in Figure 1, *L. paracasei* HII01 administration did not alter the fasting plasma glucose, insulin, leptin and adiponectin levels among the normal rats. Similarly, the oral administration of *L. paracasei* HII01 did not affect the HOMA-IR, a method used to quantify insulin resistance, in the normal rats ($p > 0.05$) (Figure 1C). These results established that the administration of *L. paracasei* HII01 in normal rats had no effect on glycemic parameters. The diabetic rats showed higher fasting plasma glucose and insulin levels as well as HOMA-IR compared with normal rats ($p < 0.05$). Remarkably, the administration of *L. paracasei* HII01, metformin alone or in combination with *L. paracasei* HII01 significantly ameliorated the fasting plasma glucose (−42.87%, −49.13% and −49.29%, respectively, $p < 0.05$) and insulin levels compared with the DMC group (−28.80%, −27.46% and −41.44%, respectively, $p < 0.05$). In accordance with these results, the HOMA-IR of the DM-L, DMM and DMM-L groups were significantly reduced compared with the DMC group (−59.38%, −62.54% and −69.29%, respectively, $p < 0.05$).

Leptin and adiponectin are two adipokine hormones that play a crucial role in metabolic regulation and are involved in insulin sensitivity. Therefore, we assessed the plasma leptin and adiponectin levels. As illustrated in Figure 1D,E, there were no significant differences in the plasma leptin and adiponectin levels between the two normal experimental groups, while the DMC group showed significantly increased plasma leptin level compared with the normal rats, indicating that leptin resistance was developed in diabetes (150.46%, $p < 0.05$). The plasma leptin level significantly dropped in the DM-L, DMM and DMM-L groups compared with the DMC group (−41.58%, −39.33% and −36.24%, respectively, $p < 0.05$). However, the plasma adiponectin level in the DMC group significantly decreased compared with the NDC group (−21.35%, $p < 0.05$). The administration of *L. paracasei* HII01, metformin, as well as the combination of *L. paracasei* HII01 and metformin, significantly increased the plasma adiponectin level (25.37%, 26.31% and 25.85%, respectively, $p < 0.05$).

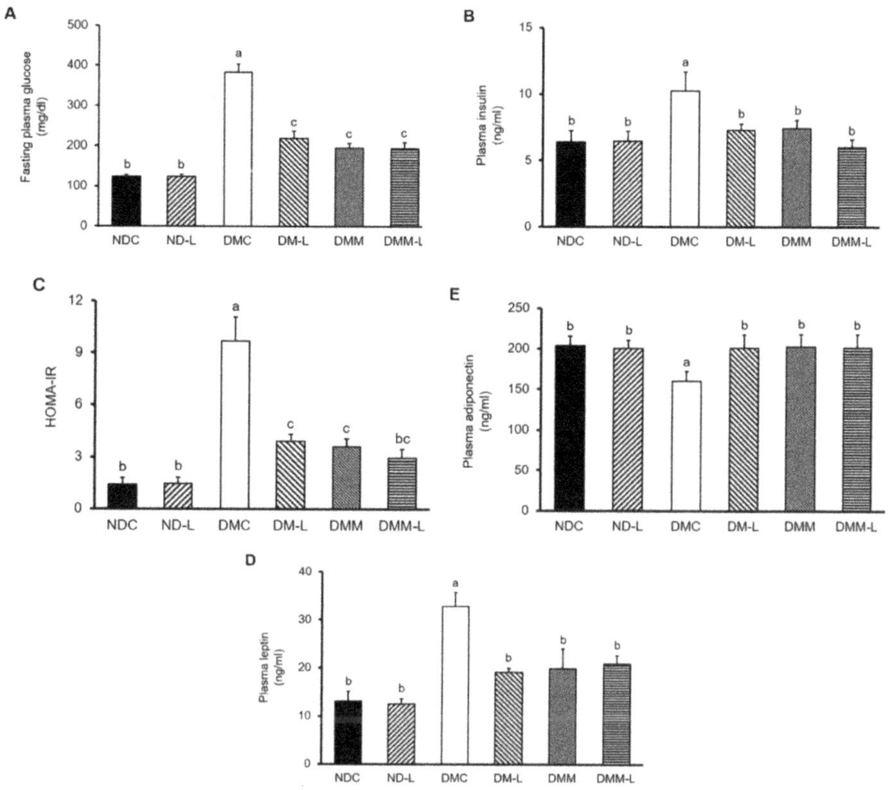


**Figure 1.** Effects of *L. paracasei* HII01 on the fasting plasma levels of (**A**) glucose, (**B**) insulin, (**C**) HOMA-IR, (**D**) leptin and (**E**) adiponectin in experimental rats. NDC, normal control rats; ND-L, normal control rats supplemented with *L. paracasei* HII01 ($10^8$ CFU/day); DMC, diabetic rats control; DM-L, diabetic rats supplemented with *L. paracasei* HII01 ($10^8$ CFU/day); DMM, diabetic rats treated with metformin 30 mg/kg; DMM-L, diabetic rats supplemented with combination of *L. paracasei* HII01 ($10^8$ CFU/day) and metformin 30 mg/kg; HOMA-IR index, homeostasis model assessment of insulin resistance. All data are expressed as mean ± SEM. Different lowercase letters indicate significant differences among different groups ($p < 0.05$).

### 3.3. Effects of *L. paracasei HII01* on the Glucose Tolerance Test

To determine whether the administration of *L. paracasei* HII01 could affect the whole-body insulin sensitivity in type 2 diabetic rats, the OGTT was conducted on the rats after 11 weeks of intervention. As shown in Figure 2A,B, there were no significant differences in the plasma glucose levels at all time points and the AUC for glucose between the NDC and ND-L groups. As expected, the plasma glucose levels after glucose loading revealed significantly higher values in the DMC group at all time points compared with the NDC group (Figure 2A, $p < 0.05$). Compared with the NDC group, the incremental area under the curve (IAUC) was markedly increased in the DMC group (Figure 2B, $p < 0.05$). These findings proved that impaired glucose tolerance was established in T2DM rats. Notably, the glucose levels at all time points in rats supplemented with *L. paracasei* HII01, metformin alone or in combination with *L. paracasei* HII01 were significantly reduced in comparison with the DMC group ($p < 0.05$). There was significant reduction in the total area under the curve (TAUC) and IAUC values in the DM-L, DMM and DMM-L groups compared with the DMC group ($p < 0.05$).

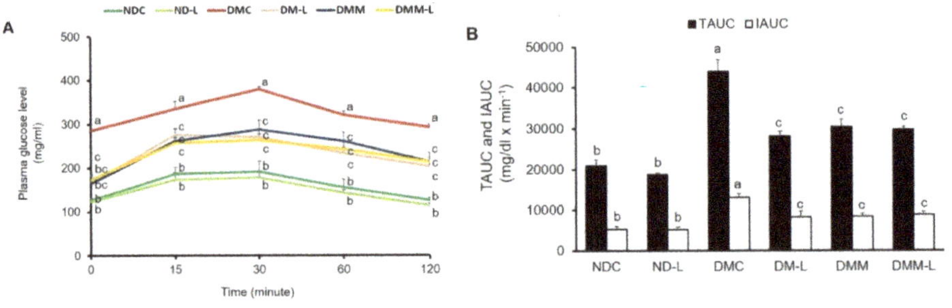


**Figure 2.** Effects of *L. paracasei* HII01 on the OGTT in experimental group. (**A**) Glucose response; (**B**) area under the curve for glucose. NDC, normal control rats; ND-L, normal control rats supplemented with *L. paracasei* HII01 ($10^8$ CFU/day); DMC, diabetic rats control; DM-L, diabetic rats supplemented with *L. paracasei* HII01 ($10^8$ CFU/day); DMM, diabetic rats treated with metformin 30 mg/kg; DMM-L, diabetic rats supplemented with combination of *L. paracasei* HII01 ($10^8$ CFU/day) and metformin 30 mg/kg; OGTT, oral glucose tolerance test; TAUC, total area under the curve; IAUC, incremental area under the curve. All data are expressed as mean ± SEM. Different lowercase letters indicate significant differences among different groups ($p < 0.05$).

### 3.4. Effects of L. paracasei HII01 on Lipid Parameters

In the present study, we also examined the hypolipidemic effect of probiotic *L. paracasei* HII01 on type 2 diabetic rats. As revealed in Table 2, the levels of plasma TG, total cholesterol and LDL in the DMC group were significantly increased compared with the normal control rats at the end of the study ($p < 0.05$). Oral administration of *L. paracasei* HII01 or in combination with metformin significantly restored the plasma TG, total cholesterol and LDL levels compared with the DMC group ($p < 0.05$), while changes in the plasma TG, cholesterol and LDL levels were not observed in normal rats treated with *L. paracasei* HII01. However, no significant change in the plasma HDL level was displayed in the DMC group compared with the NDC group. All intervention groups had significantly increased plasma HDL levels compared with the NDC group ($p < 0.05$).

**Table 2.** Effects of *L. paracasei* HII01 on lipid parameters in experimental groups.

| Parameters (mg/dL) | NDC | ND-L | DMC | DM-L | DMM | DMM-L |
|---|---|---|---|---|---|---|
| Triglyceride | 38.35 ± 1.26 $^{b}$ | 31.82 ± 1.67 $^{b}$ | 83.57 ± 7.18 $^{a}$ | 35.76 ± 2.78 $^{b}$ | 33.88 ± 1.02 $^{b}$ | 38.15 ± 3.55 $^{b}$ |
| Cholesterol | 44.35 ± 1.04 $^{b}$ | 40.53 ± 2.48 $^{b}$ | 72.92 ± 6.02 $^{a}$ | 42.85 ± 1.81$^{b}$ | 35.13 ± 2.63 $^{c}$ | 33.85 ± 3.21$^{c}$ |
| HDL | 59.75 ± 1.55 $^{b}$ | 64.00 ± 3.67 $^{ab}$ | 69.00 ± 5.98 $^{ab}$ | 72.33 ± 1.11$^{a}$ | 71.00 ± 2.64 $^{a}$ | 72.33 ± 0.76 $^{a}$ |
| LDL | 10.00 ± 1.22 $^{b}$ | 10.08 ± 1.32 $^{b}$ | 23.75 ± 3.50 $^{a}$ | 16.67 ± 3.28 $^{c}$ | 14.00 ± 1.30 $^{b}$ | 13.00 ± 1.14 $^{b}$ |

NDC, normal control rats; ND-L, normal control rats supplemented with *L. paracasei* HII01 ($10^8$ CFU/day); DMC, diabetic rats control; DM-L, diabetic rats supplemented with *L. paracasei* HII01 ($10^8$ CFU/day); DMM, diabetic rats treated with metformin 30 mg/kg; DMM-L, diabetic rats supplemented with combination of *L. paracasei* HII01 ($10^8$ CFU/day) and metformin 30 mg/kg; HDL, high-density lipoprotein; LDL, low-density lipoprotein. All data are expressed as mean ± SEM. Different lowercase letters indicate significant differences among different groups ($p < 0.05$).

### 3.5. Effects of L. paracasei HII01 on Tissue Triglyceride Accumulation

Next, we evaluated the effects of *L. paracasei* HII01 on TG accumulation in both the skeletal muscle and liver because the accumulation of lipid within target tissues of insulin is closely associated with insulin resistance and abnormal lipid metabolism. The muscle TG accumulation of the DMC group significantly increased compared with the NDC group ($p < 0.05$), as shown in Supplementary Table S3. A significant reduction in muscle TG accumulation was found in diabetic rats administered *L. paracasei* HII01, metformin alone or in combination with *L. paracasei* HII01 compared with the DMC group (−36.40%, −38.31% and −46.61%, respectively, $p < 0.05$). For the liver, the DMC group also demonstrated a significant increase in hepatic TG accumulation compared with the NDC group

($p < 0.05$). Interestingly, the administration of *L. paracasei* HII01 significantly reduced the hepatic TG accumulation compared with the DMC group (−21.65%, $p < 0.05$). The hepatic TG accumulation of the DMM and DMM-L groups tended to decrease compared with the DMC group (−14.12% and −10.95%, respectively, $p > 0.05$).

### 3.6. Effects of L. paracasei HII01 on In Vitro Skeletal Muscle Glucose Uptake

To examine whether *L. paracasei* HII01 had any effects on the skeletal muscle glucose transport system, the basal and insulin-stimulated glucose uptakes in the isolated hemi-diaphragm were determined (Figure S1). Our results found that the rate of basal glucose uptake and insulin-stimulated glucose uptake by the hemi-diaphragm in the DMC group significantly decreased compared with the NDC group ($p < 0.05$). Likewise, the insulin-stimulated glucose uptake and the delta glucose uptake, which was calculated as insulin-treated minus basal glucose uptake for paired muscles, in the DMC group were significantly reduced compared with the NDC group ($p < 0.05$). These findings implied an impairment of insulin action in skeletal muscle. In contrast, the administration of *L. paracasei* HII01 for 12 weeks significantly enhanced the rates of insulin-stimulated glucose uptake and delta glucose uptake compared with the DMC group (28.94%, $p < 0.05$). Similarly, significant increases in the rates of insulin-stimulated glucose uptake and delta glucose uptake were precisely noted in the DMM and DMM-L groups compared with the DMC group (50.58% and 43.03%, respectively, $p < 0.05$) (Figure S1).

### 3.7. Effects of L. paracasei HII01 on Protein Expressions of GLUT4, pAkt$^{Ser473}$, pAMPK$^{Thr172}$, NF-kB and TNF-$\alpha$ in Soleus Muscle

To elucidate the underlying mechanisms sustaining the possible beneficial effects of *L. paracasei* HII01 regarding improvement of insulin-stimulated glucose uptake, the expressions of key proteins involved in insulin-stimulated glucose uptake, such as GLUT4 protein expression and Akt$^{Ser473}$ phosphorylation in soleus muscle, were investigated. In normal rats, supplementation of *L. paracasei* HII01 for 12 weeks did not alter the GLUT4 protein expression compared with the NDC group (Figure 3A). As expected, the expression of GLUT4 protein of the DMC group significantly decreased compared with the NDC group ($p < 0.01$). The protein expressions of GLUT4 were markedly restored in the DM-L, DMM and DMM-L groups compared with the DMC group ($p < 0.05$). As illustrated in Figure 3B, there were no significant differences in pAkt$^{Ser473}$/Akt protein ratio between the NDC and ND-L groups. A reduction in the pAkt$^{Ser473}$/Akt protein ratio was found in the DMC group compared with the NDC group ($p < 0.05$). The administration of *L. paracasei* HII01, metformin alone or in combination with *L. paracasei* HII01 effectively reversed the activation of Akt compared with the DMC group ($p < 0.05$).

We also evaluated the effects of *L. paracasei* HII01 on AMPK activation. In addition to the insulin signaling proteins, the phosphorylation of AMPK can stimulate GLUT4 translocation for glucose uptake in the skeletal muscle via the insulin-independent pathway. As shown in Figure 4, the oral administration of *L. paracasei* HII01 had no effect on the pAMPK$^{Thr172}$/AMPK protein ratio in normal rats. The DMC group showed a significant decrease in the pAMPK$^{Thr172}$/AMPK protein ratio compared with the NDC group ($p < 0.05$). Interestingly, the administration of probiotic *L. paracasei* HII01, metformin alone or in combination with *L. paracasei* HII01 efficiently recovered the pAMPK $^{Thr172}$/AMPK protein ratio compared with the DMC group ($p < 0.05$).

It is well established that in addition to lipid accumulation-induced insulin resistance, chronic inflammation can also induce insulin resistance. Thus, we evaluated the effects of *L. paracasei* HII01 ($10^8$ CFU/day) on inflammatory cytokines, NF-kB and TNF-$\alpha$. As shown in Figure 5A, the NF-kB expression in the DMC group was significantly higher than in the NDC group ($p < 0.05$). However, the oral administration of probiotic *L. paracasei* HII01 successfully reversed that result ($p < 0.05$). However, *L. paracasei* HII01 administration in normal rats did not alter the NF-kB protein expression. The expression of TNF-$\alpha$ is shown in Figure 5B. The administration of probiotic *L. paracasei* HII01 did not affect the TNF-$\alpha$ protein expressions in the normal rats ($p > 0.05$). The protein expression of TNF-$\alpha$

was significantly higher in the DMC group than in the NDC group ($p < 0.05$). Compared with the DMC group, the protein expressions of TNF-α were significantly reduced in the DMM, DM-L and DMM-L groups ($p < 0.05$).

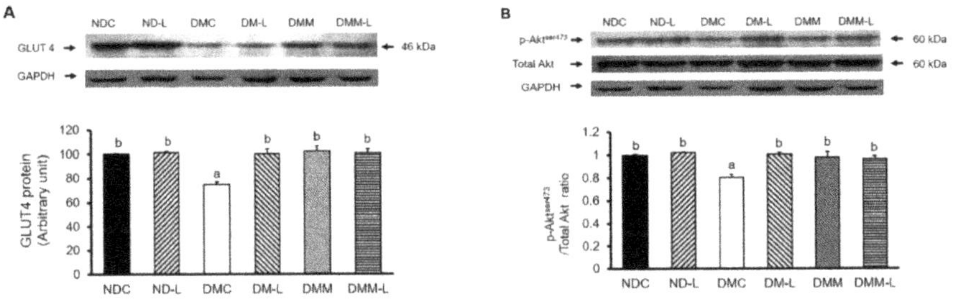

**Figure 3.** Effects of *L. paracasei* HII01 on Western blotting of insulin-stimulated glucose uptake marker proteins (**A**) GLUT4 protein (**B**) pAkt$^{ser473}$/Total Akt ratio in experimental groups. NDC, normal control rats; ND-L, normal control rats supplemented with *L. paracasei* HII01 ($10^8$ CFU/day); DMC, diabetic rats control; DM-L, diabetic rats supplemented with *L. paracasei* HII01 ($10^8$ CFU/day); DMM, diabetic rats treated with metformin 30 mg/kg; DMM-L, diabetic rats supplemented with combination of *L. paracasei* HII01 ($10^8$ CFU/day) and metformin 30 mg/kg. GLUT4, glucose transporter 4; pAkt$^{Ser473}$/total Akt ratio, phosphorylation of protein kinase B per total protein kinase B ratio. All data are expressed as mean ± SEM. Different lowercase letters indicate significant differences among different groups ($p < 0.05$).

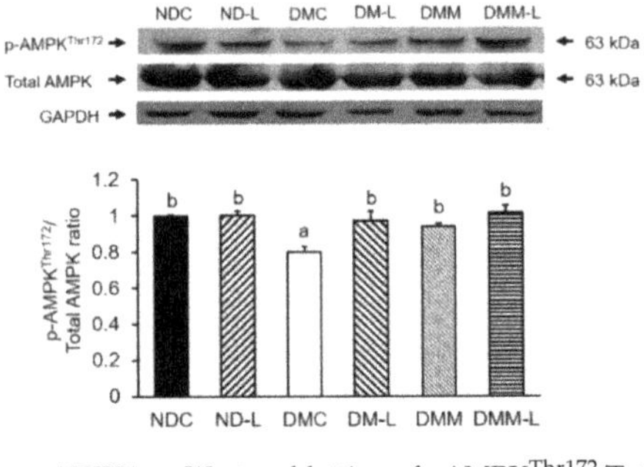

**Figure 4.** Effects of *L. paracasei* HII01 on Western blotting of pAMPK$^{Thr172}$/Total AMPK ratio in skeletal muscle of experimental groups. NDC, normal control rats; ND-L, normal control rats supplemented with *L. paracasei* HII01 ($10^8$ CFU/day); DMC, diabetic rats control; DM-L, diabetic rats supplemented with *L. paracasei* HII01 ($10^8$ CFU/day); DMM, diabetic rats treated with metformin 30 mg/kg; DMM-L, diabetic rats supplemented with combination of *L. paracasei* HII01 ($10^8$ CFU/day) and metformin 30 mg/kg. pAMPK$^{Thr172}$/ total AMPK ratio, phosphorylation of AMP-activated protein kinase per total AMPK ratio. All data are expressed as mean ± SEM. Different lowercase letters indicate significant differences among different groups ($p < 0.05$).

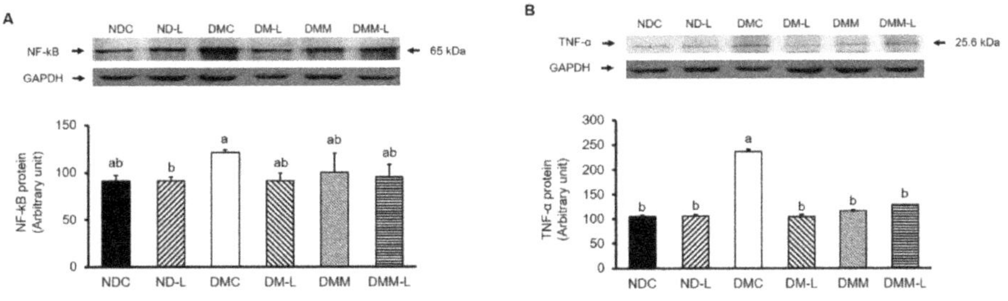


**Figure 5.** Effects of *L. paracasei* HII01 on Western blotting of inflammatory protein marker (**A**) NF-kB (**B**) TNF-α in skeletal muscle of experimental groups. NDC, normal control rats; ND-L, normal control rats supplemented with *L. paracasei* HII01 ($10^8$ CFU/day); DMC, diabetic rats control; DM-L, diabetic rats supplemented with *L. paracasei* HII01 ($10^8$ CFU/day); DMM, diabetic rats treated with metformin 30 mg/kg; DMM-L, diabetic rats supplemented with combination of *L. paracasei* HII01 ($10^8$ CFU/day) and metformin 30 mg/kg. NF-kB, nuclear factor-kappa B; TNF-α, tumor necrosis factor alpha. All data are expressed as mean ± SEM. Different lowercase letters indicate significant differences among different groups ($p < 0.05$).

### 3.8. Effects of L. paracasei HII01 on Plasma Endotoxemia

It is well accepted that endotoxemia, characterized by excess circulating bacterial wall LPS, is associated with systemic inflammation and T2DM. Consequently, we measured the plasma LPS levels. As shown in Figure 6, the DMC group had a significantly higher plasma LPS level than the NDC group ($p < 0.05$). Remarkably, the administration of metformin, *L. paracasei* HII01 alone or in combination with metformin significantly reduced the plasma LPS level compared with the DMC group ($p < 0.05$).

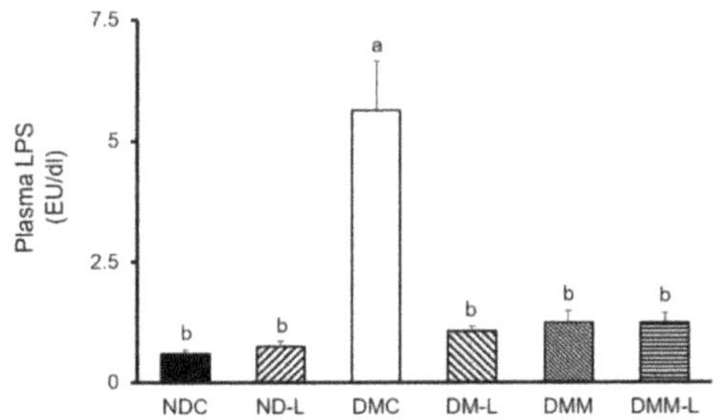


**Figure 6.** Effects of *L. paracasei* HII01 on the plasma LPS levels in experimental groups. NDC, normal control rats; ND-L, normal control rats supplemented with *L. paracasei* HII01 ($10^8$ CFU/day); DMC, diabetic rats control; DM-L, diabetic rats supplemented with *L. paracasei* HII01 ($10^8$ CFU/day); DMM, diabetic rats treated with metformin 30 mg/kg; DMM-L, diabetic rats supplemented with combination of *L. paracasei* HII01 ($10^8$ CFU/day) and metformin 30 mg/kg. LPS, lipopolysaccharide. All data are expressed as mean ± SEM. Different lowercase letters indicate significant differences among different groups ($p < 0.05$).

### 3.9. Effects of L. paracasei HII01 on Intestinal Permeability

Since the underling mechanisms behind the reinforcement of the increased plasma LPS level is expected to involve the gut permeability, we also examined the integrity of the intestinal membrane. This was carried out using an indirect method for the assessment of gut leakiness: measuring the level of DX-4000–FITC in plasma. The plasma levels of DX-4000-FITC of diabetic rats at 2.5 and

5 h were significantly higher than those of normal rats (Figure 7), indicating that an increase in intestinal permeability was found in diabetic rats. Interestingly, treatment with probiotic *L. paracasei* HII01, metformin alone or in combination with *L. paracasei* HII01 significantly reduced the plasma DX-4000-FITC level at 2.5 h compared with the DMC group ($p < 0.05$). In addition, in comparison with the DMC group, treatment with *L. paracasei* HII01 combined with metformin significantly decreased the plasma DX-4000-FITC level at 5 h ($p < 0.05$).

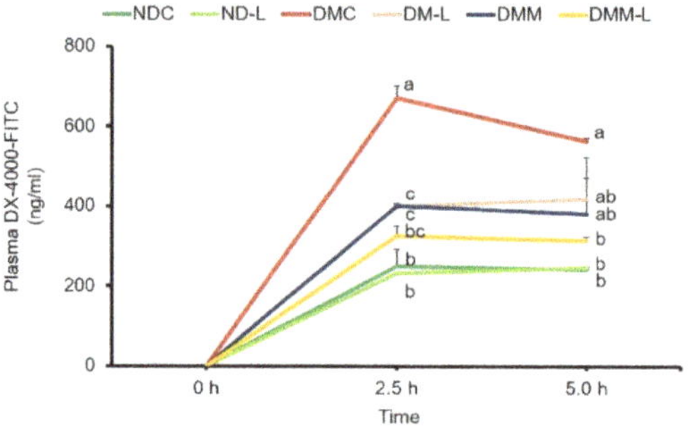


**Figure 7.** Effects of *L. paracasei* HII01 on gut permeability measured by plasma FITC-fluorescent dye levels in experimental groups. NDC, normal control rats; ND-L, normal control rats supplemented with *L. paracasei* HII01 ($10^8$ CFU/day); DMC, diabetic rats control; DM-L, diabetic rats supplemented with *L. paracasei* HII01 ($10^8$ CFU/day); DMM, diabetic rats treated with metformin 30 mg/kg; DMM-L, diabetic rats supplemented with combination of *L. paracasei* HII01 ($10^8$ CFU/day) and metformin 30 mg/kg; 4000-DX-FITC, 4000 Da Fluorescein isothiocyanate–dextran. All data are expressed as mean ± SEM. Different lowercase letters indicate significant differences among different groups ($p < 0.05$).

### *3.10. Effects of* L. paracasei *HII01 on Short-Chain Fatty Acids in Cecal Content*

SCFAs are carbon chain 1–6 organic fatty acids that are generated from the fermentation of undigested starch and fiber by lactic acid bacteria. The major SCFAs are lactic acid, propionic acid, butyric acid and acetic acid. Thus, we evaluated the effects of *L. paracasei* HII01 on the levels of SCFAs in cecal content. As shown in Supplementary Figure S2, there was no significant difference in lactic acid level in normal rats administered *L. paracasei* HII01 for 12 weeks compared with normal control rats. However, the lactic acid level in the DMC group was significantly reduced compared with the NDC group ($p < 0.05$). Interestingly, treatment with *L. paracasei* HII01, metformin alone or in combination with *L. paracasei* HII01 significantly increased the level of lactic acid compared with the DMC group ($p < 0.05$). Furthermore, the DMM group had significantly increased lactic acid level compared with the DM-L group ($p < 0.05$). The level of propionic acid in normal rats administered *L. paracasei* HII01 was similar to that of normal control rats ($p > 0.05$). The DMC group had a significantly reduced level of propionic acid compared with the NCD group ($p < 0.05$). Meanwhile, the propionic acid levels in all treatment groups were significantly increased compared with the DMC group ($p < 0.05$). Moreover, the combined treatment of *L. paracasei* HII01 and metformin was significantly enhanced compared to the DM-L group ($p < 0.05$). The butyric acid level in the DMC group did not differ from that of the NDC group ($p < 0.05$). However, treatment with *L. paracasei* HII01, metformin alone or in combination with *L. paracasei* HII01 significantly increased the level of butyric acid compared with the DMC group ($p < 0.05$). The acetic acid levels in normal rats administered *L. paracasei* HII01 were significantly higher than those of the NDC rats ($p < 0.05$). In addition, the level of acetic acid in the DMC group did not significantly differ from that of the NDC group ($p < 0.05$). Nevertheless, the administration of probiotic *L. paracasei* HII01 to diabetic rats significantly enhanced the acetic acid level compared with the DMC group ($p < 0.05$).

### 3.11. *Effects of L. paracasei HII01 on the Bacterial DNA in Feces*

The relative abundance of Lactobacillus and Bifidobacterium spp. in the gut microbiota is shown in Figure S3 (as a Supplementary). In comparison with the NDC group, the number of fecal Lactobacillus spp. was significantly altered in normal rats that received *L. paracasei* HII01 ($p < 0.05$). On the other hand, the DMC group had a significantly decreased number of fecal Lactobacillus spp. compared with the NDC group ($p < 0.05$). Oral administration of *L. paracasei* HII01, metformin alone or in combination with *L. paracasei* HII01 to diabetic rats significantly increased the number of fecal Lactobacillus spp. compared with the DMC rats ($p < 0.05$). However, the DMC group had a higher number of fecal Bifidobacterium spp. than the NDC group ($p < 0.05$). The administration of metformin, *L. paracasei* HII01 alone or in combination with metformin to diabetic rats significantly increased the number of fecal Bifidobacterium spp. compared with the DMC group ($p < 0.05$). Additionally, the number of this beneficial bacteria significantly increased in the DMM and DMM-L groups in comparison with the DM-L group ($p < 0.01$). As illustrated in Figure S3, the result revealed that the numbers of fecal *E. coli* and *C. perfringens* in the ND-L group were significantly lessened compared with the NDC group ($p < 0.05$). The DMC group had a higher number of fecal *E. coli* than the NDC group ($p < 0.05$). Interestingly, all the treatment groups had significantly reduced *E. coli* numbers compared with the DMC group ($p < 0.05$). The number of *C. perfringens*, known as bad bacteria, in the DMC group did not differ from that of the NDC group. However oral administration of metformin, *L. paracasei* HII01 alone or in combination with metformin to diabetic rats significantly reduced the number of Fecal *C. perfringens* compared with the DMC group ($p < 0.05$). Interestingly, the combination of probiotic *L. paracasei* HII01 and metformin significantly decreased the number of *C. perfringens* compared with the diabetic rats treated with probiotic alone ($p < 0.05$).

## 4. Discussion

The present study was undertaken to investigate the possible beneficial effect of *L. paracasei* HII01 on glycemia in type 2 diabetic rat model. Our results demonstrated that the administration of *L. paracasei* HII01 at a dose of $10^8$ CFU/day for 12 weeks effectively resulted in the following: (1) reduction in fasting plasma glucose, insulin, leptin and lipids levels as well as improvement in glucose intolerance; (2) improvement of PI3K/Akt signaling and AMPK activation, which are involved in enhancing the rate of insulin-stimulated glucose uptake of the isolated hemi-diaphragm; (3) modulation of gut microbiota and subsequent amelioration of plasma endotoxemia.

The type 2 diabetic rat model used in this study presented the general characteristics of T2DM, including obesity, hyperglycemia, insulin resistance, impaired glucose tolerance and dyslipidemia, similar to T2DM patients [28]. In addition, the plasma LPS level was significantly increased, which, at least in part, is linked to insulin resistance in untreated diabetic rats. Similar to humans, our findings also found that an abundance of pathogenic bacteria, *E. coli* and *C. perfringens* in diabetic rats [29,30]. At the end of the study, we found that *L. paracasei* HII01 administration significantly improved not only BW, VF/BW, and plasma lipid levels (TG, cholesterol, LDL, and HDL) but also reduced the fasting blood glucose (FBG) levels and improved glucose tolerance, demonstrating its antidiabetic effect. Nevertheless, the beneficial effect of *L. paracasei* HII01 on glycemic control in the present study is not linked to its insulinotropic action. The anti-hyperglycemic effect of *L. paracasei* HII01 might be explained by other mechanisms, such as enhanced insulin sensitivity or relevant glucose uptake, the same as the results found in the metformin treatment group. This assertion is supported by the HOMA-IR index and the outcomes of OGTT. Additionally, the results of this study revealed that gut dysbiosis was attenuated after 12 weeks of oral administration of *L. paracasei* HII01 in diabetic rats. It was interesting to note that the amelioration efficiency of *L. paracasei* HII01 on those blood metabolic parameters were close to the results of studies involving diabetic rats treated with metformin alone or combination with *L. paracasei* HII01. It was suggested that there were no synergistic or additive effects of metformin and probiotic *L. paracasei* HII01, particularly induction of hypoglycemia.

Alteration of adipokine is one of the possible mechanisms contributing to hyperglycemia and insulin resistance of diabetes. This study demonstrated that *L. paracasei* HII01 supplement effectively modulated adipokine imbalance in diabetic rats. Adiponectin acts as an insulin-sensitizer, which serves to enhance fatty acid oxidation, reduce tissue TG accumulation and, finally, improve insulin signaling [31]. While, leptin is a hormone that is important for glucose homeostasis by stimulating the PI3K signaling pathway [32]. The current study has demonstrated that the decreased leptin level is related to an improvement in insulin sensitivity and reduction in plasma glucose and lipid levels in type 2 diabetic rats treated with *L. casei* CCFM419 [33]. The reduction in visceral fat accumulation acknowledged in the present study might be involved with the suppression in plasma leptin and increasing in plasma adiponectin levels, which contribute to the improvement in insulin sensitivity, anti-hyperglycemia as well as anti-hyperlipidemia in diabetic rats treated with *L. paracasei* HII01. However, there are several mechanisms involved to the pathogenesis of insulin resistance. Among these, genes involved in adipose tissue metabolism can be considered possibly responsible for insulin sensitivity. Petrone A et al., 2007, reported that the promoter region of the adiponectin gene (+45T>G Adiponectin SNPs) could influence adiponectin levels and, consequently, insulin sensitivity in obesity and diabetes mellitus. [34].

Next, to explore more about the molecular mechanisms of *L. paracasei* HII01 in insulin sensitivity, we further evaluated the insulin signaling and glucose transport system in the skeletal muscle since the skeletal muscle is the major site of glucose uptake in the postprandial state in normal condition. Comparable with other studies, the results demonstrated that the insulin action is diminished and the ability of insulin to stimulate glucose uptake is blunted in type 2 diabetic condition [35]. Although the mechanisms of insulin resistance are not fully understood, lipid and inflammation-induced insulin resistance is one of the potential candidate mechanisms for insulin resistance [36]. Previous studies found that the TLR-4-LPS pathway can stimulate pro-inflammatory cytokines in the skeletal muscle, and pro-inflammatory cytokines, such as TNF-α, promote insulin resistance though NF-kB inflammatory signaling [37]. In addition, the rate of insulin-stimulated glucose uptake depends on the phosphorylation of Akt, the major insulin signaling protein, and the total or membrane GLUT4 protein expression [38]. Interestingly, we found that *L. paracasei* HII01 effectively increased $Akt^{Ser473}$ phosphorylation and GLUT4 protein expression as well as decreased the expression of TNF-α and NF-kB in the skeletal muscle. Similarly, the supplementation of a combination of the probiotics *L. rhamnosus*, *L. acidophilus* and *B. bifidium* enhanced $pAkt^{Ser473}$ in the muscle of diet-induced obese (DIO) mice [39]. Besides, in TNF-α treated L6 cells, probiotic *B. lactis* HY8101 treatment increased insulin-stimulated phosphorylation of $Akt^{Ser473}$ and GLUT4 protein [40]. Cumulatively, our data suggested that *L. paracasei* HII01 treatment improved the insulin signaling pathway, at least partly, through the reduction in systemic inflammation. Furthermore, the activation of AMPK, an energy-sensing enzyme, is one of the possible mechanisms linked to glucose uptake via directly activating GLUT4 translocation to membrane in skeletal muscles [41]. Metformin, a first-line drug for T2DM treatment, exerts its action mainly by activating AMPK. We also found that *L. paracasei* HII01 supplementation enhanced $AMPK^{Thr172}$ phosphorylation in the skeletal muscle of diabetic rats; this was also the case in the metformin treatment group.

Lactic, propionic, butyric and acetic acids are the most important SCFAs that affect glycemic control [42]. Importantly, the results demonstrated that the administration of *L. paracasei* HII01 for 12 weeks effectively enhanced the number of those SCFAs in cecal content. It is well known that the receptors of SCFAs are two G-protein coupled receptors (GPCRs), free fatty acid receptor 2 (FFAR2) and FFAR3, which are widely expressed in the skeletal muscle, intestinal, adipose, liver and pancreatic tissues [43]. In addition, these SCFAs are not only of importance in gut health and as signaling molecules but might also enter the systemic circulation and directly affect metabolism or the function of peripheral tissues via the AMPK activation [44]. Thus, SCFAs may partly influence glucose metabolism and insulin resistance in type 2 diabetic rats supplemented with *L. paracasei* HII01.

Recent studies reported that changes in gut microbiota composition are associated with an increase in gut LPS. The TLR4–LPS complex triggers the pro-inflammatory cytokines, which cause intestinal inflammation and decrease in tight junction proteins [45]. The loss of tight junction proteins is linked to increased gut permeability and the subsequent leakage of LPS to systemic circulation [46]. Interestingly, these abnormalities were attenuated after our probiotic *L. paracasei* HII01 administration for 12 weeks. However, the mechanisms were not investigated in this study. Lim et al., 2016, found that the supplementation of *L. sakei* OK67 suppressed TLR-4 expression and the NF-kB pathway, which are involved in the reduction in the level of intestinal TNF-$\alpha$ and interlukin-6 (IL-6) and, subsequently, the increase in tight junction protein, including zonula occludens (ZO-1), occludin and claudin expression in the colon of obese mice [15]. We further hypothesize that the gut microbiota modulation and anti-inflammatory effects of *L. paracasei* HII01 could be another probable underlying mechanism for the improvement in gut permeability and, subsequently, endotoxemia.

There are limitations to this study that have to be considered. Firstly, we showed the protective effect of *L. paracasei* HII01 administration on gut barrier integrity by measuring the concentration of LPS in the circulation and plasma levels of DX-4000-FITC, which is an indirect method for the assessment of gut leakiness. A direct assessment of intestinal tight junction permeability or levels of intestinal tight junction markers, such as occludin or ZO-1, would better confirm the role of our probiotic in preserving the intestinal epithelial barrier. Lastly, we used an experimental diabetic rat model to test our hypothesis and the results cannot be directly extrapolated to humans due to differences in gut microbiota and physiology.

## 5. Conclusions

This study provided the first evidence that *L. paracasei* HII01 administration ameliorates hyperglycemia and enhances insulin stimulated glucose uptake in HFD–STZ induced type 2 diabetic rats. These effects are associated with modulation of gut microbiota along with gut permeability, leading to improved systemic endotoxemia and inflammation-disturbed insulin sensitivity in the skeletal muscle through PI3K/Akt signaling and AMPK activation as summarized in Figure 8. Thus, *L. paracasei* HII01 has the potential for development as a complementary supplement strategy for type 2 diabetic patients.

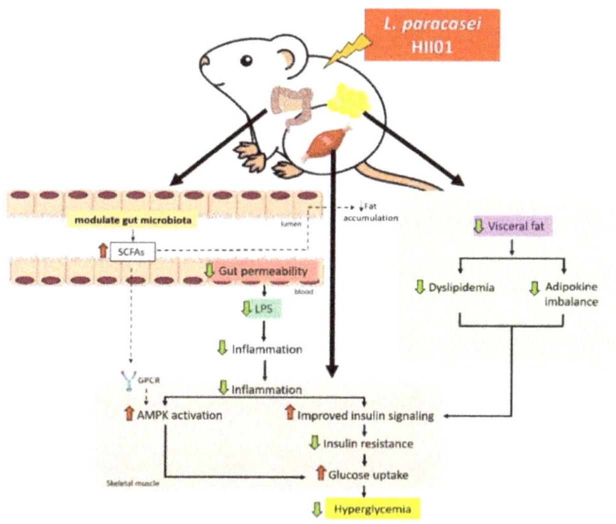


**Figure 8.** Possible mechanism of *L. paracasei* HII01 in type 2 diabetic rats.

**Supplementary Materials:** The following are available online at http://www.mdpi.com/2072-6643/12/10/3015/s1, Table S1: The composition of normal and high-fat diet., Table S2: Group-specific primers of bacterial targets based on 16 S rDNA sequences., Table S3: Effects of *L. paracasei* HII01 on tissue TG accumulation in experimental groups., Figure S1: Effects of *L. paracasei* HII01 on in vitro glucose uptake in experimental groups., Figure S2: Effects of *L. paracasei* HII01 on cecal SCFAs of the experimental groups., Figure S3: Effects of *L. paracasei* HII01 on the mean percent change in bacterial abundance from baseline analyzed in feces of the experimental groups.

**Author Contributions:** Conceptualization, C.C. and N.L.; funding acquisition, C.C. and N.L.; writing—original draft, P.T.; investigation, P.T., N.K.-U., J.I. and S.S.; methodology, P.T., N.K.-U., J.I. and S.S.; formal analysis, P.T., N.K.-U., N.L. and S.S.; writing—review and editing, C.C. and N.L.; project administration, N.L. All authors have read and agreed to the published version of the manuscript.

**Funding:** This research was funded by the Research and Researchers for Industries (RRi) grant to P.T and N.L. (PHD60I0085) under the Thailand Science Research and Innovation (TSRI), the National Research Council of Thailand (NRCT) to NL (Grant No. 270390) and Chiang Mai University for partially support.

**Acknowledgments:** We gratefully acknowledge the Research and Researchers for Industries (RRi) grant under the Thailand Science Research and Innovation (TSRI), the National Research Council of Thailand (NRCT) and Chiang Mai University.

**Conflicts of Interest:** The authors declare no conflict of interest.

## References

1. American Diabetes Association. Classification and Diagnosis of Diabetes: Standards of Medical Care in Diabetes-2018. *Diabetes Care* **2018**, *41*, 13–27. [CrossRef] [PubMed]
2. Deshpande, A.D.; Harris-Hayes, M.; Schootman, M. Epidemiology of diabetes and diabetes-related complications. *Phys. Ther.* **2008**, *88*, 1254–1264. [CrossRef] [PubMed]
3. Backhed, F.; Ding, H.; Wang, T.; Hooper, L.V.; Koh, G.Y.; Nagy, A. The gut microbiota as an environmental factor that regulates fat storage. *Proc. Natl. Acad. Sci. USA* **2004**, *101*, 15718–15723. [CrossRef] [PubMed]
4. Gill, S.R.; Pop, M.; DeBoy, R.T.; Eckburg, P.B.; Turnbaugh, P.J.; Samuel, B.S.; Gordon, J.I.; Relman, D.A.; Fraser, C.M.; Nelson, K.E. Metagenomic analysis of the human distal gut microbiome. *Science* **2006**, *312*, 1355–1359. [CrossRef]
5. Allin, K.H.; Nielsen, T.; Pedersen, O. Mechanisms in endocrinology: Gut microbiota in patients with type 2 diabetes mellitus. *Eur. J. Endocrinol.* **2015**, *172*, 167–177. [CrossRef] [PubMed]
6. Carvalho, B.M.; Saad, M.J.A. Influence of gut microbiota on subclinical inflammation and insulin resistance. *Mediat. Inflamm.* **2013**, *2013*, 986734. [CrossRef]
7. Panwar, H.; Rashmi, H.M.; Batish, V.K.; Grover, S. Probiotics as potential biotherapeutics in the management of type 2 diabetes-prospects and perspectives. *Diabetes Metab. Res. Rev.* **2013**, *29*, 103–112. [CrossRef] [PubMed]
8. Hobbs, S.; Reynoso, M.; Geddis, A.V.; Mitrophanov, A.Y.; Matheny, R.W. LPS-stimulated NF-κB p65 dynamic response marks the initiation of TNF expression and transition to IL-10 expression in RAW 264.7 macrophages. *Physiol. Rep.* **2018**, *6*, e13914. [CrossRef]
9. Khorami, S.A.H.; Movahedi, A.; Huzwah, K.; Sokhini, A.M.M. PI3K/AKT pathway in modulating glucose homeostasis and its alteration in Diabetes. *Ann. Med Biomed. Sci.* **2015**, *1*, 46–55.
10. Zhang, Q.; Wu, Y.; Fei, X. Effect of probiotics on glucose metabolism in patients with type 2 diabetes mellitus: A meta-analysis of randomized controlled trials. *Medicina (Kaunas)* **2016**, *52*, 28–34. [CrossRef]
11. Kocsis, T.; Molnár, B.; Németh, D.; Hegyi, P.; Szakács, Z.; Bálint, A.; Garami, A.; Soós, A.; Márta, K.; Solymár, M. Probiotics have beneficial metabolic effects in patients with type 2 diabetes mellitus: A meta-analysis of randomized clinical trials. *Sci. Rep.* **2020**, *10*, 11787. [CrossRef] [PubMed]
12. Gérard, C.; Vidal, H. Impact of Gut Microbiota on Host Glycemic Control. *Front. Endocrinol.* **2019**, *10*, 29. [CrossRef] [PubMed]
13. Shah, N.J.; Swami, O.C. Role of Probiotics in Diabetes: A Review of Their Rationale and Efficacy. *EMJ Diabetes* **2017**, *5*, 104–110.
14. Hsieh, F.C.; Lee, C.L.; Chai, C.Y.; Chen, W.T.; Lu, Y.C.; Wu, C.S. Oral administration of *Lactobacillus reuteri* GMNL-263 improves insulin resistance and ameliorates hepatic steatosis in high fructose-fed rats. *Nutr. Metab. (Lond.)* **2013**, *10*, 35. [CrossRef]

15. Lim, S.M.; Jeong, J.J.; Woo, K.H.; Han, M.J.; Kim, D.H. *Lactobacillus sakei* OK67 ameliorates high-fat diet-induced blood glucose intolerance and obesity in mice by inhibiting gut microbiota lipopolysaccharide production and inducing colon tight junction protein expression. *Nutr. Res.* **2016**, *36*, 337–348. [CrossRef]
16. Silva, Y.P.; Bernardi, A.; Frozza, R.L. The Role of Short-Chain Fatty Acids from Gut Microbiota in Gut-Brain Communication. *Front. Endocrinol.* **2020**, *11*, 25. [CrossRef]
17. Thiennimitr, P.; Yasom, S.; Tunapong, W.; Chunchai, T.; Wanchai, K.; Pongchaidecha, A.; Lungkaphin, A.; Sirilun, S.; Chaiyasut, C.; Chattipakorn, N.; et al. *Lactobacillus paracasei* HII01, xylooligosaccharides, and synbiotics reduce gut disturbance in obese rats. *Nutrition* **2018**, *54*, 40–47. [CrossRef]
18. Wanchai, K.; Yasom, S.; Tunapong, W.; Chunchai, T.; Eaimworawuthikul, S.; Thiennimitr, P.; Chaiyasut, C.; Pongchaidecha, A.; Chatsudthipong, V.; Chattipakorn, S.; et al. Probiotic *Lactobacillus paracasei* HII01 protects rats against obese-insulin resistance-induced kidney injury and impaired renal organic anion transporter 3 function. *Clin. Sci. (Lond.)* **2018**, *132*, 1545–1563. [CrossRef]
19. Srinivasan, K.; Viswanad, B.; Asrat, L.; Kaul, C.L.; Ramarao, P. Combination of high-fat diet-fed and low-dose streptozotocin-treated rat: A model for type 2 diabetes and pharmacological screening. *Pharm. Res.* **2005**, *52*, 313–320. [CrossRef]
20. Matthews, D.R.; Hosker, J.P.; Rudenski, A.S.; Naylor, B.A.; Treacher, D.F.; Turner, R.C. Homeostasis model assessment: Insulin resistance and beta-cell function from fasting plasma glucose and insulin concentrations in man. *Diabetologia* **1985**, *28*, 412–419. [CrossRef]
21. Matthews, J.N.; Altman, D.G.; Campbell, M.J.; Royston, P. Analysis of serial measurements in medical research. *BMJ* **1990**, *300*, 230–235. [CrossRef] [PubMed]
22. Thabet, H.S.; Saleh, N.K.; Thabet, S.S.; Abdel-Aziz, M.; Kalleny, N.K. Decreased basal non-insulin-stimulated glucose uptake by diaphragm in streptozotocin-induced diabetic mice infected with Schistosoma mansoni. *Parasitol. Res.* **2008**, *103*, 595–601. [CrossRef] [PubMed]
23. Joly Condette, C.; Khorsi-Cauet, H.; Morlière, P.; Zabijak, L.; Reygner, J.; Bach, V.; Gay-Quéheillard, J. Increased gut permeability and bacterial translocation after chronic chlorpyrifos exposure in rats. *PLoS ONE* **2014**, *14*, e102217. [CrossRef] [PubMed]
24. Christensen, E.G.; Licht, T.R.; Leser, T.D.; Bahl, M.I. Dietary xylo-oligosaccharide stimulates intestinal bifidobacteria and lactobacilli but has limited effect on intestinal integrity in rats. *BMC Res. Notes* **2014**, *19*, 660. [CrossRef] [PubMed]
25. Jayashree, B.; Bibin, Y.S.; Prabhu, D.; Shanthirani, C.S.; Gokulakrishnan, K.; Lakshmi, B.S.; Mohan, V.; Balasubramanyam, M. Increased circulatory levels of lipopolysaccharide (LPS) and zonulin signify novel biomarkers of proinflammation in patients with type 2 diabetes. *Mol. Cell Biochem.* **2014**, *388*, 203–210. [CrossRef] [PubMed]
26. Frayn, K.N.; Maycock, P.F. Skeletal muscle triacylglycerol in the rat: Methods for sampling and measurement, and studies of biological variability. *J. Lipid Res.* **1980**, *21*, 139–144.
27. Khat-udomkiri, N.; Toejing, P.; Sirilun, S.; Chaiyasut, C.; Lailerd, N. Antihyperglycemic effect of rice husk derived xylooligosaccharides in high-fat diet and low-dose streptozotocin-induced type 2 diabetic rat mod. *Food Sci. Nutr.* **2019**, *8*, 428–444. [CrossRef]
28. Kohei, K. Pathophysiology of Type 2 Diabetes and Its Treatment Policy. *JMAJ* **2010**, *53*, 41–46.
29. Wiwanitkit, V. Outbreak of Escherichia coli and diabetes mellitus. *Indian J. Endocrinol. Metab.* **2011**, *15*, 70–71. [CrossRef]
30. Ghonim, E.M.; Hindawy, G.R.E.; Motelb, T.M.A.E.; Labib, A.Z.; Ahmady, I.; Salem, E.H.H. Study of bacteremia in diabetic patients. *Menoufia Med. J.* **2016**, *29*, 846.
31. Yamauchi, T.; Kamon, J.; Minokoshi, Y.; Ito, Y.; Waki, H.; Uchida, S.; Yamashita, S.; Noda, M.; Kita, S.; Ueki, K.; et al. Adiponectin stimulates glucose utilization and fatty-acid oxidation by activating AMP-activated protein kinase. *Nat. Med.* **2002**, *8*, 1288–1295. [CrossRef] [PubMed]
32. Jose, D.J.; Renata, F.; Carol, F.E. The PI3K signaling pathway mediates the biological effects of leptin. *Arq. Bras Endocrinol. Metab.* **2010**, *54*, 591–602.
33. Li, X.; Wang, E.; Yin, B.; Fang, D.; Chen, P.; Wang, G.; Zhao, J.; Zhang, H.; Chen, W. Effects of *Lactobacillus casei* CCFM419 on insulin resistance and gut microbiota in type 2 diabetic mice. *Benef. Microbes* **2017**, *8*, 421–432. [CrossRef]

34. Petrone, A.; Zavarella, S.; Caiazzo, A.; Leto, G.; Spoletini, M.; Potenziani, S.; Osborn, J.; Vania, A.; Buzzetti, R. The promoter region of the adiponectin gene is a determinant in modulating insulin sensitivity in childhood obesity. *Obesity* **2006**, *14*, 1498–1504. [CrossRef] [PubMed]
35. Stanford, K.I.; Goodyear, L.J. Exercise and type 2 diabetes: Molecular mechanisms regulating glucose uptake in skeletal muscle. *Adv. Physiol. Educ.* **2014**, *38*, 308–314. [CrossRef]
36. Amati, F.; Dubé, J.J.; Alvarez-Carnero, E.; Edreira, M.M.; Chomentowski, P.; Coen, P.M.; Switzer, G.E.; Bickel, P.E.; Stefanovic-Racic, M.; Toledo, F.G.S.; et al. Skeletal muscle triglycerides, diacylglycerols, and ceramides in insulin resistance: Another paradox in endurance-trained athletes? *Diabetes* **2011**, *60*, 2588–2597. [CrossRef]
37. Kim, J.J.; Sears, D.D. TLR4 and Insulin Resistance. *Gastroenterol. Res. Pr.* **2010**, *2010*, 212563. [CrossRef]
38. Summers, S.A.; Garza, L.A.; Zhou, H.; Birnbaum, M.J. Regulation of Insulin-Stimulated Glucose Transporter GLUT4 Translocation and Akt Kinase Activity by Ceramide. *Mol. Cell Biol.* **1998**, *18*, 5457–5464. [CrossRef]
39. Bagarolli, R.A.; Tobar, N.; Oliveira, A.G.; Araújo, T.G.; Carvalho, B.M.; Rocha, G.Z.; Vecina, J.F.; Calisto, K.; Guadagnini, D.; Prada, P.O.; et al. Probiotics modulate gut microbiota and improve insulin sensitivity in DIO mice. *J. Nutr. Biochem.* **2017**, *50*, 16–25. [CrossRef]
40. Kim, S.H.; Huh, C.S.; Choi, I.D.; Jeong, J.W.; Ku, H.K.; Ra, J.H.; Kim, T.Y.; Kim, G.B.; Sim, J.H.; Ahn, Y.T. The anti-diabetic activity of *Bifidobacterium lactis* HY8101 in vitro and in vivo. *J. Appl. Microbiol.* **2014**, *117*, 834–845. [CrossRef]
41. Kurth-Kraczek, E.J.; Hirshman, M.F.; Goodyear, L.J.; Winder, W.W. 5′ AMP-activated protein kinase activation causes GLUT4 translocation in skeletal muscle. *Diabetes* **1999**, *48*, 1667–1671. [CrossRef] [PubMed]
42. Puddu, A.; Sanguineti, R.; Montecucco, F.; Viviani, G.L. Evidence for the gut microbiota short-chain fatty acids as key pathophysiological molecules improving diabetes. *Mediat. Inflamm.* **2014**, *2014*, 162021. [CrossRef]
43. Ang, Z.; Xiong, D.; Wu, M.; Ding, J.L. FFAR2-FFAR3 receptor heteromerization modulates short-chain fatty acid sensing. *Faseb. J.* **2018**, *32*, 289–303. [CrossRef] [PubMed]
44. Hu, G.; Chen, G.; Xu, H.; Ge, R.; Lin, J. Activation of the AMP activated protein kinase by short-chain fatty acids is the main mechanism underlying the beneficial effect of a high fiber diet on the metabolic syndrome. *Med. Hypotheses* **2010**, *74*, 123–126. [CrossRef] [PubMed]
45. Kim, K.A.; Gu, W.; Lee, I.A.; Joh, E.H.; Kim, D.H. High fat diet-induced gut microbiota exacerbates inflammation and obesity in mice via the TLR4 signaling pathway. *PLoS ONE* **2012**, *7*, e47713. [CrossRef] [PubMed]
46. Cani, P.D.; Bibiloni, R.; Knauf, C.; Waget, A.; Neyrinck, A.M.; Delzenne, N.M.; Burcelin, R. Changes in gut microbiota control metabolic endotoxemia-induced inflammation in high-fat diet-induced obesity and diabetes in mice. *Diabetes* **2008**, *57*, 1470–1481. [CrossRef]

*nutrients*

MDPI

*Article*

# Associations between Pro- and Anti-Inflammatory Gastro-Intestinal Microbiota, Diet, and Cognitive Functioning in Dutch Healthy Older Adults: The NU-AGE Study

**Annick P. M. van Soest [1,*,†], Gerben D. A. Hermes [2,†], Agnes A. M. Berendsen [1], Ondine van de Rest [1], Erwin G. Zoetendal [2], Susana Fuentes [2,3], Aurelia Santoro [4,5], Claudio Franceschi [4,6], Lisette C. P. G. M. de Groot [1,‡] and Willem M. de Vos [2,7,‡]**

1 Division of Human Nutrition and Health, Wageningen University & Research, 6708WE Wageningen, The Netherlands; agnes.berendsen@wur.nl (A.A.M.B.); ondine.vanderest@wur.nl (O.v.d.R.); lisette.degroot@wur.nl (L.C.P.G.M.d.G.)
2 Laboratory of Microbiology, Wageningen University & Research, 6708WE Wageningen, The Netherlands; gerben.hermes@wur.nl (G.D.A.H.); erwin.zoetendal@wur.nl (E.G.Z.); susana.fuentes@rivm.nl (S.F.); willem.devos@wur.nl (W.M.d.V.)
3 Center for Infectious Disease Control, National Institute for Public Health and the Environment, 3721MA Bilthoven, The Netherlands
4 Department of Experimental, Diagnostic and Specialty Medicine, University of Bologna, 40126 Bologna, Italy; aurelia.santoro@unibo.it (A.S.); claudio.franceschi@unibo.it (C.F.)
5 AlmaMater Research Institute on Global Challenges and Climate Change (Alma Climate), University of Bologna, 40126 Bologna, Italy
6 Department of Applied Mathematics, Institute of Information Technology, Mathematics and Mechanics (ITMM), Lobachevsky State University of Nizhny Novgorod-National Research University (UNN), 603950 Nizhny Novgorod, Russia
7 Human Microbiome Research Program, Faculty of Medicine, University of Helsinki, 00014 Helsinki, Finland
* Correspondence: annick.vansoest@wur.nl
† These authors contributed equally to this work.
‡ These authors contributed equally to this work.

Received: 30 September 2020; Accepted: 5 November 2020; Published: 12 November 2020

**Abstract:** Dietary modulation of the gastro-intestinal microbiota is a potential target in improving healthy ageing and age-related functional outcomes, including cognitive decline. We explored the association between diet, gastro-intestinal microbiota and cognition in Dutch healthy older adults of the 'New dietary strategies addressing the specific needs of the elderly population for healthy aging in Europe' (NU-AGE) study. The microbiota profile of 452 fecal samples from 226 subjects was determined using a 16S ribosomal RNA gene-targeted microarray. Dietary intake was assessed by 7-day food records. Cognitive functioning was measured with an extensive cognitive test battery. We observed a dietary and microbial pro- to anti-inflammatory gradient associated with diets richer in animal- or plant-based foods. Fresh fruits, nuts, seeds and peanuts, red and processed meat and grain products were most strongly associated to microbiota composition. Plant-rich diets containing fresh fruits, nuts, seeds and peanuts were positively correlated with alpha-diversity, various taxa from the Bacteroidetes phylum and anti-inflammatory species, including those related to *Faecalibacterium prausnitzii* and *Eubacterium rectale* and *E. biforme*. Animal product-rich diets associated with pro-inflammatory species, including those related to *Ruminococcus gnavus* and *Collinsella spp.*. Cognition was neither associated with microbiota composition nor alpha-diversity. In conclusion, diets richer in animal- and plant-based foods were related to a pro- and anti-inflammatory microbial profile, while cognition was associated with neither.

**Keywords:** gut microbiota; dietary intake; cognitive decline; elderly; healthy ageing; inflammation

## 1. Introduction

The ageing population is growing rapidly. Worldwide, the number of people aged 65 years or over is currently estimated at 703 million. Due to a steep rise in life expectancy, this number is expected to double to 1.5 billion in 2050 [1]. Unfortunately, as the longer lifespan is not accompanied by improvements of health outcomes [2], the increase in life expectancy poses serious challenges to the health care system, economy and society [3]. Therefore, there is an urgent need for strategies to improve healthy ageing.

The gastro-intestinal (GI) microbiota has been implicated as a potential target to enhance healthy ageing [4]. Ageing is accompanied by several physiological and lifestyle changes, including altered GI tract function, elevated inflammation levels and dietary changes, that affect the GI microbiota [5,6]. Compared to younger adults, the GI microbiota in older adults has been shown to exhibit larger inter-individual and temporal variation. It was also strongly correlated to diet, which was linked to residence location in the community [7,8]. Despite the larger variation, several universal changes in the GI microbiota that occur with ageing have been identified. Generally, the relative abundance of *Bifidobacterium* spp. was found to be lower in older adults with concomitant higher levels of Enterobacteriaceae and other pathobionts [5,6].

Changes in GI microbiota composition may influence age-related functional outcomes, such as cognitive decline. In the past decade, the link between altered GI microbiota composition and cognition has been demonstrated in various rodent models, including germ-free animals and several microbiota modulation strategies, such as antibiotics, pre- or pro-biotics, and fecal microbiota transplants [9]. For example, rodents with disrupted GI microbial homeostasis, due to infection or treatment with antibiotics, perform worse on cognitive tests compared to animals with an undisturbed GI microbiota. Restoring this homeostasis by administration of probiotics or via fecal microbiota transplantation positively influenced cognitive performance of rodents [9]. In humans, administration of *Bifidobacterium* and *Lactobacillus* species for 12 weeks has shown to positively affect cognitive functioning in older adults [10,11], providing preliminary evidence for a relation between GI microbiota and cognition in humans, thus proposing the GI microbiota as a target to prevent or delay age-related cognitive decline.

Modification of diet has been suggested as a strategy to both maintain cognition and GI homeostasis. There is special interest in the Mediterranean diet (MedDiet), which is characterized by a high intake of vegetables, fruits, legumes and olive oil and moderate to low intake of animal-based food products [12]. Greater adherence to the MedDiet has been associated to slower rates of age-related cognitive decline [13,14] and beneficial changes in GI microbiota composition [15,16].

To our knowledge, to date only one human study has investigated the relation between diet, cognition and GI microbiota. Data from all European partners of the 'New dietary strategies addressing the specific needs of the elderly population for healthy aging in Europe' (NU-AGE) study, a one-year Mediterranean-like dietary intervention, showed that individuals with better adherence to this diet had higher relative abundances of several microbial groups, including *Faecalibacterium prausnitzii*, *Anaerostipes* and *Roseburia* [16], which have previously been linked to beneficial health effects. For instance, these species exhibit anti-inflammatory properties, are able to produce the short chain fatty acid (SCFA) butyrate and have been inversely associated with diabetes mellitus type 2 and colorectal cancer [17–19]. In turn, higher relative abundances of these beneficial species were weakly, but positively, associated with cognitive function measured by BabCock memory and constructional praxis performance [16].

These results provide preliminary evidence for the potential of the MedDiet to prevent age-related cognitive decline by modulating GI microbiota. However, it remains unclear which specific food groups of the MedDiet are responsible for the potentially beneficial effects on cognition and GI microbiota composition. Moreover, in the previous study, cognitive function was measured by means of single tests [16], whereas the assessment of multiple cognitive tests representing all cognitive domains and combining these tests into composite cognitive scores is a more robust measure of cognitive functioning [20]. Therefore, the current study aims to explore the relation between diet, GI microbiota composition and cognitive function in healthy older adults (65–79 years).

## 2. Materials and Methods

### 2.1. Study Design and Participants

We used data from the Dutch cohort of the NU-AGE study, a parallel randomized one-year study investigating the effect of a dietary intervention on inflammation in European older adults [21]. Cognitive functioning and microbiota composition were determined as secondary outcomes. Information on participants, recruitment and the dietary intervention has previously been described in detail [22,23]. In short, 252 healthy Dutch older adults aged 65–79 years were randomized to the intervention or control group. Participants in the intervention group received individually tailored dietary advice to follow a Mediterranean-like diet. The control group received no specific dietary advice except for a leaflet describing the national guidelines for a healthy diet. Analyses showed that the intervention did not affect GI microbiota. Therefore, the current study has a cross-sectional design, in which data from both pre and post intervention are combined. Participants were non-frail (fried frailty $\leq 1$ [24]) and free of major diseases including cancer, dementia, diabetes mellitus type I and II and organ failure, and did not use antibiotics in the three months prior to inclusion. Dietary intake, GI microbiota composition and cognitive functioning were assessed at baseline and post intervention. Data from 26 participants were excluded due to missing GI microbiota assessments at either pre or post intervention. The NU-AGE study has been registered at clinicaltrials.gov (identifier: NCT01754012). This study was conducted according to the Declaration of Helsinki and written informed consent was obtained from all participants. The study protocol was approved by the Medical Ethics Committee of Wageningen University and Research (ABR 37818.081.11).

### 2.2. Dietary Assessment

At baseline and post intervention, dietary intake was assessed by a 7-day food record. Participants were instructed to record all consumed foods and their amounts based on household measures. All food records were reviewed by a trained research dietician during an interview. Consumed food products were coded according to standardized coding procedures. Nutrient intake data was calculated by use of the Dutch food composition table (NEVO 2011). Consumed food products with similar composition were grouped into food groups according to the EPIC-Soft Classification [25] with some local modifications. Additional groups were created for ready-to-eat meals and savory bread spreads as products in these groups were not included in the current EPIC-Soft list. Separate groups were created for low fat, and salt and sugar options within the dairy food groups based on the Dutch dietary guidelines [26]. Products containing artificial sweeteners were placed in a separate group as sweeteners have been shown to influence GI microbiota composition [27]. In addition, a separate group was made for legume-based ready to eat soups due to the relatively high fiber content. Finally, the food group meat was divided into red meat, processed meat, poultry and meat replacers instead of groups based on animal origin to limit the number of food groups.

### 2.3. Microbiota Composition Profiling

At baseline and post intervention, participants were instructed to collect a fecal sample at home with the help of a stool collection kit and store them immediately at −20 °C. Samples were transported in coolers and then stored at −20 °C and later at −80 °C before being processed. DNA extraction from fecal samples has been described in detail elsewhere [28]. In brief, DNA was extracted using a combination of column purification and repeated-bead-beating. Purity and concentration of DNA were assessed with a Nanodrop 1000 spectrophotometer (Thermo Fisher Scientific, Wilmington, USA).

The composition analysis was then performed utilizing a previously benchmarked custom made, phylogenetic microarray, the human intestinal tract chip (HITChip) [29,30]. The HITChip contains a duplicated set of 3631 probes, which target the V1 and V6 hypervariable regions of the 16S rRNA gene of 1140 intestinal bacterial phylotypes. After extraction of DNA, the full-length 16S rRNA gene was amplified by PCR using primers T7prom-Bact-27-for and Uni-1492-rev [30]. This was followed by in vitro transcription and labelling of the resulting RNA with Cy3/Cy5 before hybridization to the array. The signal intensity data from the microarray hybridizations were collected from the Agilent G2505C scanner (Agilent Technologies) using Agilent Feature Extraction software, version 10.7.3.1 and pre-processed using an in-house MySQL database and custom R scripts. Each scanner channel from the array was separately spatially normalized using polynomial regression, followed by outlier detection and filtering in each set of probes with a $\chi 2$ test. Each sample was hybridized at least twice to ensure reproducibility. Duplicate hybridizations with a Pearson correlation < 0.98 were not considered for further analysis. Microbiota profiles were summarized to genus-like 16S rRNA gene sequence groups with a sequence similarity > 90% referred to as species and relatives ('et rel.'). Measurements of probes that belong to the same phylotype were normalized with robust probabilistic averaging [31,32]. Log10-transformed hybridization signals were used as a proxy for bacterial abundance.

### 2.4. Cognitive Functioning

Cognitive functioning was assessed at baseline and post intervention with an extensive battery of cognitive tests which were administered by trained research assistants. The battery included cognitive tests from the consortium to establish a registry for Alzheimer's disease (CERAD) test battery [33] plus five additional tests.

In the verbal fluency category test [34], participants were asked to name as many animals as possible within 60 s. The number of uniquely named animals was recorded. Participants were presented with four figures in the constructional praxis test [35], and asked to copy these figures on blank paper immediately after presentation (subtest immediate) and after a few minutes (subtest recall). Scoring was based on the number of correct responses. In the word list memory test [33], participants were visually presented with ten random words. The number of correctly recalled words directly after presentation in three trials (subtest immediate) and after five minutes in one trial (subtest delayed) was recorded. Finally, the participant was asked to identify the ten words from a verbally presented list of twenty words (subtest recognition). Next, participants were read a brief story in the Babcock story recall test [36] and asked to retell the story immediately (subtest immediate) and after 20 min (subtest delayed). Scoring was based on the correctly recalled parts of the story. In the trail making test [37], participants were instructed to connect 25 numbers in chronological order (Part A) and to connect numbers and letters in chronological and alphabetical order alternately (Part B). Time to complete each task was recorded. In the number cancellation test [38], participants were presented with a list of random numbers. The number of correctly crossed out 4s in 30 s was documented. In the pattern comparison test [39], participants were asked to indicate if two patterns were similar or different. Scoring was based on the number of correct responses.

Scores for each of the cognitive tests were converted into Z-scores with baseline mean and standard deviation of the whole population. The Z-score for the trail making test was reversed as lower scores represent better cognitive functioning. The individual Z-scores for the cognitive tests were clustered into four cognitive domains:

$$\text{Episodic memory} = (zWordList_{\text{immediate}} + zWordList_{\text{delayed}} + zWordList_{\text{recognition}} + zBabcockStoryRecall_{\text{immediate}} + zBabcockStoryRecall_{\text{delayed}})/5 \quad (1)$$

$$\text{Executive functioning} = (\text{zVerbalFluency} + -\text{zTrailMakingTest}_{\text{B/A}})/2 \quad (2)$$

$$\text{Information processing speed} = (-zTrailMakingTest_{\text{A}} + zNumberCancellation + zPatternComparison)/3 \quad (3)$$

$$\text{Visuospatial ability} = (\text{zConstructionalPraxis}_{\text{immediate}} + \text{zConstructionalPraxis}_{\text{recall}})/2 \quad (4)$$

### 2.5. Assessment of Phenotypical Characteristics

Body weight and height were measured by trained research assistants. Weight was determined while wearing light clothing to the nearest 0.1 kg using a calibrated scale. Height was measured using a stadiometer to the nearest 0.1 cm. Body mass index (BMI) was calculated as weight/height$^2$. Data on age, sex, education (number of years) and smoking status (never, former or current) were collected using questionnaires. Frailty status (non-frail/pre-frail) [24] and mini-mental state examination (MMSE) [40] were assessed by trained research assistants following standardized procedures. MMSE scores from 24 to 30 are considered within the normal range [40]. Physical activity was measured using the physical activity scale for elderly (PASE). For individuals aged 70 to 75, average values for PASE are 89.1 for women and 102.4 for men [41].

### 2.6. Statistical Analyses

All microbiota analyses were performed in R version 3.4.0 [42]. Redundancy analysis (RDA) was performed to determine the multivariate effects of the explanatory variables on microbiota composition using the rda function from the vegan package [43]. RDA is a technique summarizing the linear relationships between a set of variables i.e., GI microbiota composition explained by a set of explanatory variables, i.e., dietary and host variables. The effect of an explanatory variable is defined as $R^2$, which is the percentage of variation explained from the total amount of microbiota variation. All numerical environmental variables (food groups, nutrients, phenotype and cognition) were normalized to ensure that the input variables had similar scales before performing the RDA. We first determined the simple effects of all explanatory variables on microbiota composition to help understand what was driving the interactions. Because the dietary intervention had no significant effect on microbiota composition, we performed a cross-sectional analysis with both pre- and post-intervention samples to increase power. To determine which set of food groups resulted in the most parsimonious model (i.e., explaining microbiota variation), we performed forward and reverse automatic stepwise model selection for constrained ordination methods using permutation tests with the ordistep function from the vegan package, which bases the term choice on Akaike's information criterion and $p$-value. This ordination configuration was used to test which other explanatory variables (nutrients, phenotype and cognition) significantly correlated with microbiota composition by post-hoc fitting these as vectors using the envfit function from vegan. Significance was set at $p < 0.05$. Richness, inverse Simpson and Shannon diversity were calculated to define microbial alpha-diversity using the microbiome package [44]. In ecology, alpha-diversity is defined as the species diversity within a sample. We used to commonly applied methods to determine diversity, *viz* Shannon diversity and inverse Simpson diversity. Diversity of the microbiota was based on non-logarithmic oligo-level signals and probes were counted in each sample to measure richness, by using an 80% quantile threshold for detection. To correlate microbial alpha-diversity with the significant explanatory variables we used Pearson correlations and visualized these using heatmaps with the psych package [45]. $p$-values were corrected for multiple testing using the Benjamini–Hochberg procedure [46] and $q < 0.05$ was considered significant.

## 3. Results

### 3.1. Participant Characteristics

At baseline, the mean age of participants was 70.9 ± 4.1 years and 44.4% of the study population was male (Table 1). The average body mass index (BMI) at baseline was 25.9 ± 3.6 kg/m$^2$ and mean score on the mini-mental state examination (MMSE) was 27.7 ± 1.8 points, indicating that our study population was cognitively healthy. The mean PASE score was 137 ± 54, indicating that the physical activity level was slightly higher than normal compared to a study population with similar age [41].

**Table 1.** Baseline characteristics of 226 healthy Dutch older adults.

| Characteristic | *n* = 226 |
|---|---|
| Age, years | 70.9 ± 4.1 |
| Sex, male *n* (%) | 100 (44.2%) |
| Education, years | 12.3 ± 3.7 |
| BMI, kg/m$^2$ | 25.9 ± 3.6 |
| Smoking status, *n* (%) | |
| Never | 117 (51.8%) |
| Former | 103 (45.6%) |
| Current | 6 (2.7%) |
| MMSE (score 0-30) | 27.7 ± 1.8 |
| Physical activity (PASE score) | 137 ± 54 |
| Frailty, *n* (%) | |
| Non-frail | 178 (78.8%) |
| Pre-frail | 48 (21.2%) |

Abbreviations: BMI: body mass index; MMSE: mini mental state examination; PASE: physical activity scale for the elderly. Data are presented as mean ± SD or number (%).

### 3.2. *Variables Affecting GI Microbiota Composition*

To determine how the different environmental variables impact the microbiota, we first calculated their simple effects (i.e., the effect of the environmental variable on the microbiota without any other covariates). As previously described in the methods, the dietary intervention had no significant effect on microbiota composition ($p$ = 1.0, $R^2$ = 0.08%). A total of 41 variables, existing of phenotypical characteristics, food groups and nutrients, significantly correlated to GI microbiota composition as shown in Figure 1. The largest proportion of GI microbiota variation was explained by individuals ($R^2$ = 40.0%) (Supplementary Figure S1). The phenotypical characteristics BMI ($R^2$ = 0.73%) and sex ($R^2$ = 0.22%) were both correlated with microbiota composition. BMI explained the largest proportion of microbiota variation out of all microbiota covariates. With respect to the dietary variables, 29 nutrients and 10 food groups were significantly correlated with GI microbiota composition. Concerning the food groups, fresh fruits explained the highest proportion of variation in GI microbiota composition ($R^2$ = 0.51%). Further zooming in on the fresh fruits showed that berries and grapes were the fruits most contributing to this observation. Other significant food groups were nuts, seeds and peanuts ($R^2$ = 0.45%), grain products ($R^2$ = 0.39%) and both processed and red meat ($R^2$ = 0.36% and $R^2$ = 0.25% respectively). Among the nutrients, total protein ($R^2$ = 0.46%) and protein from animal ($R^2$ = 0.62%) and plant ($R^2$ = 0.42%) sources explained the largest proportion of variation. In addition, various forms of carbohydrates, water-soluble vitamins, minerals and omega-3 fatty acids were significantly associated to GI microbiota composition, while other fatty acids and fat-soluble vitamins did not. None of the cognitive functioning domains was significantly correlated with GI microbiota composition.

To visualize the relations between dietary factors and phenotypical characteristics with microbiota composition, their conditional effects (the impact on the microbiota with the effect of other variables in the model) were calculated and plotted in two RDA bi-plots (Figure 2). We observed a gradient of participants with higher intakes of plant-based foods and participants consuming higher amounts of animal-based foods. Higher intakes of these animal-based foods, animal protein, cholesterol, vitamin B12, low fat cheese, and red and processed meat, were correlated with a higher BMI. The participants with lower intake of animal-based foods and higher intake of plant-based foods could be further divided into two groups; those consuming higher amounts of fresh fruits, nuts, seeds and peanuts and vitamin C, and those with higher intakes of grain products and digestible carbohydrates.

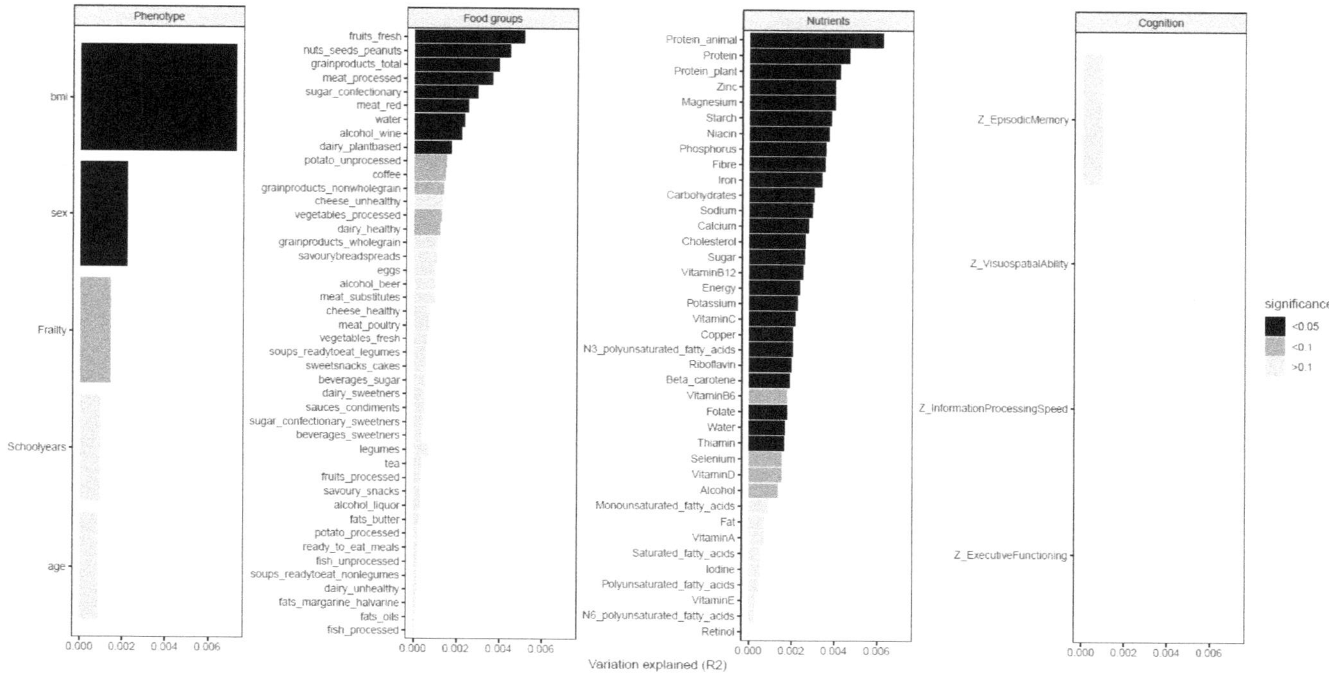


**Figure 1.** Microbiota covariates. Impact of all measured variables on microbiota composition defined as percentage variation explained ($R^2$) out of all the total microbiota variation. A higher $R^2$ implies a stronger effect size.

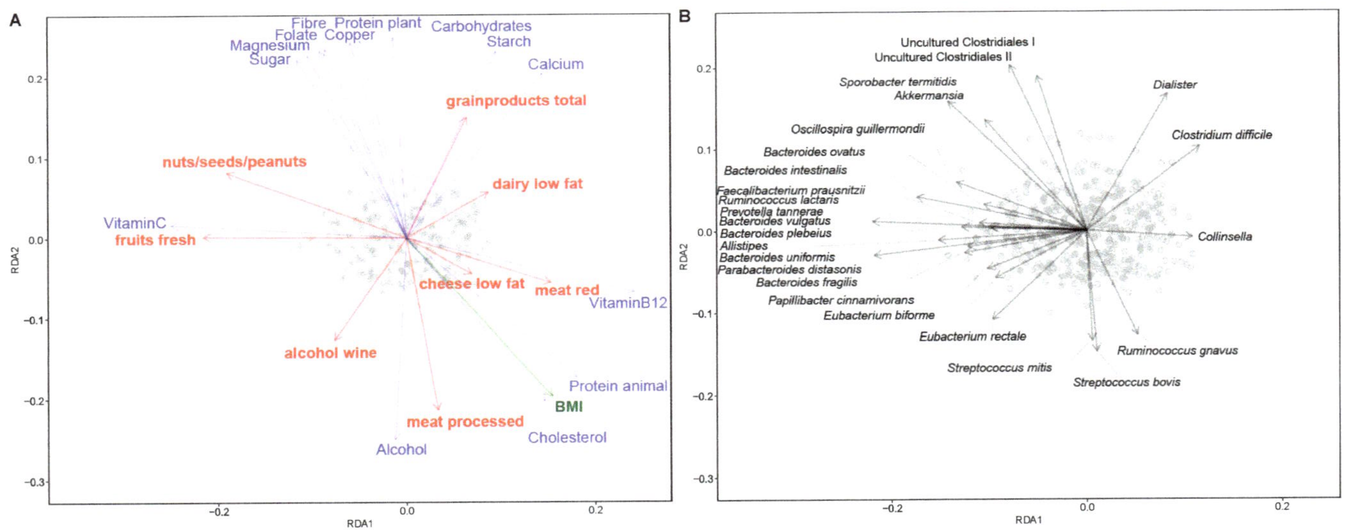


**Figure 2.** Association of microbiota with food groups, nutrients and BMI. Samples are plotted as grey circles. (**A**) Redundancy analysis (RDA) bi-plot of microbiota with explanatory variables; food groups (red), nutrients (blue) and phenotypical characteristics (green). (**B**) RDA bi-plot of samples with the associated microbial taxa (indicated as genera or species-level groups) The direction of the arrows depicts the abundance of microbial taxa. Length of the arrows is a measure of fit. The variable arrows approximate the correlation between species and explanatory variables. Samples near the coordinate origin (zero point) suggest near zero correlation. The further a sample falls in the direction indicated by the arrow, the higher the correlation.

Consumption of animal-based foods and BMI was positively associated with species related to *Ruminococcus gnavus*, *Streptococcus* spp. (*S. mitis* and *bovis*) and *Collinsella*. Conversely, animal-based foods were inversely associated with *Akkermansia muciniphila*, uncultured Clostridiales I and II and species related to *Sporobacter termitidis*. Consumption of fresh fruits, its associated nutrient vitamin C, and nuts, seeds and peanuts were associated with several genera from the Bacteroidetes phylum, including *Bacteroides* spp., *Parabacteroides*, *Alistipes* and *Prevotella*, and Firmicutes such as species related to *Faecalibacterium prausnitzii*, *Oscillospira guillermondii* and *Eubacterium rectale* and *E. biforme*. Grain products and carbohydrates were positively associated with *Dialister* and species related to *Clostridium difficile* (recently renamed to *Clostridioides difficile*). Although this group is named after *C. difficile*, the observed differences do likely not relate to this potential pathogen but probes targeting *C. bifermentans*, *C. bartlettii* and *C. glycolicum*.

### 3.3. Variables Associated with Microbial Alpha-Diversity

The relations between the significant variables in the RDA (phenotypical characteristics, nutrients, food groups) and indices that contribute to microbial alpha-diversity were calculated and visualized in Figure 3A. BMI was negatively correlated with alpha-diversity. With respect to the food groups, only fresh fruits and nuts, seeds and peanuts were positively correlated with alpha-diversity, with correlation coefficients ranging from 0.1 to 0.17. Among the fresh fruits, alpha diversity positively correlated with berries and grapes, citrus fruits and stone fruits in Supplementary Table S1. Nutrients that were positively correlated to alpha-diversity included vitamin C, various minerals, forms of carbohydrate and plant protein, with correlation coefficients between 0.09 and 0.14. None of the nutrients was negatively associated with alpha-diversity.

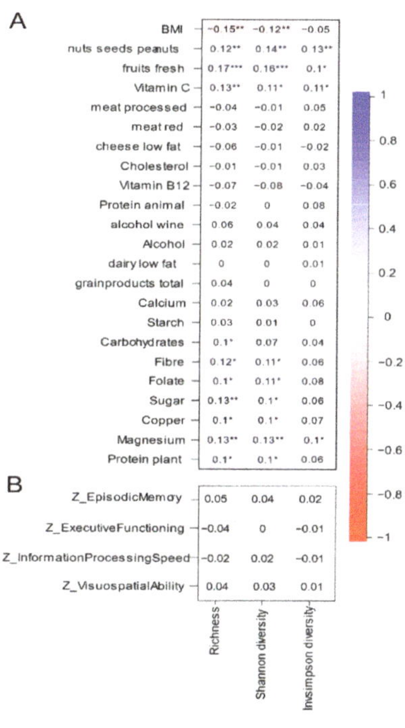


**Figure 3.** Correlation of alpha-diversity with microbiota covariates (**A**) and cognition variables (**B**). Pearson correlation of significant microbiota covariates were calculated. $p$ values are corrected for multiple testing using the Benjamini–Hochberg procedure. *** q < 0.001, ** q < 0.01 * q < 0.05.

With correlation coefficients ranging from −0.04 to 0.05, none of the cognitive domains was significantly correlated to any of the diversity indices (Figure 3B).

## 4. Discussions

By exploring associations between diet, GI microbiota and cognition in healthy Dutch older adults using food groups as the primary input, we showed that fresh fruits, nuts, seeds and peanuts, red and processed meat, grain products, low fat dairy and cheese and wine are important dietary factors in GI microbiota composition. Of these food groups, fresh fruits (berries and grapes in particular), and nuts, seeds and peanuts positively correlated with alpha-diversity. Overall, fresh fruits and nut seeds and peanuts correlated with various taxa from the Bacteroidetes phylum and species related to *Faecalibacerium prausnitzii*, grain products correlated with *Dialister,* while higher intake of animal-based foods was associated with a higher abundance of *Collinsella* and *Streptococcus* spp. as well as species related to *Ruminococcus gnavus*. Cognitive functioning was neither associated with GI microbiota composition nor alpha-diversity.

Our study is the first to investigate which food groups are related to whole GI microbiota composition and alpha-diversity in older adults. In younger adults, several studies have investigated this association before. In a large cross-sectional study with GI microbiota data from 1135 Dutch adults, 78 dietary factors, including fruit, frequency of nut consumption, red and processed meat and protein, were important dietary factors in explaining GI microbiota variation [47]. The associations of fruit and meat with GI microbiota composition were confirmed in a large cross-sectional Belgian study with adults ($n$ = 1106) [48] and the French Milieu Intérieur study ($n$ = 862) [49] showed that fruit influenced the GI microbiota. With respect to alpha-diversity, our finding that individuals with higher intakes of fresh fruit and nuts had a more diverse GI microbiota was confirmed by the studies of Dutch and French adults [47,49] and the association between nuts, seeds and peanuts and alpha-diversity was also observed in Dutch adults [47].

Despite the fact that the dietary intervention did not have a significant impact on the GI microbiota in our cohort, we could clearly identify associations between dietary variables and microbiota composition. We observed a gradient between participants consuming a diet richer in foods from animal origin and a diet richer in foods from plant origin, from now on referred to as animal- and plant-rich diets. The animal-rich diet was characterized by higher intakes of processed and red meat, low fat cheese and dairy, vitamin B12 and cholesterol. The plant-rich diet was higher in vitamin C, fresh fruits and nuts, seeds and peanuts. In addition to the classification based on origin of the food products and nutrients, these diets can also be classified as pro- and anti-inflammatory according to the dietary inflammatory index, in which various dietary factors have been scored based on their inflammatory potential [50]. Vitamin B12 and cholesterol, both associated with the animal-rich diet, were considered pro-inflammatory. With respect to the plant-rich diet, nutrients present in fresh fruits (vitamin C, flavonoids, fiber) and nuts, seeds and peanuts (polyphenols, omega-3 fatty acids, fiber) were all classified as anti-inflammatory.

Interestingly, classification of the GI microbiota based on inflammatory potential showed a similar pattern. The consumption of the pro-inflammatory diet rich in animal foods positively correlated with *Collinsella* and *Streptococcus* spp. as well as species related to *R. gnavus*. Overall, these bacteria have been classified as pro-inflammatory. Increased abundance of *Collinsella* has been observed in several inflammatory diseases, including type 2 diabetes mellitus [51,52], atherosclerosis [53] and rheumatoid arthritis [54]. Even though *Streptococcus* is a normal inhabitant of the upper GI tract, increased abundance in the colon has been associated with pro-inflammatory nutrients of animal origin [15]. Finally, higher abundance of *R. gnavus* has been linked to several inflammatory diseases as well, such as spondyloarthritis [55], eczema in infants [56] and inflammatory bowel disease, especially during active disease episodes [57]. In addition to the connection with inflammatory diseases, it has recently been shown that *R. gnavus* synthesizes an inflammatory polysaccharide that induces secretion of the inflammatory cytokine tumor necrosis factor-alpha by dendritic cells [58].

The anti-inflammatory plant-rich diet was associated with species related to *F. prausnitzii*, *E. rectale* and *E. biforme*. These species can be classified as anti-inflammatory due to their ability to produce butyrate. Butyrate has been shown to exhibit anti-inflammatory effects through their regulation of leukocyte function via inhibition of histone deacetylase and activation of G-protein coupled receptors [59]. These anti-inflammatory effects of butyrate have been demonstrated in vivo, in both animal models [60] and human clinical trials [61]. *F. prausnitzii* specifically has been shown to exhibit anti-inflammatory effects in vitro and in vivo. In peripheral blood mononuclear cells, *F. prausnitzii* led to higher levels of the anti-inflammatory cytokine IL-10 and lower production of the pro-inflammatory cytokines IL-12 and IFN-$\gamma$. In a mouse model with induced acute colitis, administration of living *F. prausnitzii* decreased colitis [62]. Moreover, in humans lower abundance of these species has been observed in several inflammatory diseases. A meta-analysis in inflammatory bowel disease patients showed that patients suffering from an active disease episode had lower abundance of *F. prausnitzii* compared to patients in remission [63] and *E. rectale* were reduced in Crohn's disease patients compared to healthy controls [64]. The plant-rich diet also positively correlated to the mucin degrading species *A. muciniphila*. Similarly, lower abundance of *A. muciniphila* has been observed in inflammatory conditions including obesity and type 2 diabetes [65–67]. Moreover, a recent human intervention trial showed that daily administration of *A. muciniphila* cells for three months increased barrier function, by decreasing the levels of pro-inflammatory lipopolysaccharides in prediabetic human subjects [68]. Overall, the links between these bacteria and inflammatory diseases and compounds, indicate that the consumption of an animal-rich diet might correlate with a more pro-inflammatory GI microbiota profile, while the plant-rich diet correlates to a more anti-inflammatory GI microbiota profile.

Moreover, several species associated to the plant-rich diet, including *F. prausnitzii* and *E. rectale*, have been previously associated with a high adherence to the MedDiet in various European countries [16]. This might imply that certain food groups that were part of the plant-rich diet, i.e., nuts, seeds and peanuts and fresh fruits, are important dietary factors in the MedDiet with respect to GI microbiota modulation. The beneficial associations of these food groups could be due to the fiber present in fruits and nuts. Fermentation of fibers in the gut leads to the production of SCFA, which have beneficial effects on health as previously discussed [69]. An additional factor underlying the beneficial associations might be the presence of polyphenols in fruit and nuts. These plant metabolites are poorly absorbed in the small intestine and reach the colon where they can interact with microbiota. Polyphenols have been shown to have prebiotic-like effects. Various types of polyphenols enhanced growth of lactobacilli and bifidobacteria as well as *Akkermansia*, in both in vitro and in vivo (animal and human) studies [70,71].

In addition to the association between the plant-rich diet and the anti-inflammatory species, the diet rich in plant foods also positively correlated with several genera from the Bacteroidetes phylum such as *Parabacteroides*, *Alistipes*, and mostly *Bacteroides* and *Prevotella* spp.. Members of the latter two maintain a complex and generally beneficial relationship with the host. Bacteroidetes are abundantly present in the human gut and many genera within this phylum respond to changes in diet. Generally, diets rich in fiber are linked with increased abundance of *Prevotella* spp. [72], while higher abundance of *Bacteroides* spp. is associated to diets rich in fat and protein from animal origin [73]. However, the latter group has also been linked to plant-based complex carbohydrates and inversely associated with dietary fat and protein [15], in line with our results. It is well known that microorganisms have context-dependent functions and a changing metabolism, depending on environmental conditions and the presence and function of other microbes. For instance, *Bacteroides* spp. contain a large repertoire of enzymes to break down complex plant carbohydrates [74], which likely underlies their association in the current study. However, several *Bacteroides* spp. are also bile resistant [75] and could thus be more prevalent in individuals consuming high fat diets with little complex carbohydrates. Additionally, different species or strains within the *Bacteroides* and *Prevotella* genera have been shown to be genetically diverse and associated with different dietary components, such as plant-based diets, while some are associated with animal-based nutrients [76,77]. Another factor in the ambiguity of the health associations of *Bacteroides* spp. is their status as a pathogen, as several species (notably *B. fragilis*) can cause significant pathology, including bacteremia and abscess formation in multiple body sites [75]. Similarly, several *Prevotella* spp. have been associated with chronic inflammatory conditions [78]. In contrast, *Bacteroides* spp. have also been linked to beneficial effects on health. This apparent duality was exemplified by the observation of a cohort specific positive or negative association with markers of insulin resistance in overweight insulin resistant males [79]. For example, *Bacteroides* spp. can contribute to the formation the SCFA propionate via the succinate pathway [80]. Propionate has been linked to several health benefits, including regulation of appetite and lipid synthesis in in vivo animal studies, and anti-colorectal cancer effects in in vitro models [81].

Specific food groups, such as berries and nuts, seeds and peanuts, were correlated with several anti-inflammatory microbial species. In addition, these food groups have been associated with slower rates of cognitive decline [82,83]. Although inflammation is a major mechanism underlying cognitive decline [84], we did not find associations between cognitive functioning and the GI microbiota composition or alpha-diversity. To our knowledge, the association between diet, gastro-intestinal microbiota and cognitive functioning in humans has only been investigated in a single other study [16]. Here, the authors showed that European individuals with high adherence to a Mediterranean-like diet had high relative abundance of several beneficial, anti-inflammatory, butyrate producing microbial groups, including *Faecalibacterium prausnitzii, Anaerostipes* and *Roseburia.* Increased relative abundance of these species was associated with improved cognitive function measured by single tests. Our approach augments this paper, but also differed in two aspects. First, we used diet as a combination of different food groups, while in the previous paper diet was only considered as adherence to the Mediterranean diet in general. Hence, it was not clear which specific food groups of the Mediterranean diet were responsible for the beneficial effect on cognition and gastro-intestinal microbiota composition. Second, we incorporated cognitive functioning outcomes using a robust measure of cognitive functioning by calculating mean scores per cognitive domain (composite cognitive scores). The previous research only considered scores of single cognitive tests. Aside from the use of a more robust measure of cognitive functioning, there are several other explanations for the apparent differing results with regard to the association of microbiota with cognitive function.

From animal studies, there is strong evidence for a relation between the gut and the brain, which has been shown with (germ-free) rodent studies, using microbiota modulating strategies such as antibiotics and fecal microbiota transplants [9]. However, there are many differences between rodents and humans, such as differences in GI tract anatomy and physiology and microbiota composition [85], severely limiting translation from rodents to humans. In addition, rodent models allow for more extreme interventions, have a very homogeneous genetic background and there is a high level of control over external factors, which allow for the demonstration of subtle effects. In contrast, we investigated cross-sectional relations in a healthy population of older adults in which diets and microbiota were relatively homogeneous. There were no extreme variations in intake of food components between participants and the dietary intervention that half of the participants underwent, resulted in small changes in dietary intake (e.g., increase of one slice of whole-wheat bread, one third of an apple, and half a serving spoon of vegetables extra per day) [23]. This may have limited the demonstration of associations between cognitive functioning and GI microbiota.

Moreover, our study population consisted of cognitively healthy older adults as shown by the mean MMSE score of 27.7 points out of 30, as scores from 24 to 30 are considered within the normal range [40]. Cognitively healthy indicates that these participants were no mild cognitive impairment or dementia patients. It is important to emphasize that cognitively healthy older individuals can benefit from the effects of diet on cognition. Cognitive health is not static, but rather a progressive phenomenon. The process of age-related cognitive decline starts from the late 20s and continuous throughout the lifespan [86]. The rate of decline can be influenced by several lifestyle factors, including nutrition. Previous research has already demonstrated that several dietary patterns can slow down cognitive decline with ageing. For example, this has been shown for the Mediterranean, Dietary Approaches to Stop Hypertension (DASH) and Mediterranean-DASH Intervention for Neurodegenerative Delay (MIND) diets [87].

Nevertheless, gastro-intestinal microbiota targeted interventions to slow down cognitive decline may be more effective in cognitively impaired individuals, i.e., mild cognitive impairment or Alzheimer's disease patients. Mild cognitive impairment and Alzheimer's disease patients have shown decreased microbial diversity and similar changes in GI microbiota compared to healthy older adults [88]. In line with this, the effectiveness of probiotic supplementation on cognition in humans likely depends on the degree of cognitive impairment. In human intervention studies, the effect of probiotic supplementation on cognitive functioning is mainly effective in cognitively impaired individuals (i.e., mild cognitive impairment or Alzheimer's disease patients), [10,89,90] while the effectiveness in relatively healthy older adults has been inconsistent [11,91,92]. Similarly, the efficacy of other dietary interventions to slow down cognitive decline has been shown to be dependent on the extent of cognitive impairment as well [93]. Therefore, our study population might have been too healthy to demonstrate the link between cognition and GI microbiota. Indeed, changes in GI microbiota in older adults seem to be more strongly associated with health status rather than with chronological age [5,94].

The study population is an important limitation of this study. We did not demonstrate associations between cognitive functioning and GI microbiota, possibly due to relatively small differences in diet and microbiota between subjects and the high cognitive health status of our study population. Further research on the association between diet, GI microbiota and cognitive ageing in humans would benefit from focusing on cognitively impaired study populations and study populations that are more heterogeneous with respect to dietary intake.

## 5. Conclusions

This cross-sectional investigation into the association between diet, GI microbiota and cognition showed that the anti-inflammatory potential of a plant-rich diet high in fresh fruits and nuts, seeds and peanuts was linked to a GI microbiota profile with a higher anti-inflammatory potential. Conversely, a pro-inflammatory animal-rich diet was associated with a more pro-inflammatory GI microbiota profile. Despite the prominent role of inflammation in cognitive decline, we did not demonstrate associations between cognitive functioning and GI microbiota.

**Supplementary Materials:** The following are available online at http://www.mdpi.com/2072-6643/12/11/3471/s1, Figure S1: microbiota covariates. Impact of the individual and all measured variables and on microbiota composition defined as percentage variation explained (R2) out of the total microbiota variation. Table S1: Pearson correlation of alpha-diversity with fresh fruit categories.

**Author Contributions:** Conceptualization, L.C.P.G.M.d.G., W.M.d.V., A.S., C.F.; formal analysis, G.D.A.H., A.P.M.v.S.; investigation, A.A.M.B, S.F.; writing—original draft preparation, A.P.M.v.S., G.D.A.H.; writing—review and editing, A.P.M.v.S., G.D.A.H., A.A.M.B., O.v.d.R., E.G.Z., S.F., L.C.P.G.M.d.G., W.M.d.V.; supervision, O.v.d.R., L.C.P.G.M.d.G., E.G.Z.; funding acquisition, L.C.P.G.M.d.G., W.M.d.V., A.S., C.F. All authors have read and agreed to the published version of the manuscript.

**Funding:** The NU-AGE study was supported by the European Union's Seventh Framework Programme, grant number No. 266486 ('NU-AGE: new dietary strategies addressing the specific needs of the elderly population for healthy aging in Europe'). G.D.A.H. was supported by the project MASTER (Microbiome Applications for Sustainable food systems through Technologies and Enterprise) funded by the European Union's H2020 research and innovation program under grant agreement No. 818368. Microbiota analysis was funded by the 2008 Spinoza Award of the Netherlands Organization for Scientific Research to W.M.d.V.

**Acknowledgments:** The authors thank all study participants and all co-workers who were involved in the data collection of the NU-AGE study. We would like to give special thanks to Wilma M. Akkermans-van Vliet for the DNA isolations and Aino Leegaard Andersen for drafting the first analysis plan.

**Conflicts of Interest:** The authors declare no conflict of interest. The funders had no role in the design of the study; in the collection, analyses, or interpretation of data; in the writing of the manuscript, or in the decision to publish the results.

## References


1. United Nations. *World Population Ageing 2019 Highlights*; UN: New York, NY, USA, 2019. [CrossRef]
2. Crimmins, E.M.; Beltrán-Sánchez, H. Mortality and Morbidity Trends: Is There Compression of Morbidity? *J. Gerontol. Ser. B* **2010**, *66*, 75–86. [CrossRef] [PubMed]
3. Harper, S. Economic and social implications of aging societies. *Science* **2014**, *346*, 587–591. [CrossRef] [PubMed]
4. Candela, M.; Biagi, E.; Brigidi, P.; O'Toole, P.W.; De Vos, W.M. Maintenance of a healthy trajectory of the intestinal microbiome during aging: A dietary approach. *Mech. Ageing Dev.* **2014**, *137*, 70–75. [CrossRef] [PubMed]
5. An, R.; Wilms, E.; Masclee, A.A.M.; Smidt, H.; Zoetendal, E.G.; Jonkers, D. Age-dependent changes in GI physiology and microbiota: Time to reconsider? *Gut* **2018**, *67*, 2213–2222. [CrossRef]
6. Biagi, E.; Nylund, L.; Candela, M.; Ostan, R.; Bucci, L.; Pini, E.; Nikkïla, J.; Monti, D.; Satokari, R.; Franceschi, C.; et al. Through Ageing, and Beyond: Gut Microbiota and Inflammatory Status in Seniors and Centenarians. *PLoS ONE* **2010**, *5*, e10667. [CrossRef]
7. Claesson, M.J.; Cusack, S.; O'Sullivan, O.; Greene-Diniz, R.; De Weerd, H.; Flannery, E.; Marchesi, J.R.; Falush, D.; Dinan, T.G.; Fitzgerald, G.F.; et al. Composition, variability, and temporal stability of the intestinal microbiota of the elderly. *Proc. Natl. Acad. Sci. USA* **2011**, *108*, 4586–4591. [CrossRef]
8. Claesson, M.J.; Jeffery, I.B.; Conde, S.; Power, S.E.; O'Connor, E.M.; Cusack, S.; Harris, H.M.B.; Coakley, M.; Lakshminarayanan, B.; O'Sullivan, O.; et al. Gut microbiota composition correlates with diet and health in the elderly. *Nature* **2012**, *488*, 178–184. [CrossRef]
9. Cryan, J.F.; Dinan, T.G. Mind-altering microorganisms: The impact of the gut microbiota on brain and behaviour. *Nat. Rev. Neurosci.* **2012**, *13*, 701–712. [CrossRef]
10. Hwang, Y.-H.; Park, S.; Paik, J.-W.; Chae, S.-W.; Kim, D.-H.; Jeong, D.-G.; Ha, E.; Kim, M.; Hong, G.; Park, S.-H.; et al. Efficacy and Safety of Lactobacillus Plantarum C29-Fermented Soybean (DW2009) in Individuals with Mild Cognitive Impairment: A 12-Week, Multi-Center, Randomized, Double-Blind, Placebo-Controlled Clinical Trial. *Nutrients* **2019**, *11*, 305. [CrossRef]
11. Kim, C.-S.; Cha, L.; Sim, M.; Jung, S.; Chun, W.Y.; Baik, H.W.; Shin, D.-M. Probiotic Supplementation Improves Cognitive Function and Mood with Changes in Gut Microbiota in Community-Dwelling Older Adults: A Randomized, Double-Blind, Placebo-Controlled, Multicenter Trial. *J. Gerontol. Ser. A Boil. Sci. Med. Sci.* **2020**. [CrossRef]
12. Trichopoulou, A.; Lagiou, P. Healthy Traditional Mediterranean Diet: An Expression of Culture, History, and Lifestyle. *Nutr. Rev.* **2009**, *55*, 383–389. [CrossRef] [PubMed]
13. Valls-Pedret, C.; Sala-Vila, A.; Serra-Mir, M.; Corella, D.; De La Torre, R.; Martínez-González, M.Á.; Martínez-Lapiscina, E.H.; Fitó, M.; Pérez-Heras, A.; Salas-Salvadó, J.; et al. Mediterranean Diet and Age-Related Cognitive Decline. *JAMA Intern. Med.* **2015**, *175*, 1094–1103. [CrossRef] [PubMed]
14. Marseglia, A.; Xu, W.; Fratiglioni, L.; Fabbri, C.; Berendsen, A.A.M.; Bialecka-Debek, A.; Jennings, A.; Gillings, R.; Meunier, N.; Caumon, E.; et al. Effect of the NU-AGE Diet on Cognitive Functioning in Older Adults: A Randomized Controlled Trial. *Front. Physiol.* **2018**, *9*, 349. [CrossRef] [PubMed]
15. De Filippis, F.; Pellegrini, N.; Vannini, L.; Jeffery, I.B.; La Storia, A.; Laghi, L.; Serrazanetti, D.I.; Di Cagno, R.; Ferrocino, I.; Lazzi, C.; et al. High-level adherence to a Mediterranean diet beneficially impacts the gut microbiota and associated metabolome. *Gut* **2015**, *65*, 1812–1821. [CrossRef] [PubMed]
16. Ghosh, T.S.; Rampelli, S.; Jeffery, I.B.; Santoro, A.; Neto, M.; Capri, M.; Giampieri, E.; Jennings, A.; Candela, M.; Turroni, S.; et al. Mediterranean diet intervention alters the gut microbiome in older people reducing frailty and improving health status: The NU-AGE 1-year dietary intervention across five European countries. *Gut* **2020**, *69*, 1218–1228. [CrossRef]
17. Machiels, K.; Joossens, M.; Sabino, J.; De Preter, V.; Arijs, I.; Eeckhaut, V.; Ballet, V.; Claes, K.; Van Immerseel, F.; Verbeke, K.; et al. Faculty Opinions recommendation of A decrease of the butyrate-producing species *Roseburia hominis* and *Faecalibacterium prausnitzii* defines dysbiosis in patients with ulcerative colitis. *Fac. Opin. Post-Public. Peer Rev. Biomed. Lit.* **2015**, *63*. [CrossRef]
18. Qin, J.; Li, Y.; Cai, Z.; Li, S.; Zhu, J.; Zhang, F.; Liang, S.; Zhang, W.; Guan, Y.; Shen, D.; et al. A metagenome-wide association study of gut microbiota in type 2 diabetes. *Nature* **2012**, *490*, 55–60. [CrossRef]

19. Wang, T.; Cai, G.; Qiu, Y.; Fei, N.; Zhang, M.; Pang, X.; Jia, W.; Cai, S.; Zhao, L. Structural segregation of gut microbiota between colorectal cancer patients and healthy volunteers. *ISME J.* **2012**, *6*, 320–329. [CrossRef]
20. Jonaitis, E.M.; Koscik, R.L.; Clark, L.R.; Ma, Y.; Betthauser, T.J.; Berman, S.E.; Allison, S.L.; Mueller, K.D.; Hermann, B.P.; Van Hulle, C.A.; et al. Measuring longitudinal cognition: Individual tests versus composites. *Alzheimer's Dement. Diagn. Assess. Dis. Monit.* **2018**, *11*, 74–84. [CrossRef]
21. Santoro, A.; Pini, E.; Scurti, M.; Palmas, G.; Berendsen, A.; Brzozowska, A.; Pietruszka, B.; Szczecinska, A.; Cano, N.; Meunier, N.; et al. Combating inflammaging through a Mediterranean whole diet approach: The NU-AGE project's conceptual framework and design. *Mech. Ageing Dev.* **2014**, *137*, 3–13. [CrossRef]
22. Berendsen, A.A.M.; Santoro, A.; Pini, E.; Cevenini, E.; Ostan, R.; Pietruszka, B.; Rolf, K.; Cano, N.J.M.; Caille, A.; Lyon-Belgy, N.; et al. Reprint of: A parallel randomized trial on the effect of a healthful diet on inflammageing and its consequences in European elderly people: Design of the NU-AGE dietary intervention study. *Mech. Ageing Dev.* **2014**, *136*, 14–21. [CrossRef] [PubMed]
23. Berendsen, A.M.; Van De Rest, O.; Feskens, E.J.M.; Santoro, A.; Ostan, R.; Pietruszka, B.; Brzozowska, A.; Stelmaszczyk-Kusz, A.; Jennings, A.; Gillings, R.; et al. Changes in Dietary Intake and Adherence to the NU-AGE Diet Following a One-Year Dietary Intervention among European Older Adults—Results of the NU-AGE Randomized Trial. *Nutrients* **2018**, *10*, 1905. [CrossRef] [PubMed]
24. Fried, L.P.; Tangen, C.M.; Walston, J.; Newman, A.B.; Hirsch, C.; Gottdiener, J.; Seeman, T.; Tracy, R.; Kop, W.J.; Burke, G.; et al. Frailty in Older Adults: Evidence for a Phenotype. *J. Gerontol. Ser. A Boil. Sci. Med. Sci.* **2001**, *56*, M146–M157. [CrossRef] [PubMed]
25. Slimani, N.; Ferrari, P.; Ocké, M.; Welch, A.; Boeing, H.; Van Liere, M.; Pala, V.; Amiano, P.; Lagiou, A.; Mattisson, I.; et al. Standardization of the 24-h diet recall calibration method used in the European Prospective Investigation into Cancer and Nutrition (EPIC): General concepts and preliminary results. *Eur. J. Clin. Nutr.* **2000**, *54*, 900–917. [CrossRef]
26. Netherlands Nutrition Centre. *Guidelines Wheel of Five;* Netherlands Nutrition Centre: The Hague, The Netherlands, 2020. (In Dutch)
27. Ruiz-Ojeda, F.J.; Plaza-Díaz, J.; Sáez-Lara, M.J.; Gil, A. Effects of Sweeteners on the Gut Microbiota: A Review of Experimental Studies and Clinical Trials. *Adv. Nutr.* **2019**, *10*, S31–S48. [CrossRef] [PubMed]
28. Salonen, A.; Nikkilä, J.; Jalanka-Tuovinen, J.; Immonen, O.; Rajilić-Stojanović, M.; Kekkonen, R.A.; Palva, A.; De Vos, W.M. Comparative analysis of fecal DNA extraction methods with phylogenetic microarray: Effective recovery of bacterial and archaeal DNA using mechanical cell lysis. *J. Microbiol. Methods* **2010**, *81*, 127–134. [CrossRef]
29. Claesson, M.J.; O'Sullivan, O.; Wang, Q.; Nikkilä, J.; Marchesi, J.R.; Smidt, H.; De Vos, W.M.; Ross, R.P.; O'Toole, P.W. Comparative Analysis of Pyrosequencing and a Phylogenetic Microarray for Exploring Microbial Community Structures in the Human Distal Intestine. *PLoS ONE* **2009**, *4*, e6669. [CrossRef]
30. Rajilić-Stojanović, M.; Heilig, H.G.H.J.; Molenaar, D.; Kajander, K.; Surakka, A.; Smidt, H.; De Vos, W.M. Development and application of the human intestinal tract chip, a phylogenetic microarray: Analysis of universally conserved phylotypes in the abundant microbiota of young and elderly adults. *Environ. Microbiol.* **2009**, *11*, 1736–1751. [CrossRef]
31. Lahti, L.; Elo, L.L.; Aittokallio, T.; Kaski, S. Probabilistic Analysis of Probe Reliability in Differential Gene Expression Studies with Short Oligonucleotide Arrays. *IEEE/ACM Trans. Comput. Biol. Bioinform.* **2011**, *8*, 217–225. [CrossRef]
32. Lahti, L.; Torrente, A.; Elo, L.L.; Brazma, A.; Rung, J. A fully scalable online pre-processing algorithm for short oligonucleotide microarray atlases. *Nucleic Acids Res.* **2013**, *41*, e110. [CrossRef]
33. Morris, J.C.; Heyman, A.; Mohs, R.C.; Hughes, J.; van Belle, G.; Fillenbaum, G.; Mellits, E.D.; Clark, C. The consortium to establish a registry for Alzheimer's disease (CERAD): I. Clinical and neuropsychological assessment of Alzheimer's disease. *Neurology* **1989**, *39*, 1159–1165.
34. Rosen, W.G. Verbal fluency in aging and dementia. *J. Clin. Neuropsychol.* **1980**, *2*, 135–146. [CrossRef]
35. Rosen, W.G.; Mohs, R.C.; Davis, K.L. A new rating scale for Alzheimer's disease. *Am. J. Psychiatry* **1984**, *141*, 1356–1364. [CrossRef] [PubMed]
36. Babcock, H.; Levy, L. *Test and Manual of Directions;* The Revised Examination for the Measurement of Efficiency of Mental Functioning; American Psychological Association: Washington, DC, USA, 1940.
37. Reitan, R.M. Validity of the Trail Making Test as an indicator of organic brain damage. *Percept. Motor Skills* **1958**, *8*, 271–276. [CrossRef]

38. Lewis, R.; Kupke, T. (Eds.) The Lafayette Clinic Repeatable Neuropsychological Test Battery: Its Development and Research Applications. In Proceedings of the Annual Meeting of the Southeastern Psychological Association, Hollywood, FL, USA, 4–7 May 1977.
39. Salthouse, T.A.; Babcock, R.L. Decomposing adult age differences in working memory. *Dev. Psychol.* **1991**, *27*, 763. [CrossRef]
40. Folstein, M.F.; Folstein, S.E.; McHugh, P.R. "Mini-mental state". A practical method for grading the cognitive state of patients for the clinician. *J. Psychiatr. Res.* **1975**, *12*, 189–198. [CrossRef]
41. Washburn, R.A.; Smith, K.W.; Jette, A.M.; Janney, C.A. The physical activity scale for the elderly (PASE): Development and evaluation. *J. Clin. Epidemiol.* **1993**, *46*, 153–162. [CrossRef]
42. R Core Team. *R: A Language and Environment for Statistical Computing*; R Foundation for Statistical Computing: Vienna, Austria, 2018.
43. Oksanen, J.; Blanchet, F.G.; Kindt, R.; Legendre, P.; Minchin, P.R.; O'Hara, R.B.; Simpson, G.L.; Solymos, P.; Stevens, M.H.H.; Wagner, H. *Vegan: Community Ecology Package*; R Package Version 2.5-2; R Foundation for Statistical Computing: Vienna, Austria, 2018.
44. Lahti, L.; Shetty, S.; Blake, T.; Salojarvi, J. *Microbiome R Package*; R Foundation for Statistical Computing: Vienna, Austria, 2017.
45. Revelle, W. *Psych: Procedures for Psychological, Psychometric, and Personality Research*; Northwestern University: Evanston, IL, USA, 2018.
46. Benjamini, Y.; Hochberg, Y. Controlling the False Discovery Rate: A Practical and Powerful Approach to Multiple Testing. *J. R. Stat. Soc. Ser. B Stat. Methodol.* **1995**, *57*, 289–300. [CrossRef]
47. Zhernakova, A.; Kurilshikov, A.; Bonder, M.J.; Tigchelaar, E.F.; Schirmer, M.; Vatanen, T.; Mujagic, Z.; Vila, A.V.; Falony, G.; Vieira-Silva, S.; et al. Population-based metagenomics analysis reveals markers for gut microbiome composition and diversity. *Science* **2016**, *352*, 565–569. [CrossRef]
48. Falony, G.; Joossens, M.; Vieira-Silva, S.; Wang, J.; Darzi, Y.; Faust, K.; Kurilshikov, A.; Bonder, M.J.; Valles-Colomer, M.; Vandeputte, D.; et al. Population-level analysis of gut microbiome variation. *Science* **2016**, *352*, 560–564. [CrossRef]
49. Partula, V.; Mondot, S.; Torres, M.J.; Kesse-Guyot, E.; Deschasaux, M.; Assmann, K.; Latino-Martel, P.; Buscail, C.; Julia, C.; Galan, P.; et al. Associations between usual diet and gut microbiota composition: Results from the Milieu Intérieur cross-sectional study. *Am. J. Clin. Nutr.* **2019**, *109*, 1472–1483. [CrossRef] [PubMed]
50. Shivappa, N.; Steck, S.E.; Hurley, T.G.; Hussey, J.R.; Hébert, J.R. Designing and developing a literature-derived, population-based dietary inflammatory index. *Public Health Nutr.* **2014**, *17*, 1689–1696. [CrossRef] [PubMed]
51. Candela, M.; Biagi, E.; Soverini, M.; Consolandi, C.; Quercia, S.; Severgnini, M.; Peano, C.; Turroni, S.; Rampelli, S.; Pozzilli, P.; et al. Modulation of gut microbiota dysbioses in type 2 diabetic patients by macrobiotic Ma-Pi 2 diet. *Br. J. Nutr.* **2016**, *116*, 80–93. [CrossRef] [PubMed]
52. Shah, V.; Lambeth, S.M.; Carson, T.; Lowe, J.; Ramaraj, T.; Leff, J.W.; Luo, L.; Bell, C.J. Composition Diversity and Abundance of Gut Microbiome in Prediabetes and Type 2 Diabetes. *J. Diabetes Obes.* **2015**, *2*, 108–114. [CrossRef] [PubMed]
53. Karlsson, F.H.; Fåk, F.; Nookaew, I.; Tremaroli, V.; Fagerberg, B.; Petranovic, D.; Bäckhed, F.; Nielsen, J.C. Symptomatic atherosclerosis is associated with an altered gut metagenome. *Nat. Commun.* **2012**, *3*, 1245. [CrossRef] [PubMed]
54. Chen, J.; Wright, K.; Davis, J.M.; Jeraldo, P.; Marietta, E.V.; Murray, J.; Nelson, H.; Matteson, E.L.; Taneja, V. An expansion of rare lineage intestinal microbes characterizes rheumatoid arthritis. *Genome Med.* **2016**, *8*, 1–14. [CrossRef] [PubMed]
55. Breban, M.; Tap, J.; Leboime, A.; Said-Nahal, R.; Langella, P.; Chiocchia, G.; Furet, J.-P.; Sokol, H. Faecal microbiota study reveals specific dysbiosis in spondyloarthritis. *Ann. Rheum. Dis.* **2017**, *76*, 1614–1622. [CrossRef]
56. Zheng, H.; Liang, H.; Wang, Y.; Miao, M.; Shi, T.; Yang, F.; Liu, E.; Yuan, W.; Ji, Z.-S.; Li, D.-K. Altered Gut Microbiota Composition Associated with Eczema in Infants. *PLoS ONE* **2016**, *11*, e0166026. [CrossRef]
57. Hall, A.B.; Yassour, M.; Sauk, J.; Garner, A.; Jiang, X.; Arthur, T.; Lagoudas, G.K.; Vatanen, T.; Fornelos, N.; Wilson, R.; et al. A novel *Ruminococcus gnavus* clade enriched in inflammatory bowel disease patients. *Genome Med.* **2017**, *9*, 1–12. [CrossRef]

58. Henke, M.T.; Kenny, D.J.; Cassilly, C.D.; Vlamakis, H.; Xavier, R.J.; Clardy, J. *Ruminococcus gnavus*, a member of the human gut microbiome associated with Crohn's disease, produces an inflammatory polysaccharide. *Proc. Natl. Acad. Sci. USA* **2019**, *116*, 12672–12677. [CrossRef]
59. Vinolo, M.A.R.; Rodrigues, H.G.; Nachbar, R.T.; Curi, R. Regulation of inflammation by short chain fatty acids. *Nutrients* **2011**, *3*, 858–876. [CrossRef] [PubMed]
60. Vieira, E.L.; Leonel, A.J.; Sad, A.P.; Beltrão, N.R.; Costa, T.F.; Ferreira, T.M.; Gomes-Santos, A.C.; Faria, A.M.; Peluzio, M.C.; Cara, D.C.; et al. Oral administration of sodium butyrate attenuates inflammation and mucosal lesion in experimental acute ulcerative colitis. *J. Nutr. Biochem.* **2012**, *23*, 430–436. [CrossRef] [PubMed]
61. Vernia, P.; Annese, V.; Bresci, G.; D'Albasio, G.; D'Incà, R.; Giaccari, S.; Ingrosso, M.; Mansi, C.; Riegler, G.; Valpiani, D.; et al. Topical butyrate improves efficacy of 5-ASA in refractory distal ulcerative colitis: Results of a multicentre trial. *Eur. J. Clin. Investig.* **2003**, *33*, 244–248. [CrossRef] [PubMed]
62. Sokol, H.; Pigneur, B.; Watterlot, L.; Lakhdari, O.; Bermúdez-Humarán, L.G.; Gratadoux, J.-J.; Blugeon, S.; Bridonneau, C.; Furet, J.-P.; Corthier, G.; et al. Faecalibacterium prausnitzii is an anti-inflammatory commensal bacterium identified by gut microbiota analysis of Crohn disease patients. *Proc. Natl. Acad. Sci. USA* **2008**, *105*, 16731–16736. [CrossRef]
63. Prosberg, M.; Bendtsen, F.; Vind, I.; Petersen, A.M.; Gluud, L.L. The association between the gut microbiota and the inflammatory bowel disease activity: A systematic review and meta-analysis. *Scand. J. Gastroenterol.* **2016**, *51*, 1407–1415. [CrossRef] [PubMed]
64. Kang, S.; Denman, S.E.; Morrison, M.; Yu, Z.; Dore, J.; Leclerc, M.; McSweeney, C.S. Dysbiosis of fecal microbiota in Crohn's disease patients as revealed by a custom phylogenetic microarray. *Inflamm. Bowel Dis.* **2010**, *16*, 2034–2042. [CrossRef] [PubMed]
65. Yassour, M.; Lim, M.Y.; Yun, H.S.; Tickle, T.L.; Sung, J.; Song, Y.-M.; Lee, K.; Franzosa, E.A.; Morgan, X.C.; Gevers, D.; et al. Sub-clinical detection of gut microbial biomarkers of obesity and type 2 diabetes. *Genome Med.* **2016**, *8*, 1–14. [CrossRef] [PubMed]
66. Zhang, X.; Shen, D.; Fang, Z.; Jie, Z.; Qiu, X.; Zhang, C.; Chen, Y.; Ji, L. Human Gut Microbiota Changes Reveal the Progression of Glucose Intolerance. *PLoS ONE* **2013**, *8*, e71108. [CrossRef]
67. Mbakwa, C.A.; Hermes, G.D.A.; Penders, J.; Savelkoul, P.H.M.; Thijs, C.; Dagnelie, P.C.; Mommers, M.; Zoetendal, E.G.; Smidt, H.; Arts, I.C.W. Gut Microbiota and Body Weight in School-Aged Children: The KOALA Birth Cohort Study. *Obesity* **2018**, *26*, 1767–1776. [CrossRef]
68. Depommier, C.; Everard, A.; Druart, C.; Plovier, H.; Van Hul, M.; Vieira-Silva, S.; Falony, G.; Raes, J.; Maiter, D.; Delzenne, N.M.; et al. Supplementation with *Akkermansia muciniphila* in overweight and obese human volunteers: A proof-of-concept exploratory study. *Nat. Med.* **2019**, *25*, 1096–1103. [CrossRef]
69. Canani, R.B. Potential beneficial effects of butyrate in intestinal and extraintestinal diseases. *World J. Gastroenterol.* **2011**, *17*, 1519–1528. [CrossRef] [PubMed]
70. Dueñas, M.; Muñoz-González, I.; Cueva, C.; Jiménez-Girón, A.; Sánchez-Patán, F.; Santos-Buelga, C.; Moreno-Arribas, M.V.; Bartolomé, B. A Survey of Modulation of Gut Microbiota by Dietary Polyphenols. *BioMed Res. Int.* **2015**, *2015*, 1–15. [CrossRef] [PubMed]
71. Roopchand, D.E.; Carmody, R.N.; Kuhn, P.; Moskal, K.; Rojas-Silva, P.; Turnbaugh, P.J.; Raskin, I. Dietary Polyphenols Promote Growth of the Gut Bacterium *Akkermansia muciniphila* and Attenuate High-Fat Diet–Induced Metabolic Syndrome. *Diabetes* **2015**, *64*, 2847–2858. [CrossRef]
72. Kovatcheva-Datchary, P.; Nilsson, A.C.; Akrami, R.; Lee, Y.S.; De Vadder, F.; Arora, T.; Hallen, A.; Martens, E.; Björck, I.; Bäckhed, F. Dietary Fiber-Induced Improvement in Glucose Metabolism Is Associated with Increased Abundance of Prevotella. *Cell Metab.* **2015**, *22*, 971–982. [CrossRef] [PubMed]
73. David, L.A.; Maurice, C.F.; Carmody, R.N.; Gootenberg, D.B.; Button, J.E.; Wolfe, B.E.; Ling, A.V.; Devlin, A.S.; Varma, Y.; Fischbach, M.A.; et al. Diet rapidly and reproducibly alters the human gut microbiome. *Nature* **2014**, *505*, 559–563. [CrossRef] [PubMed]
74. Salyers, A.A.; Vercellotti, J.R.; West, S.E.; Wilkins, T.D. Fermentation of mucin and plant polysaccharides by strains of Bacteroides from the human colon. *Appl. Environ. Microbiol.* **1977**, *33*, 319–322. [CrossRef] [PubMed]
75. Wexler, H.M.; Daya, S.; Berns, K.I. Bacteroides: The Good, the Bad, and the Nitty-Gritty. *Clin. Microbiol. Rev.* **2007**, *20*, 593–621. [CrossRef]

76. Wu, M.; McNulty, N.P.; Rodionov, D.A.; Khoroshkin, M.S.; Griffin, N.W.; Cheng, J.; Latreille, P.; Kerstetter, R.A.; Terrapon, N.; Henrissat, B.; et al. Genetic determinants of in vivo fitness and diet responsiveness in multiple human gut Bacteroides. *Science* **2015**, *350*, aac5992. [CrossRef]
77. De Filippis, F.; Pellegrini, N.; Laghi, L.; Gobbetti, M.; De Filippis, F. Unusual sub-genus associations of faecal Prevotella and Bacteroides with specific dietary patterns. *Microbiome* **2016**, *4*, 57. [CrossRef] [PubMed]
78. Larsen, J.M. The immune response to *Prevotella* bacteria in chronic inflammatory disease. *Immunology* **2017**, *151*, 363–374. [CrossRef]
79. Hermes, G.D.A.; Reijnders, D.; Kootte, R.S.; Goossens, G.H.; Smidt, H.; Nieuwdorp, M.; Blaak, E.E.; Zoetendal, E.G. Individual and cohort-specific gut microbiota patterns associated with tissue-specific insulin sensitivity in overweight and obese males. *Sci. Rep.* **2020**, *10*, 1–10. [CrossRef] [PubMed]
80. Reichardt, N.; Duncan, S.H.; Young, P.; Belenguer, A.; Leitch, C.M.; Scott, K.P.; Flint, H.J.; Louis, P. Phylogenetic distribution of three pathways for propionate production within the human gut microbiota. *ISME J.* **2014**, *8*, 1323–1335. [CrossRef] [PubMed]
81. Hosseini, E.; Grootaert, C.; Verstraete, W.; Van De Wiele, T. Propionate as a health-promoting microbial metabolite in the human gut. *Nutr. Rev.* **2011**, *69*, 245–258. [CrossRef] [PubMed]
82. Barbour, J.; Howe, P.R.C.; Buckley, J.D.; Bryan, J.; Coates, A.M. Nut consumption for vascular health and cognitive function. *Nutr. Res. Rev.* **2014**, *27*, 131–158. [CrossRef] [PubMed]
83. Devore, E.E.; Kang, J.H.; Breteler, M.M.B.; Grodstein, F. Dietary intakes of berries and flavonoids in relation to cognitive decline. *Ann. Neurol.* **2012**, *72*, 135–143. [CrossRef]
84. Kinney, J.W.; Bemiller, S.M.; Murtishaw, A.S.; Leisgang, A.M.; Salazar, A.M.; Lamb, B.T. Inflammation as a central mechanism in Alzheimer's disease. *Alzheimer's Dement. Transl. Res. Clin. Interv.* **2018**, *4*, 575–590. [CrossRef]
85. Nguyen, T.L.A.; Vieira-Silva, S.; Liston, A.; Raes, J. How informative is the mouse for human gut microbiota research? *Dis. Model. Mech.* **2015**, *8*, 1–16. [CrossRef]
86. Salthouse, T. *Major Issues in Cognitive Aging*; Oxford University Press: Oxford, UK, 2009; pp. 1–256.
87. Brink, A.C.V.D.; Brouwer-Brolsma, E.M.; Berendsen, A.A.M.; Van De Rest, O. The Mediterranean, Dietary Approaches to Stop Hypertension (DASH), and Mediterranean-DASH Intervention for Neurodegenerative Delay (MIND) Diets Are Associated with Less Cognitive Decline and a Lower Risk of Alzheimer's Disease—A Review. *Adv. Nutr.* **2019**, *10*, 1040–1065. [CrossRef]
88. Li, B.; He, Y.; Ma, J.; Huang, P.; Du, J.; Cao, L.; Wang, Y.; Xiao, Q.; Tang, H.; Chen, S. Mild cognitive impairment has similar alterations as Alzheimer's disease in gut microbiota. *Alzheimer's Dement.* **2019**, *15*, 1357–1366. [CrossRef]
89. Kobayashi, Y.; Kuhara, T.; Oki, M.; Xiao, J.-Z. Effects of Bifidobacterium breve A1 on the cognitive function of older adults with memory complaints: A randomised, double-blind, placebo-controlled trial. *Benef. Microbes* **2019**, *10*, 511–520. [CrossRef]
90. Akbari, E.; Asemi, Z.; Kakhaki, R.D.; Bahmani, F.; Kouchaki, E.; Tamtaji, O.R.; Hamidi, G.A.; Salami, M. Effect of Probiotic Supplementation on Cognitive Function and Metabolic Status in Alzheimer's Disease: A Randomized, Double-Blind and Controlled Trial. *Front. Aging Neurosci.* **2016**, *8*, 256. [CrossRef] [PubMed]
91. Benton, D.; Williams, C.S.; Brown, A.E. Impact of consuming a milk drink containing a probiotic on mood and cognition. *Eur. J. Clin. Nutr.* **2007**, *61*, 355–361. [CrossRef] [PubMed]
92. Inoue, T.; Kobayashi, Y.; Mori, N.; Sakagawa, M.; Xiao, J.-Z.; Moritani, T.; Sakane, N.; Nagai, N. Effect of combined bifidobacteria supplementation and resistance training on cognitive function, body composition and bowel habits of healthy elderly subjects. *Benef. Microbes* **2018**, *9*, 843–853. [CrossRef] [PubMed]

93. Mazereeuw, G.; Lanctôt, K.L.; Chau, S.A.; Swardfager, W.; Herrmann, N. Effects of omega-3 fatty acids on cognitive performance: A meta-analysis. *Neurobiol. Aging* **2012**, *33*, 1482.e17–1482.e29. [CrossRef]
94. An, R.; Wilms, E.; Smolinska, A.; Hermes, G.D.A.; Masclee, A.; De Vos, P.; Schols, H.; Van Schooten, F.-J.; Smidt, H.; Jonkers, D.; et al. Sugar Beet Pectin Supplementation Did Not Alter Profiles of Fecal Microbiota and Exhaled Breath in Healthy Young Adults and Healthy Elderly. *Nutrition* **2019**, *11*, 2193. [CrossRef] [PubMed]

**Publisher's Note:** MDPI stays neutral with regard to jurisdictional claims in published maps and institutional affiliations.

nutrients

MDPI

Article

# Comparison of Different Dietary Indices as Predictors of Inflammation, Oxidative Stress and Intestinal Microbiota in Middle-Aged and Elderly Subjects

**Sergio Ruiz-Saavedra [1,2,3], Nuria Salazar [1,3], Ana Suárez [2,3], Clara G. de los Reyes-Gavilán [1,3], Miguel Gueimonde [1,3] and Sonia González [2,3,*]**

1 Department of Microbiology and Biochemistry of Dairy Products, Instituto de Productos Lácteos de Asturias (IPLA-CSIC), 33300 Asturias, Spain; sergioruizsa3@gmail.com (S.R.-S.); nuriasg@ipla.csic.es (N.S.); greyes_gavilan@ipla.csic.es (C.G.d.l.R.-G.); mgueimonde@ipla.csic.es (M.G.)

2 Department of Functional Biology, University of Oviedo, 33006 Oviedo, Spain; anasua@uniovi.es

3 Metabolism Area, Instituto de Investigación Sanitaria del Principado de Asturias (ISPA), 33011 Oviedo, Spain

* Correspondence: soniagsolares@uniovi.es; Tel.: +34-985-104-209

Received: 12 November 2020; Accepted: 12 December 2020; Published: 15 December 2020

**Abstract:** During the last decades the gut microbiota has been identified as a key mediator in the diet-health interaction. However, our understanding on the impact of general diet upon microbiota is still limited. Dietary indices represent an essential approach for addressing the link between diet and health from a holistic point of view. Our aim was to test the predictive potential of seven dietary ratings on biomarkers of inflammation, oxidative stress and on the composition and metabolic activity of the intestinal microbiota. A cross-sectional descriptive study was conducted on a sample of 73 subjects aged >50 years with non-declared pathologies. Dietary inflammatory index (DII), Empirical Dietary Inflammatory Index (EDII), Healthy Eating Index (HEI), Alternative Healthy Eating Index (AHEI), Mediterranean adapted Diet Quality Index-International (DQI-I), Modified Mediterranean Diet Score (MMDS) and relative Mediterranean Diet Score (rMED) were calculated based on a Food Frequency Questionnaire. Major phylogenetic types of the intestinal microbiota were determined by real time polymerase chain reaction (qPCR) and fecal short chain fatty acids (SCFAs) by gas chromatography. While DII, HEI, DQI-I and MMDS were identified as predictors of *Faecalibacterium prausnitzii* levels, AHEI and MMDS were negatively associated with *Lactobacillus* group. HEI, AHEI and MMDS were positively associated with fecal SCFAs. In addition, DII and EDII explained lipoperoxidation level and Mediterranean scores the serum IL-8 concentrations. The lower detection of IL-8 in individuals with higher scores on Mediterranean indices may be partially explained by the increased levels of the anti-inflammatory bacterium *F. prausnitzii* in such individuals.

**Keywords:** dietary patterns; Mediterranean diet; dietary indices; microbiota; elderly

## 1. Introduction

The empirical relationship between diet and health has been recognized since the time of Hippocrates (400 BC). During the last decades, solid scientific evidence has accumulated on the protective role of certain foods, such as fruits and vegetables, on the risk of suffering non-transmissible diseases [1,2]. However, understanding the net impact of diet on health is more complex than studying isolated components. Humans do not consume single foods but a wide variety of combinations of foods forming the so-called dietary pattern. Therefore, from a physiological point of view, the analysis of the eating habits, considering the interactions between different foods and their components, is of paramount interest [3,4]. In this context, dietary indices have been developed as a useful tool for categorizing dietary practices in different populations, encouraging the comparison among different studies [5,6]. The currently available dietary indices could be clustered into three main categories: the inflammatory ones [7,8], those quantifying the adherence to dietary guidelines [9,10] and those evaluating the degree of adaptation to the Mediterranean dietary pattern [11]. Both, the number and type of components included in each index is different, depending on the purpose for which they have been created and the dietary habits of the population for which they have been designed. While inflammatory ratings, such as the Dietary Inflammatory Index (DII) [7] or the Empirical Dietary Inflammatory Index (EDII) [8], have proven to be useful in the prediction of inflammatory parameters such as C-reactive protein (CRP), interleukin 6 (IL-6) or adiponectin levels [12–14], the Healthy Eating Index (HEI) [15] or the Alternative Healthy Eating Index (AHEI) [10], based on diet quality, have been useful for assessing the risk of chronic diseases [10,16]. Among the different indicators, the Mediterranean dietary index is perhaps the one accumulating more scientific evidence about its beneficial impact on morbidity and mortality [17] through the reduction of different parameters related to oxidative stress [18,19].

Therefore, these indices represent a key tool in the assessment of the association between diet and health. Moreover, the inclusion of some novel biological parameters, such as the gut microbiota, in the study of such correlations may broaden their applicability [20,21]. In this regard, HEI and Mediterranean ratings were recently found to be associated with gut microbiota in terms of both microbial composition and diversity [22–24]. In more detail, MedDietScore index has been associated with higher fecal bifidobacteria: *Escherichia coli* ratio, total bacteria and short chain fatty acids (SCFAs) [25]. However, the studies in this area are still limited and there are no studies comparing the different indices for assessing the interaction between regular diet and microbiota.

In view of this evidence, the main objective of this work was to analyze the diet of a group of middle aged and elderly participants, without declared pathologies, through different dietary indices and to examine their predictive potential on parameters related to inflammation, oxidative stress and the composition and metabolic activity of the intestinal microbiota.

## 2. Materials and Methods

### 2.1. Participants

The sample included 73 volunteers recruited in Asturias (North of Spain), aged between 56 and 95 years and with a Body Mass Index (BMI) [weight (kg)/height ($m^2$)] from 19.9 to 37.5 kg/$m^2$. In an individual interview, volunteers were informed about the objectives of the study and an informed written consent was obtained before enrolment. Exclusion criteria were the presence of diagnosed immune or digestive related pathologies as well as consumption of corticoids, immunosuppressive drugs, monoclonal antibodies, probiotics or antibiotics in the previous month. The study was approved by the Regional Ethics Committee for Clinical Research (Servicio de Salud del Principado de Asturias n° 17/2010).

### 2.2. Nutritional Assessment

A previously validated annual semi-quantitative food frequency questionnaire (FFQ) [26,27] was used by trained personnel to assess volunteers' regular food intake in a personal interview of approximately 1 h duration. Methodological issues about dietary assessment were published elsewhere [28]. Food composition tables of CESNID (Centro de Enseñanza Superior de Nutrición Humana y Dietética) were used to transform food consumption into energy and macronutrients intake [29]. The (poly)phenol content in foods was completed using the Phenol Explorer database [30]. Fiber components were determined using the Marlett et al. food composition tables [31]. Glucosinolates levels were obtained from McNaughton et al. [32] and isothiocyanates along with aliphatic glucosinolates food content were derived from glucosinolates levels following European Prospective Investigation into Cancer and Nutrition (EPIC) criteria [33]. Glucosinolate side chains concentrations were ascertained from International

Agency for Research of Cancer (IARC) data [34]. At the time of carrying out the blood extraction, height and weight were taken by standardized protocols [35]. At the same time as the FFQ interview was conducted, a questionnaire on socio-economic factors (such as level of education or type of work) and lifestyle (physical activity, smoking and self-perception of health status) was administered.

### 2.3. Dietary Indices Calculation

A calculation of seven dietary indices was carried out, including DII, EDII, HEI, AHEI, Mediterranean adapted Diet Quality Index-International (DQI-I) [36], Modified Mediterranean Diet Score (MMDS) [37] and relative Mediterranean Diet Score (rMED) [38], as shown in Supplementary Table S1. IBM SPSS program version 24.0 (IBM SPSS, Inc., Chicago, IL, USA) was used to design a database for calculating indices scores from our FFQ data.

DII scores were calculated by evaluating 35 parameters (out of 45 possible items). Components such as eugenol, ginger, saffron, turmeric, green/black tea, isoflavones, pepper, thyme/oregano and rosemary were excluded because a lack of information about them in the FFQ recordings. First, the consumption levels of parameters were standardized by subtracting daily global consumption mean and dividing by the global standard deviation. The resulting Z-scores were then converted to percentile scores and centered by doubling and subtracting one. These centered-percentile scores were multiplied by the overall food parameter-specific inflammatory effect score to obtain the 'food parameter-specific score' (FPES). All FPES of an individual were summed to obtain the final DII Score. Values in our sample ranged from −4.62 to 4.45, with negative scores predicting lower inflammation and positive scores higher dietary-derived inflammation.

For EDII scores, 18 components were accounted. The number of servings consumed was calculated for each component. The resulting values were multiplied by the "Weight" of the components and divided by 1000. All weighted components were summed to obtain each EDII Score, with values ranging from −1.56 to 2.21 in our sample. The more positive the result, the more prone to higher concentrations of inflammatory biomarkers.

The HEI is an index comprised of 13 components. For each parameter, the amount of a dietary component, in g, cups or oz equivalents were calculated per 1000 kcal. These densities were scored according to recommended consumption values. For negative scoring components, considered to be consumed in small quantities due to their negative impact on health, the individuals get higher scores when the consumption values are lower than the established threshold. Then, all components were summed to obtain the HEI score. HEI was reflecting a total score from 29.49 to 77.76 in our sample, with higher values reflecting a healthier diet.

AHEI score calculations procedure was quite similar to that of HEI, with 11 components in total and values from 37.34 to 80.79 in our sample. In the same way, higher values represent healthier diets. For each component, the consumption in servings per day was calculated and scored according to AHEI-2010 criteria. All components were summed to obtain the total AHEI score of each individual.

Considered the most suitable index for international analyses, DQI-I accounts for 18 components. Here, we evaluated a Mediterranean adaptation of DQI-I. The amount consumed of each component was rated from 0 to 3, 5 or 6, depending on the component and all the points were summed to obtain DQI-I scores, whose final values were between 33 and 68 (minimum and maximum). Here, a total sum of 100 would show an individual with perfect adherence to the main four categories evaluated in this index, while a score of 0 is reflecting a diet far away from recommended dietary guidelines.

Composed of 9 components, rMED and MMDS were derived from the index originally developed by Trichopoulou et al. [11] to evaluate the degree of adherence to a traditional Mediterranean diet. rMED was calculated from components intake based on the density nutrients model. Once amounts of consumption per 1000 kcal were obtained, the sample was split in tertiles for each of the nine components. According to the tertile position for each component, individuals were rated. Summing all the ratings we obtained the rMED score, ranging from 2 to 12 in our sample. In the case of MMDS, each parameter was scored 0 or 1 according to the cut-off values of the sex specific-medians among the participants. Final MMDS scores ranged from 1 to 7 in our sample. In both indices, the highest scores are showing a higher adherence to Mediterranean diet patterns.

### 2.4. Blood Biochemical Analyses

Fasting blood samples were drawn by venipuncture and centrifuged (1000× *g*, 15 min). Plasma and serum aliquots were kept at −20 °C for later analyses. Serum glucose, serum total cholesterol, serum HDL-cholesterol, serum LDL-cholesterol and serum triglycerides were determined by using an automated biochemical autoanalyzer in an independent laboratory. Serum levels of CRP were determined by ELISA (CRP Human Instant ELISA, Ebioscience, San Diego, CA, USA) and malondialdehyde (MDA) by a colorimetric assay of lipid peroxidation (Byoxytech LPO-586 assay, Oxis International S.A., Paris, France) [39]. Serum leptin was determined by ELISA (Human Leptin ELISA Development Kit 900-K90, PeproTech Inc., Rocky Hill, NJ, USA) according to the manufacturer's instructions. Colorimetric assay P40117 (Innoprot, Innovative Technologies in Biological Systems, S.L., Spain) was used to determine total antioxidant capacity (TAC) in serum [40]. A multiplex immunoassay (Cytometric Bead Array, CBA, BD Biosciences) by flow cytometry allowed to quantify levels of serum IL-10, Tumor Necrosis Factor-Alpha (TNF-α), IL-8, IL-17 and IL-12, while the concentration of transforming growth factor (TGF)-β was determined by ELISA (BD OptEIATM, BD Biosciences). The phagocytic capacity was quantified in a FACSCanto II Flow Cytometer (Becton Dickinson, BD Biosciences, San Diego, CA) by using the Phagotest® kit (Orpegen Pharma, Heildelberg, Germany). Natural killer (NK) cell activity was determined by flow cytometry, using the NKtest® kit (Orpegen Pharma).

### 2.5. Fecal Collection and Microbial Analysis

Detailed instructions about fecal samples collection were given to participants who also were provided with a sterile container. After deposition, samples were immediately frozen at −20 °C and transported to the laboratory. For analyses, samples were thawed at room temperature (24 ± 2 °C), weighed, diluted 1/10 in sterile PBS and homogenized using a LabBlender 400 Stomacher (Seward Medical, London, UK) for 4 min; the DNA was extracted using the QIAamp DNA stool mini kit (Qiagen, Hilden, Germany) following previously described procedures [41]. 7500 Fast Real-Time PCR System (Applied Biosystems, Foster City, CA, USA) and SYBR Green PCR Master Mix (Applied Biosystems) were used to achieve the quantification of bacterial populations, including the major bacterial groups present in the human gut (Supplementary Table S2). Procedure instructions were published elsewhere [41]. Fecal DNA extracts were analyzed and the mean quantity per gram of fecal wet weight was calculated for each bacterial group. One milliliter of the homogenized feces were centrifuged and supernatants were analyzed by gas chromatography to determine acetate, propionate and butyrate concentrations, as previously indicated [42]. A chromatograph 6890N (Agilent Technologies Inc., Palo Alto, CA, USA) connected to a mass spectrometry detector (MS) 5973N (Agilent Technologies) and a flame ionization detector (FID) was used for identification and quantification of SCFAs, respectively, as described previously [43].

### 2.6. Statistical Analyses

Statistical analysis was performed using the IBM SPSS program version 24.0 (IBM SPSS, Inc., Chicago, IL, USA). Mean dietary scores were analyzed by a Student t-test and Bonferroni multiple comparison according to general, socio-economic and health-related characteristics, such as smoking status, educational level, mood feeling or self-health perception among others. Similar procedure was performed to analyze mean levels of microbiological and blood variables according to age group. Goodness of fit to the normal distribution was analyzed employing the Kolmogorov–Smirnov test. Scores were examined as predictors of gut microbial groups, fecal SCFAs and blood biomarkers by regression analyses controlling by age and energy intake. These variables included *Akkermansia*, *Bacteroides-Prevotella-Porphyromonas*, *Bifidobacterium*, *Clostridia* cluster XIVa, *Lactobacillus* group,

*Faecalibacterium prausnitzii*, acetic acid, propionic acid, butyric acid, glucose, triglycerides, Low density lipoprotein-high density lipoprotein (LDL-HDL) ratio, leptin, serum malondialdehyde (MDA), (total antioxidant capacity) TAC, C-Reactive protein (CRP), TGF-β, IL-10, IL-17, IL-8, IL-12, TNF-α, Phagocytosis granulocytes (%), Phagocytosis granulocytes and monocytes (%) and NK cell activity. When the distribution of variables was skewed (CRP, TGF-β, IL-10, IL-17, IL-8, IL-12, TNF-α) the values were converted to their natural logarithm. Association of microbial groups with food groups, macronutrients and micronutrients were evaluated through linear regression analyses adjusting by age and energy. The resulting data were plotted in a heatmap using the "pheatmat" function of the R program (version 3.5.1 for Windows). A Pearson correlation test was performed to elucidate closeness among dietary indices. This information was introduced in the program R, using the package "pheatmap," to clusterize and plotting indices based on Euclidean distances. To test the association among dietary indices and previously reported health-beneficial dietary compounds, linear regression analyses were performed and plotted as forehead mentioned. To indicate statistical significance in the interpretation of results, the probability value of 0.05 was used.

## 3. Results

The average score on dietary indices according to general characteristics and socio-economic status, lifestyle and health-related factors of the studied sample is presented in Table 1. Among all the variables examined, only significant differences were observed for the age and these were found in all the studied dietary indices. Subjects over 65 years presented worse dietary scores than those in the group of 50–65 years. At the time of interpreting the results it should be taken into account that unlike the rest of the indices studied, in DII and EDII a higher score is associated with a more pro-inflammatory diet. Both DII and DQI-I showed better scores in people with energy intake higher than 1994.8 kcal. EDII score was found to be higher (worse dietary quality) as BMI increases. A lower score in the AHEI, DQI-I and Mediterranean dietary indices (rMED, MMDS), associated with worse dietary quality, was found in those subjects with bad self-health perception. Therefore, age and energy intake have been introduced in further analyses carried out as a covariate.

A general description of the variables that are subsequently analyzed in the study according to age groups is shown in Table 2. The levels of the bacterial groups analyzed were in the range of those previously reported in similar populations and demonstrated the large inter-individual variability present in the human adult fecal microbiota. Significant differences were observed in most of the microbiological parameters analyzed according to age. Subjects over 65 years of age presented lower fecal levels of *Bacteroides-Prevotella-Porphyromonas* group, *Clostridia* cluster *XIVa* and *Faecalibacterium*, as well as all the short chain fatty acids determined. Blood parameters are within the normal physiological ranges and were similar between the groups evaluated except for MDA, IL-8, IL-12 and TNF-α, whose concentration is higher in subjects over 65 years of age.

In order to analyze possible linear relationships among the different dietary indices scores and fecal microbial groups, a linear regression analysis was conducted and a heatmap was plotted (Figure 1). Lower scores on indices related to diet quality (HEI, AHEI, DQI-I), suggestive of an unhealthier diet, were associated with increased *Akkermansia* levels. While DII, HEI, DQI-I and MMDS have shown potential as predictors of *F. prausnitzii*, which showed higher levels in those individuals with healthier diets, only AHEI and MMDS were negatively associated with *Lactobacillus* levels. These results were further examined including other health-related parameters in the analysis (Table 3). In relation to the production of SCFAs, higher scores in HEI, AHEI and MMDS indices were positively associated with the formation of acetic, propionic and butyric acids. Furthermore, as expected, inflammatory indices (DII, EDII) were the best determinants of lipoperoxidation blood levels, while Mediterranean ones were the best identifiers of serum IL-8 concentrations.

**Table 1.** Dietary indices score according to different characteristics of the sample.

| Variable | n | DII | EDII | HEI | AHEI | DQI-I | rMED | MMDS |
|---|---|---|---|---|---|---|---|---|
| Mean (IQR) | | −0.35 (1.07–−2.25) | 0.67 (1.17–0.19) | 58.16 (65.68–49.95) | 60.99 (67.97–55.08) | 49.03 (55–44) | 7.16 (9–6) | 3.37 (4–2) |
| Gender | | | | | | | | |
| Male | 20 | −1.04 ± 2.27 $_{a}$ | 0.64 ± 0.73 $_{a}$ | 55.92 ± 11.47 $_{a}$ | 63.48 ± 8.51 $_{a}$ | 47.15 ± 9.53 $_{a}$ | 7.80 ± 2.48 $_{a}$ | 3,65 ± 1,31 $_{a}$ |
| Female | 53 | −0.10 ± 2.54 $_{a}$ | 0.68 ± 0.78 $_{a}$ | 59.00 ± 10.00 $_{a}$ | 60.10 ± 10.52 $_{a}$ | 49.74 ± 7.04 $_{a}$ | 6.92 ± 2.28 $_{a}$ | 3.26 ± 1.46 $_{a}$ |
| Age (years) | | | | | | | | |
| 50–65 | 33 | −1.92 ± 2.05 $_{a}$ | 0.24 ± 0.62 $_{a}$ | 62.63 ± 9.01 $_{a}$ | 64.07 ± 8.94 $_{a}$ | 51.97 ± 0.83 $_{a}$ | 8.39 ± 2.14 $_{a}$ | 3.67 ± 1.29 $_{a}$ |
| >65 | 40 | 0.98 ± 2.02 $_{b}$ | 1.02 ± 0.69 $_{b}$ | 54.46 ± 10.16 $_{b}$ | 58.39 ± 10.37 $_{b}$ | 46.60 ± 5.96 $_{b}$ | 6.15 ± 2.03 $_{b}$ | 3.13 ± 1.49 $_{a}$ |
| Energy intake (kcal/day) | | | | | | | | |
| ≤1538.9 | 25 | 0.01 ± 2.3 $_{a,b}$ | 0.41 ± 0.67 $_{a}$ | 58.95 ± 11.43 $_{a}$ | 63.49 ± 8.21 $_{a}$ | 46.36 ± 7.76 $_{a}$ | 7.08 ± 2.33 $_{a}$ | 3.52 ± 1.66 $_{a}$ |
| 1539–1994.8 | 24 | 0.40 ± 2.19 $_{a}$ | 0.77 ± 0.62 $_{a}$ | 57.22 ± 9.42 $_{a}$ | 60.17 ± 9.80 $_{a}$ | 47.46 ± 7.17 $_{a}$ | 6.79 ± 1.86 $_{a}$ | 3 ± 1.02 $_{a}$ |
| >1994.9 | 24 | −1.47 ± 2.65 $_{b}$ | 0.84 ± 0.92 $_{a}$ | 58.27 ± 10.67 $_{a}$ | 59.32 ± 11.84 $_{a}$ | 53.38 ± 6.88 $_{b}$ | 7.63 ± 2.76 $_{a}$ | 3.58 ± 1.47 $_{a}$ |
| BMI (kg/m$^2$) | | | | | | | | |
| <25 | 19 | −0.99 ± 2.77 $_{a}$ | 0.37 ± 0.72 $_{a}$ | 10.02 ± 19 $_{a}$ | 62.74 ± 9.90 $_{a}$ | 48.95 ± 9.20 $_{a}$ | 7.79 ± 2.32 $_{a}$ | 3.68 ± 1.60 $_{a}$ |
| 25.0–29.9 | 37 | −0.39 ± 2.37 $_{a}$ | 0.62 ± 0.77 $_{a,b}$ | 10.10 ± 37 $_{a}$ | 61.83 ± 10.16 $_{a}$ | 48.43 ± 7.65 $_{a}$ | 7.11 ± 2.50 $_{a}$ | 3.22 ± 1.38 $_{a}$ |
| ≥30 | 17 | 0.44 ± 2.33 $_{a}$ | 1.10 ± 0.62 $_{b}$ | 10.35 ± 17 $_{a}$ | 57.28 ± 9.77 $_{a}$ | 50.41 ± 6.75 $_{a}$ | 6.59 ± 1.97 $_{a}$ | 3.35 ± 1.32 $_{a}$ |
| Depositions (times/week) | | | | | | | | |
| ≤3 | 8 | 0.51 ± 2.12 $_{a}$ | 0.85 ± 0.58 $_{a}$ | 53.09 ± 12.71 $_{a}$ | 60.29 ± 6.12 $_{a}$ | 47.50 ± 7.96 $_{a}$ | 7.00 ± 2.51 $_{a}$ | 3.00 ± 1.60 $_{a}$ |
| >3–7 | 57 | −0.11 ± 2.48 $_{a}$ | 0.71 ± 0.76 $_{a}$ | 58.40 ± 10.02 $_{a}$ | 60.74 ± 10.44 $_{a}$ | 49.19 ± 7.89 $_{a}$ | 6.93 ± 2.34 $_{a}$ | 3.35 ± 1.38 $_{a}$ |
| >7 | 8 | −2.82 ± 1.28 $_{a}$ | 0.16 ± 0.85 $_{b}$ | 61.51 ± 10.59 $_{a}$ | 63.44 ± 10.92 $_{a}$ | 49.38 ± 8.07 $_{a}$ | 9.00 ± 1.51 $_{a}$ | 3.88 ± 1.55 $_{a}$ |
| Smoking status | | | | | | | | |
| Never | 9 | −2.27 ± 1.74 $_{a}$ | 0.11 ± 0.51 $_{a}$ | 67.14 ± 8.34 $_{a}$ | 64.66 ± 7.90 $_{a}$ | 49.22 ± 10.94 $_{a}$ | 8.78 ± 2.54 $_{a}$ | 4.0 ± 1.32 $_{a}$ |
| Current | 17 | −0.40 ± 2.55 $_{a}$ | 0.63 ± 0.67 $_{a}$ | 60.79 ± 9.12 $_{a}$ | 61.14 ± 8.21 $_{a}$ | 50.35 ± 6.72 $_{a}$ | 7.53 ± 2.24 $_{a}$ | 3.65 ± 1.37 $_{a}$ |
| Former | 16 | −1.11 ± 2.04 $_{a}$ | 0.38 ± 0.59 $_{a}$ | 60.17 ± 10.28 $_{a}$ | 65.93 ± 7.43 $_{a}$ | 49.81 ± 8.23 $_{a}$ | 8.25 ± 1.77 $_{a}$ | 3.75 ± 1.18 $_{a}$ |
| Educational level | | | | | | | | |
| None or primary school | 2 | −2.46 ± 0.42 $_{a}$ | 0.45 ± 0.49 $_{a}$ | 66.19 ± 5.27 $_{a}$ | 63.13 ± 7.38 $_{a}$ | 57.00 ± 7.07 $_{a}$ | 7.00 ± 1.41 $_{a}$ | 3.50 ± 0.71 $_{a}$ |
| Technical school | 5 | −1.73 ± 1.80 $_{a}$ | −0.03 ± 0.34 $_{a}$ | 63.81 ± 6.21 $_{a}$ | 63.80 ± 1.99 $_{a}$ | 51.60 ± 6.35 $_{a}$ | 9.60 ± 1.67 $_{a}$ | 3.20 ± 1.10 $_{a}$ |
| Secondary school | 7 | −1.40 ± 2.86 $_{a}$ | 0.37 ± 0.65 $_{a}$ | 59.20 ± 8.96 $_{a}$ | 60.29 ± 8.83 $_{a}$ | 54.00 ± 8.06 $_{a}$ | 7.71 ± 2.75 $_{a}$ | 3.29 ± 1.25 $_{a}$ |
| University degree | 13 | −2.00 ± 1.61 $_{a}$ | 0.19 ± 0.46 $_{a}$ | 63.55 ± 9.67 $_{a}$ | 66.63 ± 7.77 $_{a}$ | 48.85 ± 9.14 $_{a}$ | 8.54 ± 1.90 $_{a}$ | 4.00 ± 1.22 $_{a}$ |
| Mood feeling | | | | | | | | |
| Really satisfied | 42 | −0.35 ± 2.61 $_{a}$ | 0.70 ± 0.81 $_{a}$ | 57.99 ± 10.07 $_{a}$ | 60.63 ± 9.90 $_{a}$ | 48.88 ± 8.16 $_{a}$ | 7.24 ± 2.54 $_{a}$ | 3.48 ± 1.45 $_{a}$ |
| Satisfied | 21 | −0.68 ± 2.28 $_{a}$ | 0.57 ± 0.65 $_{a}$ | 61.28 ± 10.14 $_{a}$ | 63.27 ± 10.61 $_{a}$ | 49.48 ± 7.61 $_{a}$ | 7.76 ± 1.84 $_{a}$ | 3.38 ± 1.40 $_{a}$ |
| Unsatisfied | 8 | 0.96 ± 2.17 $_{a}$ | 0.78 ± 0.58 $_{a}$ | 51.65 ± 11.64 $_{a}$ | 57.86 ± 7.67 $_{a}$ | 46.75 ± 6.90 $_{a}$ | 5.25 ± 1.98 $_{a}$ | 2.63 ± 1.30 $_{a}$ |

**Table 1.** *Cont.*

| Variable | n | DII | EDII | HEI | AHEI | DQI-I | rMED | MMDS |
|---|---|---|---|---|---|---|---|---|
| Self-health perception | | | | | | | | |
| Excellent-good | 42 | $-0.01 \pm 2.20_a$ | $0.75 \pm 0.72_a$ | $57.17 \pm 9.85_a$ | $59.24 \pm 7.88_a$ | $47.12 \pm 7.15_a$ | $6.95 \pm 2.39_{a,b}$ | $3.26 \pm 1.31_{a,b}$ |
| Normal | 18 | $-1.53 \pm 2.54_a$ | $0.51 \pm 0.68_a$ | $61.38 \pm 101.39_a$ | $66.44 \pm 10.49_b$ | $52.28 \pm 7.89_a$ | $8.33 \pm 2.22_a$ | $4.00 \pm 1.57_a$ |
| Bad | 12 | $0.56 \pm 2.76_a$ | $0.75 \pm 0.95_a$ | $56.43 \pm 11.15_a$ | $58.14 \pm 13.62_{a,b}$ | $49.92 \pm 8.40_a$ | $6.08 \pm 1.88_b$ | $2.67 \pm 1.23_b$ |

First row values are the mean of the corresponding index in the sample followed by the interquartile range (IQR = Q3 − Q1). The rest of values are shown as mean ± standard deviation (SD). For a certain variable showing two of more subgroups, only values in the same column with different subscript present a statistically significant difference ($p \leq 0.05$). Mood feeling refer to volunteer's awareness about the satisfaction with their life. Self-perception of health status refer to individual perception about health-quality degree. Inflammatory indices (DII, Dietary Inflammatory Index. EDII, Empirical Dietary Inflammatory Index): negative values favor non-inflammatory states and positives values enhance inflammation. Dietary Quality indices (HEI, Healthy Eating Index. AHEI, Alternative Healthy Eating Index. DQI-I, Diet Quality Index-International): higher scores are reflecting consumption values similar to those recommended in dietary guidelines. Mediterranean Dietary indices (rMED, Relative Mediterranean Diet Score. MMDS, Modified Mediterranean Diet Score): higher values showing a higher degree of adherence to the Mediterranean diet.

**Table 2.** Microbial levels, short chain fatty acid (SCFA) concentration and blood biomarkers according to age groups.

| Variable | Age Groups | |
|---|---|---|
| | G1 (≤65) ($n$ = 33) | G2 (>65) ($n$ = 40) |
| *Akkermansia* ($Log_{10}$ n° cells/gram feces) | 6.43 ± 1.88 $_a$ | 6.99 ± 1.77 $_a$ |
| *Bacteroides-Prevotella-Porphyromonas* ($Log_{10}$ n° cells/gram feces) | 9.32 ± 0.82 $_a$ | 8.79 ± 0.69 $_b$ |
| *Bifidobacterium* ($Log_{10}$ n° cells/gram feces) | 7.93 ± 1.53 $_a$ | 7.55 ± 1.10 $_a$ |
| *Clostridia* cluster XIVa ($Log_{10}$ n° cells/gram feces) | 7.57 ± 1.49 $_a$ | 6.45 ± 1.54 $_b$ |
| *Lactobacillus* group ($Log_{10}$ n° cells/gram feces) | 5.91 ± 1.26 $_a$ | 6.97 ± 1.83 $_b$ |
| *Faecalibacterium prausnitzii* ($Log_{10}$ n° cells/gram feces) | 7.07 ± 0.76 $_a$ | 6.42 ± 1.31 $_b$ |
| Acetic acid (mM) | 29.81 ± 9.25 $_a$ | 23.18 ± 14.45 $_b$ |
| Propionic acid (mM) | 12.94 ± 5.43 $_a$ | 9.50 ± 7.46 $_b$ |
| Butyric acid (mM) | 11.76 ± 9.39 $_a$ | 8.44 ± 7.94 $_a$ |
| Glucose (mg/dL) | 97.76 ± 12.99 $_a$ | 106.78 ± 33.85 $_a$ |
| Triglycerides (mg/dL) | 118.82 ± 54.23 $_a$ | 121.75 ± 48.41 $_a$ |
| Cholesterol (mg/dL) | 233.18 ± 39.06 $_a$ | 203.39 ± 37.48 $_b$ |
| LDL-HDL ratio | 2.76 ± 0.78 $_a$ | 2.72 ± 0.81 $_a$ |
| Leptin (ng/mL) | 9.62 ± 5.72 $_a$ | 12.01 ± 7.87 $_a$ |
| Serum MDA (μM) | 2.01 ± 0.53 $_a$ | 2.60 ± 0.49 $_b$ |
| Antioxidant capacity (mM) | 0.34 ± 0.09 $_a$ | 0.35 ± 0.09 $_a$ |
| CRP (mg/L) | 1.28 ± 1.22 $_a$ | 1.19 ± 1.03 $_a$ |
| TGF-β (ng/mL) | 4.44 ± 2.71 $_a$ | 6.25 ± 5.70 $_a$ |
| IL-10 (pg/mL) | 0.14 ± 0.78 $_a$ | 0.80 ± 3.70 $_a$ |
| IL-17 (pg/mL) | 1.42 ± 3.10 $_a$ | 2.28 ± 11.57 $_a$ |
| IL-8 (pg/mL) | 7.09 ± 6.01 $_a$ | 20.80 ± 9.91 $_b$ |
| IL-12 (pg/mL) | 0.21 ± 1.21 $_a$ | 3.54 ± 8.92 $_b$ |
| TNF-α (pg/mL) | 0.23 ± 1.23 $_a$ | 4.82 ± 7.94 $_b$ |
| Phagocytosis granulocytes (%) | 72.23 ± 22.13 $_a$ | 86.37 ± 18.53 $_a$ |
| Phagocytosis granulocytes and monocytes (%) | 71.39 ± 21.32 $_a$ | 82.35 ± 17.49 $_a$ |
| NK cell activity (%) | 53.09 ± 11.25 $_a$ | 52.70 ± 18.19 $_a$ |

Values are shown as mean ± standard deviation (SD). Values in the same row showing different subscripts present a statistically significant difference ($p \leq 0.05$).

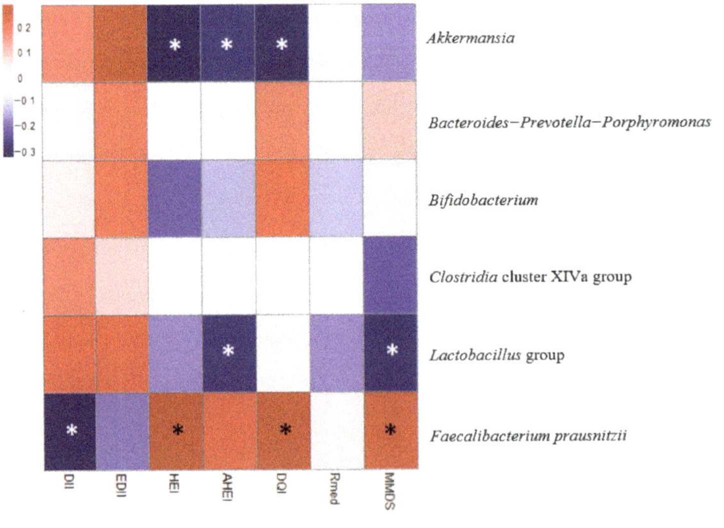


**Figure 1.** Heatmap defined by Pearson's correlations between dietary indices scores and intestinal microbial groups. Blue and red colors represent negative and positive associations, respectively. The color intensity is proportional to the degree of association. Asterisks indicate correlation significance: * $p \leq 0.05$.

**Table 3.** Results obtained from regression analyses to identify dietary indices as predictors of gut microbiota levels ($log_{10}$ n°. cells per gram of feces), fecal short chain fatty acids (mM) and blood biomarkers.

| | Dependent Variable | Independent Variable | $R^2$ | β | *p* |
|---|---|---|---|---|---|
| Model 1. Fecal microbiota groups *Akkermansia, Bacteroides-Prevotella-Porphyromonas, Bifidobacterium, Clostridia* cluster XIVa, *Faecalibacterium prausnitzii, Lactobacillus* group | *Akkermansia* | HEI | 0.080 | −0.307 | 0.026 |
| | | AHEI | 0.059 | −0.256 | 0.050 |
| | | DQI-I | 0.072 | −0.285 | 0.038 |
| | *Faecalibacterium prausnitzii* | DII | 0.124 | −0.312 | 0.030 |
| | | HEI | 0.128 | 0.284 | 0.035 |
| | | DQI-I | 0.122 | 0.265 | 0.047 |
| | | MMDS | 0.123 | 0.240 | 0.044 |
| | *Lactobacillus* group | AHEI | 0.264 | −0.256 | 0.027 |
| | | MMDS | 0.283 | −0.275 | 0.012 |
| Model 2. Fecal short chain fatty acids Acetic acid, Propionic acid, Butyric acid | Acetic acid | DII | 0.252 | −0.425 | 0.003 |
| | | EDII | 0.244 | −0.369 | 0.013 |
| | | HEI | 0.239 | 0.320 | 0.016 |
| | | AHEI | 0.335 | 0.478 | 0.000 |
| | | MMDS | 0.356 | 0.451 | 0.000 |
| | Propionic acid | DII | 0.198 | −0.316 | 0.031 |
| | | HEI | 0.246 | 0.348 | 0.009 |
| | | AHEI | 0.303 | 0.441 | 0.000 |
| | | MMDS | 0.292 | 0.378 | 0.001 |
| | Butyric acid | HEI | 0.189 | 0.289 | 0.034 |
| | | AHEI | 0.213 | 0.338 | 0.007 |
| | | MMDS | 0.211 | 0.298 | 0.012 |
| Model 3. Blood biomarkers Glucose, Triglycerides, LDL-HDL ratio, Leptin, Serum malondialdehyde (MDA), Antioxidant capacity, C-Reactive protein (CRP), Transforming growth factor-beta (TGF-β), IL-10, IL-17, IL-8, IL-12, TNF-α, % Phagocytosis granulocytes, % Phagocytosis granulocytes and monocytes, NK cell activity | MDA | DII | 0.297 | 0.373 | 0.003 |
| | | EDII | 0.318 | 0.408 | 0.002 |
| | IL-8 | rMED | 0.443 | −0.251 | 0.018 |
| | | MMDS | 0.443 | −0.221 | 0.017 |

Linear regression analyses are adjusted by age and energy. $R^2$, coefficient of multiple determination; β, standardized regression coefficient for examined variable. $p \leq 0.05$. Inflammatory indices (DII, Dietary Inflammatory Index. EDII, Empirical Dietary Inflammatory Index): negative values favor non-inflammatory states and positives values enhance inflammation. Dietary Quality indices (HEI, Healthy Eating Index. AHEI, Alternative Healthy Eating Index. DQI-I, Diet Quality Index-International): higher scores are reflecting consumption values similar to those recommended in dietary guidelines. Mediterranean Dietary indices (rMED, Relative Mediterranean Diet Score. MMDS, Modified Mediterranean Diet Score): higher values showing a higher degree of adherence to the Mediterranean diet.

To deepen into how dietary components may modulate gut microbiota and SCFAs, linear regressions models adjusting by age and energy intake were applied and β-coefficient values were plotted in the heatmap of Supplementary Figure S1. Sauces and dips seemed to be the most significant food group for determining *Clostridia* cluster XIVa and *Lactobacillus* groups. However, the consumption levels of this food group were very low in our sample (data not shown) so that this association should be considered with caution. *F. prausnitzii* showed a positive association with fruits, legumes and with fiber, whereas a negative association was found for saturated fats. SCFAs appeared to be mostly related to oils and fats, seafood, total polyphenols and saturated fats.

Heatmap showing Pearson´s correlation and clusterization among the different dietary indices is presented in Figure 2. Here, inflammatory indices revealed to be closer between them, forming the first cluster while the others integrate the second cluster. At the same time, HEI and AHEI grouped more closely with DQI-I and then with rMED and MMDS.

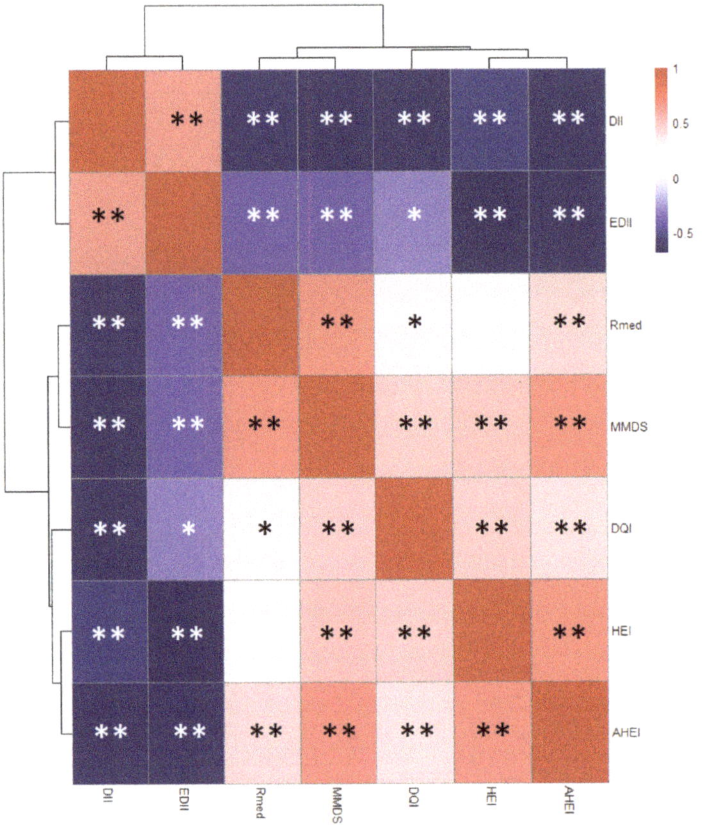


**Figure 2.** Heatmap defined by Pearson's correlations between dietary indices scores. Blue and red colors represent negative and positive association, respectively. The color intensity is proportional to the degree of association between indices. Asterisks indicate correlation significance: * $p \leq 0.05$; ** $p \leq 0.01$. Due to the scale of the Dietary inflammatory index (DII) and the Empirical Dietary Inflammatory Index (EDII), they show an inverse relationship with the rest of the indices.

To elucidate the relationship of dietary indices scores with the consumption of dietary compounds with reported anti-inflammatory and health-protective effects in the literature, a new linear regression analysis was performed adjusting by age and energy and a heatmap was plotted with the results (Figure 3). DII showed more negative correlation with vitamins, fiber, glucosinolates and isothiocyanates than EDII. All indices were associated with total polyphenols and ORAC. Except for rMED, indices had a significant correlation with flavonols, DHA, lutein + zeaxanthin, carotenoids, insoluble fiber and pectin. Glucosinolates and isothiocyanates did not show a significant association with the Mediterranean indices. In general, splitting indices into inflammatory, Mediterranean and diet quality ones, heatmap revealed similar patterns of association for the indices within each group. With the exception of flavones, all evaluated compounds revealed a negative association with inflammatory indices but positive with the Mediterranean and diet quality indices.

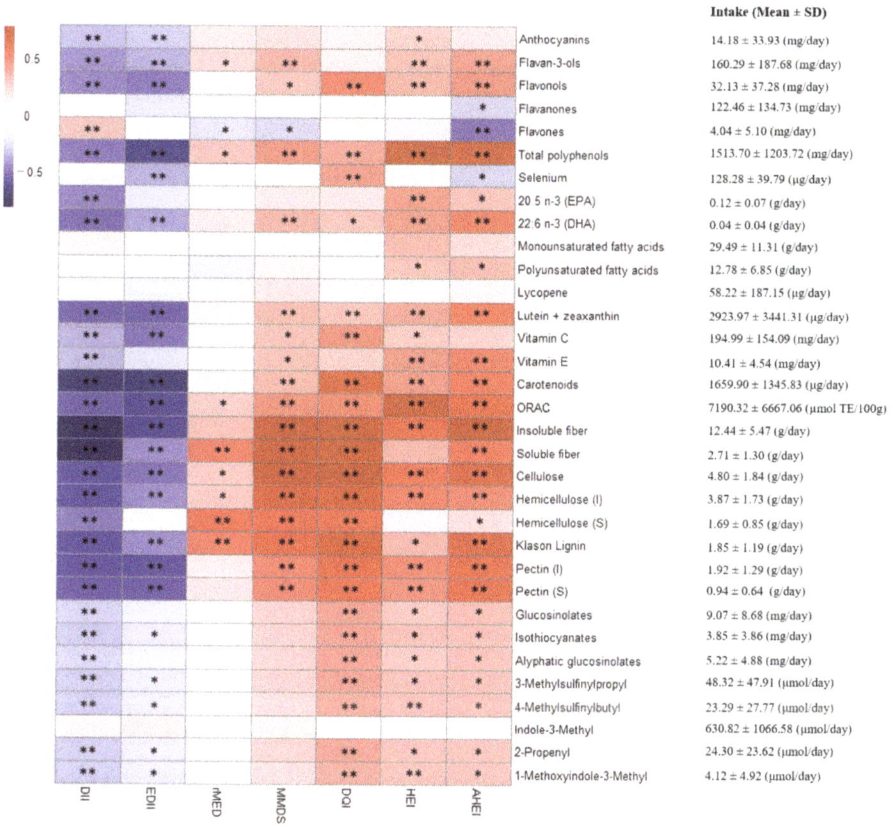


**Figure 3.** Heatmap showing β-coefficient values resulting from univariate linear regressions adjusting by age and energy among six different dietary indexes and health-related compounds. Rows include compounds with previously reported beneficial effect as (poly)phenols, fatty acids, carotenoids, vitamins, fiber, glucosinolates and isothiocyanates. Blue and red colors denote negative and positive association, respectively. Asterisks indicate the significance of the association degree * $p \leq 0.05$; ** $p \leq 0.01$.

## 4. Discussion

In recent years, dietary indices have been a major step towards addressing the diet-health binomial from a global perspective. While tailored to different populations and constructed for specific purposes, a high degree of similarity can be observed among some of them. To our knowledge, this is the first study comparing the usefulness of dietary indices as predictors of human gut microbiota, the production of fecal SCFAs and the concentration in blood of different parameters related to the immune and inflammatory status, in a sample of middle-aged and elderly subjects without diagnosed pathology. As it has been previously proposed by other authors, for some of the indices included in the study, our data showed the existence of differences according to age. Furthermore, changes in the levels of some bacterial groups such as *Akkermansia* [44] or butyrate-producing bacteria, mainly *F. prausnitzii* [45,46], have been reported in aged individuals. Therefore, the factor "age" has been introduced in the models as a covariate in order to improve the interpretation of results.

Limitations in the use of dietary indices need to be also taken into account. DII encompasses a total of 45 components, of which 35 were evaluated in this sample. The effects of lacking 10 components in the scoring system may be attenuated by the own nature of the index and the fact of having a sample that does not show extreme values of consumption for any component. Moreover, nutritional supplements, weighted in other versions of EDII to compute the final score, were not included here. Some dietary indices such as EDII, HEI or AHEI were developed in the context of almost fully "westernized" societies, which could drive to an underrating or overrating in diets of populations with mixed diets (Mediterranean, African, etc.). Indeed, in our sample, the component "Trans Fat" that is evaluated in the AHEI, showed very low values while in typical EEUU diet is probably highly present. This may entail a loss of power in the accuracy of the prediction of our scores. Furthermore, DQI-I incorporates dietary variety, adequacy and moderation as a quality criterion. Although all these parameters have been included, the accuracy of FFQ to provide an accurate measurement of variety may be one of the limitations of this work. One of the main difficulties arises in the capacity of the indices to classify the subjects under study. In this sense, the scores from dietary quality indices showed low variability inter subject, with almost the whole sample obtaining scores indicative of poor or average diet quality for DQI-I, HEI and AHEI. Thus, we propose these indices as the worse in differentiating the poor-quality Mediterranean-style diet from a middle-age-elderly sample. On the other hand, indices related to adherence to a Mediterranean-style dietary pattern have presented a wide range of scores in the sample, ranging from 2–12 points and 1–7 for rMED and MMDS, respectively.

Based on the data obtained, the identification of the best tool to predict the composition and metabolic activity of the gut microbiota as a function of diet, is a difficult task. We considered that in the present study, the HEI, AHEI and DQI-I resulted likely inappropriate as predictor variables of differences between different microbiota as poor diets, that are different but score similarly, may mask trends associated to specific dietary constituents [47]. Interestingly, we found decreased levels of the mucin degrading *Akkermansia* in better scoring individuals compared to increased levels of *Akkermansia* in the worse ones, mainly influenced by vegetable consumption (data not shown). When vegetables are included at significant levels in diet, fiber consumption increases, which could promote the rise of some fiber-degrading species at the expense of other microorganisms such *Akkermansia* [48]. This could contrast with the fact that the presence of *Akkermansia* has been associated with healthy intestine and its abundance has been inversely correlated to several disease states [49–54]. We propose Mediterranean indices and more precisely MMDS, as the most accurate and best predictor in our population sample. Probably, socio-geographical reasons do Mediterranean indices the most suitable ones to measure the quality of diet in the sample and therefore, to predict microbiological and immunological variables. Also, some dietary indices related with inflammation (DII), quality of diet (HEI and DQI-I) and adherence to the Mediterranean-diet (MMDS) seem to be predictors of *F. prausnitzii* fecal levels, which were higher in individuals with healthier diets. *F. prausnitzii*, a member of the commensal microbiota, has been related with intestinal health and gut homeostasis [55]. Several studies highlight the anti-inflammatory properties of *F. prausnitzii* and its ability for upregulating T cell production and reducing IL-8 levels by blocking the NF-kB activation [56,57]. *F. prausnitzii* is a member of *Clostridium* cluster IV, one of the main producers of butyrate in the human colon [58] during fermentation of nondigestible polysaccharides such as dietary fiber. Butyrate plays several pleiotropic effects on host physiology and enhances the protection against pathogen invasion [59,60]. Remarkably, *Faecalibacterium* was found to be at high abundance in an Irish elderly sample [61] whereas some studies showed decreased levels of *F. prausnitzii* in centenarians as compared with younger adults [62]. Further studies are needed to determine the role of these bacteria in the intestinal microbiota of elderly populations.

The indices AHEI and MMDS (related to quality of diet and adherence to Mediterranean diet, respectively) were negatively associated with intestinal *Lactobacillus* levels. Increased *Lactobacillus* levels have been correlated with a higher PUFA/SFA ratio intake, probably mediated by changes in bile acid secretion and composition [23,63,64]. Extra virgin olive oil is an important component of Mediterranean diet and is a source of unsaturated fatty acids that can be metabolized by some intestinal *Lactobacillus* species [65–67]. Dietary indices used in the present work add from 0 to different positive numerical values to the formula relating dietary fats, as depending on the type and amount of fat consumed (calculated as a percentage of total energy intake). Therefore, as higher scores in dietary indices are generally accompanied in the general population by lower consumption of all type of fats (correlation values of "Lipids" with AHEI and MMDS of −0.154 and −0.149, respectively), this could provide a rationale to the inverse association found by us between scores for AHEI and MMDS and fecal levels of *Lactobacillus*. In this regard, we recently reported increased levels of the *Lactobacillus* group in Spanish adults displaying altered profiles of serum free fatty acids, which were accompanied by subclinical metabolic alterations [68].

Regarding the fecal SCFAs evaluated in our sample, acetic and propionic acids, correlated positively with heathier dietary scores for most of the indices. Microorganisms colonizing the gastrointestinal tract can participate in beneficial interactions within the intestinal ecological niche, as modulated by external factors such as diet. This is the case of the increase of the intestinal butyrate production by cross-feeding mechanisms. In cross-feeding, fiber-degrading bacteria can produce acetate as an end-product of fermentation, which is then metabolized by other members of the intestinal microbiota, as those belonging to *Clostridia* clusters XIVa and IV, to produce butyrate as an end-product of fermentation [51]. Both, bacteria and SCFAs contribute to cell expansion, immunosuppressive functions and overall intestinal homeostasis. Therefore, better dietary scores could be related with an enhanced SCFAs production in the gut [23].

Inflammatory indices have been identified as good predictors of inflammation variables such as CRP, IL-6 and TNF-$\alpha$ receptor 2 [12–14,69,70]. MDA is considered an oxidative stress biomarker that reflects levels of lipoperoxidation in blood. However, to the best of our knowledge, there are no previous reports relating dietary inflammatory indices and MDA. The lower detection of IL-8 in those individuals in the present work showing better scores on the indices related with the adherence to the Mediterranean diet may be partially explained by the increased levels of the anti-inflammatory bacterium *F. prausnitzii* [71] in such individuals. This bacterium was able to block the production of the inflammatory interleukin IL-8 in Crohn disease patients and in a murine colitis model [72].

## 5. Conclusions

The associations found among intestinal bacterial groups, SCFAs, blood biomarkers and dietary indices are indirectly reflecting how these variables are influenced by the specific components or food groups scoring in each index. When trying to discern differences among indices by clustering them, they are split in 3 main classes: inflammatory, diet quality and adherence to Mediterranean diet. The methodologies followed to construct them, the population they target and the scoring criteria define their nature and the way they correlated with others. The extension of the usefulness of dietary indices may shed some light into how to modulate gut microbiota focusing on dietary patterns.

**Supplementary Materials:** The following are available online at http://www.mdpi.com/2072-6643/12/12/3828/s1, Figure S1: Heatmap showing β-coefficient values resulting from univariate linear regressions adjusting by age and energy among microbial groups, SCFAs and food groups and dietary compounds, Table S1: Characteristics of dietary indexes, Table S2: Primers and annealing temperatures used for the quantification of intestinal microbial groups by qPCR.

**Author Contributions:** The authors' responsibilities were as follows: Conceptualization, M.G. and S.G.; Funding acquisition, M.G.; Methodology, S.R.-S., N.S., A.S., C.G.d.l.R.-G. and M.G.; Supervision, C.G.d.l.R.-G. and S.G.; Writing—original draft, S.R.-S. and S.G.; Writing—review & editing, N.S., A.S., C.G.d.l.R.-G. and M.G. All authors have read and agreed to the published version of the manuscript.

**Funding:** This research was funded by Plan Estatal de I+D+I through projects AGL2017-83653-R (AEI/FEDER, UE) and RTI2018-098288-B-I00 (MCIU/AEI/FEDER, UE) and by contracts with Biopolis SL (Valencia, Spain), CAUCE Foundation (Oviedo, Spain) and Alimerka Foundation (Llanera, Spain). N.S. is granted a postdoctoral contract awarded by the Fundación para la Investigación Biosanitaria de Asturias (FINBA) and S.R.-S. is the recipient of a Research Training contract awarded under project RTI2018-098288-B-I00.

**Acknowledgments:** We show our greatest gratitude to all the volunteers participating in the study.

**Conflicts of Interest:** The authors declare no conflict of interest.

## References


1. Shenkin, A. Micronutrients in health and disease. *Postgrad. Med. J.* **2006**, *82*, 559–567. [CrossRef] [PubMed]
2. Howes, M.J.R.; Simmonds, M.S.J. The role of phytochemicals as micronutrients in health and disease. *Curr. Opin. Clin. Nutr. Metab. Care* **2014**, *17*, 558–566. [CrossRef] [PubMed]
3. Hu, F.B. Dietary pattern analysis: A new direction in nutritional epidemiology. *Curr. Opin. Lipidol.* **2002**, *13*, 3–9. [CrossRef] [PubMed]
4. Ocké, M.C. Evaluation of methodologies for assessing the overall diet: Dietary quality scores and dietary pattern analysis. *Proc. Nutr. Soc.* **2013**, *72*, 191–199. [CrossRef]
5. Hodge, A.M.; Bassett, J.K.; Dugué, P.A.; Shivappa, N.; Hébert, J.R.; Milne, R.L.; English, D.R.; Giles, G.G. Dietary inflammatory index or Mediterranean diet score as risk factors for total and cardiovascular mortality. *Nutr. Metab. Cardiovasc. Dis.* **2018**, *28*, 461–469. [CrossRef]
6. Roy, R.; Hebden, L.; Rangan, A.; Allman-Farinelli, M. The development, application and validation of a Healthy eating index for Australian Adults (HEIFA-2013). *Nutrition* **2016**, *32*, 432–440. [CrossRef]
7. Shivappa, N.; Steck, S.E.; Hurley, T.G.; Hussey, J.R.; Hébert, J.R. Designing and developing a literature-derived, population-based dietary inflammatory index. *Public Health Nutr.* **2014**, *17*, 1689–1696. [CrossRef]
8. Tabung, F.K.; Smith-Warner, S.A.; Chavarro, J.E.; Wu, K.; Fuchs, C.S.; Hu, F.B.; Chan, A.T.; Willett, W.C.; Giovannucci, E.L. Development and validation of an empirical dietary inflammatory index. *J. Nutr.* **2016**, *146*, 1560–1570. [CrossRef]
9. Guenther, P.M.; Reedy, J.; Krebs-Smith, S.M. Development of the Healthy Eating Index-2005. *J. Am. Diet. Assoc.* **2008**, *108*, 1896–1901. [CrossRef]
10. Chiuve, S.E.; Fung, T.T.; Rimm, E.B.; Hu, F.B.; McCullough, M.L.; Wang, M.; Stampfer, M.J.; Willett, W.C. Alternative dietary indices both strongly predict risk of chronic disease. *J. Nutr.* **2012**, *142*, 1009–1018. [CrossRef]
11. Trichopoulou, A.; Costacou, T.; Bamia, C.; Trichopoulos, D. Adherence to a Mediterranean diet and survival in a Greek population. *N. Engl. J. Med.* **2003**, *348*, 2599–2608. [CrossRef] [PubMed]
12. Cavicchia, P.P.; Steck, S.E.; Hurley, T.G.; Hussey, J.R.; Ma, Y.; Ockene, I.S.; Hébert, J.R. A new dietary inflammatory index predicts interval changes in serum high-sensitivity C-reactive protein. *J. Nutr.* **2009**, *139*, 2365–2372. [CrossRef] [PubMed]
13. Tabung, F.K.; Smith-Warner, S.A.; Chavarro, J.E.; Fung, T.T.; Hu, F.B.; Willett, W.C.; Giovannucci, E.L. An empirical dietary inflammatory pattern score enhances prediction of circulating inflammatory biomarkers in adults. *J. Nutr.* **2017**, *147*, 1567–1577. [CrossRef] [PubMed]
14. Shivappa, N.; Steck, S.E.; Hurley, T.G.; Hussey, J.R.; Ma, Y.; Ockene, I.S.; Tabung, F.; Hébert, J.R. A population-based dietary inflammatory index predicts levels of C-reactive protein in the Seasonal Variation of Blood Cholesterol Study (SEASONS). *Public Health Nutr.* **2014**, *17*, 1825–1833. [CrossRef]
15. Krebs-Smith, S.M.; Pannucci, T.R.E.; Subar, A.F.; Kirkpatrick, S.I.; Lerman, J.L.; Tooze, J.A.; Wilson, M.M.; Reedy, J. Update of the Healthy Eating Index: HEI-2015. *J. Acad. Nutr. Diet.* **2018**, *113*, 569–580. [CrossRef]
16. Onvani, S.; Haghighatdoost, F.; Surkan, P.J.; Larijani, B.; Azadbakht, L. Adherence to the Healthy Eating Index and Alternative Healthy Eating Index dietary patterns and mortality from all causes, cardiovascular disease and cancer: A meta-analysis of observational studies. *J. Hum. Nutr. Diet.* **2017**, *30*, 216–226. [CrossRef]
17. Sofi, F.; Abbate, R.; Gensini, G.F.; Casini, A. Accruing evidence on benefits of adherence to the Mediterranean diet on health: An updated systematic review and meta-analysis. *Am. J. Clin. Nutr.* **2010**, *92*, 1189–1196. [CrossRef]
18. Tosti, V.; Bertozzi, B.; Fontana, L. Health Benefits of the Mediterranean Diet: Metabolic and Molecular Mechanisms. *J. Gerontol. Ser. A Biol. Sci. Med. Sci.* **2018**, *73*, 318–326. [CrossRef]
19. Ciancarelli, M.; Massimo, C.; Amicis, D.; Ciancarelli, I. Mediterranean Diet and Health Promotion: Evidence and current concerns. *Med. Res. Arch.* **2017**, *9*, 99–114. [CrossRef]
20. Levy, M.; Kolodziejczyk, A.A.; Thaiss, C.A.; Elinav, E. Dysbiosis and the immune system. *Nat. Rev. Immunol.* **2017**, *17*, 219–232. [CrossRef]
21. Kau, A.L.; Ahern, P.P.; Griffin, N.W.; Goodman, A.L.; Gordon, J.I. Human nutrition, the gut microbiome and the immune system. *Nature* **2011**, *474*, 327–336. [CrossRef] [PubMed]
22. Gutiérrez-Díaz, I.; Fernández-Navarro, T.; Sánchez, B.; Margolles, A.; González, S. Mediterranean diet and faecal microbiota: A transversal study. *Food Funct.* **2016**, *7*, 2347–2356. [CrossRef] [PubMed]

23. Garcia-Mantrana, I.; Selma-Royo, M.; Alcantara, C.; Collado, M.C. Shifts on gut microbiota associated to mediterranean diet adherence and specific dietary intakes on general adult population. *Front. Microbiol.* **2018**, *9*, 890. [CrossRef] [PubMed]
24. Ghosh, T.S.; Rampelli, S.; Jeffery, I.B.; Santoro, A.; Neto, M.; Capri, M.; Giampieri, E.; Jennings, A.; Candela, M.; Turroni, S.; et al. Mediterranean diet intervention alters the gut microbiome in older people reducing frailty and improving health status: The NU-AGE 1-year dietary intervention across five European countries. *Gut* **2020**, *69*, 1218–1228. [CrossRef]
25. Mitsou, E.K.; Kakali, A.; Antonopoulou, S.; Mountzouris, K.C.; Yannakoulia, M.; Panagiotakos, D.B.; Kyriacou, A. Adherence to the Mediterranean diet is associated with the gut microbiota pattern and gastrointestinal characteristics in an adult population. *Br. J. Nutr.* **2017**, *117*, 1645–1655. [CrossRef]
26. Cuervo, A.; Valdés, L.; Salazar, N.; de los Reyes-Gavilán, C.G.; Ruas-Madiedo, P.; Gueimonde, M.; González, S. Pilot Study of Diet and Microbiota: Interactive Associations of Fibers and Polyphenols with Human Intestinal Bacteria. *J. Agric. Food Chem.* **2014**, *62*, 5330–5336. [CrossRef]
27. Cuervo, A.; de los Reyes-Gavilán, C.G.; Ruas-Madiedo, P.; Lopez, P.; Suarez, A.; Gueimonde, M.; González, S. Red Wine Consumption Is Associated with Fecal Microbiota and Malondialdehyde in a Human Population. *J. Am. Coll. Nutr.* **2015**, *34*, 135–141. [CrossRef]
28. Yamamoto, T.; Takahashi, H.; Suzuki, K.; Hirano, A.; Kamei, M.; Goto, T.; Takahashi, N.; Kawada, T. Theobromine enhances absorption of cacao polyphenol in rats. *Biosci. Biotechnol. Biochem.* **2014**, *78*, 2059–2063. [CrossRef]
29. Centro de Ensenanza Superior de Nutricion Humana y Dietetica (CESNID). *Tablas de Composicion de Alimentos Por Medidas Caseras de Consumo Habitual en España*; McGrawHill, Ed.; Publicaciones y Ediciones de la Universidad de Barcelona: Barcelona, Spain, 2008.
30. Neveu, V.; Perez-Jimenez, J.; Vos, F.; Crespy, V.; du Chaffaut, L.; Mennen, L.; Knox, C.; Eisner, R.; Cruz, J.; Wishart, D.; et al. Phenol-Explorer: An online comprehensive database on polyphenol contents in foods. *Database* **2010**, *2010*, bap024. [CrossRef]
31. Marlett, J.A.; Cheung, T.-F. Database and Quick Methods of Assessing Typical Dietary Fiber Intakes using data for 228 Commonly Consumed Foods. *J. Am. Diet. Assoc.* **1997**, *97*, 1139–1151. [CrossRef]
32. McNaughton, S.A.; Marks, G.C. Development of a food composition database for the estimation of dietary intakes of glucosinolates, the biologically active constituents of cruciferous vegetables. *Br. J. Nutr.* **2003**, *90*, 687–697. [CrossRef] [PubMed]
33. Agudo, A.; Ibáñez, R.; Amiano, P.; Ardanaz, E.; Barricarte, A.; Berenguer, A.; Chirlaque, M.D.; Dorronsoro, M.; Jakszyn, P.; Larrañaga, N.; et al. Consumption of cruciferous vegetables and glucosinolates in a Spanish adult population. *Eur. J. Clin. Nutr.* **2008**, *62*, 324–331. [CrossRef] [PubMed]
34. International Agency for Research of Cancer (IARC). Cruciferous Vegetables, Isothicynates and Indoles. In *Handbooks of Cancer Prevention*; IARC: Lyon, France, 2004.
35. Fernández-Navarro, T.; Díaz, I.; Gutiérrez-Díaz, I.; Rodríguez-Carrio, J.; Suárez, A.; de los Reyes-Gavilán, C.G.; Gueimonde, M.; Salazar, N.; González, S. Exploring the interactions between serum free fatty acids and fecal microbiota in obesity through a machine learning algorithm. *Food Res. Int.* **2019**, *121*, 533–541. [CrossRef] [PubMed]
36. Mariscal-Arcas, M.; Romaguera, D.; Rivas, A.; Feriche, B.; Pons, A.; Tur, J.A.; Olea-Serrano, F. Diet quality of young people in southern Spain evaluated by a Mediterranean adaptation of the Diet Quality Index-International (DQI-I). *Br. J. Nutr.* **2007**, *98*, 1267–1273. [CrossRef] [PubMed]
37. Buckland, G.; Agudo, A.; Travier, N.; María Huerta, J.; Cirera, L.; Tormo, M.J.; Navarro, C.; Dolores Chirlaque, M.; Moreno-Iribas, C.; Ardanaz, E.; et al. Adherence to the Mediterranean diet reduces mortality in the Spanish cohort of the European Prospective Investigation into Cancer and Nutrition (EPIC-Spain). *Br. J. Nutr.* **2011**, *330*, 991–995. [CrossRef] [PubMed]
38. Trichopoulou, A.; Orfanos, P.; Norat, T.; Bueno-de-Mesquita, B.; Ocké, M.C.; Peeters, P.H.M.; van der Schouw, Y.T.; Boeing, H.; Hoffmann, K.; Boffetta, P.; et al. Modified Mediterranean diet and survival: EPIC-elderly prospective cohort study. *Br. Med. J.* **2005**, *330*, 991–995. [CrossRef] [PubMed]
39. Gérard-Monnier, D.; Erdelmeier, I.; Régnard, K.; Moze-Henry, N.; Yadan, J.-C.; Chaudière, J. Reactions of 1-Methyl-2-phenylindole with Malondialdehyde and 4-Hydroxyalkenals. Analytical Applications to a Colorimetric Assay of Lipid Peroxidation. *Chem. Res. Toxicol.* **1998**, *11*, 1176–1183. [CrossRef]

40. Çelik, S.E.; Özyürek, M.; Güçlü, K.; Apak, R. CUPRAC total antioxidant capacity assay of lipophilic antioxidants in combination with hydrophilic antioxidants using the macrocyclic oligosaccharide methyl β-cyclodextrin as the solubility enhancer. *React. Funct. Polym.* **2007**, *67*, 1548–1560. [CrossRef]
41. Nogacka, A.M.; Salazar, N.; Arboleya, S.; Ruas-Madiedo, P.; Mancabelli, L.; Suarez, A.; Martinez-Faedo, C.; Ventura, M.; Tochio, T.; Hirano, K.; et al. In vitro evaluation of different prebiotics on the modulation of gut microbiota composition and function in morbid obese and normal-weight subjects. *Int. J. Mol. Sci.* **2020**, *21*, 906. [CrossRef]
42. Valdés, L.; Salazar, N.; González, S.; Arboleya, S.; Ríos-Covián, D.; Genovés, S.; Ramón, D.; de los Reyes-Gavilán, C.G.; Ruas-Madiedo, P.; Gueimonde, M. Selection of potential probiotic *bifidobacteria* and prebiotics for elderly by using in vitro faecal batch cultures. *Eur. Food Res. Technol.* **2017**, *243*, 157–165. [CrossRef]
43. Salazar, N.; Gueimonde, M.; Hernández-Barranco, A.M.; Ruas-Madiedo, P.; De Los Reyes-Gavilán, C.G. Exopolysaccharides produced by intestinal *Bifidobacterium* strains act as fermentable substrates for human intestinal bacteria. *Appl. Environ. Microbiol.* **2008**, *74*, 4737–4745. [CrossRef]
44. Collado, M.C.; Derrien, M.; Isolauri, E.; De Vos, W.M.; Salminen, S. Intestinal integrity and *Akkermansia muciniphila*, a mucin-degrading member of the intestinal microbiota present in infants, adults and the elderly. *Appl. Environ. Microbiol.* **2007**, *73*, 7767–7770. [CrossRef] [PubMed]
45. O'Toole, P.W. Changes in the intestinal microbiota from adulthood through to old age. *Clin. Microbiol. Infect.* **2012**, *18*, 44–46. [CrossRef]
46. Salazar, N.; Arboleya, S.; Fernández-Navarro, T.; de los Reyes-Gavilán, C.G.; Gonzalez, S.; Gueimonde, M. Age-associated changes in gut microbiota and dietary components related with the immune system in adulthood and old age: A cross-sectional study. *Nutrients* **2019**, *11*, 1765. [CrossRef] [PubMed]
47. Bowyer, R.C.E.; Jackson, M.A.; Pallister, T.; Skinner, J.; Spector, T.D.; Welch, A.A.; Steves, C.J. Use of dietary indices to control for diet in human gut microbiota studies. *Microbiome* **2018**, *6*, 77. [CrossRef] [PubMed]
48. Zmora, N.; Suez, J.; Elinav, E. You are what you eat: Diet, health and the gut microbiota. *Nat. Rev. Gastroenterol. Hepatol.* **2019**, *16*, 35–56. [CrossRef] [PubMed]
49. Everard, A.; Belzer, C.; Geurts, L.; Ouwerkerk, J.P.; Druart, C.; Bindels, L.B.; Guiot, Y.; Derrien, M.; Muccioli, G.G.; Delzenne, N.M.; et al. Cross-talk between *Akkermansia muciniphila* and intestinal epithelium controls diet-induced obesity. *Proc. Natl. Acad. Sci. USA* **2013**, *110*, 9066–9071. [CrossRef]
50. Plovier, H.; Everard, A.; Druart, C.; Depommier, C.; Van Hul, M.; Geurts, L.; Chilloux, J.; Ottman, N.; Duparc, T.; Lichtenstein, L.; et al. A purified membrane protein from *Akkermansia muciniphila* or the pasteurized bacterium improves metabolism in obese and diabetic mice. *Nat. Med.* **2017**, *16*, 35–56. [CrossRef]
51. Belzer, C.; Chia, L.W.; Aalvink, S.; Chamlagain, B.; Piironen, V.; Knol, J.; de Vos, W.M. Microbial metabolic networks at the mucus layer lead to diet-independent butyrate and vitamin B12 production by intestinal symbionts. *MBio* **2017**, *8*, e00770-17. [CrossRef]
52. Lyra, A.; Forssten, S.; Rolny, P.; Wettergren, Y.; Lahtinen, S.J.; Salli, K.; Cedgård, L.; Odin, E.; Gustavsson, B.; Ouwehand, A.C. Comparison of bacterial quantities in left and right colon biopsies and faeces. *World J. Gastroenterol.* **2012**, *18*, 4404–4411. [CrossRef]
53. Rajilić-Stojanović, M.; Shanahan, F.; Guarner, F.; De Vos, W.M. Phylogenetic analysis of dysbiosis in ulcerative colitis during remission. *Inflamm. Bowel Dis.* **2013**, *19*, 481–488. [CrossRef] [PubMed]
54. Swidsinski, A.; Dörffel, Y.; Loening-Baucke, V.; Theissig, F.; Rückert, J.C.; Ismail, M.; Rau, W.A.; Gaschler, D.; Weizenegger, M.; Kühn, S.; et al. Acute appendicitis is characterised by local invasion with *Fusobacterium nucleatum/necrophorum*. *Gut* **2011**, *60*, 34–40. [CrossRef] [PubMed]
55. Miquel, S.; Martín, R.; Rossi, O.; Bermúdez-Humarán, L.G.; Chatel, J.M.; Sokol, H.; Thomas, M.; Wells, J.M.; Langella, P. *Faecalibacterium prausnitzii* and human intestinal health. *Curr. Opin. Microbiol.* **2013**, *16*, 255–261. [CrossRef] [PubMed]
56. Qiu, X.; Zhang, M.; Yang, X.; Hong, N.; Yu, C. *Faecalibacterium prausnitzii* upregulates regulatory T cells and anti-inflammatory cytokines in treating TNBS-induced colitis. *J. Crohn's Colitis* **2013**, *7*, e558–e568. [CrossRef]
57. Ferreira-Halder, C.V.; de Sousa Faria, A.V.; Andrade, S.S. Action and function of *Faecalibacterium prausnitzii* in health and disease. *Best Pract. Res. Clin. Gastroenterol.* **2017**, *31*, 643–648. [CrossRef]
58. Pryde, S.E.; Duncan, S.H.; Hold, G.L.; Stewart, C.S.; Flint, H.J. The microbiology of butyrate formation in the human colon. *FEMS Microbiol. Lett.* **2002**, *217*, 133–139. [CrossRef]

59. Corrêa-Oliveira, R.; Fachi, J.L.; Vieira, A.; Sato, F.T.; Vinolo, M.A.R. Regulation of immune cell function by short-chain fatty acids. *Clin. Transl. Immunol.* **2016**, *5*, e73. [CrossRef]
60. Macfarlane, G.T.; Macfarlane, S. Fermentation in the human large intestine: Its physiologic consequences and the potential contribution of prebiotics. *J. Clin. Gastroenterol.* **2011**, *45*, S120–S127. [CrossRef]
61. Claesson, M.J.; Cusack, S.; O'Sullivan, O.; Greene-Diniz, R.; De Weerd, H.; Flannery, E.; Marchesi, J.R.; Falush, D.; Dinan, T.; Fitzgerald, G.; et al. Composition, variability and temporal stability of the intestinal microbiota of the elderly. *Proc. Natl. Acad. Sci. USA* **2011**, *108*, S4586–S4591. [CrossRef]
62. Biagi, E.; Nylund, L.; Candela, M.; Ostan, R.; Bucci, L.; Pini, E.; Nikkïla, J.; Monti, D.; Satokari, R.; Franceschi, C.; et al. Through ageing and beyond: Gut microbiota and inflammatory status in seniors and centenarians. *PLoS ONE* **2010**, *5*, e10667. [CrossRef]
63. Sandhu, K.V.; Sherwin, E.; Schellekens, H.; Stanton, C.; Dinan, T.G.; Cryan, J.F. Feeding the microbiota-gut-brain axis: Diet, microbiome and neuropsychiatry. *Transl. Res.* **2017**, *179*, 223–244. [CrossRef] [PubMed]
64. Lopez-Legarrea, P.; Fuller, N.R.; Zulet, M.A.; Martinez, J.A.; Caterson, I.D. The influence of Mediterranean, carbohydrate and high protein diets on gut microbiota composition in the treatment of obesity and associated inflammatory state. *Asia Pac. J. Clin. Nutr.* **2014**, *23*, 360–368. [CrossRef] [PubMed]
65. Druart, C.; Dewulf, E.M.; Cani, P.D.; Neyrinck, A.M.; Thissen, J.P.; Delzenne, N.M. Gut microbial metabolites of polyunsaturated fatty acids correlate with specific fecal bacteria and serum markers of metabolic syndrome in obese women. *Lipids* **2014**, *49*, 397–402. [CrossRef] [PubMed]
66. Miyamoto, J.; Igarashi, M.; Watanabe, K.; Karaki, S.I.; Mukouyama, H.; Kishino, S.; Li, X.; Ichimura, A.; Irie, J.; Sugimoto, Y.; et al. Gut microbiota confers host resistance to obesity by metabolizing dietary polyunsaturated fatty acids. *Nat. Commun.* **2019**, *10*, 1–5. [CrossRef] [PubMed]
67. Aziz, T.; Sarwar, A.; Fahim, M.; ud Din, J.; Al-Dalali, S.; Ud Din, Z.; Khan, A.A.; Jian, Z.; Yang, Z. Dose-dependent production of linoleic acid analogues in food derived *Lactobacillus plantarum* K25 and in silico characterization of relevant reactions. *Acta Biochim. Pol.* **2020**, *67*, 123–129. [CrossRef] [PubMed]
68. Rodríguez-Carrio, J.; Salazar, N.; Margolles, A.; González, S.; Gueimonde, M.; de los Reyes-Gavilán, C.G.; Suárez, A. Free fatty acids profiles are related to gut microbiota signatures and short-chain fatty acids. *Front. Immunol.* **2017**, *8*, 823. [CrossRef]
69. Shin, D.; Lee, K.W.; Brann, L.; Shivappa, N.; Hébert, J.R. Dietary inflammatory index is positively associated with serum high-sensitivity C-reactive protein in a Korean adult population. *Nutrition* **2019**, *63–64*, 155–161. [CrossRef]
70. Suzuki, K.; Shivappa, N.; Kawado, M.; Yamada, H.; Hashimoto, S.; Wakai, K.; Iso, H.; Okada, E.; Fujii, R.; Hébert, J.R.; et al. Association between dietary inflammatory index and serum C-reactive protein concentrations in the Japan collaborative cohort study. *Nagoya J. Med. Sci.* **2020**, *82*, 237–249. [CrossRef]
71. Sánchez-Tapia, M.; Tovar, A.R.; Torres, N. Diet as Regulator of Gut Microbiota and its Role in Health and Disease. *Arch. Med. Res.* **2019**, *50*, 259–268. [CrossRef]
72. Sokol, H.; Pigneur, B.; Watterlot, L.; Lakhdari, O.; Bermúdez-Humarán, L.G.; Gratadoux, J.J.; Blugeon, S.; Bridonneau, C.; Furet, J.P.; Corthier, G.; et al. *Faecalibacterium prausnitzii* is an anti-inflammatory commensal bacterium identified by gut microbiota analysis of Crohn disease patients. *Proc. Natl. Acad. Sci. USA* **2008**, *105*, 16731–16736. [CrossRef]


**Publisher's Note:** MDPI stays neutral with regard to jurisdictional claims in published maps and institutional affiliations.

*nutrients*

MDPI

*Review*

# The Microbiota–Gut–Brain Axis and Alzheimer's Disease: Neuroinflammation Is to Blame?

**Ashwinipriyadarshini Megur, Daiva Baltriukienė, Virginija Bukelskienė and Aurelijus Burokas ***

Department of Biological Models, Institute of Biochemistry, Life Sciences Center, Vilnius University, Sauletekio Ave. 7, LT-10257 Vilnius, Lithuania; avee.megur@gmail.com (A.M.); daiva.baltriukiene@bchi.vu.lt (D.B.); virginija.bukelskiene@bchi.vu.lt (V.B.)

* Correspondence: aurelijus.burokas@gmc.vu.lt; Tel.: +370-52234382

**Abstract:** For years, it has been reported that Alzheimer's disease (AD) is the most common cause of dementia. Various external and internal factors may contribute to the early onset of AD. This review highlights a contribution of the disturbances in the microbiota–gut–brain (MGB) axis to the development of AD. Alteration in the gut microbiota composition is determined by increase in the permeability of the gut barrier and immune cell activation, leading to impairment in the blood–brain barrier function that promotes neuroinflammation, neuronal loss, neural injury, and ultimately AD. Numerous studies have shown that the gut microbiota plays a crucial role in brain function and changes in the behavior of individuals and the formation of bacterial amyloids. Lipopolysaccharides and bacterial amyloids synthesized by the gut microbiota can trigger the immune cells residing in the brain and can activate the immune response leading to neuroinflammation. Growing experimental and clinical data indicate the prominent role of gut dysbiosis and microbiota–host interactions in AD. Modulation of the gut microbiota with antibiotics or probiotic supplementation may create new preventive and therapeutic options in AD. Accumulating evidences affirm that research on MGB involvement in AD is necessary for new treatment targets and therapies for AD.

**Keywords:** microbiota; Alzheimer's disease; microbiota–gut–brain axis; neuroinflammation; probiotics

**Citation:** Megur, A.; Baltriukienė, D.; Bukelskienė, V.; Burokas, A. The Microbiota–Gut–Brain Axis and Alzheimer's Disease: Neuroinflammation Is to Blame? *Nutrients* **2021**, *13*, 37. https://dx.doi.org/10.3390/nu13010037

Received: 27 November 2020
Accepted: 22 December 2020
Published: 24 December 2020

**Publisher's Note:** MDPI stays neutral with regard to jurisdictional claims in published maps and institutional affiliations.



## 1. Introduction

Dementia is a non-curable syndrome which over time leads to a progressive decrease in memory, thinking, and the capacity to perform everyday activities [1]. There are alternative forms of dementia which include vascular dementia, dementia with Lewy bodies, and frontotemporal dementia [2], which can be provoked by neurodegenerative disorders, cerebrovascular disease, brain injury [3], and infections [4]. The progression of dementia can result in a lack of consequential speech generation and inability to understand scriptural as well as phonetic language, failure to recognize and identify objects, execution of poor motor skills, and incapability to think abstractly and to execute paradoxical tasks [4,5].

Alzheimer's Disease (AD) is a persistent neurodegenerative (neuronal loss) disorder [6,7] which was first described by Alois Alzheimer in 1906 [8,9] while investigating a female patient Auguste Deter [10]. AD is known to be the major cause of dementia worldwide, mainly observed in the elderly [11], accounting for approximately 60–70% of all dementia cases [12]. The incidence of AD is higher in women than in men. AD is an extremely incapacitating disorder, progressing from slight memory impairments to a complete loss of mental function, and in the long period, resulting in death [13]. AD can affect distinct people in various ways. Most of the common warning signs include depression [14], memory loss, challenge in planning a task and problem-solving skills, confusion in recognizing time, mood swings and personality shifts, poor judgment in motor activities, difficulty in memorizing the literature, etc. [15].

Many factors can contribute to AD, but the greatest risk factors are determined to be exacerbations due to aging [16–18], degradation of anatomical pathways [12], environmental factors [19–21], mitochondrial dysfunction [22,23], immune system dysfunction [24,25], and genetic factors including mutations of amyloid precursor proteins (APP) [26,27].

In this review, we will be focusing on the role of the gut microbiota on the brain. We will be discussing the recent findings which show that a disturbance in the microbiota-brain axis can lead to neuroinflammation giving rise to AD. We will be discussing the recent studies which draw attention towards neuroinflammation in the brain, eventually leading to neuronal loss. Finally, we will be focusing on the administration of antibiotics and pre-vand probiotics modulating the brain function and used as a therapeutic agent in curing AD.

## 2. AD Pathology

The two major markers contributing to AD progression include amyloid-beta (Aβ) plaques and neurofibrillary tangles (NFTs) [28,29]. It was proposed that Aβ plaques are developed originally in the orbitofrontal, basal, and temporal neocortex regions of the human brain [30,31]. The accumulation of Aβ stimulates NFT formation [32,33]. The main constituent of NFTs is the protein tau in a hyperphosphorylated form. It is a highly soluble protein playing an essential role in maintenance of the stability of microtubules in the axons of neurons [34]. NFTs formed inside the neuron disrupt the microtubule structure and form an insoluble substance, which is detected in the locus coeruleus, and transentorhinal and entorhinal areas of the brain [35]. In the curtailed stage, it can spread to the hippocampus and neocortex [36]. The aggregation of plaques and tangles is followed by microglia recruitment surrounding the plaques [37]. This raises microglial activation and local inflammatory response which advance the neurotoxicity [25]. Aβ has been recognized as an antimicrobial peptide that activates the immune pathways recognized by toll-like receptor 2 (TLR2) leading to neuroinflammation [38].

A recent study has shown that amyloid pathogenesis begins with altered cleavage of APP β-secretase and γ-secretase to produce insoluble Aβ fibrils [22,39] (Figure 1). Aβ then oligomerizes, diffuses into synaptic clefts, and interferes with synaptic signaling [40]. Subsequently, it polymerizes into insoluble amyloid fibrils that aggregate into plaques [31]. This polymerization leads to activation of kinases [30], which can accelerate hyperphosphorylation of the microtubule-associated tau protein and its polymerization into insoluble NFTs [41].

NFTs are fragments of paired and helically wound protein filaments in the cell cytoplasm of neurons [42]. It has the proficiency of stabilizing microtubules and forging interconnections between adjoining microtubules to form a substantial network of microtubules and to hold them together [43]. The hyperphosphorylation of tau protein occurs when it comes into contact with the kinases released due to their abundance in the environment [44]. Its hyperphosphorylation leads to the formation of oligomers [45]. The microtubule becomes highly unstable due to the dissociation of tubule subunits [46] that fall apart and then get converted into enormous chunks of tau filaments, which further aggregate into NFTs [40]. The appearance of NFTs are straight, fibrillary, and highly insoluble patches [27] in the neuronal cytoplasm [47]. The major property known causes an abnormal loss of communication between neurons and signal processing and finally apoptosis of neurons [32]. Phosphorylation of tau is regulated by several kinases, including glycogen synthase kinase-3 (GSK3) and cyclin-dependent kinase 5 activated by extracellular Aβ [48]. Even GSK3 beta and cell division protein kinase 5 are primarily responsible kinases for tau hyperphosphorylation [13], and other kinases like protein kinase C, protein kinase A [49], ERK2, serine/threonine kinase, caspase 3, and caspase 9 also have a prominent role, which may be activated by Aβ [50].

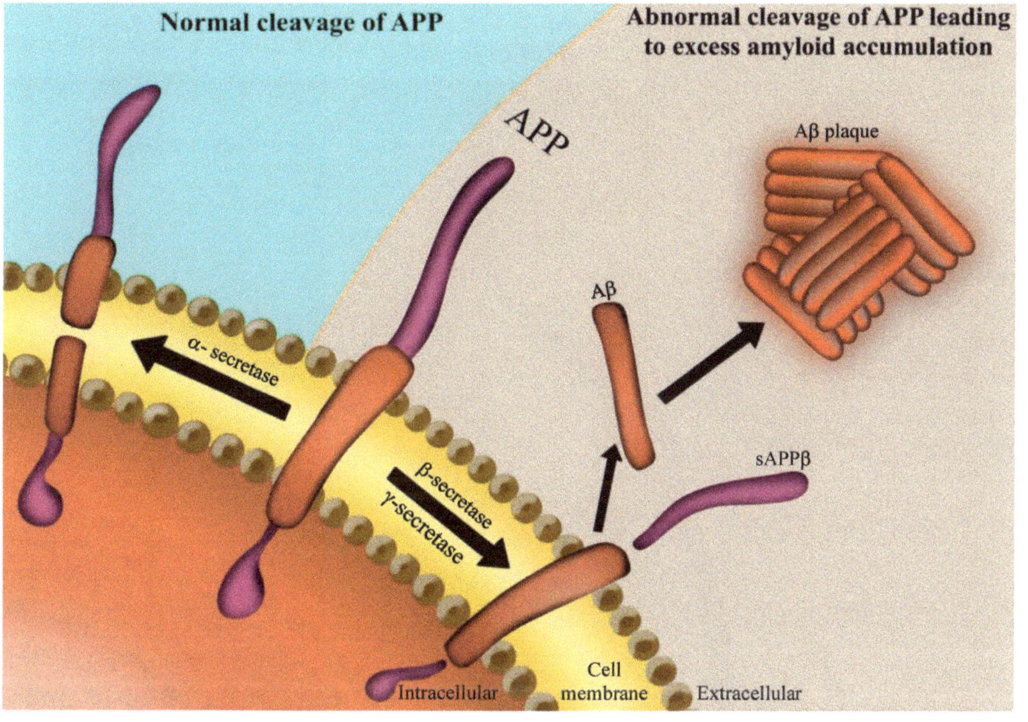


**Figure 1.** Aβ formation: the amyloid precursor protein (APP) is a transmembrane protein of the neuronal cell. In the case when it is cleaved by α-secretase, the formed soluble aggregates can be digested by microglial cells. When APP is cleaved by β-secretase and γ-secretase, it leads to formation of Aβ insoluble aggregates. Such protein aggregation results in amyloid plaques, one of the hallmarks of AD.

## 3. The Microbiota–Gut–Brain Axis

A microbiota is an ecological community of commensal microorganisms that live symbiotically and pathogenically in our body [5] and plays a vital role in regulatory functions in health and disease [51,52] (Figure 2). At the level of bacterial strains, the gut microbiota demonstrates tremendous diversity and variation in microorganisms related to the age of the person and can be different in the individuals [53]. To date, it was considered that microbial colonization in the gut was only involved in colon-specific activities, which includes fermentation of carbohydrates, vitamin synthesis, and metabolism of xenobiotics [54,55]. Furthermore, it was also found that the role of the gut microbiota is to act as a barrier for the pathogenic bacteria invading the gastrointestinal tract (GIT) [56].

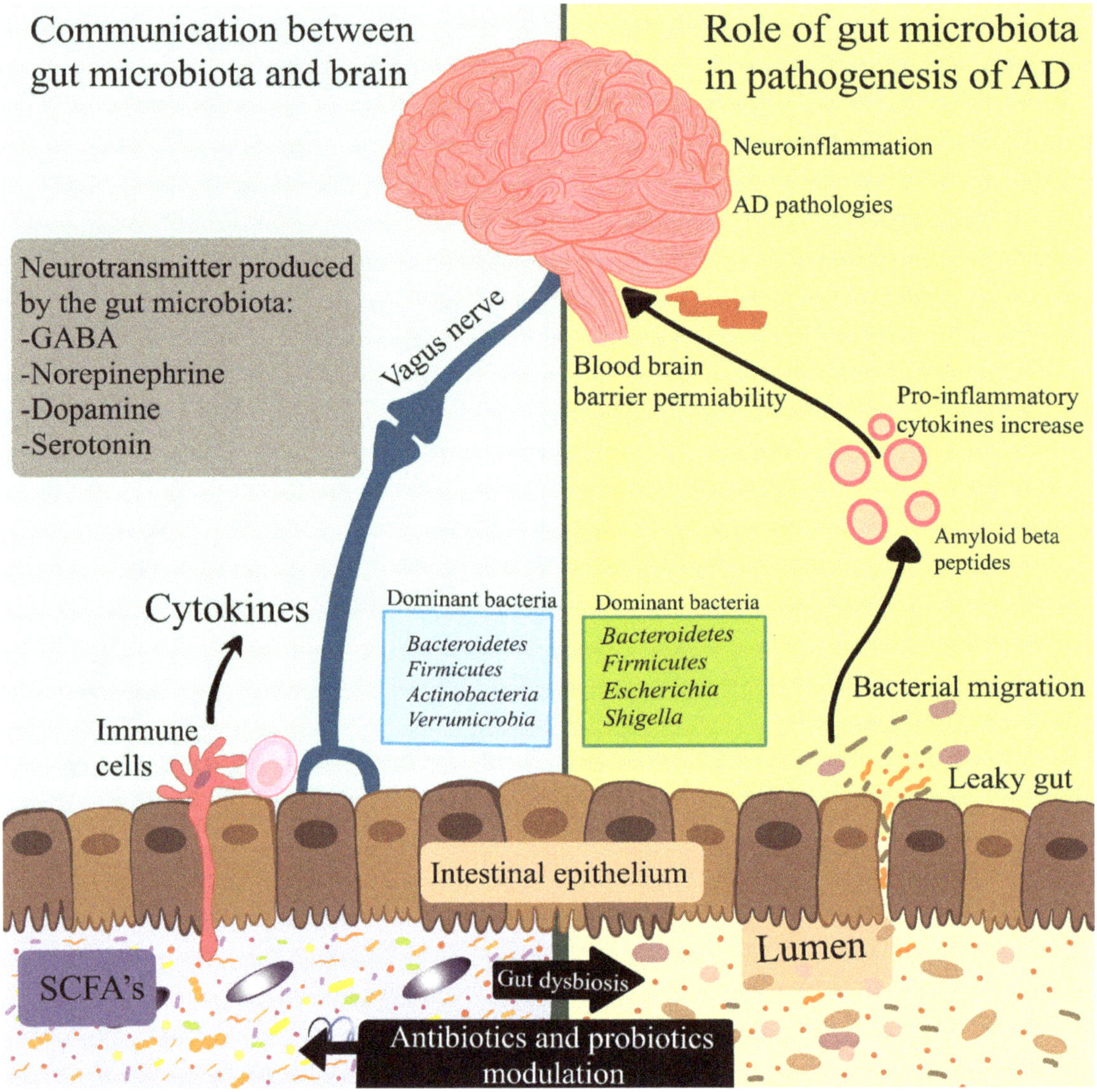


**Figure 2.** Modulation of the microbiota–gut–brain axis by antibiotics and probiotics. The communication between the gut microbiota and the brain includes neuronal, immune-mediated, and metabolite-mediated pathways. Gut dysbiosis leads to activation of the immune response and alters the production of neurotransmitters as well as bacterial metabolites. These may have a contribution to abnormal signaling through the vagus nerve. Reduction in the integrity of the gastrointestinal barrier causes bacterial migration and inflammation. Pro-inflammatory cytokines induce disruption of the blood–brain barrier permeability. Antibiotics can hinder the growth of certain bacteria, and probiotics have the potential to normalize the gut microbiota in microbiota–gut–brain processes.

The microbial colonization in humans is estimated to begin at birth. The new born infant is initially colonized by microorganisms common to its mother, which are *Lactobacillus* and *Prevotella* spp. [57]. When compared with healthy and preterm infants, usually delivered by caesarean section, preterm infants seem to have variations in the microbiota [58].

As well, further comparison with elderly people in nursing homes and in the community showed large differences. The individuals in the nursing home had less microbiota attributed to a limited diet [59]. Alterations of the composition of microorganisms due to dietary changes can result in augmentation of several diseases such as obesity, colorectal cancer, inflammatory bowel disease, heart failure, type 2 diabetes, and neurodegenerative disorders (AD, Parkinson's disease, multiple sclerosis, etc.) [52,57,60,61]. Furthermore, antibiotic treatment in early life can modulate the composition of microbiota in the gut later in life and can have a negative impact on the brain functions [62,63].

Numerous studies indicate that gut microbiota can have an influence in synthesizing various neurotransmitters and neuromodulators, which affect gut–brain communication and brain function [64–66]. Signal transduction is complex and can have the propensity to include neural, endocrine, immune, and metabolic pathways. However, its detailed mechanism and signals still have to be elucidated [53,67,68]. Clinical and preclinical studies have shown that gut microorganisms can produce metabolites, which affect brain functioning (Table 1).

**Table 1.** Effect of metabolites on brain produced by gut microbiota.

| No. | Gut Microorganisms | Metabolites | Effects of Metabolites on Brain | Subjects | References |
|---|---|---|---|---|---|
| 1 | *Lactobacillus* | Short chain fatty acids (SCFA), Serotonin, Acetylcholine | Increases emotional level | Wistar rats | [69,70] |
| | | | Improves attention, memory and motivation | Humans | [71] |
| | | | Improves sleep | C57BL/6J mice | [72] |
| 2 | *Bifidobacterium* | Gamma-aminobutyric acid | Reduces anxiety, stress, and fear Improves ADHD | Humans | [69,73,74] |
| | | Tryptophan | Improves behaviors relevant to depression | Pregnant Sprague–Dawley dams, rats | [75] |
| 3 | *Escherichia* | Dopamine, Norepinephrine, Endotoxin and Serotonin | Improves mood, blood flow, sleep regulation, cognition and concentration, hormonal activity | Human | [76–79] |
| 4 | *Bacillus* | Tryptophan | Improves cognitive function | Pigs | [80–82] |
| 5 | *Saccharomyces* | Norepinephrine | Enhances formation of retrieval of memory | Wistar rats | [77] |
| 6 | *Enterococcus* | Histamine, Serotonin | Promotes wakefulness, cognition orchestrates desperate behavior | C57BL/6J | [83] |

Bacterial strains such as *Escherichia*, *Lactobacillus*, *Saccharomyces*, and *Bacillus* can synthesize amino acids including gamma-aminobutyric acid, 5-hydroxytryptamine, dopamine, butyrate, histamine, and serotonin, which can play a significant role in emphasizing the brain activity of the individuals [84,85]. These neurotransmitters synthesized can cross the mucosal layer of the intestine and are capable of entering the blood stream [61,86]. It was found that the microbiota of aged individuals with AD have a lower level of bacteria that resulted in decreased butyrate levels [87], which, in turn, could lead to increased inflammation in the brain and the progression of cognitive loss [27,86]. These findings suggest that the microbiota performs numerous vital functions in our body, including releasing biochemical by-products such as SCFA and gases [88]. Moreover, animal studies conducted on pigs and rats showed an effect on memory due to microbiota, *bacillus* and *saccharomyces* [85–87]. Interestingly, a recent study has shown that microbiota transfer from human subjects with obesity led to reduced memory scores in mice, aligning this trait in humans with that of recipient mice [89], where RNA sequencing of the medial prefrontal cortex of those mice uncovered that short-term memory is associated with aromatic amino acid pathways, inflammatory genes, and clusters of bacterial species [89].

As the GIT of humans are inhabited by numerous microorganisms essential for by-product formation, it has been recently reevaluated in functional terms and different important mechanisms have been established in the bidirectional connection with the brain [90–92]. This bidirectional connection with the brain is termed as the "microbiota–gut–brain (MGB) axis". MGB refers to a crosstalk between the brain and the gut involving multiple overlapping pathways, including the autonomic, neuroendocrine, vagus nerve, the immune system, or the metabolic processes of gut microorganisms and immune system as well as bacterial metabolites and neuromodulatory molecules [93,94]. The MGB axis mirrors the constant connection between the central nervous system (CNS) and the GIT [95]. A number of rodent studies suggest potential involvement of the gut microbiota in behavioral changes [75,96–98]. The sympathetic and parasympathetic arms of the autonomic nervous system, including the neuroendocrine and neuroimmune systems, are known to be vital pathways in MGB [99]. The precise mechanism that arbitrates gut–brain interplay is not fully comprehended, yet it is suggested that it entails immune, endocrine, and neural pathways, leading to a possible alteration in AD patients or aggravation of inflammation (Table 2). The results from a rat study showed that *Bifidobacterium infantis*, an intestinal resident microorganism, has a link to immune response in the brain [75]. An augmentation in the number of *Lactobacillus casei*, *Bacteroides fragilis*, and *Streptococcus thermophilus* in the rodent intestine showed a positive effect on brain activity and performance [75,98–102]. On the other hand, *Eubacterium rectale*, *Porphyromonas gingivalis*, and *Lactobacillus rhamnosus* can play a vital role in the onset of AD [103–107].

Consideration of the human microbiota as a substantial correspondent to nutrition, health, and disease is a relatively fairly contemporary study, and currently, peer-reviewed studies relating modifications in the microbiota to the etiopathology of human diseases are few [108]. Claims on the potential involvement of the gut microbiota in brain function are made, in part, due to the well-described pathways of communication between the brain and the GIT which has been intensively studied in the area of food intake, satiety, and regulation of the digestive tract [109].

**Table 2.** Roles played by different microorganisms residing in the gut.

| No. | Organism | Positive ↑/ Negative ↓ Effects | Subjects | Role | Reference |
|---|---|---|---|---|---|
| 1. | *Bacteroides fragilis* | ↑ | AD patients | Protected against CNS demyelinating disease | [100,101] |
| | | | C57BL/6 mice | In pregnant mice showed an immediate significant diminished autistic behavior | [102,110,111] |
| 2. | *Lactobacillus casei* | ↑ | SAMP8 mice | A decreased in anxiety symptoms | [112] |
| 3. | *Lactobacillus rhamnosus* | ↑ | Wistar rats | Ameliorated the inflammation level in the brain | [103] |
| 4. | *Streptococcus thermophilus* | ↑ | SJL/J mice | • Robust effects on brain regions that control the central processing of emotions and sensation<br>• Degradation of Aβ 42 load | [113,114] |
| 5. | *Bifidobacterium infantis* | ↑ | Sprague–Dawley dams rats | Normalized the immune response | [75] |
| 6. | *Campylobacter jejuni* | ↓ | AD patients | Induced anxiety-like behavior<br>Impaired memory | [104] |

**Table 2.** *Cont.*

| No. | Organism | Positive ↑/ Negative ↓ Effects | Subjects | Role | Reference |
|---|---|---|---|---|---|
| 7. | *Campylobacter rodentium* | ↓ | C57BL/6 mice | Led to stress and contributed to behavioral abnormalities | [105] |
| 8. | *Porphyromonas gingivalis* | ↓ | AD patients | Caused an inflammatory response in the liver, which subsequently led to neuroinflammation and causes neurodegenerative disease | [106] |
| 9. | *Eubacterium rectale* | ↓ | AD patients | Leads to amyloidosis | [107] |
| 10. | *Lactobacillus acidophilus* | ↑ | SAMP8 mice | Improved the impairment in neural proteolysis | [112,113] |
| 11. | *Lactobacillus johnsonii* | ↑ | BB-DR rats Healthy humans | Improved gastric vagus nerve activity | [115,116] |

Incorporation of certain microorganisms, such as probiotics, in diet intake can be used as a therapeutic strategy to reduce neurological disorders. *Bifidobacterium* and *Lactobacillus casei* are two microorganisms which show a beneficial effect on neurological disorders [75,112].

## 4. Gut Microbiota in AD

Changes altering the gut microbiota can activate proinflammatory cytokines and increase intestinal permeability, which lead to the development of insulin resistance that is associated with AD [117] (Figure 2). Interestingly, recent work has shown that AD development could start even in the gut and then spread to the brain [118]. In this study, the gastric wall of mice was injected with A$\beta_{1-42}$ oligomers. Over 1 year, it was observed that the amyloid migrated from the intestine to the brain. Consequently, the translocation of A$\beta$ oligomers from the gut to the brain can have a major contribution in causing AD and neuroinflammation [118].

*Escherichia coli*, *Salmonella enterica*, *Bacillus subtilis*, *Mycobacterium Tuberculosis*, and *Staphylococcus aureus* are some of the bacterial strains that can produce functional extracellular amyloid fibers [107]. These amyloid proteins help the bacterial strains to form biofilms and to strongly bind to each other to resist destruction by physical and immune factors [119]. The amyloids formed by bacteria are different from the CNS amyloids in the primary structure but show resemblance in their tertiary structure [120]. The appearance of bacterial amyloid in the gut can trigger the immune system, which could lead to enhanced immune responses with endogenous formation of neuronal amyloid in the brain [119]. Studies of AD patient's blood and cerebrospinal fluid showed an escalated inflammatory response when compared to healthy adults [107]. In the latter case, the clearance of amyloid is very precise [121].

In a recent study, aged Fischer 344 rats were orally exposed to transgenic *E. coli* producing the extracellular bacterial amyloid protein curli (a type of amyloid fiber protein). The data showed an enhanced alpha-synuclein production in the gut and intensified aggregation of alpha-synuclein in the brain, leading to enhanced microgliosis and astrogliosis. Elevated expressions of TLR2, IL-6, and TNF-$\alpha$ in the brain of animals exposed to curli-producing bacteria were determined. This suggested that bacterial amyloid functions as a trigger initiating alpha-synuclein aggregation through cross-seeding and prime responses of the innate immune system [122].

A profound experiment conducted on the APP transgenic mouse model for AD suggested that variation in the number of microbial strains could lead to amyloid deposition. These APPPS1 mice showed reduced numbers of *Firmicutes* and an increased number of *Bacteroides* in the intestine. The germ-free APP transgenic mice demonstrated a reduction in cerebral Aβ pathology [123]. This finding strongly points towards the intestinal microbiota forming amyloid-triggering immune responses that can lead to hallmarks of AD.

Clinical studies of the gut microbiota of AD patients as well as microbiota from AD model mice revealed decreased microbial diversity when compared with controls (Table 3). These include decreased levels of *Fusobacteriaceae*, *Firmicutes*, *Actinobacteria*, and *Bifidobacterium* and increased levels of *Bacteroidetes* [54,124]. *Cyanobacteria*, one of the gut-residing bacteria, produces a neurotoxin β-N-methylamino-L-alanine, which interferes with the N-methyl-D-aspartate glutamate receptor and leads to signal dysfunction in AD [125].

**Table 3.** Investigation of microbiota in the gut of human as well as animal models of AD.

| No. | Microorganisms | Increase ↑/ Decrease ↓ | Animal Model | Location | Reference |
|---|---|---|---|---|---|
| 1. | *Firmicutes/Actinobacteria* | ↓ | CONVR-APP/PS1 | Intestine | [54,124] |
| 2. | *Bacteroides/tenericutes* | ↑ | | | |
| 3. | *E. coli/B. subtilis* | ↑ | AD patient | Brain tissues/Stool | [69,126–128] |
| 4. | *E. rectale* | ↓ | AD patient | Stool | [107,129] |
| 5. | *Escherichia/shigella* | ↑ | | | |
| 6. | *B. fragilis* | ↓ | | | |
| 7. | *Lactobacilli/Bifidobacteria* | ↑ | SAMP-8 mice | Intestine | [71] |
| 8. | *Fusobacteriaceae* | ↓ | AD patients | Stool | [123] |
| 9. | *Prevotellaceae* | ↑ | | Stool | |
| 10. | *Verrucomicrobia* | ↑ | $APP_{SWE}/PS1_{\Delta E9}$ (PAP) transgenic mice | Stool | [130] |

Not only the bacterial strains residing in the gut can lead to neurodegeneration but also the invading pathogens, such as *Mycobacterium leprae*, are known to be responsible for demyelination and nerve damage. *M. leprae* assists in initiation of the pathogen by changing the internal environment of Schwann cells and stimulation of apoptotic pathways in cells [131]. *Chlamydia pneumoniae* causing respiratory tract infection has been reported in CNS disorders, including AD [132]. *C. pneumoniae* antigens were also found in the neocortex of AD in association with NFTs and senile plaques [133]. Moreover, *Cladosporium*, *Malassezia*, *Phoma*, *Saccharomyces*, and *Candida* species DNA, polysaccharide, and proteins were observed in the CNS samples of AD patients [134]. Fungal footprints were identified in the cerebrospinal fluid by using PCR and slot bolt assay techniques [135].

Upon infection, various cell signaling pathways can occur in the body, which can activate inflammation. When infectious microorganisms cross the blood–brain barrier, it leads to neuronal death due to inflammation and forms similar hallmarks to AD. Lipopolysaccharide (LPS) is found in many gram-negative bacteria [136], exclusively on the outer membrane [137]. An experiment conducted on animal models has shown that bacterial LPS injection in the fourth ventricle of the brain produced inflammatory and pathological characteristics as observed in AD [138] and the peritoneal cavity led to extended elevation of Aβ in the hippocampal regions of mice resulting in cognitive decline [139]. An in vitro study conducted on *E. coli* confirmed that bacterial LPS advanced amyloid fibrillogenesis [127]. Studies conducted on AD patients confirmed LPS presence in the hippocampus and neocortex brain lysates in which most of the LPS aggregation has been observed in the perinuclear region [129,140]. The LPSs are located near Aβ 1-40/42 in amyloid plaques as well as blood vessels [128], and in AD patients, its levels are slightly higher compared with healthy adults [141]. When microglial cells come in contact with LPS, the TLRs present on the cell membrane of microglia gets activated through interaction with glycosylphosphatidylinositol-anchored receptor CD14 and MD-2 protein promoting inflammatory responses [110,142]. CD14-activated receptor TLR4 mediates responses to Aβ [143]. This activation affects the immune response and induces neuroinflammation.

## 5. Neuroinflammation

Our brain sustains the immune cells that protect against infection and injury, also supporting neurons in plasticity and circuit efficient connectivity. Inflammation is a response necessary for protection and regulation of the process which is associated with managing and reducing damage of the organism: protection against microorganisms, tissue repair, and removal of debris from the body [144]. Various studies currently indicate the involvement of neuroinflammation playing a crucial role in the progression of neuropathological changes that are observed in AD [145] (Figure 2). A broad variety of cellular and molecular mechanisms, assumedly identical in aging and chronic metabolic diseases such as hypertension, diabetes, metabolic syndrome, dementia, depression, or traumatic brain injury, are currently considered silent contributors to neuroinflammation [146]. The key players responsible for induction of neuroinflammation are known to be activated microglia and astrocytes [24,147].

Microglia which originate from myeloids are known as immunocompetent cells in the brain. Microglia cells are considered to be the most important player in the development and progression of neuroinflammation [25]. Microglia are immensely plastic cells that can transform into complex phenotypes depending on specific microenvironmental signals within the brain [148]. On the membrane, these cells express a diverse range of innate immune receptors that belong to the pattern recognition receptors family [147]. When pattern recognition receptors get activated on microglia, activation of the cell and the production of inflammatory mediators occur in the presence of a distinct signaling cascade [149]. Repeatedly activated microglia release a broad range of proinflammatory [150] and toxic products and, among them, reactive oxygen species, nitric oxide, and cytokines. In addition, endothelial cells and perivascular macrophages are also important in interpreting and propagating these inflammatory signals within the CNS [24]. A threat to the CNS, such as invasion, injury, or disease, activates microglia, induces morphological changes, and increases motility of cells.

In AD, there are studies conducted that the primary initiator of activation of microglia is the accumulation of Aβ [151]. The activated microglia respond to Aβ, resulting in migration to the plaques and phagocytosis of Aβ. It initiates a microglial-mediated inflammatory response by binding to various pattern recognition receptors [152], which, in turn, results in cell activation and release of proinflammatory factors (iNOS, TNF-α, IL-1, and IL-6) [152–155]. In the case of AD, the receptors present on the surface of the microglia bind to Aβ oligomers and Aβ fibrils. In the process of phagocytosis, microglia begin to clean up Aβ fibrils; hence, fibrils undergo an endolysosomal pathway.

Other than microglia, astrocytes are also major participants in neuroinflammation [156]. They are fivefold more than neurons in the CNS [157] and are known to have functions in the maintenance of CNS integrity, such as control of blood perfusion in the cerebrum, maintenance of blood–brain barrier stability, and modulation of neuron or nutrient transmission [158]. In AD patient brains, there have been observed alterations in the morphology of astrocytes, their protein composition, gene expression, and function [150]. The accumulation of activated astrocytes is often present in clusters around amyloid plaques. Aβ deposit can activate the astrocytes which lead to overexpression of cytokines, such as IL-1β and IL-6, resulting in oxidative stress [24,159]. It was recently shown that neurodegeneration presumably associates astrocytes, which, by taking on a microglia-induced A1 proinflammatory phenotype, would encourage neuronal cell death, with TNF-α as the most eminent arbitrator [160,161].

On the other hand, the activated microglia lose their phagocytic effect, thus decreasing the degree of Aβ phagocytosis, inevitably developing its accumulation [162]. Moreover, such discoveries are supported by the results of an association between an increase in AD risk and alterations in genes encoding immune receptors such as TREM2, CD33, and CR1 (myeloid cell surface antigen) [163]. Since they are all expressed on myeloid cells, it is a more convincing demonstration that alterations in microglial biology are linked to AD pathogenesis. Worth mentioning, a variety of transcriptomic and proteomic analysis of inflammatory cells might provide biomarkers for preclinical detection as well as insights on the progression from mild cognitive impairment to AD condition [164–166].

A relatively close connection has also been reported between microglia and cognitive dysfunction [167]. Importantly, in healthy tissue, microglia have a ramified morphology and prolongations that continuously look after the synaptic activity. However, phagocytic microglia have a salient role in synaptic pruning and honing in the developing nervous system [168]. The most fascinating mechanism describing memory dysfunction in AD suggests that Aβ oligomers lead to microglial activation, which, in turn, excessively engulfs and accelerates the termination of synapses through complement factors such as C1q and C3 [169]. It has also been reported that Aβ oligomer arbitrates memory problems which are closely connected with glial activation [100,170].

Recent evidences now shed light on a dangerous dialogue between central immune cells and the gut microbiota, potentially leading to AD in humans.

## 6. The Link between Microbiota and Neuroinflammation

The immune system modulates the gut microbiota framework and issuance [171], while in return, the microbial symbionts control immune system maturation and function [172,173]. Numerous rodent studies have affirmed that there is an interaction between the gut microbiota and various immune cell populations [174,175] or the expression of genes related to neuroinflammation [176,177].

The study furnished evidence stating that microbiota residing in the gut predisposes the development of the immune system by administering hematopoiesis of primary immune cells. It was shown that germ-free (GF) mice have a lower ratio and less distinction capability of myeloid cell progenitors of both yolk sac and bone marrow origin. This supports the idea of the widespread effects of gut microbiota on the immune system, microglia included [175]. Microglia from antibiotic-treated mice or GF mice showed an immature profile and impaired immune response. The absence of gut microbiota alters microglial mRNA profiles and suppresses various microglial genes involved in cell activation, pathogen recognition, and host defense. Microglia transcription and survival factors, normally suppressed in mature adult microglia, were increased in GF mice [178]. The experiment was conducted to examine the transcriptional profiles of different microglial development stages, referring to the genes related to the adult phase of microglial maturation and immune response that are abnormally regulated in GF mice [179].

A number of studies have coined a protective association between dietary polyphenols and the prevention of age-related chronic diseases such as diabetes, cancer, and neurodegenerative diseases [180–182]. Dietary flavonoids and nonsteroidal anti-inflammatory agents modulate the nuclear factor-kappa β signaling pathway and therefore are termed as a potential therapeutic target for AD [182–184]. Polyphenols make an impact on microbiota-related metabolism and have a potential to improve neurological health, including their ability to interact with intracellular neuronal and glial signaling, to modulate peripheral and cerebrovascular blood flow, and to reduce neuronal damage and loss induced by neurotoxins and neuroinflammation [185–187]. Flavonoids, a subclass of polyphenols, are more likely to combat neuronal dysfunction and toxicity by recruiting antiapoptotic pro-survival signaling pathways, increasing antioxidant gene expression and reducing Aβ pathology [182,188,189]. Flavonoids that are not absorbed in the small intestine and other sugars are then broken down by the gut microbiota into phenolic acids and other metabolites that inhibit the growth of *Ruminococcus gauvreauii*, *Bacteroides galacturonicus*, and *Lactobacillus* sp. strains [190] and flavonoids present in berries have also shown inhibitory actions against *Bacillus cereus*, *Campylobacter jejuni*, *Clostridium perfingens*, *Helicobacter pylori*, *Staphylococcus aureus*, *Staphylococcus epidermidis*, and *Candida albicans* [191]. Recently, it was reported that anthocyanins (one of the flavonoids) could significantly ameliorate the expression of proinflammatory cytokines and ROS/JNK, thus preventing neuroinflammation and AD pathology [192–194]. In an experiment conducted on aged rodents, blueberry supplementations have shown improved spatial memory, object recognition memory, and inhibitory fear conditioning learning [195–197]. In another study on blueberry anthocyanins given to adults aged 40–74 years over 3 weeks, plasma concentrations of NF-kB-related proinflammatory cytokines and chemokines (IL-4, IL-13, IL-8, and IFN-$\alpha$) were significantly reduced [198]. However, a study conducted by Spilsbury et al. did not reveal any remarkable effect of lower concentrations of flavonoids on NF-$\kappa$B activity in astrocytes [199]. Nevertheless, the literature date supports that the dietary supplementation of flavonoids might be implicated in the regulation of NF-$\kappa$B in neurons [199].

Flavonoids are important players in the prevention of neuroinflammation via several anti-inflammatory mechanisms, inhibiting the microglial activation of inflammatory cytokines (TNF-$\alpha$ and IL-1$\beta$), inhibiting iNOS and ROS generation in activated glia, and downregulating the activity of pro- inflammatory transcription factors such as NF-$\kappa$B through modulation of glial and neuronal signaling pathways [182].

Chicory root, known for its high content of fibers (galacto-oligosaccharides and fructans, such as inulin) and beneficial for the MGB axis modulation [64,177,200], recently also has received attention due to its sesquiterpene lactones (a class of sesquiterpenoids that contain a lactone ring) [201]. Interestingly, it has been shown that different sesquiterpene lactones from chicory root have the potential to influence anti-inflammatory responses through modulation of the nuclear factor of the activated T-cells pathway [201].

Bacterial metabolites such as SCFAs were considered the key mediators for microbiota–microglia interaction. These compounds have the potential to translocate from the mucosa to systemic circulation and to cross the blood–brain barrier affecting the CNS and their function [68,202]. Oral administration of SCFA for 4 weeks restored many facets of the immature microglial morphology of GF mice. SCFA claimed to reestablish microglial density and normalized CSF1R surface expression [203]. It is crucial to accentuate that the gut microbiota–microglia interaction is extremely dynamic as many of the defects noticed in the microglia of GF mice could be partially restored by recolonization with conventional gut microbiota or SCFA supplementation [203].

## 7. Role of Antibiotics on Microbiota in AD

Antibiotics or antimicrobial substances are typically used to remove or prevent bacterial colonization in the human body [204]. These can alter the bacteria without any specific target or type [205]. As a consequence, a broad spectrum of antibiotics can immensely affect the composition of the gut microbiota, lower its biodiversity, and withhold colonization for a long period after administration. Various studies with distinct antibiotic treatments resulted in long-/or short-term changes in the gut microbiota in both animals as well as humans [206]. Numerous studies have demonstrated that the use of antibiotics has an association with changes in behavior and brain chemistry [207–209]. Studies conducted in vivo with long-term broad spectrum antibiotic treatment have shown a decreased A$\beta$ plaque deposition, attenuation of plaque localization in glial reactivity, and alteration in microglial morphology in the $APP_{SWE}/PS1_{\Delta E9}$ mouse model of AD [210]. Another study conducted on 68 patients with advanced AD demonstrated a correlation among usage of antibiotics and prolonged survival. Of the patients who survived for more than 6 months, 31% were on antibiotic care and 14% were on palliative care [211]. Another study in humans showed that antibiotics, i.e., cefepime, can cross the blood–brain barrier, causing altered mental status, along altered consciousness and confusion without mediation of the gut microbiota [212]. Below, some of the preclinical studies of antibiotics in animals and humans have been described briefly.

The patients suffering from infection caused by *Helicobacter pylori* were administered with a cocktail of antibiotics consisting of proton pump inhibitor and clarithromycin, along with amoxicillin or metronidazole. The outcome of this treatment showed an association with neurological disorders, including panic attacks due to major depression and anxiety, delirium, and psychosis [213]. On the other hand, the elimination of pathogenic bacteria such as *Helicobacter pylori* in AD patients by the triple eradication antibiotic regimen (clarithromycin, amoxicillin, and omeprazole) led to positive results for cognitive and functional status parameters [214].

Antibiotic administration with rifampicin and minocycline in AD animal models reduced the Aβ levels in the brain and abbreviates inflammation cytokines [215]. Oral administration of rifampicin to three different mouse models of Alzheimer's disease and tauopathy showed that this antibiotic reduced the accumulation of Aβ oligomers and tau oligomers and enhanced the memory of the mice. These results suggested that rifampicin could prevent AD [216]. In 6 months, AD patients' improvement in the Standardized AD Assessment Scale cognitive subscale was observed when treated with a combination of doxycycline and rifampicin [217].

A pilot study conducted on the TgCRND8 transgenic mouse model showed that 3 months of treatment with erythromycin in drinking water at 0.1 g/L reduced the $A\beta_{1-42}$ levels in the cortex by 54% when compared to vehicle-treated mice [218].

Several studies conducted on minocycline suggested that it has neuroprotective and anti-inflammatory actions in many animal models. In microglial cell cultures, it was remarkably able to reduce the oligomeric Aβ-induced neuroinflammatory response and enhancement of fibrillar Aβ phagocytosis [219]. Minocycline treatment at 50 mg/kg for 4 weeks in a transgenic hAPP mouse model of AD exhibited attenuated behavioral abnormalities, neuroinflammatory markers, and Aβ [220]. In another study, 4 months of treatment with minocycline at 55 mg/kg/day in food in 3×Tg-AD mice showed a reduction in brain levels of insoluble Aβ, decreased neuroinflammatory markers, and reversed cognitive deficit [221].

A contrary effect of antibiotics was also observed after administration of ampicillin in the Sprague–Dawley rats. In this case, an elevated level of corticosterone in serum, intensified anxiety-like behavior, impairment of memory due to elevated glucocorticoids, and reduction in hippocampal brain-derived neurotrophic factor were determined [222]. Distinct studies demonstrated that administration of intracerebroventricular streptozotocin into the brain of wild-type mice and rats can cause learning impairment and memory loss [223–227].

An experiment conducted on $APP_{SWE}/PS1_{\Delta E9}$ transgenic mice administered with antibiotics demonstrated that it led to an alteration in several circulating inflammatory cytokines and chemokines in the blood. It also showed an elevated level of CCL11 (which has a link to age-related deficits in hippocampal neurogenesis) [228] in the blood serum of mice [210]. A recent study conducted on $APP_{SWE}/PS1_{L166P}$ mice treated with a cocktail of antibiotics revealed a selective, microbiome-dependent, sex-specific effect on brain Aβ amyloidosis of Aβ and microglial physiology [229]. Interestingly, the transplants of fecal microbiota from age-matched $APP_{SWE}/PS1_{L166P}$ mice into antibiotic-treated $APP_{SWE}/PS1_{L166P}$ mice restores the gut microbiota and partially restores AD pathology along with microglial morphology [229].

## 8. Role of Probiotics on Microbiota in AD

Probiotics are defined as living microbial feed supplements which show a beneficial effect on the host, resulting in improved intestinal microbial balance [230]. The most commonly used probiotics are lactic acid bacteria, particularly *Lactobacilli*, *Streptococci*, *Pediococcus*, *Enterococcus*, and *Bifidobacteria* and some yeast like *Saccharomyces boulardii*. However, not all microorganisms can be probiotic, as they need to be strain-specific (Table 4).

**Table 4.** Effects of probiotics on neurological disorders.

| No. | Probiotic Supplementation | Subject | Effect | Reference |
|---|---|---|---|---|
| 1. | *L. helveticus R0052* | WT mice IL-10 deficient 129/SvEv mice | Prevented from anxiety-like behavior and memory impairment | [231] |
| 2. | *Lactobacillus plantarum MTCC 1325* | AD rat model (IP injection of D-galactosea) | Reestablished acetylcholine levels, debilitated Aβ plaque formation, and ameliorated cognitive function | [232] |
| 3. | *L. helveticus, L. rhamnosus* | Streptozocin injected rats (diabetes rats) | Improved spatial memory impairment and recovered declined basic synaptic transmission | [233] |
| 4. | *Lactobacillus casei* strain Shirota (LcS) | In vivo mouse model of EAE | Reduced neuroinflammation | [234] |
| 5. | *Lactobacillus* and *Bifidobacterium* | AD rat model (intrahippocampal injection of Aβ) | Ameliorated memory, learning deficits, and oxidative stress | [235] |
| 6. | *Clostridium butyricum* | Mouse model of vascular dementia | Reduced neuronal apoptosis and attenuated cognitive dysfunction and histopathological changes | [236] |
| 7. | SLAB51 probiotic formulation | 3×Tg-AD mice | Altered plasma concentration of inflammatory cytokines and gut hormones induced also a decrease in brain damage and accumulation of Aβ aggregates | [113] |
| 8. | *Bifidobacterium breve* strain A1 | AD mouse model (ICV injection of Aβ) | Blocked Aβ-induced cognitive dysfunction and suppressed Aβ-induced changes in gene expression in the hippocampus | [237] |
| 9. | oligosaccharides from *Morinda officinalis* | APP/PS1 mice | Ameliorated brain tissue swelling and neuronal apoptosis and downregulated the expression of Aβ | [238] |
| 10. | *Bifidobacterium longum* 1714 | Healthy humans | Reduced stress and improved memory | [239] |
| 11. | *Lactobacillus brevis FPA3709* | Sprague–Dawley rats | Similar effects to a generally used antidepressant drugs | [240] |

A broad range of probiotics have been used in an animal study and in the models of AD. In rats, *Bifidobacterium* and *Lactobacillus* administration have shown a positive effect on AD treatment [235]. In an AD mouse model, *Bifidobacterium breve* strain A1 prevented cognitive function, making it one of the effective treatments for AD [237]. A reduction in neuroinflammation in mouse models due to *Lactobacillus casei* strain *Shirota* can be effective against AD [234]. Despite the fact that there are few human clinical studies compared to animals, there is increasing indication that probiotics can be used for reducing depression and anxiety-like symptoms [241].

A study with thirty-six healthy women assigned to three groups showed the importance of probiotics in the modulation of brain activity [242]. In this experiment, the group which was treated with fermented milk products containing *Bifidobacterium animalis sub. lactis*, *Streptococcus thermophilus*, *Lactobacillus bulgarigaricus*, and *Lactococcus lactis subs. lactis* showed a compelling reduction in the activity of the specific area in the brain. This region of the brain is involved in sensory/affective tasks when compared to the activation of other cortical regulatory brain areas. The experiment confirmed that probiotic supplementation has a major contribution in activating specific areas in the brain involved in the central control of emotion and sensation [242].

In another study conducted to understand the probiotic application in AD, sixty patients with AD were randomly assigned into two groups [243]. The first group received 200 mL/day milk enriched with *Lactobacillus acidophilus, Lactobacillus casei, Bifidobacterium bifidum,* and *Lactobacillus fermentum* for weeks, whereas the control group received plain milk of the same amount. The subjects, which were on probiotic supplementation showed a significant improvement in the mini-mental state examination test when compared with controls. The study revealed a beneficial effect on cognitive function and metabolic status of patients with AD. However, the treatment with probiotics was ineffective on oxidative stress and inflammation [243].

A study conducted by Leblhuber et al. showed an increased level of serum kynurenine, which was observed after probiotic administration, potentially caused by macrophage activation. The stimulation of immune cells could induce mechanisms that can be helpful in removing amyloid aggregates and damaged cells or on the other perspective. On the other hand, the intensive activating events could negatively affect gut barrier function and further stimulate neurodegenerative events [244].

When taken together, these human and animal studies prove that probiotics can have a major role in the bidirectional communication between the gut microbiota and the brain, modulating brain function. The exact mechanism of probiotics on the MGB axis is not yet well defined. Therefore, the data suggest that the proper dose of probiotics in AD treatment would be a new way to eliminate amyloid deposition in the brain by the MGB axis and to reduce neuroinflammation (Figure 2).

## 9. Conclusions

Accumulating all information from the human as well as animal studies, it can be suggested that GIT microbiota has an important role in the bidirectional communication between the brain and the gut. There is increasing evidence stating that the gut microbiota has a contribution to the pathogenesis of AD. As the gut microbiota is known as the source of a large number of amyloid, LPS, and other toxins, it can contribute to systemic inflammation and disruption of physiological barriers. The products formed by bacteria can move from the GIT to the CNS, especially in aging. Bacterial amyloid can trigger misfolding and can enhance native amyloid aggregation. The gut microbiota products can activate microglia, augmenting inflammatory response in the CNS, which in turn results in microglial function. Triggered microglia start neuroinflammation in the brain, causing loss of neurons, a major factor in AD. Modulation of the gut microbiota composition can be used as a therapeutic target in AD. Some antibiotics as well as probiotics can be implemented as a preventive measure that successfully targets ongoing inflammation. The role of antibiotics and probiotics in modulating the microbiota is under intense debate. The certain microbiota profile also strongly depends on the host's genetics and diet. This only confirms that research on MGB involvement in AD is crucial for new treatment targets and therapies for AD.

**Author Contributions:** Conceptualization, A.M. and A.B.; investigation, A.M.; writing—original draft preparation, A.M., D.B., and A.B.; writing—review and editing, A.M., A.B., and V.B.; visualization, A.M.; supervision, A.B.; project administration, A.B.; All authors have read and agreed to the published version of the manuscript.

**Funding:** This project received funding from the European Regional Development Fund (project No. 01.2.2-LMT-K-718-02-0014) under grant agreement with the Research Council of Lithuania (LMTLT).

**Conflicts of Interest:** No conflict of interest declared.

## Abbreviations

| | |
|---|---|
| AD | Alzheimer's disease |
| Aβ | Amyloid-beta |
| NFTs | Neurofibrillary tangles |
| MGB | Microbiota–gut–brain |
| GIT | Gastrointestinal tract |
| CNS | Central nervous system |
| LPS | Lipopolysaccharides |
| TLR | Toll-like receptor |
| SCFA | Short chain fatty acids |
| APP | Amyloid precursor protein |

## References


1. Kolanowski, A.; Fortinsky, R.H.; Calkins, M.; Devanand, D.P.; Gould, E.; Heller, T.; Hodgson, N.A.; Kales, H.C.; Kaye, J.; Lyketsos, C.; et al. Advancing Research on Care Needs and Supportive Approaches for Persons with Dementia: Recommendations and Rationale. *J. Am. Med. Dir. Assoc.* **2018**, *19*, 1047–1053. [CrossRef] [PubMed]
2. Toepper, M.; Falkenstein, M. Driving Fitness in Different Forms of Dementia: An Update. *J. Am. Geriatr. Soc.* **2019**, *67*, 2186–2192. [CrossRef] [PubMed]
3. Annear, M.J.; Toye, C.; McInerney, F.; Eccleston, C.; Tranter, B.; Elliott, K.E.; Robinson, A. What should we know about dementia in the 21st century? A Delphi consensus study. *BMC Geriatr.* **2015**, *15*, 1–13. [CrossRef] [PubMed]
4. Sibbett, R.A.; Russ, T.C.; Deary, I.J.; Starr, J.M. Risk factors for dementia in the ninth decade of life and beyond: A study of the Lothian birth cohort 1921. *BMC Psychiatry* **2017**, *17*, 1–10. [CrossRef] [PubMed]
5. Mancuso, C.; Santangelo, R. Alzheimer's disease and gut microbiota modifications: The long way between preclinical studies and clinical evidence. *Pharmacol. Res.* **2018**, *129*, 329–336. [CrossRef] [PubMed]
6. Burns, A.; Iliffe, S. Alzheimer's disease. *BMJ* **2009**, *338*, 467–471. [CrossRef]
7. Perl, D.P. Neuropathology of Alzheimer's disease. *Mt. Sinai J. Med.* **2010**, *77*, 32–42. [CrossRef]
8. Grøntvedt, G.R.; Schröder, T.N.; Sando, S.B.; White, L.; Bråthen, G.; Doeller, C.F. Alzheimer's disease. *Curr. Biol.* **2018**, *28*, R645–R649. [CrossRef]
9. Armstrong, R.A. Review article what causes alzheimer's disease? *Folia Neuropathol.* **2013**, *3*, 169–188. [CrossRef]
10. Dage, J.L.; Wennberg, A.M.V.; Airey, D.C.; Hagen, C.E.; David, S.; Machulda, M.M.; Roberts, R.O.; Ronald, C.; Mielke, M.M.; Lilly, E.; et al. Levels of tau protein in plasma are associated with neurodegeneration and cognitive function in a population-based elderly cohort. *Alzheimer's Dement.* **2017**, *12*, 1226–1234. [CrossRef]
11. Di Resta, C.; Ferrari, M. New molecular approaches to Alzheimer's disease. *Clin. Biochem.* **2019**, *72*, 81–86. [CrossRef] [PubMed]
12. Frigerio, C.S.; Wolfs, L.; Fattorelli, N.; Perry, V.H.; Fiers, M.; Strooper, B.D.; Frigerio, C.S.; Wolfs, L.; Fattorelli, N.; Thrupp, N.; et al. The major risk factors for Alzheimer's disease: Age, sex, and genes modulate the microglia response to Aβ plaques. *Cell Rep.* **2019**, *27*, 1293–1306.e6. [CrossRef] [PubMed]
13. Evans, D.A.; Bienias, J.L.; Schneider, J.A.; Wilson, R.S.; Bennett, D.A. Mild cognitive impairment is related to Alzheimer disease pathology and cerebral infarctions. *Neurology* **2011**, *64*, 834–841. [CrossRef]
14. Cortés, N.; Andrade, V.; Maccioni, R.B. Behavioral and Neuropsychiatric Disorders in Alzheimer's Disease. *J. Alzheimer's Dis.* **2018**, *63*, 899–910. [CrossRef]
15. Souza, R.K.M.d.; Barboza, A.F.; Gasperin, G. Prevalence of dementia in patients seen at a private hospital in the Southern Region of Brazil. *Einstein (São Paulo)* **2020**, *18*, 1–7. [CrossRef]
16. Hara, Y.; McKeehan, N.; Fillit, H.M. Translating the biology of aging into novel therapeutics for Alzheimer disease. *Neurology* **2019**, *92*, 84–93. [CrossRef]
17. Bostancıklıoğlu, M. The role of gut microbiota in pathogenesis of Alzheimer's disease. *J. Appl. Microbiol.* **2019**, *127*, 954–967. [CrossRef]
18. Mathay, M.T.; Ito, K.; Boppana, S.; Ito, N.; Yadav, S.K.; Mindur, J.E.; Patel, A.; Dhib-Jalbut, S. Gut dysbiosis breaks immunological tolerance toward the central nervous system during young adulthood. *Proc. Natl. Acad. Sci. USA* **2017**, *114*, E9318–E9327. [CrossRef]
19. Tan, S.H.; Karri, V.; Tay, N.W.R.; Chang, K.H.; Ah, H.Y.; Ng, P.Q.; Ho, H.S.; Keh, H.W.; Candasamy, M. Emerging pathways to neurodegeneration: Dissecting the critical molecular mechanisms in Alzheimer's disease, Parkinson's disease. *Biomed. Pharm.* **2019**, *111*, 765–777. [CrossRef]
20. Wojtunik-Kulesza, K.; Oniszczuk, A.; Waksmundzka-Hajnos, M. An attempt to elucidate the role of iron and zinc ions in development of Alzheimer's and Parkinson's diseases. *Biomed. Pharmacother.* **2019**, *111*, 1277–1289. [CrossRef]
21. Goschorska, M.; Baranowska-Bosiacka, I.; Gutowska, I.; Metryka, E.; Skórka-Majewicz, M.; Chlubek, D. Potential role of fluoride in the etiopathogenesis of alzheimer's disease. *Int. J. Mol. Sci.* **2018**, *19*, 3965. [CrossRef] [PubMed]
22. Correia, S.C.; Perry, G.; Moreira, P.I. Mitochondrial traffic jams in Alzheimer's disease—pinpointing the roadblocks. *Biochim. Biophys. Acta Mol. Basis Dis.* **2016**, *1862*, 1909–1917. [CrossRef] [PubMed]

23. Flannery, P.J.; Trushina, E. Mitochondrial dynamics and transport in Alzheimer's disease. *Mol. Cell. Neurosci.* **2019**, *98*, 109–120. [CrossRef]
24. Ahmad, M.H.; Fatima, M.; Mondal, A.C. Influence of microglia and astrocyte activation in the neuroinflammatory pathogenesis of Alzheimer's disease: Rational insights for the therapeutic approaches. *J. Clin. Neurosci.* **2019**, *59*, 6–11. [CrossRef] [PubMed]
25. Cowan, M.; Petri, W.A. Microglia: Immune regulators of neurodevelopment. *Front. Immunol.* **2018**, *9*, 1–8. [CrossRef]
26. Konijnenberg, E.; den Braber, A.; ten Kate, M.; Tomassen, J.; Mulder, S.D.; Yaqub, M.; Teunissen, C.E.; Lammertsma, A.A.; van Berckel, B.N.M.; Scheltens, P.; et al. Association of amyloid pathology with memory performance and cognitive complaints in cognitively normal older adults: A monozygotic twin study. *Neurobiol. Aging* **2019**, *77*, 58–65. [CrossRef]
27. Naveed, M.; Mubeen, S.; Khan, A.; Ibrahim, S.; Meer, B. Plasma Biomarkers: Potent Screeners of Alzheimer's Disease. *Am. J. Alzheimer's Dis. Other Dementias®* **2019**, *34*, 290–301. [CrossRef]
28. Olsson, B.; Lautner, R.; Andreasson, U.; Öhrfelt, A.; Portelius, E.; Bjerke, M.; Hölttä, M.; Rosén, C.; Olsson, C.; Strobel, G.; et al. CSF and blood biomarkers for the diagnosis of Alzheimer's disease: A systematic review and meta-analysis. *Lancet Neurol.* **2016**, *15*, 673–684. [CrossRef]
29. Blennow, K.; Zetterberg, H. Biomarkers for Alzheimer's disease: Current status and prospects for the future. *J. Intern. Med.* **2018**, *284*, 643–663. [CrossRef]
30. Reiss, A.B.; Arain, H.A.; Stecker, M.M.; Siegart, N.M.; Kasselman, L.J. Amyloid toxicity in Alzheimer's disease. *Rev. Neurosci.* **2018**, *29*, 613–627. [CrossRef]
31. Mroczko, B.; Groblewska, M.; Litman-Zawadzka, A.; Kornhuber, J.; Lewczuk, P. Amyloid β oligomers (AβOs) in Alzheimer's disease. *J. Neural Transm.* **2018**, *125*, 177–191. [CrossRef] [PubMed]
32. Saha, P.; Sen, N. Tauopathy: A common mechanism for neurodegeneration and brain aging. *Mech. Ageing Dev.* **2019**, *178*, 72–79. [CrossRef] [PubMed]
33. Ahmadian, N.; Hejazi, S.; Mahmoudi, J.; Talebi, M. Tau pathology of Alzheimer disease: Possible role of sleep deprivation. *Basic Clin. Neurosci.* **2018**, *9*, 307–316. [CrossRef] [PubMed]
34. Goedert, M.; Spillantini, M.G.; Jakes, R.; Rutherford, D.; Crowther, R.A. Multiple isoforms of human microtubule-associated protein tau: Sequences and localization in neurofibrillary tangles of Alzheimer's disease. *Neuron* **1989**, *3*, 519–526. [CrossRef]
35. Cortés, N.; Andrade, V.; Guzmán-Martínez, L.; Estrella, M.; Maccioni, R.B. Neuroimmune tau mechanisms: Their role in the progression of neuronal degeneration. *Int. J. Mol. Sci.* **2018**, *19*, 956. [CrossRef]
36. Goedert, M. Alzheimer's and Parkinson's diseases: The prion concept in relation to assembled Aβ, tau, and α-synuclein. *Science* **2015**, *349*, 61–69. [CrossRef]
37. Penke, B.; Bogár, F.; Fülöp, L. β-amyloid and the pathomechanisms of Alzheimer's disease: A comprehensive view. *Molecules* **2017**, *22*, 1692. [CrossRef]
38. Allen, H.B. Alzheimer's Disease: Assessing the Role of Spirochetes, Biofilms, the Immune System, and Amyloid-β with Regard to Potential Treatment and Prevention. *J. Alzheimer's Dis.* **2016**, *53*, 1271–1276. [CrossRef]
39. Yndart, A. Alzheimer's disease: Pathogenesis, diagnostics, and therapeutics. *Int. J. Nanomed.* **2019**, *14*, 5541–5554.
40. Ihara, M.; Washida, K. Linking atrial fibrillation with Alzheimer's disease: Epidemiological, pathological, and mechanistic evidence. *J. Alzheimer's Dis.* **2018**, *62*, 61–72. [CrossRef]
41. Crews, L.; Masliah, E. Molecular mechanisms of neurodegeneration in Alzheimer's disease. *Hum. Mol. Genet.* **2010**, *19*, 12–20. [CrossRef] [PubMed]
42. Quinn, J.P.; Corbett, N.J.; Kellett, K.A.B.; Hooper, N.M. Tau Proteolysis in the Pathogenesis of Tauopathies: Neurotoxic Fragments and Novel Biomarkers. *J. Alzheimer's Dis.* **2018**, *63*, 13–33. [CrossRef] [PubMed]
43. Leuzy, A.; Heurling, K.; Ashton, N.J.; Schöll, M.; Zimmer, E.R. In vivo detection of alzheimer's disease. *Yale J. Biol. Med.* **2018**, *91*, 291–300.
44. Weller, J.; Budson, A. Current understanding of Alzheimer's disease diagnosis and treatment. *F1000Research* **2018**, *7*, 1–9. [CrossRef] [PubMed]
45. Jeong, S. Molecular and cellular basis of neurodegeneration in alzheimer's disease. *Mol. Cells* **2017**, *40*, 613–620. [CrossRef]
46. Laurent, C.; Buée, L.; Blum, D. Tau and neuroinflammation: What impact for Alzheimer's Disease and Tauopathies? *Biomed. J.* **2018**, *41*, 21–33. [CrossRef]
47. Castellani, R.J.; Perry, G.; Tabaton, M. Tau biology, tauopathy, traumatic brain injury, and diagnostic challenges. *J. Alzheimer's Dis.* **2019**, *67*, 447–467. [CrossRef]
48. Gao, Y.; Wang, N.; Sun, F.; Cao, X.; Zhang, W.; Yu, J. Tau in neurodegenerative disease. *Ann. Transl. Med.* **2018**, *21*, 1–13. [CrossRef]
49. Zetterberg, H.; Wilson, D.; Andreasson, U.; Minthon, L.; Blennow, K.; Randall, J. Plasma tau levels in Alzheimer's disease Plasma tau levels in Alzheimer' s disease. *Alzheimer's Res. Ther.* **2013**, *5*, 9. [CrossRef]
50. Leuzy, A.; Chiotis, K.; Lemoine, L.; Gillberg, P.G.; Almkvist, O.; Rodriguez-Vieitez, E.; Nordberg, A. Tau PET imaging in neurodegenerative tauopathies—Still a challenge. *Mol. Psychiatry* **2019**, *24*, 1112–1134. [CrossRef]
51. De-Paula, V.d.J.R.; Forlenza, A.S.; Forlenza, O.V. Relevance of gutmicrobiota in cognition, behaviour and Alzheimer's disease. *Pharmacol. Res.* **2018**, *136*, 29–34. [CrossRef] [PubMed]
52. Burokas, A.; Moloney, R.D.; Dinan, T.G.; Cryan, J.F. Microbiota Regulation of the Mammalian Gut-Brain Axis. *Adv. Appl. Microbiol.* **2015**, *91*, 1–62. [CrossRef]
53. Quigley, E.M.M. Microbiota-Brain-Gut Axis and Neurodegenerative Diseases. *Curr. Neurol. Neurosci. Rep.* **2017**, *17*, 94. [CrossRef]

54. Salminen, S.; Bouley, C.; Boutron, M.-C.; Cummings, J.H.; Franck, A.; Gibson, G.R.; Isolauri, E.; Moreau, M.-C.; Roberfroid, M.; Rowland, I. Functional food science and gastrointestinal physiology and function. *Br. J. Nutr.* **1998**, *80*, S147–S171. [CrossRef] [PubMed]
55. Schmidt, T.S.B.; Raes, J.; Bork, P. The Human Gut Microbiome: From Association to Modulation. *Cell* **2018**, *172*, 1198–1215. [CrossRef] [PubMed]
56. Jandhyala, S.M.; Talukdar, R.; Subramanyam, C.; Vuyyuru, H.; Sasikala, M.; Nageshwar Reddy, D. Role of the normal gut microbiota. *World J. Gastroenterol.* **2015**, *21*, 8787–8803. [CrossRef]
57. Kowalski, K.; Mulak, A.; Words, K. Brain-Gut-Microbiota Axis in Alzheimer's Disease. *J. Neurogastroenterol. Motil.* **2019**, *25*, 48. [CrossRef]
58. Penders, J.; Thijs, C.; Vink, C.; Stelma, F.F.; Snijders, B.; Kummeling, I.; van den Brandt, P.A.; Stobberingh, E.E. Factors influencing the composition of the intestinal microbiota in early infancy. *Pediatrics* **2006**, *118*, 511–521. [CrossRef]
59. Claesson, M.J.; Jeffery, I.B.; Conde, S.; Power, S.E.; O'Connor, E.M.; Cusack, S.; Harris, H.M.B.; Coakley, M.; Lakshminarayanan, B.; O'Sullivan, O.; et al. Gut microbiota composition correlates with diet and health in the elderly. *Nature* **2012**, *488*, 178–184. [CrossRef]
60. Bermúdez-Humarán, L.G.; Salinas, E.; Ortiz, G.G.; Ramirez-Jirano, L.J.; Morales, J.A.; Bitzer-Quintero, O.K. From probiotics to psychobiotics: Live beneficial bacteria which act on the brain-gut axis. *Nutrients* **2019**, *11*, 890. [CrossRef]
61. Mayer, X.E.A.; Knight, R.; Mazmanian, S.K.; Cryan, X.J.F.; Tillisch, K. Gut Microbes and the Brain: Paradigm Shift in Neuroscience. *J. Neurosci.* **2014**, *34*, 15490–15496. [CrossRef] [PubMed]
62. Angelucci, F.; Cechova, K.; Amlerova, J.; Hort, J. Antibiotics, gut microbiota, and Alzheimer's disease. *J. Neuroinflamm.* **2019**, *16*, 1–10. [CrossRef] [PubMed]
63. Hao, W.-Z.; Li, X.-J.; Zhang, P.-W.; Chen, J.-X. A review of antibiotics, depression, and the gut microbiome. *Psychiatry Res.* **2020**, *284*, 112691. [CrossRef] [PubMed]
64. Burokas, A.; Arboleya, S.; Moloney, R.D.; Peterson, V.L.; Murphy, K.; Clarke, G.; Stanton, C.; Dinan, T.G.; Cryan, J.F. Targeting the Microbiota-Gut-Brain Axis: Prebiotics Have Anxiolytic and Antidepressant-like Effects and Reverse the Impact of Chronic Stress in Mice. *Biol. Psychiatry* **2017**, *82*, 472–487. [CrossRef] [PubMed]
65. Strandwitz, P. Neurotransmitter modulation by the gut microbiota. *Brain Res.* **2018**, *1693*, 128–133. [CrossRef] [PubMed]
66. Fjell, A.M.; McEvoy, L.; Holland, D.; Dale, A.M.; Walhovd, K.B. What is normal in normal aging? Effects of aging, amyloid and Alzheimer's disease on the cerebral cortex and the hippocampus. *Prog. Neurobiol.* **2014**, *117*, 20–40. [CrossRef]
67. Golubeva, A.V.; Joyce, S.A.; Moloney, G.; Burokas, A.; Sherwin, E.; Arboleya, S.; Flynn, I.; Khochanskiy, D.; Moya-Pérez, A.; Peterson, V.; et al. Microbiota-related Changes in Bile Acid & Tryptophan Metabolism are Associated with Gastrointestinal Dysfunction in a Mouse Model of Autism. *EBioMedicine* **2017**, *24*, 166–178. [CrossRef]
68. Borre, Y.E.; O'Keeffe, G.W.; Clarke, G.; Stanton, C.; Dinan, T.G.; Cryan, J.F. Microbiota and neurodevelopmental windows: Implications for brain disorders. *Trends Mol. Med.* **2014**, *20*, 509–518. [CrossRef]
69. Alkasir, R.; Li, J.; Li, X.; Jin, M.; Zhu, B. Human gut microbiota: The links with dementia development. *Protein Cell* **2017**, *8*, 90–102. [CrossRef]
70. Ranuh, R.; Athiyyah, A.F.; Darma, A.; Risky, V.P.; Riawan, W.; Surono, I.S.; Sudarmo, S.M. Effect of the probiotic lactobacillus plantarum is-10506 on bdnf and 5ht stimulation: Role of intestinal microbiota on the gut-brain axis. *Iran. J. Microbiol.* **2019**, *11*, 145–150. [CrossRef]
71. Ma, D.; Forsythe, P.; Bienenstock, J. Live Lactobacillus reuteri is essential for the inhibitory effect on tumor necrosis factor alpha-induced interleukin-8 expression. *Infect. Immun.* **2004**, *72*, 5308–5314. [CrossRef] [PubMed]
72. Lin, A.; Shih, C.T.; Huang, C.L.; Wu, C.C.; Lin, C.T.; Tsai, Y.C. Hypnotic effects of lactobacillus fermentum PS150TM on pentobarbital-induced sleep in mice. *Nutrients* **2019**, *11*, 2409. [CrossRef]
73. Roy Sarkar, S.; Banerjee, S. Gut microbiota in neurodegenerative disorders. *J. Neuroimmunol.* **2019**, *328*, 98–104. [CrossRef] [PubMed]
74. Bull-Larsen, S.; Hasan Mohajeri, M. The potential influence of the bacterial microbiome on the development and progression of adhd. *Nutrients* **2019**, *11*, 2805. [CrossRef] [PubMed]
75. Desbonnet, L.; Garrett, L.; Clarke, G.; Kiely, B.; Cryan, J.F.; Dinan, T.G. Effects of the probiotic Bifidobacterium infantis in the maternal separation model of depression. *Neuroscience* **2010**, *170*, 1179–1188. [CrossRef] [PubMed]
76. Lopes, J.G.; Sourjik, V. Chemotaxis of Escherichia coli to major hormones and polyamines present in human gut. *ISME J.* **2018**, *12*, 2736–2747. [CrossRef]
77. Asano, Y.; Hiramoto, T.; Nishino, R.; Aiba, Y.; Kimura, T.; Yoshihara, K.; Koga, Y.; Sudo, N. Critical role of gut microbiota in the production of biologically active, free catecholamines in the gut lumen of mice. *Am. J. Physiol. Gastrointest. Liver Physiol.* **2012**, *303*, 1288–1295. [CrossRef]
78. Agus, A.; Planchais, J.; Sokol, H. Gut Microbiota Regulation of Tryptophan Metabolism in Health and Disease. *Cell Host Microbe* **2018**, *23*, 716–724. [CrossRef]
79. Freestone, P.P.; Williams, P.H.; Haigh, R.D.; Maggs, A.F.; Neal, C.P.; Lyte, M. Growth stimulation of intestinal commensal Escherichia coli by catecholamines: A possible contributory factor in trauma-induced sepsis. *Shock* **2002**, *18*, 465–470. [CrossRef]
80. Johnson, K.V.A.; Foster, K.R. Why does the microbiome affect behaviour? *Nat. Rev. Microbiol.* **2018**, *16*, 647–655. [CrossRef]

81. Bjerre, K.; Cantor, M.D.; Nørgaard, J.V.; Poulsen, H.D.; Blaabjerg, K.; Canibe, N.; Jensen, B.B.; Stuer-Lauridsen, B.; Nielsen, B.; Derkx, P.M.F. Development of Bacillus subtilis mutants to produce tryptophan in pigs. *Biotechnol. Lett.* **2017**, *39*, 289–295. [CrossRef] [PubMed]
82. Sheng, Q.K.; Yang, Z.J.; Zhao, H.B.; Wang, X.L.; Guo, J.F. Effects of L-tryptophan, fructan, and casein on reducing ammonia, hydrogen sulfide, and skatole in fermented swine manure. *Asian Australas. J. Anim. Sci.* **2015**, *28*, 1202–1208. [CrossRef] [PubMed]
83. Kambe, J.; Watcharin, S.; Makioka-Itaya, Y.; Inoue, R.; Watanabe, G.; Yamaguchi, H.; Nagaoka, K. Heat-killed Enterococcus fecalis (EC-12) supplement alters the expression of neurotransmitter receptor genes in the prefrontal cortex and alleviates anxiety-like behavior in mice. *Neurosci. Lett.* **2020**, *720*, 134753. [CrossRef] [PubMed]
84. Baj, A.; Moro, E.; Bistoletti, M.; Orlandi, V.; Crema, F.; Giaroni, C. Glutamatergic Signaling along the Microbiota-Gut-Brain Axis. *Int. J. Mol. Sci.* **2019**, *20*, 1482. [CrossRef]
85. Jameson, K.G.; Hsiao, E.Y. Linking the Gut Microbiota to a Brain Neurotransmitter. *Trends Neurosci.* **2018**, *41*, 413–414. [CrossRef]
86. Franceschi, F.; Ojetti, V.; Candelli, M.; Covino, M.; Cardone, S. Microbes and Alzheimer' disease: Lessons from H pylori and GUT microbiota. *Eur. Rev. Med. Pharmacol. Sci.* **2019**, *23*, 426–430.
87. Rieder, R.; Wisniewski, P.J.; Alderman, B.L.; Campbell, S.C. Microbes and mental health: A review. *Brain. Behav. Immun.* **2017**, *66*, 9–17. [CrossRef]
88. Rowland, I.; Gibson, G.; Heinken, A.; Scott, K.; Swann, J.; Thiele, I.; Tuohy, K. Gut microbiota functions: Metabolism of nutrients and other food components. *Eur. J. Nutr.* **2018**, *57*, 1–24. [CrossRef]
89. Arnoriaga-Rodríguez, M.; Mayneris-Perxachs, J.; Burokas, A.; Contreras-Rodríguez, O.; Blasco, G.; Coll, C.; Biarnés, C.; Miranda-Olivos, R.; Latorre, J.; Moreno-Navarrete, J.M.; et al. Obesity Impairs Short-Term and Working Memory through Gut Microbial Metabolism of Aromatic Amino Acids. *Cell Metab.* **2020**, *32*, 548–560.e7. [CrossRef]
90. Martin, C.R.; Osadchiy, V.; Kalani, A.; Mayer, E.A. The Brain-Gut-Microbiome Axis. *Cell. Mol. Gastroenterol. Hepatol.* **2018**, *6*, 133–148. [CrossRef]
91. Giau, V.V.; Wu, S.Y.; Jamerlan, A.; An, S.S.A.; Kim, S.Y.; Hulme, J. Gut Microbiota and Their Neuroinflammatory Implications in Alzheimer's Disease. *Nutrients* **2018**, *10*, 1765. [CrossRef] [PubMed]
92. Grenham, S.; Clarke, G.; Cryan, J.F.; Dinan, T.G.; Makharia, G.K. Brain—gut—microbe communication in health and disease. *Front. Physiol.* **2011**, *2*, 1–15. [CrossRef] [PubMed]
93. Luca, M.; Di Mauro, M.; Di Mauro, M.; Luca, A. Gut Microbiota in Alzheimer's Disease, Depression, and Type 2 Diabetes Mellitus: The Role of Oxidative Stress. *Oxid. Med. Cell. Longev.* **2019**, *2019*, 1–10. [CrossRef] [PubMed]
94. Liu, R.; Han, Y.; Du, J.; Yi, W.; Jin, K.; Zhu, X. Microbiota-gut-brain axis and the central nervous system. *Oncotarget* **2017**, *8*, 53829–53838. [CrossRef]
95. Bravo, J.A.; Julio-Pieper, M.; Forsythe, P.; Kunze, W.; Dinan, T.G.; Bienenstock, J.; Cryan, J.F. Communication between gastrointestinal bacteria and the nervous system. *Curr. Opin. Pharmacol.* **2012**, *12*, 667–672. [CrossRef]
96. van de Wouw, M.; Boehme, M.; Lyte, J.M.; Wiley, N.; Strain, C.; O'Sullivan, O.; Clarke, G.; Stanton, C.; Dinan, T.G.; Cryan, J.F. Short-chain fatty acids: Microbial metabolites that alleviate stress-induced brain-gut axis alterations. *J. Physiol.* **2018**, *596*, 4923–4944. [CrossRef]
97. Singh, K.; Loreth, D.; Pöttker, B.; Hefti, K.; Innos, J.; Schwald, K.; Hengstler, H.; Menzel, L.; Sommer, C.J.; Radyushkin, K.; et al. Neuronal Growth and Behavioral Alterations in Mice Deficient for the Psychiatric Disease-Associated Negr1 Gene. *Front. Mol. Neurosci.* **2018**, *11*, 30. [CrossRef]
98. Webster, S.J.; Bachstetter, A.D.; Nelson, P.T.; Schmitt, F.A.; Van Eldik, L.J. Using mice to model Alzheimer's dementia: An overview of the clinical disease and the preclinical behavioral changes in 10 mouse models. *Front. Genet.* **2014**, *5*, 88. [CrossRef]
99. Farzi, A.; Fröhlich, E.E.; Holzer, P. Gut Microbiota and the Neuroendocrine System. *Neurother. J. Am. Soc. Exp. Neurother.* **2018**, *15*, 5–22. [CrossRef]
100. Cerovic, M.; Forloni, G.; Balducci, C. Neuroinflammation and the Gut Microbiota: Possible Alternative Therapeutic Targets to Counteract Alzheimer's Disease? *Front. Aging Neurosci.* **2019**, *11*, 284. [CrossRef]
101. Ochoa-Repáraz, J.; Kasper, L.H. The Microbiome and Neurologic Disease: Past and Future of a 2-Way Interaction. *Neurother. J. Am. Soc. Exp. Neurother.* **2018**, *15*, 1–4. [CrossRef] [PubMed]
102. Colpitts, S.L.; Kasper, E.J.; Keever, A.; Liljenberg, C.; Kirby, T.; Magori, K.; Kasper, L.H.; Ochoa-Repáraz, J. A bidirectional association between the gut microbiota and CNS disease in a biphasic murine model of multiple sclerosis. *Gut Microbes* **2017**, *8*, 561–573. [CrossRef] [PubMed]
103. Mehrabadi, S.; Sadr, S.S. Assessment of probiotics mixture on memory function, inflammation markers, and oxidative stress in an Alzheimer's disease model of rats. *Iran. Biomed. J.* **2020**, *24*, 220–228. [CrossRef] [PubMed]
104. Chang, C.; Lin, C.; Lane, H.Y. D-glutamate and Gut Microbiota in Alzheimer's Disease. *Int. J. Mol. Sci.* **2020**, *21*, 2676. [CrossRef]
105. Gareau, M.G.; Wine, E.; Rodrigues, D.M.; Cho, J.H.; Whary, M.T.; Philpott, D.J.; MacQueen, G.; Sherman, P.M. Bacterial infection causes stress-induced memory dysfunction in mice. *Gut* **2011**, *60*, 307–317. [CrossRef]
106. Singhrao, S.K.; Harding, A.; Poole, S.; Kesavalu, L.; Crean, S.J. Porphyromonas gingivalis periodontal infection and its putative links with Alzheimer's disease. *Mediat. Inflamm.* **2015**, *2015*. [CrossRef]
107. Cattaneo, A.; Cattane, N.; Galluzzi, S.; Provasi, S.; Lopizzo, N.; Festari, C.; Ferrari, C.; Guerra, U.P.; Paghera, B.; Muscio, C.; et al. Association of brain amyloidosis with pro-inflammatory gut bacterial taxa and peripheral inflammation markers in cognitively impaired elderly. *Neurobiol. Aging* **2017**, *49*, 60–68. [CrossRef]

108. Hill, J.M.; Lukiw, W.J.; Clement, C.; Bhattacharjee, S.; Zhao, Y. Alzheimer'S Disease and the Microbiome. *Alzheimer's Dement.* **2014**, *10*, P873. [CrossRef]
109. Vogt, N.M.; Kerby, R.L.; Dill-McFarland, K.A.; Harding, S.J.; Merluzzi, A.P.; Johnson, S.C.; Carlsson, C.M.; Asthana, S.; Zetterberg, H.; Blennow, K.; et al. Gut microbiome alterations in Alzheimer's disease. *Sci. Rep.* **2017**, *7*, 1–11. [CrossRef]
110. Lukiw, W.J. Bacteroides fragilis lipopolysaccharide and inflammatory signaling in alzheimer's disease. *Front. Microbiol.* **2016**, *7*, 1–6. [CrossRef]
111. Deng, H.; Yang, S.; Zhang, Y.; Qian, K.; Zhang, Z.; Liu, Y.; Wang, Y.; Bai, Y.; Fan, H.; Zhao, X.; et al. Bacteroides fragilis Prevents Clostridium difficile Infection in a Mouse Model by Restoring Gut Barrier and Microbiome Regulation. *Front. Microbiol.* **2018**, *9*, 2976. [CrossRef] [PubMed]
112. Yang, X.; Yu, D.; Xue, L.; Li, H.; Du, J. Probiotics modulate the microbiota-gut-brain axis and improve memory deficits in aged SAMP8 mice. *Acta Pharm. Sin. B* **2020**, *10*, 475–487. [CrossRef] [PubMed]
113. Bonfili, L.; Cecarini, V.; Berardi, S.; Scarpona, S.; Suchodolski, J.S.; Nasuti, C.; Fiorini, D.; Boarelli, M.C.; Rossi, G.; Eleuteri, A.M. Microbiota modulation counteracts Alzheimer' s disease progression influencing neuronal proteolysis and gut hormones plasma levels. *Sci. Rep.* **2017**, *7*, 1–21. [CrossRef]
114. Dargahi, N.; Matsoukas, J.; Apostolopoulos, V. Streptococcus thermophilus ST285 alters pro-inflammatory to anti-inflammatory cytokine secretion against multiple sclerosis peptide in mice. *Brain Sci.* **2020**, *10*, 126. [CrossRef] [PubMed]
115. Tanida, M.; Yamano, T.; Maeda, K.; Okumura, N.; Fukushima, Y.; Nagai, K. Effects of intraduodenal injection of Lactobacillus johnsonii La1 on renal sympathetic nerve activity and blood pressure in urethane-anesthetized rats. *Neurosci. Lett.* **2005**, *389*, 109–114. [CrossRef] [PubMed]
116. Marcial, G.E.; Ford, A.L.; Haller, M.J.; Gezan, S.A.; Harrison, N.A.; Cai, D.; Meyer, J.L.; Perry, D.J.; Atkinson, M.A.; Wasserfall, C.H.; et al. Lactobacillus johnsonii N6.2 modulates the host immune responses: A double-blind, randomized trial in healthy adults. *Front. Immunol.* **2017**, *8*, 655. [CrossRef]
117. Ma, Q.; Xing, C.; Long, W.; Wang, H.Y.; Liu, Q.; Wang, R.F. Impact of microbiota on central nervous system and neurological diseases: The gut-brain axis. *J. Neuroinflamm.* **2019**, *16*, 1–14. [CrossRef]
118. Sun, Y.; Sommerville, N.R.; Liu, J.Y.H.; Ngan, M.P.; Poon, D.; Ponomarev, E.D.; Lu, Z.; Kung, J.S.C.; Rudd, J.A. Intra-gastrointestinal amyloid-β1-42 oligomers perturb enteric function and induce Alzheimer's disease pathology. *J. Physiol.* **2020**, *598*, 4209–4223. [CrossRef]
119. Friedland, R.P.; Chapman, M.R. The role of microbial amyloid in neurodegeneration. *PLoS Pathog.* **2017**, *13*, e1006654. [CrossRef]
120. Friedland, R.P. Mechanisms of Molecular Mimicry Involving the Microbiota in Neurodegeneration. *J. Alzheimer's Dis.* **2015**, *45*, 349–362. [CrossRef]
121. McIntee, F.L.; Giannoni, P.; Blais, S.; Sommer, G.; Neubert, T.A.; Rostagno, A.; Ghiso, J. In vivo differential brain clearance and catabolism of monomeric and oligomeric alzheimer's aβ protein. *Front. Aging Neurosci.* **2016**, *8*, 1–15. [CrossRef] [PubMed]
122. Chen, S.G.; Stribinskis, V.; Rane, M.J.; Demuth, D.R.; Gozal, E.; Roberts, A.M.; Jagadapillai, R.; Liu, R.; Choe, K.; Shivakumar, B.; et al. Exposure to the Functional Bacterial Amyloid Protein Curli Enhances Alpha-Synuclein Aggregation in Aged Fischer 344 Rats and Caenorhabditis elegans. *Sci. Rep.* **2016**, *6*, 1–10. [CrossRef] [PubMed]
123. Harach, T.; Marungruang, N.; Duthilleul, N.; Cheatham, V.; Mc Coy, K.D.; Frisoni, G.; Neher, J.J.; Fåk, F.; Jucker, M.; Lasser, T.; et al. Reduction of Abeta amyloid pathology in APPPS1 transgenic mice in the absence of gut microbiota. *Sci. Rep.* **2017**, *7*, 41802. [CrossRef] [PubMed]
124. Westfall, S.; Lomis, N.; Kahouli, I.; Dia, S.Y.; Singh, S.P.; Prakash, S. Microbiome, probiotics and neurodegenerative diseases: Deciphering the gut brain axis. *Cell. Mol. Life Sci.* **2017**, *74*, 3769–3787. [CrossRef]
125. Aziz, Q.; Doré, J.; Emmanuel, A.; Guarner, F.; Quigley, E.M.M. Gut microbiota and gastrointestinal health: Current concepts and future directions. *Neurogastroenterol. Motil.* **2013**, *25*, 4–15. [CrossRef]
126. Salazar, N.; Arboleya, S.; Valdés, L.; Stanton, C.; Ross, P.; Ruiz, L.; Gueimonde, M.; de los Reyes-Gavilán, C.G. The human intestinal microbiome at extreme ages of life. Dietary intervention as a way to counteract alterations. *Front. Genet.* **2014**, *5*, 1–9. [CrossRef]
127. Asti, A.; Gioglio, L. Can a Bacterial Endotoxin be a Key Factor in the Kinetics of Amyloid Fibril Formation? *J. Alzheimer's Dis.* **2014**, *39*, 169–179. [CrossRef]
128. Zhan, X. Author response: Gram-negative bacterial molecules associate with Alzheimer disease pathology. *Neurology* **2017**, *88*, 2338. [CrossRef]
129. Zhao, Y.; Jaber, V.; Lukiw, W.J. Secretory products of the human GI tract microbiome and their potential impact on Alzheimer's disease (AD): Detection of lipopolysaccharide (LPS) in AD hippocampus. *Front. Cell. Infect. Microbiol.* **2017**, *7*, 1–9. [CrossRef]
130. Zhang, L.; Wang, Y.; Xiayu, X.; Shi, C.; Chen, W.; Song, N.; Fu, X.; Zhou, R.; Xu, Y.-F.; Huang, L.; et al. Altered Gut Microbiota in a Mouse Model of Alzheimer's Disease. *J. Alzheimer's Dis.* **2017**, *60*, 1241–1257. [CrossRef]
131. Vardhini, D.; Suneetha, S.; Ahmed, N.; Joshi, D.S.M.; Karuna, S.; Magee, X.; Vijayalakshmi, D.S.R.; Sridhar, V.; Karunakar, K.V.; Archelos, J.J.; et al. Comparative proteomics of the Mycobacterium leprae binding protein myelin P0: Its implication in leprosy and other neurodegenerative diseases. *Infect. Genet. Evol.* **2004**, *4*, 21–28. [CrossRef] [PubMed]
132. Wunderink, R.G.; Waterer, G.W. Community-acquired pneumonia. *N. Engl. J. Med.* **2014**, *370*, 543–551. [CrossRef] [PubMed]
133. Choroszy-Król, I.; Frej-Madrzak, M.; Hober, M.; Sarowska, J.; Jama-Kmiecik, A. Infections caused by Chlamydophila pneumoniae. *Adv. Clin. Exp. Med.* **2014**, *23*, 123–126. [CrossRef] [PubMed]

134. Pisa, D.; Alonso, R.; Juarranz, A.; Rábano, A.; Carrasco, L. Direct visualization of fungal infection in brains from patients with Alzheimer's disease. *J. Alzheimer's Dis.* **2015**, *43*, 613–624. [CrossRef] [PubMed]
135. Alonso, R.; Pisa, D.; Rábano, A.; Carrasco, L. Alzheimer's disease and disseminated mycoses. *Eur. J. Clin. Microbiol. Infect. Dis.* **2014**, *33*, 1125–1132. [CrossRef]
136. Galloway, S.M.; Raetz, C.R.H. A mutant of Escherichia coli defective in the first step of endotoxin biosynthesis. *J. Biol. Chem.* **1990**, *265*, 6394–6402.
137. Whitfield, C.; Stephen Trent, M. Biosynthesis and export of bacterial lipopolysaccharides. *Annu. Rev. Biochem.* **2014**, *83*, 99–128. [CrossRef]
138. Hauss-Wegrzyniak, B.; Vraniak, P.D.; Wenk, G.L. LPS-induced neuroinflammatory effects do not recover with time. *Neuroreport* **2000**, *11*, 1759–1763. [CrossRef]
139. Kahn, M.S.; Kranjac, D.; Alonzo, C.A.; Haase, J.H.; Cedillos, R.O.; McLinden, K.A.; Boehm, G.W.; Chumley, M.J. Prolonged elevation in hippocampal Aβ and cognitive deficits following repeated endotoxin exposure in the mouse. *Behav. Brain Res.* **2012**, *229*, 176–184. [CrossRef]
140. Zhao, Y.; Cong, L.; Jaber, V.; Lukiw, W.J. Microbiome-derived lipopolysaccharide enriched in the perinuclear region of Alzheimer's disease brain. *Front. Immunol.* **2017**, *8*, 1–6. [CrossRef]
141. Zhang, R.; Miller, R.G.; Gascon, R.; Champion, S.; Katz, J.; Lancero, M.; Narvaez, A.; Honrada, R.; Ruvalcaba, D.; McGrath, M.S. Circulating endotoxin and systemic immune activation in sporadic amyotrophic lateral sclerosis (sALS). *J. Neuroimmunol.* **2009**, *206*, 121–124. [CrossRef] [PubMed]
142. Zhao, Y.; Dua, P.; Lukiw, W.J. Microbial Sources of Amyloid and Relevance to Amyloidogenesis and Alzheimer's Disease (AD). *J. Alzheimer's Dis. Park.* **2015**, *5*, 1–13. [CrossRef]
143. Fassbender, K.; Walter, S.; Kühl, S.; Landmann, R.; Ishii, K.; Bertsch, T.; Stalder, A.K.; Muehlhauser, F.; Liu, Y.; Ulmer, A.J.; et al. The LPS receptor (CD14) links innate immunity with Alzheimer's disease. *FASEB J. Off. Publ. Fed. Am. Soc. Exp. Biol.* **2004**, *18*, 203–205. [CrossRef] [PubMed]
144. Dansokho, C.; Heneka, M.T. Neuroinflammatory responses in Alzheimer's disease. *J. Neural Transm.* **2018**, *125*, 771–779. [CrossRef] [PubMed]
145. Finneran, D.J.; Nash, K.R. Neuroinflammation and fractalkine signaling in Alzheimer's disease. *J. Neuroinflamm.* **2019**, *16*, 1–8. [CrossRef] [PubMed]
146. Parimisetty, A.; Dorsemans, A.C.; Awada, R.; Ravanan, P.; Diotel, N.; Lefebvre d'Hellencourt, C. Secret talk between adipose tissue and central nervous system via secreted factors-an emerging frontier in the neurodegenerative research. *J. Neuroinflamm.* **2016**, *13*, 1–13. [CrossRef] [PubMed]
147. Venneti, S.; Wiley, C.A.; Kofler, J. Imaging microglial activation during neuroinflammation and Alzheimer's disease. *J. Neuroimmune Pharmacol.* **2009**, *4*, 227–243. [CrossRef]
148. Heneka, M.T.; McManus, R.M.; Latz, E. Inflammasome signalling in brain function and neurodegenerative disease. *Nat. Rev. Neurosci.* **2018**, *19*, 610–621. [CrossRef]
149. Olson, J.K.; Miller, S.D. Microglia Initiate Central Nervous System Innate and Adaptive Immune Responses through Multiple TLRs. *J. Immunol.* **2004**, *173*, 3916–3924. [CrossRef]
150. Bagyinszky, E.; Giau, V.V.; Shim, K.; Suk, K.; An, S.S.A.; Kim, S.Y. Role of inflammatory molecules in the Alzheimer's disease progression and diagnosis. *J. Neurol. Sci.* **2017**, *376*, 242–254. [CrossRef]
151. Yu, Y.; Ye, R.D. Microglial Aβ Receptors in Alzheimer's Disease. *Cell. Mol. Neurobiol.* **2014**, *35*, 71–83. [CrossRef] [PubMed]
152. Griffin, W.S.T.; Liu, L.; Li, Y.; Mrak, R.E.; Barger, S.W. Interleukin-1 mediates Alzheimer and Lewy body pathologies. *J. Neuroinflamm.* **2006**, *3*, 1–9. [CrossRef]
153. Ojala, J.O.; Sutinen, E.M.; Salminen, A.; Pirttilä, T. Interleukin-18 increases expression of kinases involved in tau phosphorylation in SH-SY5Y neuroblastoma cells. *J. Neuroimmunol.* **2008**, *205*, 86–93. [CrossRef] [PubMed]
154. Park, K.M.; Bowers, W.J. Tumor necrosis factor-alpha mediated signaling in neuronal homeostasis and dysfunction. *Cell. Signal.* **2010**, *22*, 977–983. [CrossRef] [PubMed]
155. Hickman, S.E.; Allison, E.K.; El Khoury, J. Microglial dysfunction and defective β-amyloid clearance pathways in aging alzheimer's disease mice. *J. Neurosci.* **2008**, *28*, 8354–8360. [CrossRef] [PubMed]
156. Medeiros, R.; LaFerla, F.M. Astrocytes: Conductors of the Alzheimer disease neuroinflammatory symphony. *Exp. Neurol.* **2013**, *239*, 133–138. [CrossRef] [PubMed]
157. Avila-Muñoz, E.; Arias, C. When astrocytes become harmful: Functional and inflammatory responses that contribute to Alzheimer's disease. *Ageing Res. Rev.* **2014**, *18*, 29–40. [CrossRef]
158. González-Reyes, R.E.; Nava-Mesa, M.O.; Vargas-Sánchez, K.; Ariza-Salamanca, D.; Mora-Muñoz, L. Involvement of astrocytes in Alzheimer's disease from a neuroinflammatory and oxidative stress perspective. *Front. Mol. Neurosci.* **2017**, *10*, 1–20. [CrossRef] [PubMed]
159. Söllvander, S.; Nikitidou, E.; Brolin, R.; Söderberg, L.; Sehlin, D.; Lannfelt, L.; Erlandsson, A. Accumulation of amyloid-β by astrocytes result in enlarged endosomes and microvesicle-induced apoptosis of neurons. *Mol. Neurodegener.* **2016**, *11*, 1–19. [CrossRef]
160. Chang, R.; Knox, J.; Chang, J.; Derbedrossian, A.; Vasilevko, V.; Cribbs, D.; Boado, R.J.; Pardridge, W.M.; Sumbria, R.K. Blood-Brain Barrier Penetrating Biologic TNF-α Inhibitor for Alzheimer's Disease. *Mol. Pharm.* **2017**, *14*, 2340–2349. [CrossRef]

161. Decourt, B.; Lahiri, D.K.; Sabbagh, M.N. Targeting Tumor Necrosis Factor Alpha for Alzheimer's Disease. *Curr. Alzheimer Res.* **2017**, *14*, 412–425. [CrossRef] [PubMed]
162. Krabbe, G.; Halle, A.; Matyash, V.; Rinnenthal, J.L.; Eom, G.D.; Bernhardt, U.; Miller, K.R.; Prokop, S.; Kettenmann, H.; Heppner, F.L. Functional impairment of microglia coincides with Beta-amyloid deposition in mice with Alzheimer-like pathology. *PLoS ONE* **2013**, *8*, e60921. [CrossRef] [PubMed]
163. Villegas-Llerena, C.; Phillips, A.; Garcia-Reitboeck, P.; Hardy, J.; Pocock, J.M. Microglial genes regulating neuroinflammation in the progression of Alzheimer's disease. *Curr. Opin. Neurobiol.* **2016**, *36*, 74–81. [CrossRef]
164. Fiala, M.; Veerhuis, R. Biomarkers of inflammation and amyloid-beta phagocytosis in patients at risk of Alzheimer disease. *Exp. Gerontol.* **2010**, *45*, 57–63. [CrossRef]
165. Bonham, L.W.; Sirkis, D.W.; Yokoyama, J.S. The Transcriptional Landscape of Microglial Genes in Aging and Neurodegenerative Disease. *Front. Immunol.* **2019**, *10*, 1170. [CrossRef]
166. Rangaraju, S.; Dammer, E.B.; Raza, S.A.; Gao, T.; Xiao, H.; Betarbet, R.; Duong, D.M.; Webster, J.A.; Hales, C.M.; Lah, J.J.; et al. Quantitative proteomics of acutely-isolated mouse microglia identifies novel immune Alzheimer's disease-related proteins. *Mol. Neurodegener.* **2018**, *13*, 34. [CrossRef]
167. Bosch, T.C.G. Rethinking the role of immunity: Lessons from Hydra. *Trends Immunol.* **2014**, *35*, 495–502. [CrossRef] [PubMed]
168. Weinhard, L.; di Bartolomei, G.; Bolasco, G.; Machado, P.; Schieber, N.L.; Neniskyte, U.; Exiga, M.; Vadisiute, A.; Raggioli, A.; Schertel, A.; et al. Microglia remodel synapses by presynaptic trogocytosis and spine head filopodia induction. *Nat. Commun.* **2018**, *9*, 1228. [CrossRef]
169. Hong, S.; Beja-Glasser, V.F.; Nfonoyim, B.M.; Frouin, A.; Li, S.; Ramakrishnan, S.; Merry, K.M.; Shi, Q.; Rosenthal, A.; Barres, B.A.; et al. Complement and microglia mediate early synapse loss in Alzheimer mouse models. *Science* **2016**, *352*, 712–716. [CrossRef]
170. Balducci, C.; Frasca, A.; Zotti, M.; La Vitola, P.; Mhillaj, E.; Grigoli, E.; Iacobellis, M.; Grandi, F.; Messa, M.; Colombo, L.; et al. Toll-like receptor 4-dependent glial cell activation mediates the impairment in memory establishment induced by β-amyloid oligomers in an acute mouse model of Alzheimer's disease. *Brain. Behav. Immun.* **2017**, *60*, 188–197. [CrossRef]
171. Sigal, M.; Meyer, T.F. Coevolution between the Human Microbiota and the Epithelial Immune System. *Dig. Dis.* **2016**, *34*, 190–193. [CrossRef] [PubMed]
172. Belkaid, Y.; Hand, T.W. Role of the microbiota in immunity and inflammation. *Cell* **2014**, *157*, 121–141. [CrossRef] [PubMed]
173. Schluter, J.; Peled, J.U.; Taylor, B.P.; Markey, K.A.; Smith, M.; Taur, Y.; Niehus, R.; Staffas, A.; Dai, A.; Fontana, E.; et al. The gut microbiota is associated with immune cell dynamics in humans. *Nature* **2020**, *588*, 1–5. [CrossRef] [PubMed]
174. Kamada, N.; Seo, S.-U.; Chen, G.Y.; Núñez, G. Role of the gut microbiota in immunity and inflammatory disease. *Nat. Rev. Immunol.* **2013**, *13*, 321–335. [CrossRef]
175. Khosravi, A.; Yáñez, A.; Price, J.G.; Chow, A.; Merad, M.; Helen, S. Gut microbiota promote hematopoiesis to control bacterial infection. *Cell Host Microbe* **2015**, *15*, 374–381. [CrossRef]
176. Fülling, C.; Lach, G.; Bastiaanssen, T.F.S.; Fouhy, F.; O'Donovan, A.N.; Ventura-Silva, A.-P.; Stanton, C.; Dinan, T.G.; Cryan, J.F. Adolescent dietary manipulations differentially affect gut microbiota composition and amygdala neuroimmune gene expression in male mice in adulthood. *Brain. Behav. Immun.* **2020**, *87*, 666–678. [CrossRef]
177. Boehme, M.; van de Wouw, M.; Bastiaanssen, T.F.S.; Olavarría-Ramírez, L.; Lyons, K.; Fouhy, F.; Golubeva, A.V.; Moloney, G.M.; Minuto, C.; Sandhu, K.V.; et al. Mid-life microbiota crises: Middle age is associated with pervasive neuroimmune alterations that are reversed by targeting the gut microbiome. *Mol. Psychiatry* **2019**, *25*, 2567–2583. [CrossRef]
178. Kierdorf, K.; Prinz, M. Factors regulating microglia activation. *Front. Cell. Neurosci.* **2013**, *7*, 44. [CrossRef]
179. Matcovitch-Natan, O.; Winter, D.R.; Giladi, A.; Vargas Aguilar, S.; Spinrad, A.; Sarrazin, S.; Ben-Yehuda, H.; David, E.; Zelada González, F.; Perrin, P.; et al. Microglia development follows a stepwise program to regulate brain homeostasis. *Science* **2016**, *353*, aad8670. [CrossRef]
180. Arts, I.C.W.; Hollman, P.C.H. Polyphenols and disease risk in epidemiologic studies. *Am. J. Clin. Nutr.* **2005**, *81*, 317S–325S. [CrossRef]
181. Scalbert, A.; Manach, C.; Morand, C.; Rémésy, C.; Jiménez, L. Dietary polyphenols and the prevention of diseases. *Crit. Rev. Food Sci. Nutr.* **2005**, *45*, 287–306. [CrossRef]
182. Flanagan, E.; Müller, M.; Hornberger, M.; Vauzour, D. Impact of Flavonoids on Cellular and Molecular Mechanisms Underlying Age-Related Cognitive Decline and Neurodegeneration. *Curr. Nutr. Rep.* **2018**, *7*, 49–57. [CrossRef] [PubMed]
183. Rodriguez-Mateos, A.; Vauzour, D.; Krueger, C.G.; Shanmuganayagam, D.; Reed, J.; Calani, L.; Mena, P.; Del Rio, D.; Crozier, A. Bioavailability, bioactivity and impact on health of dietary flavonoids and related compounds: An update. *Arch. Toxicol.* **2014**, *88*, 1803–1853. [CrossRef] [PubMed]
184. Jha, N.K.; Jha, S.K.; Kar, R.; Nand, P.; Swati, K.; Goswami, V.K. Nuclear factor-kappa β as a therapeutic target for Alzheimer's disease. *J. Neurochem.* **2019**, *150*, 113–137. [CrossRef] [PubMed]
185. Del Rio, D.; Rodriguez-Mateos, A.; Spencer, J.P.E.; Tognolini, M.; Borges, G.; Crozier, A. Dietary (poly)phenolics in human health: Structures, bioavailability, and evidence of protective effects against chronic diseases. *Antioxid. Redox Signal.* **2013**, *18*, 1818–1892. [CrossRef]
186. Godos, J.; Currenti, W.; Angelino, D.; Mena, P.; Castellano, S.; Caraci, F.; Galvano, F.; Rio, D.D.; Ferri, R.; Grosso, G. Diet and mental health: Review of the recent updates on molecular mechanisms. *Antioxidants* **2020**, *9*, 346. [CrossRef]

187. De Bruyne, T.; Steenput, B.; Roth, L.; De Meyer, G.R.Y.; Dos Santos, C.N.; Valentová, K.; Dambrova, M.; Hermans, N. Dietary polyphenols targeting arterial stiffness: Interplay of contributing mechanisms and gut microbiome-related Metabolism. *Nutrients* **2019**, *11*, 578. [CrossRef]
188. Williams, R.J.; Spencer, J.P.E.; Rice-Evans, C. Flavonoids: Antioxidants or signalling molecules? *Free Radic. Biol. Med.* **2004**, *36*, 838–849. [CrossRef]
189. Williams, R.J.; Spencer, J.P.E. Flavonoids, cognition, and dementia: Actions, mechanisms, and potential therapeutic utility for Alzheimer disease. *Free Radic. Biol. Med.* **2012**, *52*, 35–45. [CrossRef]
190. Duda-Chodak, A. The inhibitory effect of polyphenols on human gut microbiota. *J. Physiol. Pharmacol. Off. J. Polish Physiol. Soc.* **2012**, *63*, 497–503.
191. Nohynek, L.J.; Alakomi, H.-L.; Kähkönen, M.P.; Heinonen, M.; Helander, I.M.; Oksman-Caldentey, K.-M.; Puupponen-Pimiä, R.H. Berry phenolics: Antimicrobial properties and mechanisms of action against severe human pathogens. *Nutr. Cancer* **2006**, *54*, 18–32. [CrossRef]
192. Khan, M.S.; Ali, T.; Kim, M.W.; Jo, M.H.; Jo, M.G.; Badshah, H.; Kim, M.O. Anthocyanins protect against LPS-induced oxidative stress-mediated neuroinflammation and neurodegeneration in the adult mouse cortex. *Neurochem. Int.* **2016**, *100*, 1–10. [CrossRef] [PubMed]
193. Zhao, L.; Chen, S.; Liu, T.; Wang, X.; Huang, H.; Liu, W. Callistephin enhances the protective effects of isoflurane on microglial injury through downregulation of inflammation and apoptosis. *Mol. Med. Rep.* **2019**, *20*, 802–812. [CrossRef]
194. Shukitt-Hale, B.; Kelly, M.E.; Bielinski, D.F.; Fisher, D.R. Tart cherry extracts reduce inflammatory and oxidative stress signaling in microglial cells. *Antioxidants* **2016**, *5*, 33. [CrossRef] [PubMed]
195. Williams, C.M.; El Mohsen, M.A.; Vauzour, D.; Rendeiro, C.; Butler, L.T.; Ellis, J.A.; Whiteman, M.; Spencer, J.P.E. Blueberry-induced changes in spatial working memory correlate with changes in hippocampal CREB phosphorylation and brain-derived neurotrophic factor (BDNF) levels. *Free Radic. Biol. Med.* **2008**, *45*, 295–305. [CrossRef] [PubMed]
196. Goyarzu, P.; Malin, D.H.; Lau, F.C.; Taglialatela, G.; Moon, W.D.; Jennings, R.; Moy, E.; Moy, D.; Lippold, S.; Shukitt-Hale, B.; et al. Blueberry supplemented diet: Effects on object recognition memory and nuclear factor-kappa B levels in aged rats. *Nutr. Neurosci.* **2004**, *7*, 75–83. [CrossRef] [PubMed]
197. Barros, D.; Amaral, O.B.; Izquierdo, I.; Geracitano, L.; do Carmo Bassols Raseira, M.; Henriques, A.T.; Ramirez, M.R. Behavioral and genoprotective effects of Vaccinium berries intake in mice. *Pharmacol. Biochem. Behav.* **2006**, *84*, 229–234. [CrossRef] [PubMed]
198. Karlsen, A.; Retterstøl, L.; Laake, P.; Paur, I.; Bøhn, S.K.; Sandvik, L.; Blomhoff, R. Anthocyanins inhibit nuclear factor-kappaB activation in monocytes and reduce plasma concentrations of pro-inflammatory mediators in healthy adults. *J. Nutr.* **2007**, *137*, 1951–1954. [CrossRef]
199. Spilsbury, A.; Vauzour, D.; Spencer, J.P.E.; Rattray, M. Regulation of NF-κB activity in astrocytes: Effects of flavonoids at dietary-relevant concentrations. *Biochem. Biophys. Res. Commun.* **2012**, *418*, 578–583. [CrossRef]
200. Walsh, J.; Gheorghe, C.E.; Lyte, J.M.; van de Wouw, M.; Boehme, M.; Dinan, T.G.; Cryan, J.F.; Griffin, B.T.; Clarke, G.; Hyland, N.P. Gut microbiome-mediated modulation of hepatic cytochrome P450 and P-glycoprotein: Impact of butyrate and fructo-oligosaccharide-inulin. *J. Pharm. Pharmacol.* **2020**, *72*, 1072–1081. [CrossRef]
201. Matos, M.S.; Anastácio, J.D.; Allwood, J.W.; Carregosa, D.; Marques, D.; Sungurtas, J.; McDougall, G.J.; Menezes, R.; Matias, A.A.; Stewart, D.; et al. Assessing the intestinal permeability and anti-inflammatory potential of sesquiterpene lactones from chicory. *Nutrients* **2020**, *12*, 3547. [CrossRef] [PubMed]
202. Schönfeld, P.; Wojtczak, L. Short- and medium-chain fatty acids in energy metabolism: The cellular perspective. *J. Lipid Res.* **2016**, *57*, 943–954. [CrossRef] [PubMed]
203. Erny, D.; Hrabě de Angelis, A.L.; Jaitin, D.; Wieghofer, P.; Staszewski, O.; David, E.; Keren-Shaul, H.; Mahlakoiv, T.; Jakobshagen, K.; Buch, T.; et al. Host microbiota constantly control maturation and function of microglia in the CNS. *Nat. Neurosci.* **2015**, *18*, 965–977. [CrossRef] [PubMed]
204. Mohr, K.I. History of Antibiotics Research. *Curr. Top. Microbiol. Immunol.* **2016**, *398*, 237–272. [CrossRef] [PubMed]
205. Roberts, J.A.; Lipman, J. Pharmacokinetic issues for antibiotics in the critically ill patient. *Crit. Care Med.* **2009**, *37*, 840–851. [CrossRef]
206. Ianiro, G.; Tilg, H.; Gasbarrini, A. Antibiotics as deep modulators of gut microbiota: Between good and evil. *Gut* **2016**, *65*, 1906–1915. [CrossRef]
207. Hoban, A.E.; Moloney, R.D.; Golubeva, A.V.; McVey Neufeld, K.A.; O'Sullivan, O.; Patterson, E.; Stanton, C.; Dinan, T.G.; Clarke, G.; Cryan, J.F. Behavioural and neurochemical consequences of chronic gut microbiota depletion during adulthood in the rat. *Neuroscience* **2016**, *339*, 463–477. [CrossRef]
208. Neuman, H.; Forsythe, P.; Uzan, A.; Avni, O.; Koren, O. Antibiotics in early life: Dysbiosis and the damage done. *FEMS Microbiol. Rev.* **2018**, *42*, 489–499. [CrossRef]
209. Fröhlich, E.E.; Farzi, A.; Mayerhofer, R.; Reichmann, F.; Jačan, A.; Wagner, B.; Zinser, E.; Bordag, N.; Magnes, C.; Fröhlich, E.; et al. Cognitive impairment by antibiotic-induced gut dysbiosis: Analysis of gut microbiota-brain communication. *Brain. Behav. Immun.* **2016**, *56*, 140–155. [CrossRef]
210. Minter, M.R.; Zhang, C.; Leone, V.; Ringus, D.L.; Zhang, X.; Oyler-Castrillo, P.; Musch, M.W.; Liao, F.; Ward, J.F.; Holtzman, D.M.; et al. Antibiotic-induced perturbations in gut microbial diversity influences neuro-inflammation and amyloidosis in a murine model of Alzheimer's disease. *Sci. Rep.* **2016**, *6*, 1–12. [CrossRef]

211. Volicer, B.J.; Hurley, A.; Fabiszewski, K.J.; Montgomery, P.; Volicer, L. Predicting short-term survival for patients with advanced Alzheimer's disease. *J. Am. Geriatr. Soc.* **1993**, *41*, 535–540. [CrossRef] [PubMed]
212. Payne, L.E.; Gagnon, D.J.; Riker, R.R.; Seder, D.B.; Glisic, E.K.; Morris, J.G.; Fraser, G.L. Cefepime-induced neurotoxicity: A systematic review. *Crit. Care* **2017**, *21*, 1–8. [CrossRef] [PubMed]
213. Neufeld, N.H.; Mohamed, N.S.; Grujich, N.; Shulman, K. Acute Neuropsychiatric Symptoms Associated with Antibiotic Treatment of Helicobacter Pylori Infections. *J. Psychiatr. Pract.* **2017**, *23*, 25–35. [CrossRef] [PubMed]
214. Kountouras, J.; Boziki, M.; Gavalas, E.; Zavos, C.; Grigoriadis, N.; Deretzi, G.; Tzilves, D.; Katsinelos, P.; Tsolaki, M.; Chatzopoulos, D.; et al. Eradication of Helicobacter pylori may be beneficial in the management of Alzheimer's disease. *J. Neurol.* **2009**, *256*, 758–767. [CrossRef] [PubMed]
215. Yulug, B.; Hanoglu, L.; Ozansoy, M.; Isık, D.; Kilic, U.; Kilic, E.; Schabitz, W.R. Therapeutic role of rifampicin in Alzheimer's disease. *Psychiatry Clin. Neurosci.* **2018**, *72*, 152–159. [CrossRef]
216. Umeda, T.; Ono, K.; Sakai, A.; Yamashita, M.; Mizuguchi, M.; Klein, W.L.; Yamada, M.; Mori, H.; Tomiyama, T. Rifampicin is a candidate preventive medicine against amyloid-β and tau oligomers. *Brain* **2016**, *139*, 1568–1586. [CrossRef] [PubMed]
217. Loeb, M.B.; Molloy, D.W.; Smieja, M.; Standish, T.; Goldsmith, C.H.; Mahony, J.; Smith, S.; Borrie, M.; Decoteau, E.; Davidson, W.; et al. A Randomized, Controlled Trial of Doxycycline and Rifampin for Patients with Alzheimer's Disease. *J. Am. Geriatr. Soc.* **2004**, *52*, 381–387. [CrossRef]
218. Tucker, S.; Ahl, M.; Bush, A.; Westaway, D.; Huang, X.; Rogers, J.T. Pilot study of the reducing effect on amyloidosis in vivo by three FDA pre-approved drugs via the Alzheimer's APP 5′ untranslated region. *Curr. Alzheimer Res.* **2005**, *2*, 249–254. [CrossRef]
219. El-Shimy, I.A.; Heikal, O.A.; Hamdi, N. Minocycline attenuates Aβ oligomers-induced pro-inflammatory phenotype in primary microglia while enhancing Aβ fibrils phagocytosis. *Neurosci. Lett.* **2015**, *609*, 36–41. [CrossRef]
220. Cuello, A.C.; Ferretti, M.T.; Leon, W.C.; Iulita, M.F.; Melis, T.; Ducatenzeiler, A.; Bruno, M.A.; Canneva, F. Early-stage inflammation and experimental therapy in transgenic models of the Alzheimer-like amyloid pathology. *Neurodegener. Dis.* **2010**, *7*, 96–98. [CrossRef]
221. Parachikova, A.; Vasilevko, V.; Cribbs, D.H.; Laferla, F.M.; Green, K.N. Reductions in amyloid-β-derived neuroinflammation, with minocycline, restore cognition but do not significantly affect tau hyperphosphorylation. *J. Alzheimer's Dis.* **2010**, *21*, 527–542. [CrossRef] [PubMed]
222. Wang, T.; Hu, X.; Liang, S.; Li, W.; Wu, X.; Wang, L.; Jin, F. Lactobacillus fermentum NS9 restores the antibiotic induced physiological and psychological abnormalities in rats. *Benef. Microbes* **2015**, *6*, 707–717. [CrossRef] [PubMed]
223. Ravelli, K.G.; dos Anjos Rosário, B.; Camarini, R.; Hernandes, M.S.; Britto, L.R. Intracerebroventricular Streptozotocin as a Model of Alzheimer's Disease: Neurochemical and Behavioral Characterization in Mice. *Neurotox. Res.* **2017**, *31*, 327–333. [CrossRef] [PubMed]
224. Muller, A.P.; Zimmer, E.R.; Haas, C.B.; Oses, J.P.; Martimbianco De Assis, A.; Galina, A.; Souza, D.O.; Portela, L.V. Physical exercise exacerbates memory deficits induced by intracerebroventricular stz but improves insulin regulation of $H_2O_2$ production in mice synaptosomes. *J. Alzheimer's Dis.* **2012**, *30*, 889–898. [CrossRef]
225. Santos, D.B.; Colle, D.; Moreira, E.L.G.; Peres, K.C.; Ribeiro, R.P.; Dos Santos, A.A.; de Oliveira, J.; Hort, M.A.; de Bem, A.F.; Farina, M. Probucol mitigates streptozotocin-induced cognitive and biochemical changes in mice. *Neuroscience* **2015**, *284*, 590–600. [CrossRef]
226. Sharma, B.; Singh, N.; Singh, M.; Jaggi, A.S. Exploitation of HIV protease inhibitor Indinavir as a memory restorative agent in experimental dementia. *Pharmacol. Biochem. Behav.* **2008**, *89*, 535–545. [CrossRef]
227. Shoham, S.; Bejar, C.; Kovalev, E.; Schorer-Apelbaum, D.; Weinstock, M. Ladostigil prevents gliosis, oxidative-nitrative stress and memory deficits induced by intracerebroventricular injection of streptozotocin in rats. *Neuropharmacology* **2007**, *52*, 836–843. [CrossRef]
228. Villeda, S.A.; Luo, J.; Mosher, K.I.; Zou, B.; Britschgi, M.; Bieri, G.; Stan, T.M.; Fainberg, N.; Ding, Z.; Eggel, A.; et al. The ageing systemic milieu negatively regulates neurogenesis and cognitive function. *Nature* **2011**, *477*, 90–96. [CrossRef]
229. Dodiya, H.B.; Kuntz, T.; Shaik, S.M.; Baufeld, C.; Leibowitz, J.; Zhang, X.; Gottel, N.; Zhang, X.; Butovsky, O.; Gilbert, J.A.; et al. Sex-specific effects of microbiome perturbations on cerebral Ab amyloidosis and microglia phenotypes. *J. Exp. Med.* **2019**, *216*, 1542–1560. [CrossRef]
230. Fuller, R. Probiotics in man and animals. *J. Appl. Bacteriol.* **1989**, *66*, 365–378.
231. Ohland, C.L.; Kish, L.; Bell, H.; Thiesen, A.; Hotte, N.; Pankiv, E.; Madsen, K.L. Effects of Lactobacillus helveticus on murine behavior are dependent on diet and genotype and correlate with alterations in the gut microbiome. *Psychoneuroendocrinology* **2013**, *38*, 1738–1747. [CrossRef] [PubMed]
232. Nimgampalle, M.; Kuna, Y. Anti-Alzheimer Properties of Probiotic, Lactobacillus plantarum MTCC 1325 in Alzheimer's Disease induced Albino Rats. *J. Clin. Diagn. Res.* **2017**, *11*, KC01–KC05. [CrossRef] [PubMed]
233. Davari, S.; Talaei, S.A.; Alaei, H.; Salami, M. Probiotics treatment improves diabetes-induced impairment of synaptic activity and cognitive function: Behavioral and electrophysiological proofs for microbiome-gut-brain axis. *Neuroscience* **2013**, *240*, 287–296. [CrossRef] [PubMed]
234. Kobayashi, T.; Suzuki, T.; Kaji, R.; Serata, M.; Nagata, T.; Ando, M.; Iizuka, R.; Tsujibe, S.; Murakami, J.; Kiyoshima-Shibata, J.; et al. Probiotic upregulation of peripheral IL-17 responses does not exacerbate neurological symptoms in experimental autoimmune encephalomyelitis mouse models. *Immunopharmacol. Immunotoxicol.* **2012**, *34*, 423–433. [CrossRef]

235. Athari Nik Azm, S.; Djazayeri, A.; Safa, M.; Azami, K.; Ahmadvand, B.; Sabbaghziarani, F.; Sharifzadeh, M.; Vafa, M. Lactobacilli and bifidobacteria ameliorate memory and learning deficits and oxidative stress in β-amyloid (1-42) injected rats. *Appl. Physiol. Nutr. Metab.* **2018**, *43*, 718–726. [CrossRef]
236. Liu, J.; Sun, J.; Wang, F.; Yu, X.; Ling, Z.; Li, H.; Zhang, H.; Jin, J.; Chen, W.; Pang, M.; et al. Neuroprotective Effects of Clostridium butyricum against Vascular Dementia in Mice via Metabolic Butyrate. *Biomed Res. Int.* **2015**, *2015*. [CrossRef]
237. Kobayashi, Y.; Sugahara, H.; Shimada, K.; Mitsuyama, E.; Kuhara, T.; Yasuoka, A.; Kondo, T.; Abe, K.; Xiao, J.Z. Therapeutic potential of Bifidobacterium breve strain A1 for preventing cognitive impairment in Alzheimer's disease. *Sci. Rep.* **2017**, *7*, 1–10. [CrossRef]
238. Xin, Y.; Diling, C.; Jian, Y.; Ting, L.; Guoyan, H.; Hualun, L.; Xiaocui, T.; Guoxiao, L.; Ou, S.; Chaoqun, Z.; et al. Effects of oligosaccharides from morinda officinalis on gut microbiota and metabolome of APP/PS1 transgenic mice. *Front. Neurol.* **2018**, *9*, 1–14. [CrossRef]
239. Allen, A.P.; Hutch, W.; Borre, Y.E.; Kennedy, P.J.; Temko, A.; Boylan, G.; Murphy, E.; Cryan, J.F.; Dinan, T.G.; Clarke, G. Bifidobacterium longum 1714 as a translational psychobiotic: Modulation of stress, electrophysiology and neurocognition in healthy volunteers. *Transl. Psychiatry* **2016**, *6*, e939. [CrossRef]
240. Ko, C.Y.; Lin, H.T.V.; Tsai, G.J. Gamma-aminobutyric acid production in black soybean milk by Lactobacillus brevis FPA 3709 and the antidepressant effect of the fermented product on a forced swimming rat model. *Process Biochem.* **2013**, *48*, 559–568. [CrossRef]
241. Scriven, M.; Dinan, T.; Cryan, J.; Wall, M. Neuropsychiatric Disorders: Influence of Gut Microbe to Brain Signalling. *Diseases* **2018**, *6*, 78. [CrossRef] [PubMed]
242. Tillisch, K.; Labus, J.; Kilpatrick, L.; Jiang, Z.; Stains, J.; Ebrat, B.; Guyonnet, D.; Legrain-Raspaud, S.; Trotin, B.; Naliboff, B.; et al. Consumption of fermented milk product with probiotic modulates brain activity. *Gastroenterology* **2013**, *144*, 1394–1401. [CrossRef] [PubMed]
243. Akbari, E.; Asemi, Z.; Kakhaki, R.D.; Bahmani, F.; Kouchaki, E.; Tamtaji, O.R.; Hamidi, G.A.; Salami, M. Effect of probiotic supplementation on cognitive function and metabolic status in Alzheimer's disease: A randomized, double-blind and controlled trial. *Front. Aging Neurosci.* **2016**, *8*, 256. [CrossRef] [PubMed]
244. Leblhuber, F.; Steiner, K.; Schuetz, B.; Fuchs, D.; Gostner, J.M. Probiotic Supplementation in Patients with Alzheimer's Dementia—An Explorative Intervention Study. *Curr. Alzheimer Res.* **2018**, *15*, 1106–1113. [CrossRef]

*nutrients*

Article

# Application of *Ligilactobacillus salivarius* CECT5713 to Achieve Term Pregnancies in Women with Repetitive Abortion or Infertility of Unknown Origin by Microbiological and Immunological Modulation of the Vaginal Ecosystem

**Leónides Fernández [1,*], Irma Castro [2], Rebeca Arroyo [2], Claudio Alba [2], David Beltrán [3] and Juan M. Rodríguez [2,*]**

[1] Department of Galenic Pharmacy and Food Technology, Complutense University of Madrid, 28040 Madrid, Spain
[2] Department of Nutrition and Food Science, Complutense University of Madrid, 28040 Madrid, Spain; irmacastro@ucm.es (I.C.); rebecaa@vet.ucm.es (R.A.); c.alba@ucm.es (C.A.)
[3] Centro de Diagnóstico Médico, Ayuntamiento de Madrid, 28006 Madrid, Spain; beltrangine@gmail.com
* Correspondence: leonides@vet.ucm.es (L.F.); jmrodrig@ucm.es (J.M.R.); Tel.: +34-913943745 (L.F.); +34-913943745 (J.M.R.)

**Abstract:** In this study, the cervicovaginal environment of women with reproductive failure (repetitive abortion, infertility of unknown origin) was assessed and compared to that of healthy fertile women. Subsequently, the ability of *Ligilactobacillus salivarius* CECT5713 to increase pregnancy rates in women with reproductive failure was evaluated. Vaginal pH and Nugent score were higher in women with reproductive failure than in fertile women. The opposite was observed regarding the immune factors TGF-β 1, TFG-β 2, and VEFG. Lactobacilli were detected at a higher frequency and concentration in fertile women than in women with repetitive abortion or infertility. The metataxonomic study revealed that vaginal samples from fertile women were characterized by the high abundance of *Lactobacillus* sequences, while DNA from this genus was practically absent in one third of samples from women with reproductive failure. Daily oral administration of *L. salivarius* CECT5713 (~9 $\log_{10}$ CFU/day) to women with reproductive failure for a maximum of 6 months resulted in an overall successful pregnancy rate of 56%. The probiotic intervention modified key microbiological, biochemical, and immunological parameters in women who got pregnant. In conclusion, *L. salivarius* CECT5713 has proved to be a good candidate to improve reproductive success in women with reproductive failure.

**Keywords:** infertility; repetitive abortion; implantation failure; *Lactobacillus salivarius*; probiotics; vaginal microbiome; TGF-β; VEGF

**Citation:** Fernández, L.; Castro, I.; Arroyo, R.; Alba, C.; Beltrán, D.; Rodríguez, J.M. Application of *Ligilactobacillus salivarius* CECT5713 to Achieve Term Pregnancies in Women with Repetitive Abortion or Infertility of Unknown Origin by Microbiological and Immunological Modulation of the Vaginal Ecosystem. *Nutrients* **2021**, *13*, 162. https://doi.org/10.3390/nu13010162

Received: 27 November 2020
Accepted: 30 December 2020
Published: 6 January 2021



## 1. Introduction

Increasing evidence has highlighted the relevance of the microbiota of the female genital tract for human reproduction [1,2]. Under physiological conditions, and in contrast to the gut, the human vaginal microbiota is usually characterized by a low microbial diversity and the dominance of bacteria from the genus *Lactobacillus* [3,4]. In fact, a low diversity in the gut has been linked to a variety of gastrointestinal processes, including inflammatory bowel disease [5], while a high diversity in the vagina has been associated to vaginosis [6].

The vaginal microbiota in healthy reproductive-age women is mainly composed of one or a few *Lactobacillus* species, which represent more than 90% of the total microbiota [7,8]. In a seminal study, the bacterial communities of 396 asymptomatic women were classified into five distinct vaginotypes; four of them were dominated by *Lactobacillus crispatus*, *Lactobacillus gasseri*, *Lactobacillus iners*, and *Lactobacillus jensenii*, respectively; in contrast, the fifth one had lower proportions of lactobacilli and was predominantly composed of strictly anaerobic bacterial genera, such as *Gardnerella*, *Prevotella*, *Megasphaera*, *Atopobium*, or *Dialister* [3]. This last vaginotype was associated to high Nugent scores, a Gram-staining based technique used for the diagnosis of bacterial vaginosis (BV).

Several factors are known to contribute to interindividual and intraindividual changes in the vaginal microbiota [9]. Although shifts between different vaginotypes may occur naturally, increase of diversity and colonization by strict anaerobes and decrease or depletion of lactobacilli are considered as risk factors for BV. In fact, vaginal microbiota dysbiosis has been associated with higher rates of intra-amniotic infection, premature delivery, spontaneous abortion, and infertility [10–15].

Different studies have shown that infertile women harbor a differential vaginal microbiota when compared to fertile women [16–19]. Therefore, the composition of the vaginal microbiota (and, particularly, any deviation from the *Lactobacillus*-dominated, low-diversity vaginal microbiome) may play a key role in fertility and in the outcomes of assisted reproduction treatments (ARTs) [20–22]. Abundant isolation of enterococci, streptococci, staphylococci, and/or Gram-negative bacteria (*Escherichia coli*, *Klebsiella pneumoniae*) from the tip of the catheter used for embryo transfer has been correlated with lower implantation and pregnancy rates and increased miscarriage rates [23], while abundant isolation of lactobacilli and low density or no isolation of the aforementioned bacteria has been correlated with better reproductive outcomes [24–28]. Metataxonomic studies of endometrial samples have also revealed that an abnormal endometrial bacterial profile (with a low percentage of sequences of the genus *Lactobacillus*) is a common feature in a high percentage of infertile women subjected to ART [21,29]. Although at least a part of the bacterial DNA detected in endometrial samples may arise from vaginal contamination during sampling, these studies suggest that an abundant presence of *Lactobacillus* DNA in such samples may be a predictor of implantation success [29,30].

As a consequence, the assessment of the microbial communities in the reproductive tract should be considered as a relevant part of the evaluation and personalized care in cases of reproductive failure of unknown cause or origin. When this happens, the use of probiotics may be a possible strategy to modulate the reproductive tract microbiome and to increase the success rates [31]. However, such a combined strategy (assessment of vaginal communities together with use of a target-selected probiotic) has not been explored yet, and commercially available probiotics are being empirically prescribed for repopulation of the female reproductive tract with *Lactobacillus* strains [2], without a proper scientific evidence of their actual usefulness.

Lactobacilli may have different biological activities that contribute to fertility and to a healthy pregnancy, including, among others: (a) the inhibition of the colonization and growth of potentially harmful microbes, including viruses, bacteria, yeast, and protozoa that may compromise fertility [32,33]; (b) contribution to angiogenesis and vasculogenesis that may favor the implantation of the embryo [34]; and, (c) induction of immunomodulation activities, such as those involved in implantation and in tolerance towards the embryo, first, and the fetus, later [35,36]. However, those properties might be strain-specific and, therefore, a strain-by-strain evaluation has to be performed for this specific target.

*Lactobacillus salivarius* CECT5713 [37] has been shown to be a probiotic strain suitable for applications in the mother–infant dyad due to a wide repertoire of desirable phenotypic and genotypic properties [38]. This includes a high survival rate when exposed to gastrointestinal tract conditions, a high acidifying activity, and antimicrobial, anti-inflammatory, and immunomodulatory properties, which have been demonstrated *in vitro*, in animal models, and in human clinical trials [38–44]. Therefore, and after evaluating some vaginal-related properties in this study, it was selected to be administered in a clinical trial in order to assess its efficacy for the infertility target. It must be highlighted that this species has been renamed as *Ligilactobacillus salivarius* in the recent proposal for reclassification of the genus *Lactobacillus* [45].

In this context, the first objective of this study was to assess the differences in several vaginal parameters (pH, Nugent score, microbiota composition as determined through culture and metataxonomic methods, and soluble immune factor levels) between women with reproductive failure (because of repetitive abortion during the first 12 weeks of pregnancy or infertility of unknown origin) and fertile women. The second objective was to evaluate the ability of *L. salivarius* CECT5713 to modulate those vaginal parameters and to increase pregnancy rates (currently ~29% after IVF procedures in this setting) in the group of women with reproductive failure.

## 2. Materials and Methods

### 2.1. Characterization of Vaginal-Related Properties of L. salivarius CECT5713

An overlay method [46] was used to determine the ability of *L. salivarius* CECT5713 to inhibit the growth of various species of bacteria and yeasts. It was performed as described previously [37]. All indicator strains had been previously isolated from clinical cases of vaginal or cervical infections, and included five strains of *G. vaginalis*, three of *Streptococcus agalactiae* and of *Candida albicans*, and two of *Candida glabrata*, *Candida parapsilosis*, and *Ureaplasma urealyticum* (our own culture collection). All inhibitory activity assays were performed in triplicate.

The ability of *L. salivarius* CECT5713 to aggregate with cells of the indicator strains cited above was investigated following the procedure of Younes et al. [47]. The suspensions were observed under a phase-contrast microscope. Adherence to vaginal epithelial cells collected from healthy premenopausal women was performed and interpreted as described previously [48]. Adherence was measured as the number of lactobacilli adhered to the vaginal cells in 20 random microscopic fields. *L. salivarius* CECT9145 was used as a control strain because of its high adherence to vaginal cells [49]. The assay was performed in triplicate.

Initially, the α-amylase activity of *L. salivarius* CECT5713 was qualitatively assessed using the method described by Padmavathi et al. [50]. Briefly, the strain was inoculated into a modified MRS media containing starch (0.5% peptone, 0.7% yeast extract, 0.2% NaCl, 2% starch, and 1.5% agar). The plates were incubated at 37 °C for 48 h in anaerobiosis and, then, the zone of clearance was observed by adding Gram's iodine as detecting agent. Quantitation of the cell-bound α-amylase activity of *L. salivarius* CECT5713 was done with a kit (Kikkoman Co., Tokyo, Japan) using 2-chloro-4-nitrophenyl $6^5$-azido-$6^5$-deoxy-β-maltopentaoside as substrate and using conditions described previously [51]. One unit of activity was defined as the amount of enzyme needed to release 1 μmol 2-chloro-4-nitrophenol from 2-chloro-4-nitrophenyl $6^5$-azido-$6^5$-deoxy-β-maltopentaoside per min at 37 °C.

### 2.2. Participants, Sampling, and Design of the Human Study

A total of 58 women, aged 28–45, participated in this study (Table 1). Volunteers were classified into 3 groups. All women in the RA group ($n$ = 21) had a history of recurrent miscarriage with three or more pregnancy losses during the first 12 weeks of pregnancy. All women of the INF group ($n$ = 23) had a history of infertility (inability to conceive) despite being the recipients of ART for at least three times, including two cycles, at least, of in vitro fertilization (IVF). Finally, the control group ($n$ = 14) included fertile women having at least two children after uncomplicated term pregnancies. None of the women of the RA and INF groups received ART during the whole period of the study. None of the RA group components were diagnosed of antiphospholipidic syndrome and, therefore, they did not receive either heparin and/or salicylic acid during the study. None of the participants had received hormonal therapy, antibiotics or probiotics in the 4 weeks previous to sampling. Vaginal samples were taken at least 7 days after coitus to avoid or minimize the impact of the partner's semen on the vaginal pH, microbiota composition or immunoprofile (in the latter case, particularly in relation to the concentration of the two isoforms of the transforming growth factors beta 1 and 2 (TGF-β 1 and TGF-β 2)). Women with lactose intolerance or cow's milk protein allergy were excluded because of the excipient used to administer the strain in the subsequent pilot trial (see below). Informed consent was obtained from all subjects involved in the study.

**Table 1.** Characteristics of the participants ($N = 58$) which included fertile women (Control group), women with a history of repetitive abortion (RA group), and women with infertility of unknown origin (INF group).

| Characteristic | Group | | | |
|---|---|---|---|---|
| | Control ($n = 14$) | RA ($n = 21$) | INF ($n = 23$) | *p*-Value |
| **Age (years)** | | | | |
| Mean (95% CI) | 34.6 (33.5–35.8) [a] | 39.4 (38.5–40.4) [b] | 38.0 (37.1–38.9) [b] | <0.001 [2] |
| Range (min–max) | (28.0–45.0) | (36.0–44.0) | (34.0–44.0) | |
| **Weight (kg)** | | | | |
| Mean (95% CI) | 62.4 (59.7–65.0) | 68.3 (66.1–70.4) | 66.5 (64.5–68.6) | 0.054 [2] |
| Range (min–max) | (46.0–87.0) | (50.0–87.0) | (51.0–78.0) | |
| **Height (cm)** | | | | |
| Mean (95% CI) | 166 (164–168) | 167 (165–169) | 168 (166–169) | 0.761 [2] |
| Range (min–max) | (156–175) | (152–190) | (160–182) | |
| **Regularity of the menstrual cycle** | | | | |
| Yes, *n* (%) | 10 (71) | 10 (48) | 11 (48) | 0.337 [3] |
| No, *n* (%) | 4 (29) | 11 (52) | 12 (52) | |
| **Duration of the menstrual cycle (days)** | | | | |
| Mean (95% CI) | 28.0 (27.4–28.7) | 27.4 (26.9–27.9) | 27.5 (27.0–28.0) | 0.502 [2] |
| Range (min–max) | (25.0–32.5) | (24.0–30.0) | (24.0–30.0) | |
| **History of infections** | | | | |
| Vaginal, *n* (%) | 2 (14) | 13 (62) | 8 (35) | 0.017 [3] |
| Urinary tract, *n* (%) | 2 (14) | 13 (62) | 15 (65) | 0.006 [3] |
| Otorhinolaryngology, *n* (%) | 3 (21) | 13 (62) | 12 (52) | 0.057 [3] |
| Lower respiratory tract, *n* (%) | 2 (14) | 7 (33) | 7 (30) | 0.490 [3] |
| Skin, *n* (%) | 1 (7) | 3 (14) | 4 (17) | 0.800 [3] |
| Gastrointestinal, *n* (%) | 0 (0) | 1 (5) | 1 (4) | 1.000 |
| **Antibiotic usage** [1] | | | | |
| In infancy, *n* (%) | 4 (29) | 19 (90) | 14 (61) | <0.001 [3] |
| In adulthood, *n* (%) | 4 (29) | 16 (76) | 19 (83) | 0.003 [3] |
| **History of other conditions** | | | | |
| Allergies, *n* (%) | 2 (14) | 5 (24) | 4 (17) | 0.835 [3] |
| Food intolerance, *n* (%) | 0 (0) | 8 (38) | 13 (57) | 0.001 [3] |
| Thyroid disease, *n* (%) | 0 (0) | 5 (24) | 3 (13) | 0.125 [3] |

[1] Antibiotic usage means ≥4 annual treatments due to recurrent infections. [2] One-way ANOVA tests were used to evaluate differences in mean values of women age, weight, and height and duration of the menstrual cycle between groups. Values followed by different superscript letters within the same row indicate statistically significant differences between groups according to Scheffé post hoc comparison tests. [3] Freeman–Halton extension of the Fisher exact probability tests for a 2 × 3 contingency table were used to compute the (two-tailed) probability of obtaining a distribution of values of categorical variables (regularity of the menstrual cycle, history of infections, antibiotic usage and history of other conditions).

At recruitment (within the first three days post-ovulation; day 0), two samples were collected: A vaginal swab specimen for in fresh determination of the Nugent score, and a cervicovaginal lavage (CVL) of the cervical and the vaginal walls with 10 mL of sterile normal saline for all the other analysis. Aliquots of the CVL samples were used for culture-based analysis. Subsequently, CVL samples were clarified by centrifugation at 800× *g* for 10 min at 4 °C. Aliquots of CVL supernatants and cell pellets were stored at −80 °C until the immunological and metataxonomic analyses were performed. Demographic, anthropometric, and health data (including a past or present history of recurrent infections at different body locations and use of antibiotics) were recorded at recruitment (Supplementary Figure S1). High use of antibiotics was defined as receiving ≥4 antibiotic treatments per year because of recurrent infections while a range between 0 and 2 annual treatments was considered as a low use of antibiotics.

Starting at day 0, women of the RA and INF groups consumed (oral route) a daily sachet with ~50 mg of freeze-dried probiotic (~9 $\log_{10}$ CFU of *L. salivarius* CECT5713) for 6 months or until a diagnosis of pregnancy (whatever happened first). At that point, the same two samples described above were collected from each woman. After a diagnosis of pregnancy, oral administration of the probiotic strain was maintained until the 15th week of pregnancy. All the spontaneous pregnancies that occurred within the first year after day 0 were recorded in this study.

Probiotic-containing sachets were kept at 4–8 °C throughout the study. All volunteers signed a written consent and were provided with diaries to record compliance with the study product intake. Minimum compliance rate (% of the total treatment doses) was set at 86%. This study was conducted according to the guidelines laid down in the Declaration of Helsinki and it was approved by the Ethical Committee of Biomedical Research of Consejería de Salud y Familias (Junta de Andalucía, Granada, Spain) (P050/19, Act 11/19). The study was registered in the ClinicalTrials.gov database (NCT04446572).

### 2.3. Measurement of Vaginal pH and Nugent Score

At each of the two study visits, the pH of the lateral vaginal wall was measured (Whatman pH paper, pH 3.8–5.5 and pH 6.0–8.1). Nugent scoring was performed as described previously [52]. Briefly, the swab material was transferred to a glass slide, heat fixed, and Gram stained. Gram-positive, Gram-negative, and Gram-variable bacterial morphotypes were quantified. A Nugent score of 0–3 was considered normal, 4–6 was considered intermediate, and 7–10 was considered consistent with bacterial vaginosis [52].

### 2.4. Culture-Dependent Analysis

CVL samples collected during the trial were serially diluted and plated onto Columbia Nalidixic Acid (CNA), Gardnerella (GAR), CHROMagar StrepB (CHR), Mac Conkey (MCK), Mycoplasma (MYC), and Sabouraud Dextrose Chloramphenicol (SDC) agar plates (BioMerieux, Marcy l'Etoile, France) for selective isolation and quantification of the main cultivable non-*Lactobacillus* bacteria and yeasts that may be found in the vagina, including the agents most frequently involved in vaginal infections. They were also inoculated onto agar plates of MRS (Oxoid, Basingstoke, UK) supplemented with either L-cysteine (2.5 g/L) (MRS-C) or horse blood (5%) (MRS-B) for isolation of lactobacilli, including *L. iners* (MRS-B). All media were incubated for 48 h at 37 °C under aerobic conditions, with the exception of the MRS-C and MRS-B plates, which were incubated anaerobically (85% nitrogen, 10% hydrogen, 5% carbon dioxide) in an anaerobic workstation (DW Scientific, Shipley, UK) for up to 72 h. After incubation, the colonies were recorded and at least one representative of each colony morphology was selected from the agar plates. The isolates were identified by Matrix Assisted Laser Desorption Ionization-Time of Flight (MALDI-TOF) mass spectrometry (Bruker, Bremen, Germany). When the identification by MALDI-TOF was not possible at the species level (particularly in the case of lactobacilli isolates), the identification was carried out by 16S ribosomal RNA (rRNA) gene sequencing as described by Mediano et al. [53].

### 2.5. DNA Extraction from the Samples

Approximately 1 mL of each CVL sample was used for DNA extraction following a method described previously [54]. Extracted DNA was eluted in 22 μL of nuclease-free water and stored at −20 °C until further analysis. Purity and concentration of each extracted DNA was initially estimated using a NanoDrop 1000 spectrophotometer (NanoDrop Technologies, Inc., Rockland, DE, USA). Negative controls (blanks) were processed in parallel.

### 2.6. Real-Time Quantitative PCR (qPCR) Assay for the Detection and Quantification of *L. salivarius* DNA

Primers and conditions for quantification of *L. salivarius* DNA have been described previously [55]. The DNA concentration of all samples was adjusted to 5 ng $\mu L^{-1}$. A commercial real-time PCR thermocycler (CFX96™, Biorad Laboratories, Hercules, CA, USA) was used for all experiments. Standard curves using 1:10 DNA dilutions (ranging from 2 ng to 0.2 pg) from *L. salivarius* CECT5713 were used to calculate the concentrations of the unknown bacterial genomic targets. Threshold cycle (Ct) values between 15.29 and 20.07 were obtained for this range of *L. salivarius* DNA ($R^2$ = 0.9915). The Ct values measured for DNA extracted from non-target species (*L. reuteri* MP07 and *Lactobacillus plantarum* MP02; our own collection) were $\geq$39.27 $\pm$ 0.64. These two control strains were selected because they belong to the *L. salivarius* taxonomically closest species [56]. All samples and standards were run in triplicate.

### 2.7. Metataxonomic Analysis

The V3-V4 hypervariable region of the 16S rDNA was amplified by PCR using the universal primers S-D-Bact-0341-b-S-17 (CCTACGGGNGGCWGCAG) and S-D-Bact-129 0785-a-A-21 (GACTACHVGGGTATCTAATCC) [57] and sequenced in the MiSeq system of Illumina at the facilities of Parque Científico de Madrid (Tres Cantos, Spain). Barcodes appended to 3′ and 5′ terminal ends of the PCR amplicons allowed separation of forward and reverse sequences in a second PCR-reaction. DNA concentration of the PCR products was quantified in a 2100 Bioanalyzer system (Agilent, Santa Clara, CA, USA). After pooling the PCR products at about equal molar ratios, DNA amplicons were purified by using a QIAEX II Gel Extraction Kit (Qiagen, Hilden, Germany) from the excised band having the correct size after running on an agarose gel. DNA concentration was then quantified with PicoGreen (BMG Labtech, Jena, Germany). The pooled, purified and barcoded DNA amplicons were sequenced using the Illumina MiSeq pair-end protocol (Illumina Inc., San Diego, CA, USA) following the manufacturer's protocols.

### 2.8. Bioinformatic Analysis

Raw sequence data were demultiplexed and quality filtered using Illumina MiSeq Reporter analysis software. Microbiome bioinformatics was done with QIIME 2 2019.1 [58]. Denoising was performed with DADA2 [59]. Taxonomy was assigned to ASVs using the q2-feature-classifier [60] and the naïve Bayes classifier *classify-sklearn* against the SILVA database version 132 [61]. Posterior bioinformatic analysis was conducted using the R version 3.5.1 (https://www.R-project.org) [62]. A table of Operational Taxonomic Units (OTUs) counts per sample was generated, and bacterial taxa abundances were normalized to the total number of sequences in each sample. The relative abundance values of the different bacterial taxa in the three groups of CVL samples (control, RA and INF) were analyzed using the linear discriminant analysis (LDA) effect size (LEfSe) algorithm [63] in an online version (http://huttenhower.sph.harvard.edu/galaxy/). Alpha diversity was studied with the Shannon and Simpson diversity indexes with the R Vegan package (Version 2.5.6) (https://github.com/vegandevs/vegan/). Beta diversity was studied using principal coordinates analysis (PCoA) to visually display patterns of bacterial profiles at the genus level through a distance matrix containing a dissimilarity value for each pairwise sample comparison. The Bray–Curtis and binary Jaccard indices were used for quantitative (relative abundance) and qualitative analyses (presence/absence), respectively. Analysis of variance of the distance matrices was performed with the "nonparametric *MANOVA* test" Adonis with 999 or permutational multivariate ANOVA (PERMANOVA) with 999 permutations with the R Vegan package. The heatmap graph was generated by using *gplots* package. Dendogram linkages were based on the relative abundance of the 20 most abundant bacterial genera within the samples and on the complete linkage method for hierarchical clustering (*hclust* function).

### 2.9. Immunological Analysis

The concentrations of several soluble immune factors (IL1β, IL1ra, IL2, IL4, IL5, IL6, IL7, IL8, IL9, IL10, IL12, IL13, IL15, IL17, IL6, basic FGF, eotaxin, GCSF, GMCSF, IFNγ, MCP1, MIP1α, MIP1β, PDGF-BB, RANTES, TNFα, VEGF) were determined by magnetic bead-based multiplex immunoassays, using a Bioplex 200 instrument (Bio-Rad, Hercules, CA, USA) and the Bio-Plex Pro™ Human Cytokine 27-plex Assay (ref. M500KCAF0Y, Bio-Rad). In parallel, the levels of TGF-β 1 and TGF-β 2 were measured by ELISA with the RayBio® Human TGF-β 1 and Human TGF-β 2 ELISA kits, respectively (RayBiotech, Norcross, GA, USA). All determinations were carried out following the manufacturer's protocols and standard curves were performed for each analyte.

### 2.10. Statistical Analysis

Microbiological data were recorded as CFU/mL and transformed to logarithmic values before statistical analysis. The normality of data distribution was analyzed using the Shapiro–Wilks test. Then, the quantitative variables were expressed as means and 95% confidence intervals (CI) or standard deviations (SD) when normally distributed and as medians and interquartile ranges (IQR) if they did not follow a normal distribution. The qualitative values were presented as total number of events and percentages. One-way ANOVA tests were used to compare the means of the experimental groups and Scheffé post hoc tests were used to identify which pairs of means were statistically different. The effect of the probiotic intervention on several vaginal parameters in each group of women with reproductive failure was analyzed using one-way ANOVA repeated measures tests. The Fisher's exact probability test, or the Freeman–Halton extension of the Fisher exact probability test for a $2 \times 3$ contingency table, was used for comparison of proportions and frequencies. For non-parametric analyses, differences between groups were assessed using Kruskal–Wallis tests and Wilcoxon–Mann–Whitney tests to identify which pair of groups were different, with Bonferroni correction for multiple comparisons when indicated. Correlations between the 20-major relative abundant bacterial genera were visualized using R package *qgraph* [64]. Statistical analysis and plotting were performed either using Statgraphics Centurion XVIII version 18.1.06 (Statgraphics Technologies, Inc., The Plains, VA, USA) or in the R environment (version 3.5.1; R-project, http://www.r-project.org) and *ggplot2* [Wickham, 2016]. Differences were considered statistically significant at $p < 0.05$.

## 3. Results

### 3.1. Characterization of Vaginal-Relevant Properties of L. salivarius CECT5713

*L. salivarius* CECT5713 showed inhibitory antimicrobial activity (inhibition zone > 2 mm around the streak) against all the *G. vaginalis*, *S. agalactiae*, *C. albicans*, *C. glabrata*, *C. parapsilosis*, and *U. urealyticum* strains used as indicators in this study. The strain was able to form large, well defined co-aggregates with all the selected vaginal and cervical pathogens. Co-aggregation was particularly intense with *G. vaginalis*, *S. agalactiae*, and *C. albicans* strains. In this study, the strain tested was strongly adhesive to vaginal epithelial cells, a mean (±SD) of 329 (±46) adherent lactobacilli in 20 random microscopic fields. The mean (±SD) value for *L. salivarius* CECT9145, a control strain with a high adherence to vaginal cells, was 336 (±52) adherent lactobacilli in 20 microscopic fields. Extracellular amylase production by *L. salivarius* CECT5713 was observed by the zone of clearance around the colonies (~2.0 mm) when flooded with iodine solution. Later, when the α-amylase activity was measured, this strain showed a high level of α-amylase activity (0.83 U/mL) at 16 h (concentration of *L. salivarius* CECT5713: ~8.6 $\log_{10}$ CFU/mL), and could be detected in supernatants at a similar level for up to 48 h (when the assay was finished).

### 3.2. Demographic, Anthropometric, and Clinical Characteristics of the Participants in the Human Study

The characteristics of the 58 women that participated in this study are presented in Table 1. The mean (95% CI) age in the control group was 34.6 years (33.5–35.8), while in those of repetitive abortions (RA) and with infertility of unknown origin (INF) was 39.4 (38.5–40.4) and 38.0 (37.1–38.9) years, respectively (Table 1). Women in the control group were significantly younger than other participants ($p < 0.001$; one-way ANOVA), but there were no differences in mean values of body weight and height between the three groups of women (Table 1).

About 71% of the women in the control group had a regular menstrual cycle, while in the other two (RA and INF) this percentage was 48%, although this difference was not statistically significant ($p$ = 0.337; Fisher exact probability tests). No differences were observed in the mean duration of the menstrual cycle that was 28, 27.4, and 27.5 days for women in the control, RA, and INF groups, respectively (Table 1).

Interestingly, statistically significant differences were found between the control women and those in the other two groups regarding a history of recurrent vaginal and urinary tract infections ($p$ = 0.017 and $p$ = 0.006, respectively; Fisher exact probability tests) and the use of antibiotics both during infancy ($p$ < 0.001) and adulthood ($p$ = 0.003), which were higher in the last two groups (Table 1; Supplementary Figure S1). A trend to a higher rate of ORL infections (pharyngitis, otitis) among women with repetitive abortion or infertility was also observed but it did not reach statistical significance ($p$ = 0.057). In contrast, no differences were observed among the three groups in relation to the rates of skin, lower respiratory tract and gastrointestinal infections (Table 1).

### *3.3. Baseline Vaginal Health Parameters*

The vaginal pH values of the control group (4.53; range 4.38–4.68) were statistically different from those of the two study groups: 5.67 (5.55–5.79) and 5.96 (5.84–6.07) for RA and INF, respectively ($p$ = 0.000; one-vay ANOVA). Similarly, the Nugent scores of the two study groups were significantly higher (5.95 (5.54–6.37) and 6.30 (5.91–6.70), respectively), than those from controls (1.79 (1.27–2.30); $p$ = 0.000; one-way ANOVA) (Table 2). The CVL concentrations of the growth factors TGF-β 1, TFG-β 2 and VEFG of the control group were 4.83 (4.65–5.01) pg/mL, 3.22 (3.10–3.34) pg/mL, and 406.0 (322.0–490.0) pg/mL, respectively, while they appeared to be halved in both study groups (RA and INF), the differences being statistically significant (Table 2). No differences were observed among the three groups in relation to the remaining soluble immune factors analyzed in this work, which showed a high degree of interindividual variability (data not shown).

All women of the control group harbored lactobacilli in their vaginas ($n$ = 14), the mean (95% CI) value being 7.24 (6.89–7.60) $\log_{10}$ CFU/mL using culture-dependent assessment. The frequency of lactobacilli detection was lower in the RA and INF groups: 57% and 26%, respectively ($p$ < 0.001; Fisher exact probability tests). In addition, mean lactobacilli concentrations were 2.20 and 1.46 $\log_{10}$ units lower in CVL samples from lactobacilli-positive women in the RA and INF groups, respectively. The lactobacilli profile was also different (Figure 1). Seven species were identified in the samples from women of the control group, including *L. crispatus* (the dominant species), *L. jensenii*, *L. gasseri*, *L. iners*, *Limosilactobacillus* (formely *Lactobacillus*) *fermentum*, *L. salivarius*, and *Limosilactobacillus vaginalis*. However, the lactobacilli species profiles in the study groups (RA and INF) were narrower than in controls and *L. fermentum*, *L. salivarius*, and *L. vaginalis* were not detected. *L. crispatus* was the dominant species in 6 samples (43%) from fertile women, 5 samples (24%) from women with repetitive abortion and only 1 sample (4%) from infertile women. It is interesting to note that *L. iners* was isolated only from one CVL sample of the control group while it was isolated from about one-third (5 out of a total of 18 lactobacilli positive samples) from samples of RA and INF groups. *L. salivarius* was detected in the sample of a unique woman from the control group as determined by species-specific qPCR (7.29 $\log_{10}$ copies/mL) and culture (7.3 $\log_{10}$ CFU/mL) (Table 2). The strain was genetically different from *L. salivarius* CECT5713 (results not shown).

**Table 2.** Comparison of baseline vaginal parameters (pH, Nugent score, cytokines, and microbiology) of the participants ($n$ = 58) which included fertile women (Control group), women with a history of repetitive abortion (RA group), and women with infertility of unknown origin (INF group).

| | Group | | | |
|---|---|---|---|---|
| **Vaginal Parameter** | **Control ($n$ = 14)** | **RA ($n$ = 21)** | **INF ($n$ = 23)** | ***p*-Value** |
| **pH** | | | | |
| Mean (95% CI) | 4.53 (4.38–4.68) [a] | 5.67 (5.55–5.79) [b] | 5.96 (5.84–6.07) [b] | 0.000 [1] |
| Range (min–max) | (4.20–5.00) | (4.70–6.50) | (4.90–6.30) | |
| **Nugent score** | | | | |
| Mean (95% CI) | 1.79 (1.27–2.30) [a] | 5.95 (5.54–6.37) [b] | 6.30 (5.91–6.70) [b] | 0.000 [1] |
| Range (min–max) | (0.00–4.00) | (3.00–8.00) | (4.00–8.00) | |
| **TGF-β 1,** pg/mL | | | | |
| Mean (95% CI) | 4.83 (4.65–5.01) [a] | 2.62 (2.47–2.76) [b] | 2.19 (2.05–2.33) [c] | 0.000 [1] |
| Range (min–max) | (4.20–5.30) | (1.70–3.80) | (1.50–2.90) | |
| **TGF-β 2,** pg/mL | | | | |
| Mean (95% CI) | 3.22 (3.10–3.34) [a] | 1.52 (1.43–1.62) [b] | 1.33 (1.24–1.43) [b] | 0.000 [1] |
| Range (min–max) | (2.70–3.70) | (0.90–2.20) | (0.80–2.00) | |
| **VEGF,** pg/mL | | | | |
| Mean (95% CI) | 406.0 (322.0–490.0) [a] | 274.8 (206.0–343.0) [a,b] | 181.2 (116.0–247.0) [b] | 0.016 [1] |
| Range (min–max) | (1.4–929.0) | (95.0–562.0) | (38.0–431.0) | |
| **Lactobacilli** | | | | |
| **Positive women** | 14 (100) | 12 (57) | 6 (26) | <0.001 [3] |
| **Viable counts,** $\log_{10}$ CFU/mL [2] | | | | |
| Mean (95% CI) | 7.24 (6.89–7.60) [a] | 5.04 (4.66–5.42) [b] | 5.78 (5.24–6.32) [b] | 0.000 [1] |
| Range (min–max) | (6.80–7.70) | (3.60–6.70) | (3.70–7.50) | |
| ***L. salivarius* qPCR,** $\log_{10}$ copies/mL | | | | |
| $n$ (%) | 1 (7) | 0 | 0 | |
| Mean (95% CI) | 7.29 | | | |

[1] One-way ANOVA tests were used to evaluate differences in mean values between groups. Values followed by different superscript letters within the same row indicate statistically significant differences between groups according to Scheffé post hoc comparison tests. [2] Mean (95% CI) and range (min–max) values in lactobacilli-positive women. [3] Freeman–Halton extension of the Fisher exact probability test for a 2 × 3 contingency table were used to compute the (two-tailed) probability of obtaining a distribution of values of lactobacilli positive women. Abbreviations: TGF-β 1, transforming growth factor β 1; TGF-β 2, transforming growth factor β 2; VEGF, vascular endothelial growth factor.

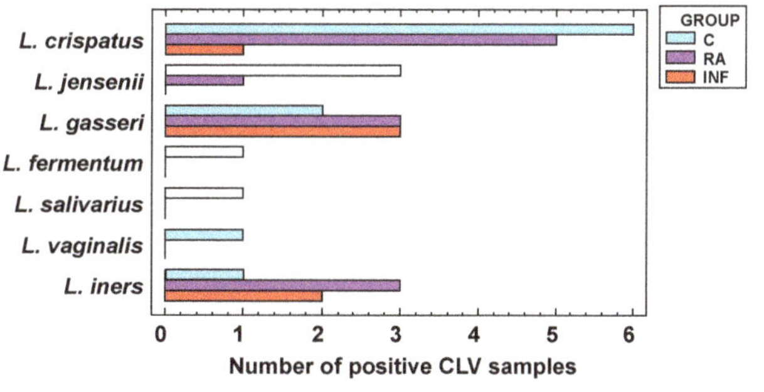


**Figure 1.** Dominant lactobacilli species (when lactobacilli could be isolated) in cervovaginal lavage (CVL) samples of fertile women (C, bluish green), women with repetitive abortion (RA, purple) and women with infertility of unknown origin (INF, red).

Globally, the comparison of RA and INF groups at the beginning of the study revealed some statistically relevant differences (Figure 2). The mean of the vaginal pH values was 0.29 units higher in the INF group, but the opposite was observed for TGF-β 1 and VEGF, which had mean concentrations 0.43 pg/mL and 94 pg/mL higher, respectively, in the RA group. No differences were observed regarding other characteristics, including age, weight, height, Nugent score, TGF-β 2, and lactobacilli viable counts (Figure 2).

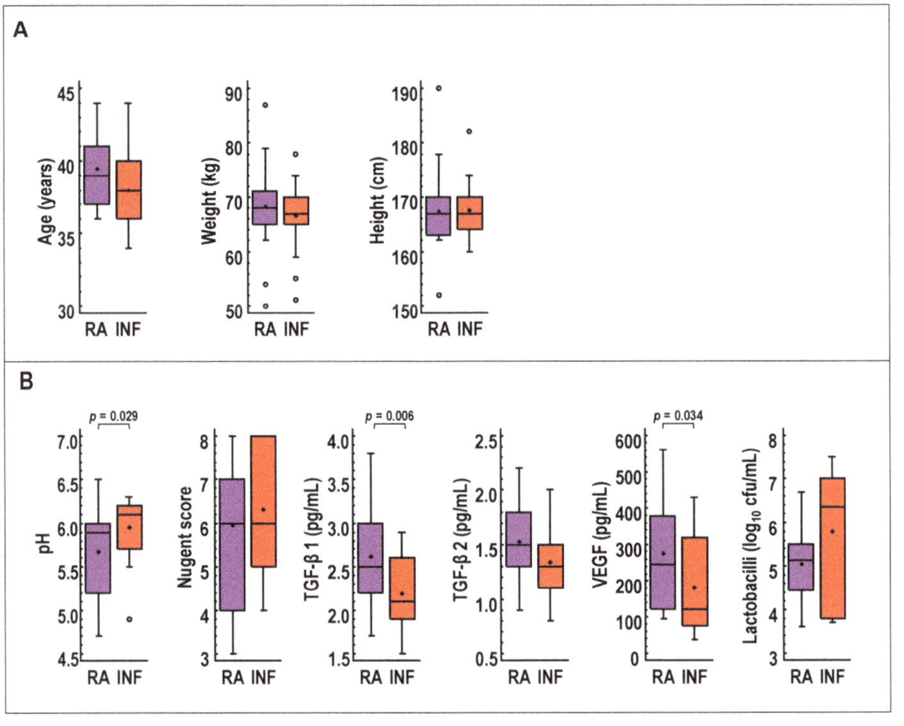


**Figure 2.** Comparison of selected baseline (**A**) demographic characteristics (age, weight and height) and (**B**) vaginal parameters (pH, Nugent score, TGF-β 1, TGF-β 2, and VEGF concentrations, and viable *Lactobacillus* counts) in CVL samples of women with repetitive abortion (RA, purple) and women with infertility of unknown origin (INF, red) at recruitment. For each boxplot, the line and the cross within the box represent the median and mean, respectively. The bottom and top boundaries of each box indicate the first and third quartiles (the 25th and 75th percentiles), respectively. The whiskers represent the lowest and highest values within the 1.5 interquartile range (IQR) and the dots outside the rectangles are suspected outliers (>1.5 × IQR). One-way ANOVA tests were used to compare both groups.

The 16S rRNA gene sequencing analysis of the CVL samples ($n = 58$) yielded 4,363,364 high quality filtered sequences, ranging from 33,160 to 139,044 per sample (median [IQR] = 73,383 [66,587–82,821] sequences per sample). Sequences were assigned to a total of 23 phyla and 453 genera, and Figure 3 shows the 5 most abundant phyla and the 20 most abundant genera in CVL samples from the fertile control group and from the RA and INF groups. The comparison of the relative abundance (% of total) of sequences at the phylum level from the three groups revealed statistically significant differences with regard to the 4 dominant phyla: *Firmicutes*, *Actinobacteria*, *Proteobacteria*, and *Bacteroidetes* (Table 3). The most frequent (present in all samples) and abundant phylum was *Firmicutes* (Figure 3). The relative abundance of *Firmicutes* in samples provided by fertile controls (median [IQR] = 99.60% [99.18–99.80%]) was higher than in samples from women of RA and INF groups (median [IQR] = 97.29% [72.34–99.35%] and 89.96% [52.46–98.85%], respectively) ($p < 0.001$; Kruskal–Wallis rank test with Bonferroni correction) (Table 3). In contrast, the median (IQR) values of the relative abundance of *Actinobacteria*, *Proteobacteria*, and *Bacteroidetes* were higher in women of the RA and INF groups ($p < 0.012$, $p < 0.003$, and $p < 0.006$, respectively; Kruskal–Wallis rank tests with Bonferroni correction) (Table 3).

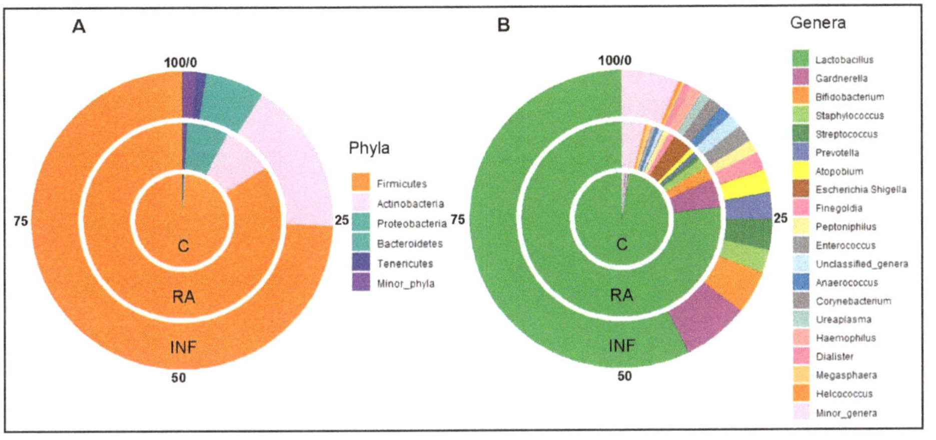


**Figure 3.** Pie charts showing the percentages of the relative abundances of the 5 most abundant phyla (**A**) and the 20 most abundant genera (**B**) in the CVL samples from healthy fertile women (inner pie charts; C group), women with a history of repetitive abortion (middle pie charts; RA group), and women with infertility of unknown origin (outer pie charts; INF group).

The only bacterial genus that was detected in all samples was *Lactobacillus*, but there were significant differences in its relative abundance in samples from the three groups (Table 3; Figure 3). The median [IQR] relative abundance of *Lactobacillus* in CVL samples from women of RA and INF groups (93.49% [67.18–97.53%] and 71.95% [0.76–94.09%], respectively) was lower than in samples from fertile control women (97.88% [96.92–99.31%]) ($p = 0.001$; Kruskal–Wallis rank test with Bonferroni correction) (Table 3). In fact, the only bacterial genus that characterized and differentially explained the greatest difference between the microbial communities in CVL samples between fertile control women and women of RA and INF groups was *Lactobacillus*, according to the LEfSe analysis (Figure 4).

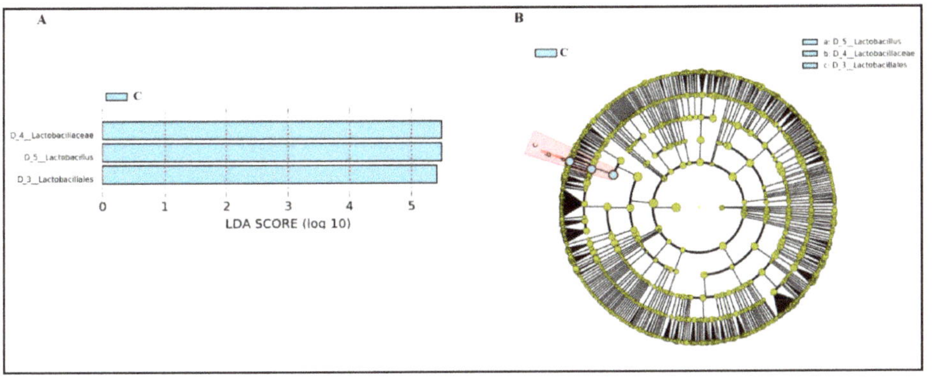


**Figure 4.** LEfSe analysis identifying taxonomic differences in the microbiota of CVL samples from healthy fertile women (C, bluish green) and women with repetitive abortion (RA) and with infertility of unknown origin (INF). Differentially abundant bacterial taxa were identified using linear discriminant analysis (LDA) and the effect size (LEfSe) algorithm. (**A**) Histogram of LDA scores (absolute LDA ($\log_{10}$) score > 2.0, $p < 0.05$) showing the substantial enrichment of *Lactobacillus* in the microbiota profile of the CVL samples from healthy fertile women. (**B**) Cladogram showing LEfSe comparison of differential bacterial taxa in CVL samples. The central point represents the root of the bacterial tree and each ring the next lower taxonomic level from phylum to genus (from the inner to the outer ring: phylum, class, order, family, and genus). The color node (other than yellow) indicates which taxa are significantly higher in relative abundance, and the diameter of the node is proportional to the relative abundance of the taxon.

**Table 3.** Relative frequencies, medians and interquartile range (IQR) of the most abundant bacterial phyla and genera detected in CVL samples from fertile women (Control group), women with a history of repetitive abortion (RA group), and women with infertility of unknown origin (INF group).

| | Control (*n* = 14) | | RA (*n* = 21) | | INF (*n* = 23) | | |
|---|---|---|---|---|---|---|---|
| **Phylum** Genus | *n* (%) [1] | Median (IQR) | *n* (%) | Median (IQR) | *n* (%) | Median (IQR) | *p*-Value [2] |
| ***Firmicutes*** | 14 (100) | 99.60 (99.18–99.80) | 21 (100) | 97.29 (72.34–99.35) | 23 (100) | 89.96 (52.46–98.85) | 0.001 |
| *Lactobacillus* | 14 (100) | 97.88 (96.92–99.31) | 21 (100) | 93.49 (67.18–97.53) | 23 (100) | 71.95 (0.76–94.09) | 0.001 |
| *Staphylococcus* | 13 (93) | 0.31 (0.11–0.66) | 19 (90) | 0.45 (0.03–1.51) | 22 (96) | 0.75 (0.14–5.40) | 0.260 |
| *Streptococcus* | 9 (64) | 0.02 (<0.01–0.03) | 14 (67) | 0.01 (<0.01–0.34) | 16 (70) | 0.06 (<0.01–2.04) | 0.180 |
| *Finegoldia* | 13 (93) | 0.17 (0.03–0.28) | 18 (86) | 0.16 (0.07–0.61) | 17 (74) | 0.12 (0.02–1.24) | 0.760 |
| *Peptoniphilus* | 11 (79) | 0.06 (0.01–0.21) | 16 (76) | 0.10 (0.02–0.49) | 17 (74) | 0.09 (<0.01–1.45) | 0.670 |
| *Enterococcus* | 2 (14) | <0.01 (<0.01–<0.01) | 6 (29) | <0.01 (<0.01–0.04) | 12 (52) | 0.01 (<0.01–0.19) | 0.044 |
| *Anaerococcus* | 11 (79) | 0.03 (0.01–0.16) | 18 (86) | 0.10 (0.05–0.30) | 18 (78) | 0.12 (0.01–1.71) | 0.220 |
| ***Actinobacteria*** | 12 (86) | 0.09 (0.02–0.20) | 21 (100) | 0.32 (0.08–7.87) | 23 (100) | 4.84 (0.1–34.36) | 0.012 |
| *Gardnerella* | 4 (29) | <0.01 (<0.01–0.01) | 11 (52) | 0.01 (<0.01–0.12) | 9 (39) | <0.01 (<0.01–0.04) | 0.300 |
| *Bifidobacterium* | 3 (21) | <0.01 (<0.01–<0.01) | 9 (43) | <0.01 (<0.01–0.07) | 9 (39) | <0.01 (<0.01–0.03) | 0.300 |
| *Atopobium* | 2 (14) | <0.01 (<0.01–<0.01) | 7 (33) | <0.01 (<0.01–0.01) | 13 (57) | 0.02 (<0.01–0.12) | 0.015 |
| ***Proteobacteria*** | 1 (93) | 0.07 (0.02–0.10) | 21 (100) | 0.28 (0.09–0.69) | 22 (96) | 0.23 (0.09–0.64) | 0.003 |
| *Escherichia/Shigella* | 1 (7) | <0.01 (<0.01–<0.01) | 9 (43) | <0.01 (<0.01–0.02) | 8 (35) | <0.01 (<0.01–0.01) | 0.084 |
| ***Bacteroidetes*** | 10 (71) | 0.03 (<0.01–0.08) | 18 (86) | 0.16 (0.06–1.33) | 22 (96) | 0.80 (0.05–3.19) | 0.006 |
| *Prevotella* | 8 (57) | 0.02 (<0.01–0.08) | 15 (71) | 0.06 (<0.01–0.45) | 19 (83) | 0.70 (0.01–2.55) | 0.660 |
| *Tenericutes* | 6 (43) | <0.01 (<0.01–0.16) | 5 (24) | <0.01 (<0.01–<0.01) | 10 (43) | <0.01 (<0.01–0.97) | 0.290 |
| **Minor phyla** | 14 (100) | 0.13 (0.07–0.18) | 21 (100) | 0.16 (0.07–0.65) | 23 (100) | 0.17 (0.09–1.29) | 0.280 |
| Minor genera | 14 (100) | 0.30 (0.09–0.70) | 21 (100) | 0.91 (0.27–2.54) | 23 (100) | 2.26 (0.40–8.35) | 0.038 |
| Unclassified_genera | 14 (100) | 0.09 (0.05–0.12) | 21 (100) | 0.13 (0.07–0.66) | 23 (100) | 0.14 (0.04–0.36) | 0.170 |

[1] *n* (%): Number of samples in which the phylum/genus was detected (relative frequency of detection). [2] Kruskal–Wallis rank tests with Bonferroni correction.

Other genera were present in a variable number of samples, ranging from 96% (*Staphylococcus* in the INF group) to 7% (*Escherichia/Shigella* in the control group), but the median relative abundance of any of these genera was <1% (Table 3). The bacterial profile at the genus level in some individual samples from women in the RA and INF groups did not differ from that of samples from women from the fertile control group, which were highly homogenous (Figure 5). However, aberrant profiles with reduced content or even complete absence of *Lactobacillus* were registered in some samples from women of the RA and INF groups (Figure 5).

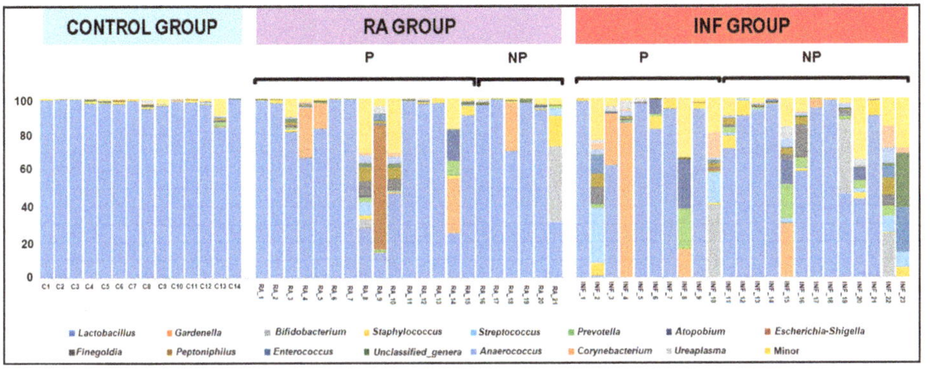


**Figure 5.** Relative abundance of the predominant bacterial genera in CVL samples of healthy fertile women (C), women with repetitive abortion (RA) and women with infertility of unknown origin (INF). In women with a history of reproductive failure, because either of recurrent miscarriage (RA group) or infertility (INF groups), P indicates the group of women who got pregnant after the probiotic intervention with *L. salivarius* CECT5713 and NP those women who did not.

The analysis of alpha diversity at the genus level, calculated either by the Shannon or the Simpson's indices, revealed significant differences between the vaginal microbiota of women in the fertile and INF groups ($p < 0.001$; Kruskal–Wallis tests with Bonferroni correction) (Figure 6A,B).

The analysis of the beta diversity, calculated according to the relative abundance of bacterial genera (Bray–Curtis distance) and the presence/absence of bacterial genera (Binnary Jaccard distance matrix), indicated that the profiles of bacterial genera of CVL samples of the 3 groups clustered apart ($p$ = 0.004 and $p$ = 0.002, respectively; PERMANOVA) (Figure 6C,D). In addition, samples from fertile controls clustered closer (shorter distance to centroid) according to the relative abundance of bacterial genera (Bray–Curtis distance) than those from RA and INF groups, indicating that the bacterial profiles in CVL samples from controls were highly uniform (Figure 6E,F).

An initial assessment of potentially dominant patterns in the bacteriological profile of the CVL samples is shown in the heatmap plot presented in Figure 6G. There was a clear separation of samples based on the presence of *Lactobacillus*. One cluster was characterized by the marked and almost exclusively presence of *Lactobacillus* in CVL samples. This cluster comprised all the samples from fertile women although not exclusively, because it included also some samples from the RA and INF groups. The second cluster was characterized by the absence or reduced presence of *Lactobacillus* and the presence of multiple bacterial genera, such as *Gardenella* and *Bifidobacterium*. This second cluster contained exclusively CVL samples from the RA and INF groups. Although globally there was no clear separation between the CVL samples from the three groups, it was perceived a higher similarity between samples from the fertile control group and women with a history of repetitive abortion than between the fertile control group and women with infertility of unknown origin (Figure 6G).

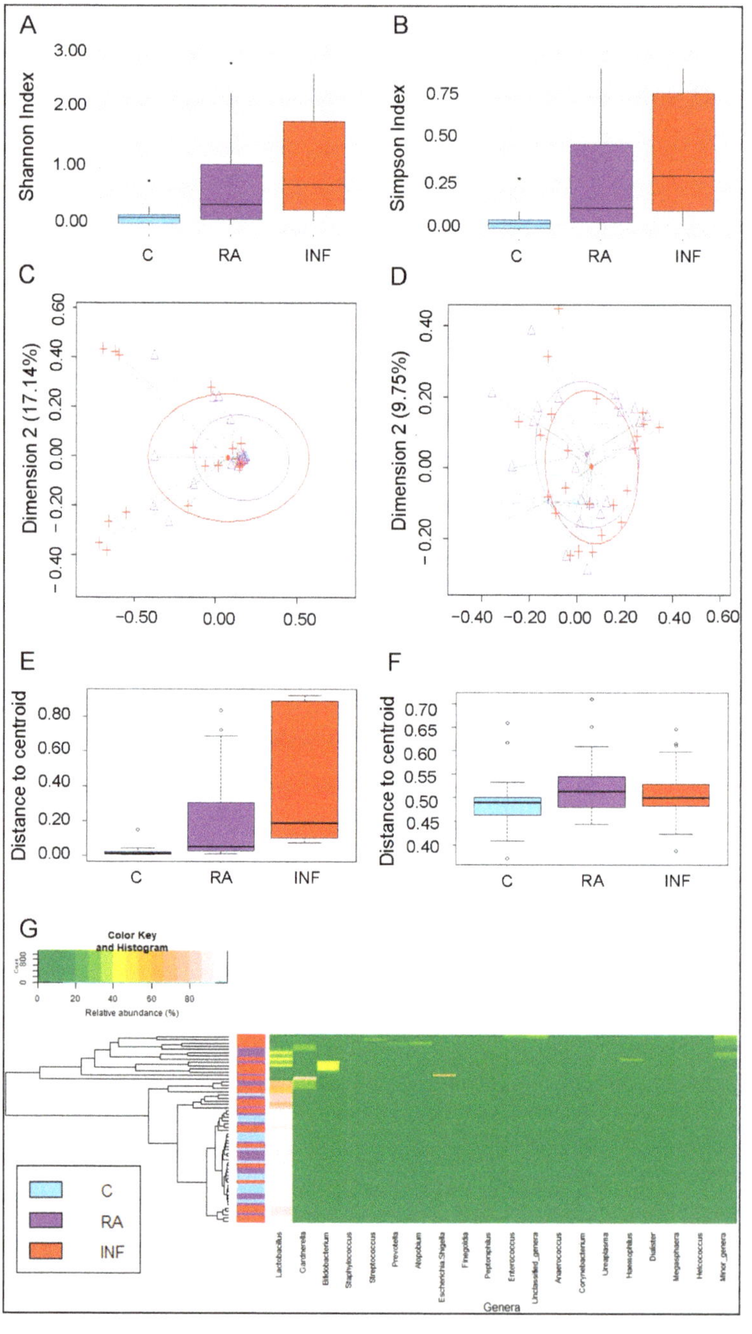


**Figure 6.** Metataxonomic profiles of CVL samples of healthy fertile women (C; bluish green), women with repetitive abortion (RA; purple) and women with infertility of unknown origin (INF; red). (**A**) Comparison of alpha diversity at genus level calculated using the Shannon index between the three groups of women. (**B**) Comparison of alpha diversity at genus level calculated using the Simpson index between the three groups of women. (**C**) Principal coordinate analysis (PCoA)

plots of bacterial profiles at the genus level based on the Bray–Curtis dissimilarity analysis (relative abundance). (**D**) Principal coordinate analysis (PCoA) plots of bacterial profiles at the genus level based on the Jaccard's coefficient for binary data (presence or absence). The values on each axis label in graphs C and D represent the percentage of the total variance explained by that axis. The differences between groups of CVL samples were analyzed using the PERMANOVA test with 999 permutations. (**E**) Comparison of the mean distances of samples to the centroids in the PCoA plots based on the Bray–Curtis dissimilarity index in each group. (**F**) Comparison of the mean distances of samples to the centroids in the PCoA plots based on the Jaccard's coefficient (graph D) in each group. (**G**) Heatmap showing the relative abundance of the 20 most abundant bacterial genera (x axis) detected in CVL samples. The relative abundance of each bacterial genus within each sample is indicated by the color of the scale ranging from white (high relative abundance) to green (low relative abundance) as indicated in the scale shown at the left down corner. Dendrogram linkages are based upon relative abundance of the genus within the samples and *hclust* was used as the clustering algorithm. The column between the dendrogram of the vaginal samples and the individual values of the relative abundance of bacterial genera indicates the study group (control fertile women: C, in bluish green; women with repetitive abortion: RA, in purple; women with infertility of unknown origin: INF, in red). The differences between groups (C, healthy fertile women; RA, women with repetitive abortion; INF, women with infertility of unknown origin) were analyzed using Kruskal–Wallis tests with Bonferroni correction for data in panels A and B, and with one-way ANOVA tests for data in panels E and F.

### 3.4. Main Outcome of the Clinical Trial: Pregnancies and Successful Pregnancies

Administration of *L. salivarius* CECT5713 (~9 $\log_{10}$ CFU/day) for 6 months (or until a diagnosis of pregnancy if this happened first) to the women of the RA and INF groups led to 29 pregnancies out of the 44 participating patients. This means a pregnancy effectiveness of 66% with a 95% CI of 52–80% (Table 4). Among them, there were 25 successful pregnancies and 4 abortions. This means an effectiveness for reproductive success of 57% with a 95% CI of 42–72% (Table 4). Interestingly, all successful pregnancies led to full-term singletons (gestational age $\geq$ 38 weeks).

**Table 4.** Main outcomes after the probiotic treatment with *L. salivarius* CECT5713 in women with repetitive abortion (RA) and women with infertility of unknown origin (INF).

| | Group | | Total | Ratio (95% CI) |
|---|---|---|---|---|
| **Outcome** | **RA** | **INF** | **RA + INF** | **(RA/INF)** |
| Pregnancy (events/total events) | 17/21 | 12/23 | 29/44 | |
| Pregnancy effectiveness (95% CI) | 81% (64–98%) | 52% (32–73%) | 66% (52–80%) | 1.55 (1.00–2.42) |
| Successful pregnancy [1] (events/total events) | 15/21 | 10/23 | 25/44 | |
| Reproductive success (95% CI) | 71% (52–91%) | 43% (23–64%) | 57% (42–72%) | 1.64 (0.96–2.82) |

[1] Two women in each group end up in abortion.

Women of the RA group had the highest rate of reproductive success (15 full term pregnancies and 2 abortions out of 21 participants) (Table 4). The rate in the INF group was lower although still noticeable: 12 pregnancies (10 full term and 2 abortions) out of 23 enrolled. Therefore, the pregnancy effectiveness and successful pregnancy rates (95% CI) tended to be higher in RA group that in INF group (RR [95% CI] = 1.55 [1.00–2.42] and 1.64 [0.96–2.82], respectively), although the difference between both groups did not reach statistical significance (Table 4). It must be highlighted that all women of these groups had been unsuccessfully subjected to ART interventions in previous attempts to avoid spontaneous miscarriage (RA group) or to get pregnant (INF group).

### 3.5. Secondary Outcomes Associated with the Probiotic Treatment: RA Group

There were no differences in age, weight, or height between women in the RA group that ended up having a successful pregnancy ($n$ = 15) and those who did not ($n$ = 6) after the probiotic intervention. However, differential changes in their vaginal parameters were observed (Table 5). The vaginal pH of women who delivered was about 0.9 units lower than in those who did not ($p < 0.001$; one-way ANOVA). Similar results were noted for the Nugent score (a mean [95% CI]) reduction of 3.33 [3.73–2.93] units in women who got pregnant after the probiotic intervention versus a mean [95% CI] reduction of 0.67 [1.29–0.04] units in those who did not complete a full-term pregnancy; $p = 0.000$ one-way ANOVA) (Table 5, Supplementary Figure S2). In fact, the probiotic treatment did not modify the Nugent score in those women that did not get pregnant ($p = 0.102$; one-way repeated measures ANOVA) (Table 5).

**Table 5.** Effect of the probiotic intervention with *L. salivarius* CECT5713 on the vaginal parameters of women who were able to complete a full-term pregnancy ($n = 15$) and of those who did not ($n = 6$) among the women that had a history of repetitive abortion (RA group; $n = 21$).

| Vaginal Parameter | Probiotic Intervention Resulted in Pregnancy<br>Yes ($n = 15$)<br>Mean (95% CI) | No ($n = 6$)<br>Mean (95% CI) | $p$-Value [1] |
|---|---|---|---|
| **pH** | | | |
| Baseline | 5.58 (5.39–5.77) | 5.88 (5.58–6.18) | 0.221 |
| Post-intervention | 4.45 (4.34–4.57) | 5.65 (0.13–5.46) | 0.000 |
| Change | −1.13 (−1.27–−0.99) | −0.23 (−0.45–−0.01) | <0.001 |
| $p$-value [3] | 0.000 | 0.002 | |
| **Nugent score** | | | |
| Baseline | 5.87 (5.24–6.49) | 6.17 (5.18–7.15) | 0.708 |
| Post-intervention | 2.53 (2.13–2.94) | 5.50 (4.86–6.14) | 0.000 |
| Change | −3.33 (−3.73–−2.93) | −0.67 (−1.29–−0.04) | 0.000 |
| $p$-value [3] | 0.000 | 0.102 | |
| **TGF-β 1**, pg/mL | | | |
| Baseline | 2.81 (2.62–3.00) | 2.15 (1.85–2.45) | 0.014 |
| Post-intervention | 4.21 (4.05–4.36) | 2.47 (2.22–2.71) | 0.000 |
| Change | 1.40 (1.18–1.62) | 0.32 (−0.02–0.66) | <0.001 |
| $p$-value [3] | 0.000 | 0.098 | |
| **TGF-β 2**, pg/mL | | | |
| Baseline | 1.67 (1.57–1.78) | 1.15 (0.99–1.31) | <0.001 |
| Post-intervention | 2.93 (2.81–3.05) | 1.30 (1.11–1.49) | 0.000 |
| Change | 1.25 (1.12–1.38) | 0.15 (−0.05–0.35) | 0.000 |
| $p$-value [3] | 0.000 | 0.328 | |
| **VEGF**, pg/mL | | | |
| Baseline | 341 (300–382) | 109 (44–173) | <0.001 |
| Post-intervention | 743 (640–846) | 138 (−25–301) | <0.001 |
| Change | 402 (319–485) | 29 (−102–160) | 0.002 |
| $p$-value [3] | 0.000 | 0.189 | |
| **Lactobacilli presence**, *n* (%) | | | |
| Baseline | 9 (60) | 3 (50) | 0.523 [2] |
| Post-intervention | 15 (100) | 4 (67) | 0.071 [2] |
| Change | 6 (40) | 1 (17) | 0.613 [2] |
| **Lactobacilli counts**, $log_{10}$ CFU/mL | | | |
| Initial | 4.99 (4.48–5.50) | 5.20 (4.31–6.09) | 0.752 |
| Final | 6.52 (6.22–6.81) | 4.74 (4.17–5.31) | <0.001 |
| Change | 2.44 (1.84–3.04) | 0.16 (−0.99–1.32) | 0.019 |
| $p$-value [3] | <0.001 | 0.697 | |

**Table 5.** *Cont.*

| Vaginal Parameter | Probiotic Intervention Resulted in Pregnancy<br>Yes ($n$ = 15)<br>Mean (95% CI) | No ($n$ = 6)<br>Mean (95% CI) | $p$-Value [1] |
|---|---|---|---|
| ***L. salivarius*** **qPCR**, $n$ (%) | | | |
| Initial | nd | nd | |
| Final | 15 (100) | 3 (50) | 0.015 [2] |
| ***L. salivarius*** **qPCR**, $\log_{10}$ copies/mL [4] | | | |
| Initial | - | - | |
| Final | 6.85 (6.58–7.12) | 2.63 (0.41–3.24) | <0.000 |

[1] One-way ANOVA tests were used to evaluate differences in mean values between groups, except for lactobacilli presence. [2] Fisher exact probability test for a 2 × 2 contingency table. [3] One-way repeated measures ANOVA tests were used to determine whether there was a change in each group of participants when comparing the baseline and post-intervention parameters. [4] Mean (95% CI) of *L. salivarius* qPCR (copies/mL) in positive samples.

The vaginal cytokine concentrations also differed in both subgroups of women (with successful pregnancy or not) in the RA group after the probiotic treatment. There was no modification in the vaginal TGF-β 1, TGF-β 2, and VEGF concentrations with respect to the baseline in the women who did not become pregnant, but there was a mean (95% CI) significant increase of 1.40 (1.18–1.62) pg/mL, 1.25 (1.12–1.38) pg/mL, and 402 (319–485) pg/mL, respectively, in those who did ($p$ = 0.000; one-way repeated measures ANOVA) (Table 5, Supplementary Figure S2). In addition, it should be noted that there were already differences in the concentration of these cytokines even before starting the treatment between those that became and those that did not become pregnant (Table 5).

On the other hand, the probiotic treatment resulted in a mean (95% CI) increase in lactobacilli counts of 2.12 (1.66–2.59) $\log_{10}$ CFU/mL in women that finally got pregnant, but there was no change in those that did not (Table 5, Supplementary Figure S2). The presence of *L. salivarius* (mean [95% CI] = 6.85 [6.58–7.12] $\log_{10}$ copies/mL) was confirmed by qPCR in all women that got pregnant, but only in 50% of the women with unsuccessful pregnancies, their concentration being significantly lower (mean [95% CI] = 2.63 [0.41–3.24] copies/mL) (Table 5). The lactobacilli profile in CVL samples obtained at the beginning of the probiotic treatment and after 6 months or until a diagnosis of pregnancy is presented in Figure 7. The most noticeable difference was the presence of viable *L. salivarius* in most women (17/21) after the probiotic treatment. In addition, *L. iners*, which was present in 3 women at the beginning of the study, was isolated at the end of the treatment only from 2 women who did not end up in pregnancy. There were no differences in the metataxonomic profile at the genus level of CVL samples from women of the RA group regarding the pregnancy outcome (Figure 5; Supplementary Table S1).

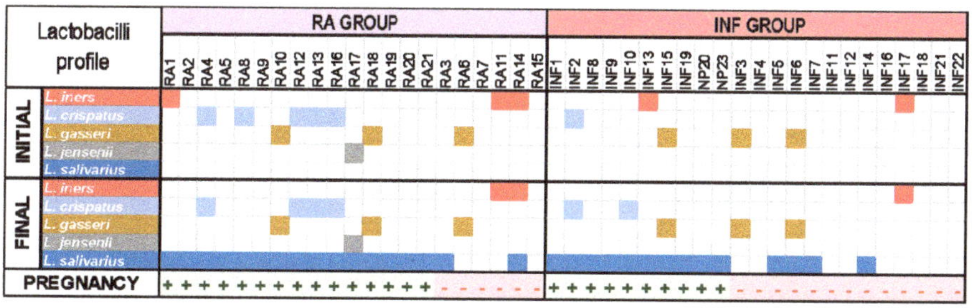


**Figure 7.** Changes in the profile of dominant *Lactobacillus* species in CVL samples from women with a history of repetitive abortion (RA group) and women with infertility of unknown origin (INF group) after the probiotic intervention with *L. salivarius* CECT5713. The outcome is indicated in the last file: +, successful full-term pregnancy and -, no pregnancy. The presence of isolates from a given species is indicated by a colored square.

### 3.6. Secondary Outcomes Associated with the Probiotic Treatment: INF Group

The women in the INF group that got pregnant after the probiotic intervention ($n$ = 10) and those who did not ($n$ = 13) did not differ in age, weight and height. The CVL pH and the Nugent score decreased significantly in all members of the INF group after the probiotic treatment ($p < 0.05$; one-way repeated measures ANOVA), although the magnitude of the change was smaller in the women that did not get pregnant when compared to those that got pregnant (Table 6; Supplementary Figure S2). Specifically, the mean (95% CI) reductions in CVL pH and Nugent score in women that got pregnant were −1.32 (−1.43–−1.21) and −3.90 (−4.25–−3.55), respectively, and in women that did not get pregnancy these reductions were only −0.19 (−0.29–−0.09) and −0.54 (−0.85–−0.23), respectively (Table 6; Supplementary Figure S2).

**Table 6.** Effect of the probiotic intervention with *L. salivarius* CECT5713 on the vaginal parameters of women who were able to complete a full-term pregnancy ($n$ = 15) and of those who did not ($n$ = 6) among the women with infertility of unknown origin (INF group; $n$ = 23).

| Vaginal Parameter | Probiotic Intervention Resulted in Pregnancy Yes ($n$ = 15) Mean (95% CI) | No ($n$ = 6) Mean (95% CI) | $p$-Value [1] |
|---|---|---|---|
| **pH** | | | |
| Baseline | 5.85 (5.70–6.00) | 6.04 (5.58–6.17) | 0.190 |
| Post-intervention | 4.53 (4.42–4.64) | 5.85 (5.75–5.95) | 0.000 |
| Change | −1.32 (−1.43–−1.21) | −0.19 (−0.29–−0.09) | 0.000 |
| $p$-value [3] | 0.000 | 0.002 | |
| **Nugent score** | | | |
| Baseline | 6.00 (5.40–6.60) | 6.54 (6.01–7.07) | 0.334 |
| Post-intervention | 2.10 (1.61–2.59) | 6.00 (5.57–6.43) | 0.000 |
| Change | −3.90 (−4.25–−3.55) | −0.54 (−0.85–−0.23) | 0.000 |
| $p$-value [3] | 0.000 | 0.028 | |
| **TGF-β 1**, pg/mL | | | |
| Baseline | 2.29 (2.10–2.48) | 2.11 (1.94–2.28) | 0.308 |
| Post-intervention | 4.58 (4.41–4.75) | 2.18 (2.04–2.33) | 0.000 |
| Change | 2.29 (2.16–2.42) | 0.08 (−0.04–0.19) | 0.000 |
| $p$-value [3] | 0.000 | 0.281 | |
| **TGF-β 2**, pg/mL | | | |
| Baseline | 1.56 (1.46–1.66) | 1.16 (1.07–1.25) | <0.001 |
| Post-intervention | 2.81 (2.68–2.94) | 1.26 (1.15–1.38) | 0.000 |
| Change | 1.25 (1.13–1.37) | 0.10 (<−0.01–0.20) | 0.000 |
| $p$-value [3] | 0.000 | 0.203 | |
| **VEGF**, pg/mL | | | |
| Baseline | 311 (279–343) | 81 (53–109) | 0.000 |
| Post-intervention | 773 (695–850) | 87 (19–155) | 0.000 |
| Change | 462 (411–513) | 6 (−39–50) | 0.000 |
| $p$-value [3] | 0.000 | 0.165 | |
| **Lactobacilli presence**, $n$ (%) | | | |
| Baseline | 3 (30) | 3 (23) | 0.537 [2] |
| Post-intervention | 10 (100) | 6 (46) | 0.007 [2] |
| Change | 7 (70) | 3 (23) | 0.040 [2] |

**Table 6.** *Cont.*

| Vaginal Parameter | Probiotic Intervention Resulted in Pregnancy<br>Yes ($n$ = 15)<br>Mmean (95% CI) | No ($n$ = 6)<br>Mean (95% CI) | $p$-Value [1] |
|---|---|---|---|
| **Lactobacilli counts**, $\log_{10}$ CFU/mL | | | |
| Initial | 5.00 (3.22–6.78) | 6.57 (4.78–8.35) | 0.290 |
| Final | 6.46 (5.94–6.98) | 4.95 (4.28–5.62) | 0.017 |
| Change | 3.05 (2.45–3.64) | 0.32 (−0.46–1.09) | <0.001 |
| $p$-value [3] | <0.001 | 0.451 | |
| ***L. salivarius* qPCR**, $n$ (%) | | | |
| Initial | nd | nd | |
| Final | 10 (100) | 4 (31) | 0.002 [2] |
| ***L. salivarius* qPCR**, $\log_{10}$ copies/mL [4] | | | |
| Initial | - | - | |
| Final | 6.48 (6.28–6.68) | 3.55 (3.24–3.86) | 0.000 |

[1] One-way ANOVA tests were used to evaluate differences in mean values between groups, except for lactobacilli presence. [2] Fisher exact probability test for a 2 × 2 contingency table. [3] One-way repeated measures ANOVA tests were used to determine whether there was a change in each group of participants when comparing the baseline and post-intervention parameters. [4] Mean (95% CI) of *L. salivarius* qPCR (copies/mL) in positive samples.

The change in the vaginal cytokine concentrations after the probiotic treatment was similar to that described in the RA group: There was no modification in the vaginal TGF-β 1, TGF-β 2, and VEGF levels of women who did not become pregnant, but there was a mean (95% CI) significant increase of 2.29 (2.16–2.42) pg/mL, 1.25 (1.13–1.37) pg/mL, and 462 (411–513) pg/mL, respectively, in those who did (Table 6; Supplementary Figure S2). In this INF group, there were already differences in the concentrations of TGF-β 2 and VEGF, but not in that of TGF-β 1, between those that became and those that did not became pregnant even before starting the treatment (Table 6).

The probiotic intervention resulted in a high degree of vaginal colonization by lactobacilli (6.46 [5.94–6.98] $\log_{10}$ CFU/mL) of all women that got pregnant, while this only happened in 46% of those that experienced a treatment failure, the density of lactobacilli reached being significantly lower (4.95 [4.28–5.62] $\log_{10}$ CFU/mL) (Table 6). Similarly to the RA group, the presence of *L. salivarius* (mean [95% CI] = 6.48 [6.28–6.68] copies/mL) was confirmed by qPCR in all women that got pregnant, but only in 31% of the women with unsuccessful pregnancies and, then, at a lower concentration (mean [95% CI] = 3.55 [3.24–3.86] copies/mL) (Table 6). The main difference in the lactobacilli profile of CVL samples of women in the INF group registered after the probiotic intervention was the detection of viable *L. salivarius* in all women who got pregnant, but only in 4 out of 13 of those women that failed to get pregnant. There were no differences in the metataxonomic profile at the genus level of CVL samples from women of the RA group regarding the pregnancy outcome (Figure 5; Supplementary Table S2).

### 3.7. Comparison of Vaginal Parameters between Women Who Became Pregnant and Those Who Did Not from Both the RA and INF Groups

The mean [95% CI] pH value in CVL samples was slightly but significantly more acidic in the women who become pregnant (5.69 [5.57–5.81] units) than in those who did not (5.99 [5.85–6.13] units) ($p$ = 0.024; one-way ANOVA) (Figure 8; Supplementary Table S3). There were also differences in the concentration of vaginal cytokines TGF-β 2 and VEFG at the beginning of the study according to the final pregnancy outcome, but the differences were similar to those described already separately for RA and INF groups (Figure 8; Supplementary Table S3). The only parameters that did not differed initially between both groups were the Nugent score, TGF-β 1 concentration, and the frequency of detection and counts of lactobacilli (Figure 8; Supplementary Table S3). Globally, *Lactobacillus* was detected in all women who became pregnant, but only in half of those that did not ($p$ < 0.001; Fisher exact probability test).

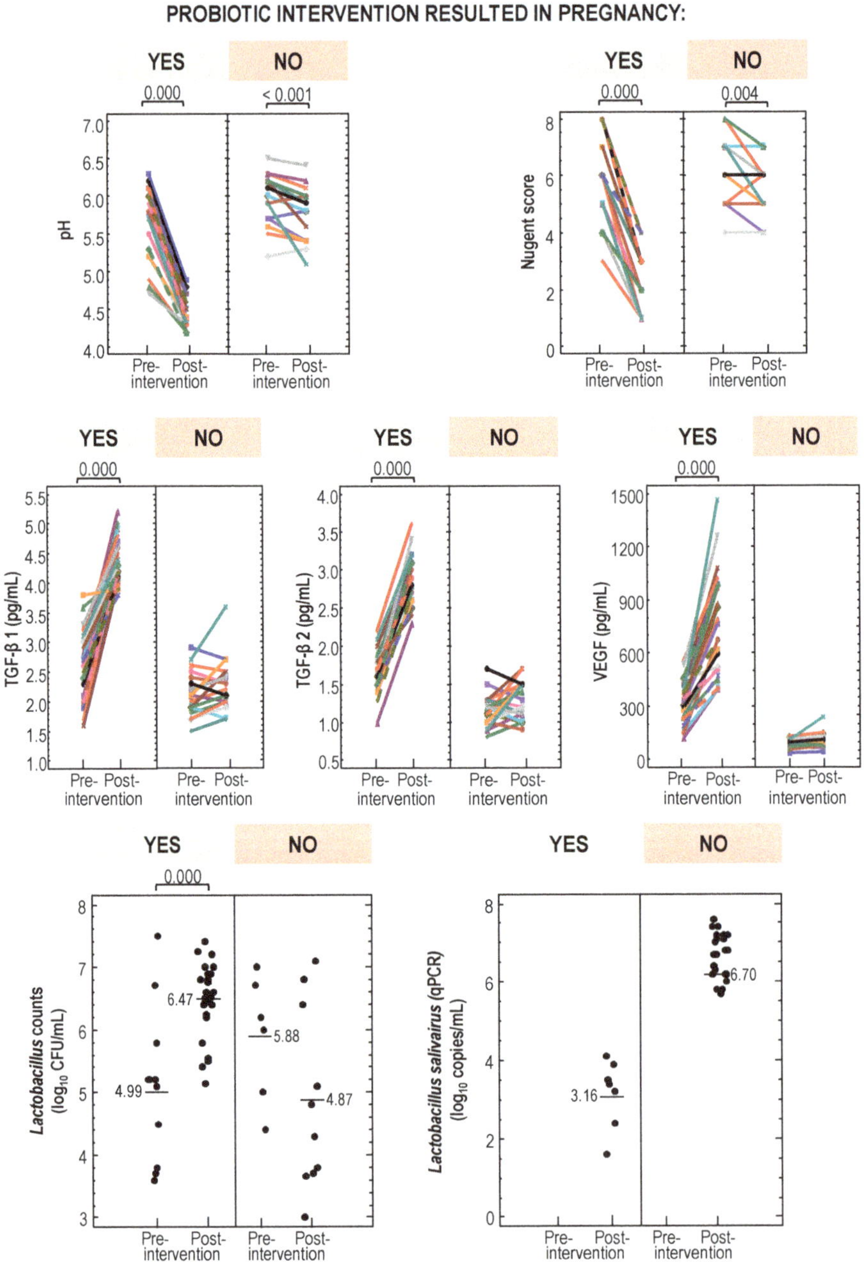


**Figure 8.** Changes in vaginal parameters (pH, Nugent score, TGF-β 1, TGF-β 2, and VEGF concentrations, viable *Lactobacillus* counts and *L. salivarius* copies in CVL samples) in women with a history of reproductive failure, because either of recurrent miscarriage (RA group) or infertility (INF groups), after the probiotic intervention with *L. salivarius* CECT5713 according to their outcome (pregnancy versus no pregnancy).

The probiotic intervention resulted in differential and remarkable changes in the vaginal parameters in those women who became pregnant but not in those who did not (Figure 8; Supplementary Table S3). First, the probiotic administration of *L. salivarius* CECT5713 resulted to be more effective regarding the change in the vaginal pH and Nugent score in women who got pregnant, which recorded mean (95% CI) decreases of −1.20 (−1.29–−1.12) and −3.56 (−3.82–−3.30) units, respectively ($p$ = 0.000; one-way repeated measures ANOVA). In contrast, the change in these two parameters was smaller (−0.21 (−0.31–−0.10) and −0.58 (−0.88–−0.28) units, respectively) in the group of women who did not get pregnant (Figure 8; Supplementary Table S3). Second, the probiotic intervention led to a significant increase in the concentrations of vaginal cytokines TGF-β 1, TGF-β 2 and VEFG (mean [95% CI] increase of 1.76 [1.60–1.91] pg/mL, 1.25 [1.17–1.33] pg/mL, and 426 [378–473] pg/mL, respectively) in women who got pregnant but no change was registered in the group that did not (Figure 8; Supplementary Table S3). Third, regarding the lactobacilli profile of CVL samples, there was a mean (95% CI) increase of 2.67 (2.26–3.08) $\log_{10}$ CFU/mL units in viable *Lactobacillus* counts after the probiotic treatment in the group of women who became pregnant as opposed to those that did not. Differences were also noted on the *L. salivarius* content in CVL samples. This lactobacilli species was detected, and at a high concentration (mean [95% CI] = 6.70 [6.52–6.89] $\log_{10}$ copies/mL), in CVL samples from all women having a successful pregnancy unlike women who did not become pregnant (Figure 8; Supplementary Table S3). The metataxonomic profile at the genus level of CVL samples from women of the INF group was equal in women that did or did not become pregnant, except for a slightly higher relative frequency of *Escherichia/Shighella* in women that got pregnant (Figure 5; Supplementary Table S4).

### 3.8. Comparison of Vaginal Parameters between Control Women, All Women Who Became Pregnant and Those Who Did Not from Both RA and INF Groups

The analysis of post-intervention vaginal parameters (pH, Nugent score, TGF-β 1, TGF-β 2, VEGF, lactobacilli counts) revealed that the pH value of CVL samples and Nugent score in women who became pregnant after the probiotic intervention were similar to those of fertile control women (Table 7; Supplementary Figure S3). The concentrations of TGF-β 1, TGF-β 2, and VEGF in post- intervention CVL samples of women who became pregnant were closer to those found in fertile control women, although statistically significant differences were found between them (Table 7; Supplementary Figure S3). Besides, it is remarkable to note that the post-intervention concentration of VEGF in women that became pregnant was about twice that registered in fertile control women (mean [95% CI] = 755.0 [637.1–872.5] pg/mL and 406.0 [322.0–490.0] pg/mL, respectively). There was a high interindividual variation in lactobacilli counts varying from undetectable (in 57% of the women who did not become pregnant) to 7.5 $\log_{10}$ CFU/mL in CVL samples of women who did not become pregnant after the probiotic intervention, but the mean [95% CI] value (4.87 [3.83–5.90] $\log_{10}$ CFU/mL) was lower than in samples of the other participants (Table 7; Supplementary Figure S3). There was less than 1 $\log_{10}$ CFU/mL difference between the lactobacilli viable counts in CVL samples of women who enjoyed a full term pregnancy after the probiotic intervention and those of fertile controls (mean [95% CI] = 6.47 [6.22–6.72] $\log_{10}$ CFU/mL and 7.24 [6.89–7.60] $\log_{10}$ CFU/mL, respectively) (Table 7; Supplementary Figure S3).

**Table 7.** Comparison of vaginal parameters (pH, Nugent score, TGF-β 1, TGF-β 2, and VEGF concentrations, and *Lactobacillus* counts) of all women who were able to complete a full-term pregnancy ($n$ = 25) and of those who did not ($n$ = 19) among all women with a history of repetitive abortion and with infertility of unknown origin (RA and INF groups) after the probiotic intervention with *L. salivarius* CECT5713 and vaginal parameters of fertile women (Control group; $n$ = 14).

| | | Probiotic Intervention Resulted in Pregnancy | | |
|---|---|---|---|---|
| **Vaginal Parameter** | **Control ($n$ = 14) Mean (95% CI)** | **Yes ($n$ = 25) Mean (95% CI)** | **No ($n$ = 23) Mean (95% CI)** | ***p*-Value** |
| pH | 4.53 (4.38–4.68) [a] | 4.48 (4.39–4.58) [a] | 5.78 (5.62–5.95) [b] | 0.000 [1] |
| Nugent score | 1.79 (1.27–2.30) [a] | 2.36 (1.92–2.80) [a] | 5.84 (5.35–6.33) [b] | 0.000 [1] |
| TGF-β 1, pg/mL | 4.83 (4.65–5.01) [a] | 4.36 (4.20–4.52) [b] | 2.27 (2.06–2.48) [c] | 0.000 [1] |
| TGF-β 2, pg/mL | 3.22 (3.10–3.34) [a] | 2.88 (2.75–3.01) [b] | 1.27 (1.15–1.40) [c] | 0.000 [1] |
| VEGF, pg/mL | 406.0 (322.0–490.0) [a] | 755.0 (637.1–872.5) [b] | 103.3 (82.4–124.1) [c] | 0.000 [1] |
| Lactobacilli | | | | |
| Positive women, $n$ (%) | 14 (100) | 25 (100) | 10 (43) | <0.001 [2] |
| Viable counts [3], $\log_{10}$ CFU/mL | 7.24 (6.89–7.60) [a] | 6.47 (6.22–6.72) [b] | 4.87 (3.83–5.90) [c] | 0.000 [1] |

[1] One-way ANOVA tests were used to evaluate differences in mean values between groups. Values followed by different superscript letters within the same row indicate statistically significant differences between groups according to Scheffé post hoc comparison tests. [2] Freeman–Halton extension of the Fisher exact probability tests for a 2 × 3 contingency table were used to compute the (two-tailed) probability of obtaining a distribution of values of lactobacilli positive women. [3] Mean (95% CI) of *L. salivarius* qPCR (copies/mL) in lactobacilli-positive women. TGF-β 1, transforming growth factor-β 1; TGF-β 2, transforming growth factor-β 2; VEGF, vascular endothelial growth factor.

Additionally, a network structure of the baseline vaginal bacterial genera communities on the three different groups of women (fertile controls, women who got pregnant after the probiotic intervention and women who did not get pregnant after the probiotic intervention) was constructed based on the genus-genus correlations (Figure 9). In the group of fertile women, the strongest correlation was observed between two minority genera, *Escherichia/Shigella* and *Enterococcus*; the most abundant genera, *Lactobacillus*, established negative and weak relationship with other Firmicutes (*Finegoldia* and *Peptoniphilus*) and *Prevotella*. In contrast, in the group of women with either repeated abortions or infertility of unknown origin, *Lactobacillus* showed strong negative association with two genera of the *Actinobacteria*, *Gardenella*, and *Bifidobacterium*. However, in the group of women that responded to the probiotic intervention and ended up in a successful pregnancy, the strongest negative association was between *Lactobacillus* and *Gardenella*, while in those women that did not get pregnant this negative association was weaker than that registered between *Lactobacillus* and *Bifidobacterium*, indicating that indeed the bacterial profile in CVL samples may indicate different fertility problems (Figure 9).

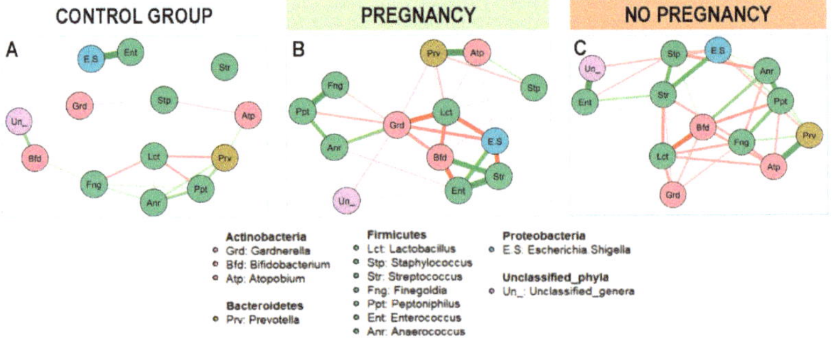


**Figure 9.** Estimated network structures based on a sample of 58 vaginal samples: 14 from healthy fertile women (**A**, Control group), 25 from women with a successful reproductive outcome after the probiotic intervention *L. salivarius* CECT5713 (**B**, Pregnancy) and 19 from women who did not have a successful pregnancy after the probiotic intervention with *L. salivarius* CECT5713 (**C**, No pregnancy). The 14 most abundant genera were represented. Red lines indicate negative correlation and green lines indicates positive correlation. The thickness and the intensity of the line reflects the intensity of the correlation.

## 4. Discussion

In this study, the comparison between the vaginal microbiota of women with a history of reproductive failure, due to recurrent miscarriage or infertility, and healthy fertile women confirmed that dominance of specific species of *Lactobacillus* in the vaginal microbiota plays a determinant role in the success of human reproduction. Overall, the lowest vaginal pH values and Nugent scores were associated with vaginal communities dominated by lactobacilli, while those with the highest pH values and Nugent scores were associated with a depletion of lactobacilli. Close associations between low pH, low Nugent score and a high concentration and dominance of lactobacilli in the human vagina has been repeatedly reported [3,4,65]. In this study, the frequency of detection of lactobacilli in the vaginal samples was much higher in fertile women (100%) than in women with repetitive miscarriage (57%). Interestingly, infertile women showed the lowest percentage of women from whom lactobacilli could be isolated (26%). Use of antibiotics in both infancy and adulthood was significantly higher among women of the RA and INF groups than among women of the control group. It has been long known that opportunistic vaginal infections may arise as an adverse effect to the use of antibiotics because of their negative effect on the lactobacilli population [66]. The results obtained in this study suggest, for the first time, that an antibiotic-associated depletion of vaginal lactobacilli may have long-term health consequences by impairing fertility or embryo implantation and that such effect may be contrasted reversed by microbiological modulation of the vaginal ecosystem.

The species most frequently isolated from vaginal samples in this study belonged to *L. crispatus*, *L. gasseri*, *L. iners*, and *L. jensenii*, which are particularly common and abundant in the human vagina and absent or infrequently found in other human habitats [3,32,67]. Stable codominance of multiple *Lactobacillus* species is rarely observed in the same vaginal community [67]. Initial presence of *L. crispatus* seemed to be positively correlated with a successful reproductive outcome after the intervention with the probiotic assayed in this study. In contrast, initial presence of *L. iners* and *L. gasseri* seemed to be negatively correlated with a successful reproductive outcome after the probiotic intervention unless the *L. salivarius* strain provided in the trial was able to become dominant in the vaginal samples. *L. crispatus* and *L. iners* are probably the most common inhabitants of the healthy human vagina and are able to perform relevant ecological functions in the vaginal environment. Transitions from a vaginal community dominated by *L. iners* to one dominated by *L. crispatus*, and viceversa, seems to be relatively frequent [68]. The relationships between these two species and their potential functions have received an increasing scientific interest in the last years [67–71]. However, while there is a general agreement that a *L. crispatus*-dominated vaginotype promotes vaginal and reproductive health [72–74], the role of *L. iners* is very controversial since this peculiar species has been associated to beneficial roles for vaginal health [8,75] but, also, to dysbiosis, vaginal infections and a variety of gynecological conditions, including adverse pregnant outcomes [69,71,76–78]. Functional studies are required to investigate its roles in vaginal bacterial communities and whether, under certain circumstances, it can be used as a biomarker of reproductive failure.

A characterization of some properties of *L. salivarius* CECT5713 that may be relevant for vaginal and reproductive health showed that this strain was able to inhibit all the clinical isolates of *G. vaginalis*, *S. agalactiae*, *C. albicans*, *C. glabrata*, *C. parapsilosis*, and *U. urealyticum* tested in this study. This antimicrobial activity is relevant since vaginal infections are associated with an increased risk of adverse urogenital and reproductive health outcomes [79]. *L. salivarius* CECT5713 has a high acidifying ability by producing high amounts of L-lactic acid and small amounts of acetic acid [37]. Eubiosis and dysbiosis in the vaginal communities are distinguished by the high concentration of lactic acid and the high acidity that characterize the eubiosis state [79–81], as a direct result of the metabolic activity of the local lactobacilli, which is enough to inactivate reproductive tract pathogens, including viruses, bacteria and yeasts [49,82–87]. The capability and rate of production of lactic acid by lactobacilli is strain-specific and only high levels of lactic acid and a concomitant very low pH can inhibit microbial growth efficiently in the local vaginal biofilm [88,89].

From this point of view, *L. salivarius* CECT5713 seems a suitable candidate as a probiotic for the cervicovaginal target. In addition, this strain encodes an α-amylase in its genome (GenBank: ADJ79335.1), which is fully functional as revealed in the activity assays performed in this work. This enzyme might contribute, together with host α-amylase, to degradation of vaginal glycogen and, therefore, to increase lactic acid production and to maintain the vaginal pH at $\leq$4.5, promoting the desired lactobacilli dominance in the vaginal ecosystem [90].

Other properties of *L. salivarius* CECT5713 that are interesting in relation to the control of harmful vaginal microbes include a high rate of adhesion to vaginal cells and co-aggregation with the vaginal pathogens used in this study. High adherence of *L. salivarius* strains to vaginal cells has been previously observed and related to the prevention of vaginal colonization by *S. agalactiae* [49]. Both adhesion and co-aggregation activities seem to be highly strain-specific traits [48,49,91,92]. Cell-dependent reduction of *Candida* spp. adhesion by *Lactobacillus* species has been related to co-aggregation and competition for binding sites [93,94]. Overall, *L. salivarius* CECT5713 seems to be a strain suitable for applications involving vaginal homeostasis. This strain was isolated from human milk and infant feces of a healthy mother–child pair [37], and has been shown to be a good probiotic strain due to its extensive repertoire of desirable properties and safety, being particularly suited for application in the mother–infant dyad [38].

In this work, oral administration of *L. salivarius* CECT5713 to women of the RA and INF groups led to a relevant number of pregnancies. Women of such groups who had term pregnancies experienced significant changes in some key microbiological, biochemical and immunological parameters in the vaginal samples, such as concentration of cultivable lactobacilli, concentration of *L. salivarius* specific DNA, pH, Nugent score, and concentrations of VEGF, TGF-β 1 and TGF-β 2. The fact that all of them had high concentrations of *L. salivarius* in the vaginal samples and that DNA from this species was also detected by the qPCR assay reveals that the strain was able to reach and colonize the vaginal mucosa. The significant reductions of the pH values after the treatment indicate that the strain was metabolically active and suggests a good agreement between the in vitro potential of the strain and its in vivo capabilities.

The changes induced by *L. salivarius* CECT5713 in the concentrations of the growth factors VEGF, TGF-β 1 and TGF-β 2 seem to be particularly relevant and can be considered as biomarkers of the efficacy of the strain for the target pursued in the clinical trial. VEGF is a 45-kDa homodimeric heparin-binding glycoprotein with angiogenic activity that plays a key role as regulator of vasculogenesis, angiogenesis and vascular function in the human endometrium [95,96]. Vasculogenesis and angiogenesis are crucial steps for embryogenesis and particularly for embryo implantation (vessel formation and trophoblastic invasion) and both processes have been correlated with an increased expression of VEGF and VEGF receptors [97–101]; otherwise, endometrial angiogenesis may be impaired and result in a lethal phenotype, ranging from failed implantation to first-trimester miscarriage [95,102–105].

TGF-β 1 and TGF-β 2 also promote angiogenesis in vivo [106], and participate in implantation, trophoblast differentiation, and immunoregulation at the maternal-fetal interface [100,107]. Transcription of TGF-β 1 increases notably in human uterine endometrium during the first trimester of pregnancy [108], while recurrent pregnancy is associated with a decrease in the decidual TGF-β [109–111]. Expression of both VEGF and TGF-β 1 is highly regulated in a temporal and spatial manner during the early stages of implantation, a fact that underlines their critical role in the evolving pregnancy [109–111]. In addition, TGF-β 1 increases expression of VEGF in the trophoblast [111–115] suggesting a link between the action of both growth factors.

TGF-β 1 and TGF-β 2 are also of particular interest in this field because of their well-known roles in regulating the inflammatory response and inducing active immune tolerance in mucosal tissues [116,117]. Interestingly, both are present at very high concentrations in human seminal fluid [118,119], acting as male-female signaling agents that regulate the female immune response to sperm after coitus and promote maternal immune tolerance for embryo implantation and subsequent pregnancy [120–123]. Although studies in mouse models have shown that exposure to the high concentrations of TGF-β present in seminal fluid is absolutely required to boost uterine Treg cells prior to embryo implantation [124–129], this fact is not taken into account in many current ARTs, including IVF techniques, where such exposure is absent. Most TGF-β present in human semen is latent and requires activation to bind to receptors on cervical cells [130,131]. Interestingly, activation after coitus is facilitated by the acid pH of the vaginal environment [123] and, in this study, administration of *L. salivarius* CECT5713 led to an increase of TGF-β 1 and TGF-β 2 concentrations and, concomitantly, to a significant decrease in the vaginal pH values.

Our study has some limitations. First, the microbiota of the genitourinary tract of the partner was not evaluated and some studies have shown that male microbiota may also play a fundamental role in reproductive outcomes [132,133]. In fact, the couple (when applicable) should be considered as a single entity to achieve the best reproductive outcomes [134]. This approach will be taken into account in our future studies in this field. In addition, the metataxomomic analysis included in this study was carried at the genus level since the 16S rRNA gene approach has poor discriminatory power at the species level [135,136]. Other approaches, such as shotgun sequencing, should be used in the future to solve such limitation and to have a broader view of the vaginal microbiome.

Although our knowledge of the mechanisms that these early embryo–maternal interactions has increased in recent years, implantation remains as a rate-limiting step in human ART and the currently available treatments for infertility or recurrent pregnancy loss of unknown etiology have a rather limited efficacy [137,138]. Therefore, the possibility of enhancing angiogenic and tolerance activities in the endometrium by modifying the reproductive microbiota using bacterial strains specifically tailored for these targets provides a novel strategy to improve reproductive functions and deserves future basic and clinical research efforts.

**Supplementary Materials:** The following are available online at https://www.mdpi.com/2072-6643/13/1/162/s1, Supplementary Figure S1. History of recurrent infections and use of antibiotics in infancy and adulthood among the women recruited in this study. Supplementary Figure S2. Changes in vaginal parameters (pH, Nugent score, TGF-β 1, TGF-β 2, and VEGF concentrations, viable *Lactobacillus* counts in CVL samples) in women with a history of reproductive failure, because either of recurrent miscarriage (RA group) or infertility of unknown origin (INF groups), after the probiotic intervention with *L. salivarius* CECT5713 according to their outcome (pregnancy versus no pregnancy). Supplementary Figure S3. Comparison of vaginal parameters (pH, Nugent score, TGF-β 1, TGF-β 2, and VEGF concentrations, viable *Lactobacillus* counts in CVL samples) in healthy fertile women (C, control group) and those of women with a history of reproductive failure, because either of recurrent miscarriage (RA group) or infertility of unknown origin (INF groups), after the probiotic intervention with *L. salivarius* CECT5713. One-way ANOVA tests followed by Scheffé *post hoc* comparison tests were used to compare the groups; different letters above the boxplots indicate significant differences. Supplementary Table S1. Relative frequencies, medians and interquartile ranges (IQR) of the most abundant bacterial phyla (grey shadow) and genera detected in CVL samples from women who were able to complete a full-term pregnancy ($n$ = 15) and of those who did not ($n$ = 6) among the women that had a history of repetitive abortion (RA group; $n$ = 21). Supplementary Table S2. Relative frequencies, medians and interquartile ranges (IQR) of the most abundant bacterial phyla (grey shadow) and genera detected in CVL samples from women who were able to complete a full-term pregnancy ($n$ = 10) and of those who did not ($n$ = 13) among the women with infertility of unknown origin (INF group; $n$ = 23). Supplementary Table S3. Differences in the baseline characteristics and effect of the probiotic intervention with *L. salivarius* CECT5713 on the vaginal parameters of all women who were able to complete a full-term pregnancy ($n$ = 25) and of those who did not ($n$ = 19) among all participants from both RA and INF groups ($n$ = 44). Supplementary Table S4. Relative frequencies, medians and interquartile ranges (IQR) of the most abundant bacterial phyla (grey shadow) and genera detected in CVL samples from women who were able to complete a full-term pregnancy ($n$ = 25) and of those who did not ($n$ = 19) among women with a history of reproductive failure, because either of recurrent miscarriage (RA group) or infertility of unknown origin (INF groups) ($n$ = 44).

**Author Contributions:** D.B., L.F., and J.M.R. designed and coordinated the study. D.B. recruited participants and recorded samples-associated metadata. I.C. and R.A. processed the samples and performed the microbiological and immunological analyses. C.A. executed bioinformatic analysis. L.F. and J.M.R. drafted the manuscript. All authors have read and agreed to the published version of the manuscript.

**Funding:** This work was partly funded by contract 291-2018 established between Complutense University of Madrid and Biosearch Life S. A. (Granada, Spain). Irma Castro is the recipient of a predoctoral contract (BES-2017-080713) from the Ministerio de Ciencia, Innovación y Universidades (Spain).

**Institutional Review Board Statement:** The study was conducted according to the guidelines of the Declaration of Helsinki, and approved by the Ethics Committee of Consejería de Salud y Familias (Junta de Andalucía, Granada, Spain) (protocol code P050/19, Act 11/19, 10th December 2019).

**Informed Consent Statement:** Informed consent was obtained from all subjects involved in the study.

**Data Availability Statement:** The data presented in this study are available on request from the corresponding author. The data are not publicly available due to ethical restrictions.

**Acknowledgments:** We sincerely thank all the women that participated in the assay. We also thank Evaristo Suárez (University of Oviedo, Spain) for critical reading of the manuscript and fruitful discussions.

**Conflicts of Interest:** Biosearch Life S.A., the company that partly funded the study, is the proprietary of the strain *L. salivarius* CECT 5713.

## References

1. Reid, G.; Brigidi, P.; Burton, J.P.; Contractor, N.; Duncan, S.; Fargier, E.; Hill, C.; Lebeer, S.; Martín, R.; McBain, A.J.; et al. Microbes Central to Human Reproduction. *Am. J. Reprod. Immunol.* **2015**, *73*, 1–11. [CrossRef] [PubMed]
2. Moreno, I.; Simón, C. Deciphering the effect of reproductive tract microbiota on human reproduction. *Reprod. Med. Biol.* **2019**, *18*, 40–50. [CrossRef] [PubMed]
3. Ravel, J.; Gajer, P.; Abdo, Z.; Schneider, G.M.; Koenig, S.S.K.; McCulle, S.L.; Karlebach, S.; Gorle, R.; Russell, J.; Tacket, C.O.; et al. Vaginal microbiome of reproductive-age women. *Proc. Natl. Acad. Sci. USA* **2011**, *108*, 4680–4687. [CrossRef] [PubMed]
4. The Human Microbiome Project Consortium. Structure, function and diversity of the healthy human microbiome. *Nature* **2012**, *486*, 207–214. [CrossRef]
5. Qin, J.; Li, R.; Raes, J.; Arumugam, M.; Burgdorf, K.S.; Manichanh, C.; Nielsen, T.; Pons, N.; Levenez, F.; Yamada, T.; et al. A human gut microbial gene catalogue established by metagenomic sequencing. *Nature* **2010**, *464*, 59–65. [CrossRef]
6. Fredricks, D.; Fiedler, T.L.; Marrazzo, J.M. Molecular Identification of Bacteria Associated with Bacterial Vaginosis. *N. Engl. J. Med.* **2005**, *353*, 1899–1911. [CrossRef]
7. Delaney, M.L. Nugent score related to vaginal culture in pregnant women. *Obstet. Gynecol.* **2001**, *98*, 79–84. [CrossRef]
8. Srinivasan, S.; Liu, C.; Mitchell, C.M.; Fiedler, T.L.; Thomas, K.K.; Agnew, K.J.; Marrazzo, J.M.; Fredricks, D.N. Temporal Variability of Human Vaginal Bacteria and Relationship with Bacterial Vaginosis. *PLoS ONE* **2010**, *5*, e10197. [CrossRef]
9. Kroon, S.J.; Ravel, J.; Huston, W.M. Cervicovaginal microbiota, women's health, and reproductive outcomes. *Fertil. Steril.* **2018**, *110*, 327–336. [CrossRef]
10. Krohn, M.A.; Hillier, S.L.; Nugent, R.P.; Cotch, M.F.; Carey, J.C.; Gibbs, R.S.; Eschenbach, D.A. Vaginal Infection and Prematurity Study Group The Genital Flora of Women with Intraamniotic Infection. *J. Infect. Dis.* **1995**, *171*, 1475–1480. [CrossRef]
11. Newton, E.R.; Piper, J.; Peairs, W. Bacterial vaginosis and intraamniotic infection. *Am. J. Obstet. Gynecol.* **1997**, *176*, 672–677. [CrossRef]
12. Leitich, H.; Bodner-Adler, B.; Brunbauer, M.; Kaider, A.; Egarter, C.; Husslein, P.W. Bacterial vaginosis as a risk factor for preterm delivery: A meta-analysis. *Am. J. Obstet. Gynecol.* **2003**, *189*, 139–147. [CrossRef] [PubMed]
13. Eckert, L.O.; Moore, D.E.; Patton, D.L.; Agnew, K.J.; Eschenbach, D.A. Relationship of Vaginal Bacteria and Inflammation With Conception and Early Pregnancy Loss Following In-Vitro Fertilization. *Infect. Dis. Obstet. Gynecol.* **2003**, *11*, 11–17. [CrossRef] [PubMed]
14. Yudin, M.H. Bacterial Vaginosis in Pregnancy: Diagnosis, Screening, and Management. *Clin. Perinatol.* **2005**, *32*, 617–627. [CrossRef]
15. Van Oostrum, N.; De Sutter, P.; Meys, J.; Verstraelen, H. Risks associated with bacterial vaginosis in infertility patients: A systematic review and meta-analysis. *Hum. Reprod.* **2013**, *28*, 1809–1815. [CrossRef]
16. Sirota, I.; Zarek, S.M.; Segars, J.H. Potential Influence of the Microbiome on Infertility and Assisted Reproductive Technology. *Semin. Reprod. Med.* **2014**, *32*, 035–042. [CrossRef]
17. Wilson, J.D.; Ralph, S.G.; Rutherford, A.J. Rates of bacterial vaginosis in women undergoing in vitro fertilisation for different types of infertility. *BJOG Int. J. Obstet. Gynaecol.* **2002**, *109*, 714–717. [CrossRef]

18. Campisciano, G.; Florian, F.; D'Eustacchio, A.; Stanković, D.; Ricci, G.; De Seta, F.; Comar, M. Subclinical alteration of the cervical-vaginal microbiome in women with idiopathic infertility. *J. Cell. Physiol.* **2017**, *232*, 1681–1688. [CrossRef]
19. Wee, B.A.; Thomas, M.; Sweeney, E.L.; Frentiu, F.D.; Samios, M.; Ravel, J.; Gajer, P.; Myers, G.S.A.; Timms, P.; Allan, J.A.; et al. A retrospective pilot study to determine whether the reproductive tract microbiota differs between women with a history of infertility and fertile women. *Aust. New Zealand J. Obstet. Gynaecol.* **2018**, *58*, 341–348. [CrossRef]
20. Hyman, R.W.; Herndon, C.N.; Jiang, H.; Palm, C.; Fukushima, M.; Bernstein, D.; Vo, K.C.; Zelenko, Z.; Davis, R.W.; Giudice, L.C. The dynamics of the vaginal microbiome during infertility therapy with in vitro fertilization-embryo transfer. *J. Assist. Reprod. Genet.* **2012**, *29*, 105–115. [CrossRef]
21. Moreno, I.; Codoñer, F.M.; Vilella, F.; Valbuena, D.; Martinez-Blanch, J.F.; Jimenez-Almazán, J.; Alonso, R.; Alamá, P.; Remohí, J.; Pellicer, A.; et al. Evidence that the endometrial microbiota has an effect on implantation success or failure. *Am. J. Obstet. Gynecol.* **2016**, *215*, 684–703. [CrossRef] [PubMed]
22. Haahr, T.; Jensen, J.; Thomsen, L.; Duus, L.; Rygaard, K.; Humaidan, P. Abnormal vaginal microbiota may be associated with poor reproductive outcomes: A prospective study in IVF patients. *Hum. Reprod.* **2016**, *31*, 795–803. [CrossRef] [PubMed]
23. Egbase, P.; Al-Sharhan, M.; Al-Othman, S.; Al-Mutawa, M.; Udo, E.; Grudzinskas, J. Fertilization and early embryology: Incidence of microbial growth from the tip of the embryo transfer catheter after embryo transfer in relation to clinical pregnancy rate following in-vitro fertilization and embryo transfer. *Hum. Reprod.* **1996**, *11*, 1687–1689. [CrossRef] [PubMed]
24. Fanchin, R.; Harmas, A.; Benaoudia, F.; Lundkvist, U.; Olivennes, F.; Frydman, R. Microbial flora of the cervix assessed at the time of embryo transfer adversely affects in vitro fertilization outcome. *Fertil. Steril.* **1998**, *70*, 866–870. [CrossRef]
25. Egbase, P.E.; Udo, E.E.; Al-Sharhan, M.; Grudzinskas, J.G. Prophylactic antibiotics and endocervical microbial inoculation of the endometrium at embryo transfer. *Lancet* **1999**, *354*, 651–652. [CrossRef]
26. Moore, D.E.; Soules, M.R.; Klein, N.A.; Fujimoto, V.Y.; Agnew, K.J.; Eschenbach, D.A. Bacteria in the transfer catheter tip influence the live-birth rate after in vitro fertilization. *Fertil. Steril.* **2000**, *74*, 1118–1124. [CrossRef]
27. Salim, R.; Ben-Shlomo, I.; Colodner, R.; Keness, Y.; Shalev, E. Bacterial colonization of the uterine cervix and success rate in assisted reproduction: Results of a prospective survey. *Hum. Reprod.* **2002**, *17*, 337–340. [CrossRef]
28. Selman, H.; Mariani, M.; Barnocchi, N.; Mencacci, A.; Bistoni, F.; Arena, S.; Pizzasegale, S.; Brusco, G.F.; Angelini, A. Examination of bacterial contamination at the time of embryo transfer, and its impact on the IVF/pregnancy outcome. *J. Assist. Reprod. Genet.* **2007**, *24*, 395–399. [CrossRef]
29. Riganelli, L.; Iebba, V.; Piccioni, M.; Illuminati, I.; Bonfiglio, G.; Neroni, B.; Calvo, L.; Gagliardi, A.; Levrero, M.; Merlino, L.; et al. Structural Variations of Vaginal and Endometrial Microbiota: Hints on Female Infertility. *Front. Cell. Infect. Microbiol.* **2020**, *10*, 350. [CrossRef]
30. Peric, A.; Weiss, J.; Vulliemoz, N.; Baud, D.; Stojanov, M. Bacterial Colonization of the Female Upper Genital Tract. *Int. J. Mol. Sci.* **2019**, *20*, 3405. [CrossRef]
31. Reid, G.; Younes, J.A.; Van Der Mei, H.C.; Gloor, G.B.; Knight, R.; Busscher, H.J. Microbiota restoration: Natural and supplemented recovery of human microbial communities. *Nat. Rev. Genet.* **2011**, *9*, 27–38. [CrossRef] [PubMed]
32. Martín, R.; Soberón, N.; Vaneechoutte, M.; Flórez, A.B.; Vázquez, F.; Suárez, J.E. Characterization of indigenous vaginal lactobacilli from healthy women as probiotic candidates. *Int. Microbiol.* **2008**, *11*, 261–266. [PubMed]
33. Amabebe, E.; Anumba, D.O. The Vaginal Microenvironment: The Physiologic Role of Lactobacilli. *Front. Med.* **2018**, *5*, 181. [CrossRef] [PubMed]
34. Halper, J.; Leshin, L.; Lewis, S.; Li, W. Wound Healing and Angiogenic Properties of Supernatants from *Lactobacillus* Cultures. *Exp. Biol. Med.* **2003**, *228*, 1329–1337. [CrossRef]
35. Witkin, S.S.; Linhares, I.M. Why do lactobacilli dominate the human vaginal microbiota? *BJOG Int. J. Obstet. Gynaecol.* **2017**, *124*, 606–611. [CrossRef]
36. Kovachev, S. Defence factors of vaginal lactobacilli. *Crit. Rev. Microbiol.* **2018**, *44*, 31–39. [CrossRef]
37. Martín, R.; Jiménez, E.; Olivares, M.; Marín, M.; Fernández, L.; Xaus, J.; Rodríguez, J. *Lactobacillus salivarius* CECT 5713, a potential probiotic strain isolated from infant feces and breast milk of a mother–child pair. *Int. J. Food Microbiol.* **2006**, *112*, 35–43. [CrossRef]
38. Langa, S.; Maldonado-Barragán, A.; Delgado, S.; Martín, R.; Martín, V.; Jiménez, E.; Ruíz-Barba, J.L.; Mayo, B.; Connor, R.I.; Suárez, J.E.; et al. Characterization of *Lactobacillus salivarius* CECT 5713, a strain isolated from human milk: From genotype to phenotype. *Appl. Microbiol. Biotechnol.* **2012**, *94*, 1279–1287. [CrossRef]
39. Díaz-Ropero, M.; Martin, R.; Sierra, S.; Lara-Villoslada, F.; Rodríguez, J.M.; Xaus, J.; Olivares, M. Two *Lactobacillus* strains, isolated from breast milk, differently modulate the immune response. *J. Appl. Microbiol.* **2007**, *102*, 337–343. [CrossRef]
40. Olivares, M.; Diaz-Ropero, M.; Martin, R.; Rodriguez, J.; Xaus, J. Antimicrobial potential of four *Lactobacillus* strains isolated from breast milk. *J. Appl. Microbiol.* **2006**, *101*, 72–79. [CrossRef]
41. Jiménez, E.; Fernández, L.; Maldonado, A.; Martín, R.; Olivares, M.; Xaus, J.; Rodríguez, J.M. Oral administration of lactobacilli strains isolated from breast milk as an alternative for the treatment of infectious mastitis during lactation. *Appl. Environ. Microbiol.* **2008**, *74*, 4650–4655. [CrossRef] [PubMed]
42. Pérez-Cano, F.J.; Dong, H.; Yaqoob, P. In vitro immunomodulatory activity of *Lactobacillus fermentum* CECT5716 and *Lactobacillus salivarius* CECT5713: Two probiotic strains isolated from human breast milk. *Inmunobiology* **2010**, *12*, 996–1004. [CrossRef] [PubMed]

43. Arroyo, R.; Martín, V.; Maldonado, A.; Jiménez, E.; Fernández, L.; Rodríguez, J.M. Treatment of Infectious Mastitis during Lactation: Antibiotics versus Oral Administration of Lactobacilli Isolated from Breast Milk. *Clin. Infect. Dis.* **2010**, *50*, 1551–1558. [CrossRef] [PubMed]
44. Maldonado, J.; Lara-Villoslada, F.; Sierra, S.; Sempere, L.; Gómez, M.; Rodríguez, J.M.; Boza, J.; Xaus, J.; Olivares, M. Safety and tolerance of the human milk probiotic strain *Lactobacillus salivarius* CECT5713 in 6-month-old children. *Nutrition* **2010**, *26*, 1082–1087. [CrossRef] [PubMed]
45. Zheng, J.; Wittouck, S.; Salvetti, E.; Franz, C.M.A.P.; Harris, H.M.B.; Mattarelli, P.; O'Toole, P.W.; Pot, B.; Vandamme, P.; Walter, J.; et al. A taxonomic note on the genus *Lactobacillus*: Description of 23 novel genera, emended description of the genus *Lactobacillus* Beijerinck 1901, and union of *Lactobacillaceae* and *Leuconostocaceae*. *Int. J. Syst. Evol. Microbiol.* **2020**, *70*, 2782–2858. [CrossRef] [PubMed]
46. Magnusson, J.; Schnürer, J. *Lactobacillus coryniformis* subsp.*coryniformis* Strain Si3 Produces a Broad-Spectrum Proteinaceous Antifungal Compound. *Appl. Environ. Microbiol.* **2001**, *67*, 1–5. [CrossRef]
47. Younes, J.A.; Van Der Mei, H.C.; Heuvel, E.V.D.; Busscher, H.J.; Reid, G. Adhesion Forces and Coaggregation between Vaginal Staphylococci and Lactobacilli. *PLoS ONE* **2012**, *7*, e36917. [CrossRef]
48. Boris, S.; Suárez, J.E.; Vázquez, F.; Barbés, C. Adherence of Human Vaginal Lactobacilli to Vaginal Epithelial Cells and Interaction with Uropathogens. *Infect. Immun.* **1998**, *66*, 1985–1989. [CrossRef]
49. Martin, V.S.; Cárdenas, N.; Ocaña, S.; Marín, M.; Arroyo, R.; Beltrán, D.; Badiola, C.; Fernández, L.; Rodríguez, J.M. Rectal and Vaginal Eradication of *Streptococcus agalactiae* (GBS) in Pregnant Women by Using *Lactobacillus salivarius* CECT 9145, A Target-specific Probiotic Strain. *Nutrients* **2019**, *11*, 810. [CrossRef]
50. Padmavathi, T.; Bhargavi, R.; Priyanka, P.R.; Niranjan, N.R.; Pavitra, P.V. Screening of potential probiotic lactic acid bacteria and production of amylase and its partial purification. *J. Genet. Eng. Biotechnol.* **2018**, *16*, 357–362. [CrossRef]
51. Narita, J.; Okano, K.; Kitao, T.; Ishida, S.; Sewaki, T.; Sung, M.-H.; Fukuda, H.; Kondo, A. Display of α-Amylase on the Surface of *Lactobacillus casei* Cells by Use of the PgsA Anchor Protein, and Production of Lactic Acid from Starch. *Appl. Environ. Microbiol.* **2006**, *72*, 269–275. [CrossRef] [PubMed]
52. Nugent, R.P.; Krohn, M.A.; Hillier, S.L. Reliability of diagnosing bacterial vaginosis is improved by a standardized method of gram stain interpretation. *J. Clin. Microbiol.* **1991**, *29*, 297–301. [CrossRef] [PubMed]
53. Mediano, P.; Fernández, L.; Jiménez, E.; Arroyo, R.; Espinosa-Martos, I.; Rodríguez, J.M.; Marín, M. Microbial Diversity in Milk of Women With Mastitis: Potential Role of Coagulase-Negative Staphylococci, Viridans Group Streptococci, and Corynebacteria. *J. Hum. Lact.* **2017**, *33*, 309–318. [CrossRef]
54. Lackey, K.A.; Williams, J.E.; Meehan, C.L.; Zachek, J.A.; Benda, E.D.; Price, W.J.; Foster, J.A.; Sellen, D.W.; Kamau-Mbuthia, E.W.; Kamundia, E.W.; et al. What's Normal? Microbiomes in Human Milk and Infant Feces Are Related to Each Other but Vary Geographically: The INSPIRE Study. *Front. Nutr.* **2019**, *6*, 45. [CrossRef]
55. Harrow, S.A.; Ravindran, V.; Butler, R.C.; Marshall, J.W.; Tannock, G.W. Real-Time Quantitative PCR Measurement of Ileal *Lactobacillus salivarius* Populations from Broiler Chickens To Determine the Influence of Farming Practices. *Appl. Environ. Microbiol.* **2007**, *73*, 7123–7127. [CrossRef]
56. Salvetti, E.; Harris, H.M.B.; Felis, G.E.; O'Toole, P.W. Comparative Genomics of the Genus *Lactobacillus* Reveals Robust Phylogroups That Provide the Basis for Reclassification. *Appl. Environ. Microbiol.* **2018**, *84*, 00993-18. [CrossRef]
57. Klindworth, A.; Pruesse, E.; Schweer, T.; Peplies, J.; Quast, C.; Horn, M.; Glöckner, F.O. Evaluation of general 16S ribosomal RNA gene PCR primers for classical and next-generation sequencing-based diversity studies. *Nucleic Acids Res.* **2012**, *41*, e1. [CrossRef] [PubMed]
58. Bolyen, E.; Rideout, J.R.; Dillon, M.R.; Bokulich, N.A.; Abnet, C.C.; Al-Ghalith, G.A.; Alexander, H.; Alm, E.J.; Arumugam, M.; Asnicar, F.; et al. Reproducible, interactive, scalable and extensible microbiome data science using QIIME 2. *Nat. Biotechnol.* **2019**, *37*, 852–857. [CrossRef] [PubMed]
59. Callahan, B.J.; McMurdie, P.J.; Rosen, M.J.; Han, A.W.; Johnson, A.J.A.; Holmes, S.P. DADA2: High-resolution sample inference from Illumina amplicon data. *Nat. Methods* **2016**, *13*, 581–583. [CrossRef]
60. Bokulich, N.A.; Kaehler, B.; Rideout, J.R.; Dillon, M.; Bolyen, E.; Knight, R.; Huttley, G.A.; Caporaso, J.G. Optimizing taxonomic classification of marker-gene amplicon sequences with QIIME 2's q2-feature-classifier plugin. *Microbiome* **2018**, *6*, 1–17. [CrossRef]
61. Quast, C.; Pruesse, E.; Yilmaz, P.; Gerken, J.; Schweer, T.; Yarza, P.; Peplies, J.; Glöckner, F.O. The SILVA ribosomal RNA gene database project: Improved data processing and web-based tools. *Nucleic Acids Res.* **2013**, *41*, D590–D596. [CrossRef]
62. R Core Team. *R: A Language and Environment for Statistical Computing*; R Foundation for Statistical Computing: Vienna, Austria, 2018; Available online: https://www.R-project.org (accessed on 10 November 2020).
63. Segata, N.; Izard, J.; Waldron, L.; Gevers, D.; Miropolsky, L.; Garrett, W.S.; Huttenhower, C. Metagenomic biomarker discovery and explanation. *Genome Biol.* **2011**, *12*, R60. [CrossRef]
64. Epskamp, S.; Cramer, A.O.J.; Waldorp, L.J.; Schmittmann, V.D.; Borsboom, D. qgraph: Network Visualizations of Relationships in Psychometric Data. *J. Stat. Softw.* **2012**, *48*, 1–18. [CrossRef]
65. O'Hanlon, D.E.; Lanier, B.R.; Moench, T.R.; Cone, R. Cervicovaginal fluid and semen block the microbicidal activity of hydrogen peroxide produced by vaginal lactobacilli. *BMC Infect. Dis.* **2010**, *10*, 120. [CrossRef]
66. Macklaim, J.M.; Clemente, J.C.; Knight, R.; Gloor, G.B.; Reid, G. Changes in vaginal microbiota following antimicrobial and probiotic therapy. *Microb. Ecol. Heal. Dis.* **2015**, *26*, 27799. [CrossRef]

67. Mendes-Soares, H.; Suzuki, H.; Hickey, R.J.; Forney, L.J. Comparative Functional Genomics of *Lactobacillus* spp. Reveals Possible Mechanisms for Specialization of Vaginal Lactobacilli to Their Environment. *J. Bacteriol.* **2014**, *196*, 1458–1470. [CrossRef]
68. France, M.T.; Mendes-Soares, H.; Forney, L.J. Genomic Comparisons of *Lactobacillus crispatus* and *Lactobacillus iners* Reveal Potential Ecological Drivers of Community Composition in the Vagina. *Appl. Environ. Microbiol.* **2016**, *82*, 7063–7073. [CrossRef]
69. Macklaim, J.M.; Gloor, G.B.; Anukam, K.C.; Cribby, S.; Reid, G. At the crossroads of vaginal health and disease, the genome sequence of *Lactobacillus iners* AB-1. *Proc. Natl. Acad. Sci. USA* **2010**, *108*, 4688–4695. [CrossRef]
70. Vaneechoutte, M. *Lactobacillus iners*, the unusual suspect. *Res. Microbiol.* **2017**, *168*, 826–836. [CrossRef]
71. Borgdorff, H.; Armstrong, S.D.; Tytgat, H.L.P.; Xia, D.; Ndayisaba, G.F.; Wastling, J.M.; Van De Wijgert, J.H.H.M. Unique Insights in the Cervicovaginal *Lactobacillus iners* and *L. crispatus* Proteomes and Their Associations with Microbiota Dysbiosis. *PLoS ONE* **2016**, *11*, e0150767. [CrossRef]
72. Petricevic, L.; Domig, K.J.; Nierscher, F.J.; Sandhofer, M.J.; Fidesser, M.; Krondorfer, I.; Husslein, P.; Kneifel, W.; Kiss, H. Characterisation of the vaginal *Lactobacillus* microbiota associated with preterm delivery. *Sci. Rep.* **2015**, *4*, 5136. [CrossRef]
73. Lepargneur, J.-P. *Lactobacillus crispatus* as biomarker of the healthy vaginal tract. *Ann. Biol. Clin.* **2016**, *74*, 421–427. [CrossRef]
74. Anton, L.; Sierra, L.-J.; Devine, A.; Barila, G.; Heiser, L.; Brown, A.G.; Elovitz, M.A. Common Cervicovaginal Microbial Supernatants Alter Cervical Epithelial Function: Mechanisms by Which *Lactobacillus crispatus* Contributes to Cervical Health. *Front. Microbiol.* **2018**, *9*, 2181. [CrossRef]
75. Feng, Y.; Yao, Z.; Klionsky, D.J. How to control self-digestion: Transcriptional, post-transcriptional, and post-translational regulation of autophagy. *Trends Cell Biol.* **2015**, *25*, 354–363. [CrossRef]
76. Petrova, M.I.; Reid, G.; Vaneechoutte, M.; Lebeer, S. *Lactobacillus iners*: Friend or Foe? *Trends Microbiol.* **2017**, *25*, 182–191. [CrossRef] [PubMed]
77. Kindinger, L.M.; Bennett, P.R.; Lee, Y.S.; Marchesi, J.R.; Smith, A.; Cacciatore, S.; Holmes, E.; Nicholson, J.K.; Teoh, T.G.; MacIntyre, D.A. The interaction between vaginal microbiota, cervical length, and vaginal progesterone treatment for preterm birth risk. *Microbiome* **2017**, *5*, 1–14. [CrossRef] [PubMed]
78. Zheng, N.; Guo, R.; Yao, Y.; Jin, M.; Cheng, Y.; Ling, Z. *Lactobacillus iners* Is Associated with Vaginal Dysbiosis in Healthy Pregnant Women: A Preliminary Study. *BioMed Res. Int.* **2019**, *2019*, 6079734. [CrossRef] [PubMed]
79. Tachedjian, G.; Aldunate, M.; Bradshaw, C.S.; Cone, R. The role of lactic acid production by probiotic *Lactobacillus* species in vaginal health. *Res. Microbiol.* **2017**, *168*, 782–792. [CrossRef]
80. Ealdunate, M.; Esrbinovski, D.; Hearps, A.C.; Latham, C.F.; Ramsland, P.A.; Egugasyan, R.; Cone, R.A.; Tachedjian, G. Antimicrobial and immune modulatory effects of lactic acid and short chain fatty acids produced by vaginal microbiota associated with eubiosis and bacterial vaginosis. *Front. Physiol.* **2015**, *6*, 164. [CrossRef]
81. O'Hanlon, D.E.; Cone, R.; Moench, T.R. Vaginal pH measured in vivo: Lactobacilli determine pH and lactic acid concentration. *BMC Microbiol.* **2019**, *19*, 1–8. [CrossRef]
82. Boskey, E.; Cone, R.; Whaley, K.; Moench, T. Origins of vaginal acidity: High D/L lactate ratio is consistent with bacteria being the primary source. *Hum. Reprod.* **2001**, *16*, 1809–1813. [CrossRef] [PubMed]
83. O'Hanlon, D.E.; Moench, T.R.; Cone, R. In vaginal fluid, bacteria associated with bacterial vaginosis can be suppressed with lactic acid but not hydrogen peroxide. *BMC Infect. Dis.* **2011**, *11*, 200. [CrossRef] [PubMed]
84. O'Hanlon, D.E.; Moench, T.R.; Cone, R.A. Vaginal pH and Microbicidal Lactic Acid When Lactobacilli Dominate the Microbiota. *PLoS ONE* **2013**, *8*, e80074. [CrossRef] [PubMed]
85. Ruíz, F.O.; Gerbaldo, G.; Garcia, M.J.; Giordano, W.; Pascual, L.; Barberis, I.L. Synergistic Effect Between Two Bacteriocin-like Inhibitory Substances Produced by Lactobacilli Strains with Inhibitory Activity for *Streptococcus agalactiae*. *Curr. Microbiol.* **2012**, *64*, 349–356. [CrossRef]
86. Aldunate, M.; Tyssen, D.; Johnson, A.; Zakir, T.; Sonza, S.; Moench, T.; Cone, R.; Tachedjian, G. Vaginal concentrations of lactic acid potently inactivate HIV. *J. Antimicrob. Chemother.* **2013**, *68*, 2015–2025. [CrossRef] [PubMed]
87. Tyssen, D.; Wang, Y.-Y.; Hayward, J.A.; Agius, P.A.; Delong, K.; Aldunate, M.; Ravel, J.; Moench, T.R.; Cone, R.A.; Tachedjian, G. Anti-HIV-1 Activity of Lactic Acid in Human Cervicovaginal Fluid. *mSphere* **2018**, *3*, e00055-18. [CrossRef]
88. Chew, S.Y.; Cheah, Y.K.; Seow, H.F.; Sandai, D.; Than, L.T.L. In vitro modulation of probiotic bacteria on the biofilm of *Candida glabrata*. *Anaerobe* **2015**, *34*, 132–138. [CrossRef]
89. Matsubara, V.H.; Wang, Y.; Bandara, H.M.H.N.; Mayer, M.P.A.; Samaranayake, L. Probiotic lactobacilli inhibit early stages of *Candida albicans* biofilm development by reducing their growth, cell adhesion, and filamentation. *Appl. Microbiol. Biotechnol.* **2016**, *100*, 6415–6426. [CrossRef]
90. Nasioudis, D.; Beghini, J.; Bongiovanni, A.M.; Giraldo, P.C.; Linhares, I.M.; Witkin, S.S. α-Amylase in vaginal fluid: Association with conditions favorable to dominance of *Lactobacillus*. *Reprod. Sci.* **2015**, *22*, 1393–1398. [CrossRef]
91. Hütt, P.; Lapp, E.; Štšepetova, J.; Smidt, I.; Taelma, H.; Borovkova, N.; Oopkaup, H.; Ahelik, A.; Rööp, T.; Hoidmets, D.; et al. Characterisation of probiotic properties in human vaginal lactobacilli strains. *Microb. Ecol. Heal. Dis.* **2016**, *27*, 30484. [CrossRef]
92. Cárdenas, N.; Martin, V.S.; Arroyo, R.; López, M.; Carrera, M.; Badiola, C.; Jiménez, E.; Rodríguez, J.M. Prevention of Recurrent Acute Otitis Media in Children Through the Use of *Lactobacillus salivarius* PS7, a Target-Specific Probiotic Strain. *Nutrients* **2019**, *11*, 376. [CrossRef]

93. Chew, S.Y.; Cheah, Y.K.; Seow, H.F.; Sandai, D.; Than, L.T.L. Probiotic *Lactobacillus rhamnosus* GR-1 and *Lactobacillus reuteri* RC-14 exhibit strong antifungal effects against vulvovaginal candidiasis-causing *Candida glabrata* isolates. *J. Appl. Microbiol.* **2015**, *118*, 1180–1190. [CrossRef]
94. Aarti, C.; Khusro, A.; Varghese, R.; Arasu, M.V.; Agastian, P.; Al-Dhabi, N.A.; Ilavenil, S.; Choi, K.C. In vitro investigation on probiotic, anti-*Candida*, and antibiofilm properties of *Lactobacillus pentosus* strain LAP1. *Arch. Oral Biol.* **2018**, *89*, 99–106. [CrossRef]
95. Nardo, L.G. Vascular endothelial growth factor expression in the endometrium during the menstrual cycle, implantation window and early pregnancy. *Curr. Opin. Obstet. Gynecol.* **2005**, *17*, 419–423. [CrossRef]
96. Demir, R.; Yaba, A.; Huppertz, B. Vasculogenesis and angiogenesis in the endometrium during menstrual cycle and implantation. *Acta Histochem.* **2010**, *112*, 203–214. [CrossRef]
97. Gordon, J.D.; Shifren, J.L.; Foulk, R.A.; Taylor, R.N.; Jaffe, R.B. Angiogenesis in the Human Female Reproductive Tract. *Obstet. Gynecol. Surv.* **1995**, *50*, 688–697. [CrossRef] [PubMed]
98. Licht, P.; Russu, V.; Lehmeyer, S.; Wissentheit, T.; Siebzehnrübl, E.; Wildt, L. Cycle dependency of intrauterine vascular endothelial growth factor levels is correlated with decidualization and corpus luteum function. *Fertil. Steril.* **2003**, *80*, 1228–1233. [CrossRef]
99. Malamitsi-Puchner, A.; Sarandakou, A.; Tziotis, J.; Stavreus-Evers, A.; Tzonou, A.; Landgren, B.-M. Circulating angiogenic factors during periovulation and the luteal phase of normal menstrual cycles. *Fertil. Steril.* **2004**, *81*, 1322–1327. [CrossRef]
100. Torry, D.S.; Leavenworth, J.; Chang, M.; Maheshwari, V.; Groesch, K.; Ball, E.R.; Torry, R.J. Angiogenesis in implantation. *J. Assist. Reprod. Genet.* **2007**, *24*, 303–315. [CrossRef]
101. Kaczmarek, M.M.; Waclawik, A.; Blitek, A.; Kowalczyk, A.E.; Schams, D.; Ziecik, A.J. Expression of the vascular endothelial growth factor-receptors ystem in the porcine endometrium throughout the estrous cycle and early pregnancy. *Mol. Reprod. Dev.* **2008**, *75*, 362–372. [CrossRef]
102. Meegdes, B.H.; Ingenhoes, R.; Peeters, L.L.; Exalto, N. Early pregnancy wastage: Relationship between chorionic vascularization and embryonic development. *Fertil. Steril.* **1988**, *49*, 216–220. [CrossRef]
103. Fong, G.-H.; Rossant, J.; Gertsenstein, M.; Breitman, M.L. Role of the Flt-1 receptor tyrosine kinase in regulating the assembly of vascular endothelium. *Nat. Cell Biol.* **1995**, *376*, 66–70. [CrossRef] [PubMed]
104. Vuorela, P.; Carpén, O.; Tulppala, M.; Halmesmäki, E. VEGF, its receptors and the Tie receptors in recurrent miscarriage. *Mol. Hum. Reprod.* **2000**, *6*, 276–282. [CrossRef] [PubMed]
105. Reynolds, L.P.; Caton, J.S.; Redmer, D.A.; Grazul-Bilska, A.T.; Vonnahme, K.A.; Borowicz, P.P.; Luther, J.S.; Wallace, J.M.; Wu, G.; Spencer, T.E. Evidence for altered placental blood flow and vascularity in compromised pregnancies. *J. Physiol.* **2006**, *572*, 51–58. [CrossRef]
106. Relf, M.; Lejeune, S.; Scott, P.A.; Fox, S.; Smith, K.; Leek, R.; Moghaddam, A.; Whitehouse, R.; Bicknell, R.; Harris, A.L. Expression of the angiogenic factors vascular endothelial cell growth factor, acidic and basic fibroblast growth factor, tumor growth factor beta-1, platelet-derived endothelial cell growth factor, placenta growth factor, and pleiotrophin in human primary breast cancer and its relation to angiogenesis. *Cancer Res.* **1997**, *57*, 963–969.
107. Ingman, W.V.; Robertson, S.A. Defining the actions of transforming growth factor beta in reproduction. *BioEssays* **2002**, *24*, 904–914. [CrossRef]
108. Giudice, L.C. Growth factors and growth modulators in human uterine endometrium: Their potential relevance to reproductive medicine. *Fertil. Steril.* **1994**, *61*, 1–17. [CrossRef]
109. Bao, S.H.; Wang, X.P.; De Lin, Q.; Wang, W.J.; Yin, G.J.; Qiu, L.H. Decidual CD4+CD25+CD127dim/- regulatory T cells in patients with unexplained recurrent spontaneous miscarriage. *Eur. J. Obstet. Gynecol. Reprod. Biol.* **2011**, *155*, 94–98. [CrossRef]
110. Xu, L.; Qiu, T.; Wang, Y.; Chen, Y.; Cheng, W. Expression of C-type lectin receptors and Toll-like receptors in decidua of patients with unexplained recurrent spontaneous abortion. *Reprod. Fertil. Dev.* **2017**, *29*, 1613–1624. [CrossRef]
111. Qian, J.; Zhang, N.; Lin, J.; Wang, C.; Pan, X.; Chen, L.; Li, D.; Wang, L. Distinct pattern of Th17/Treg cells in pregnant women with a history of unexplained recurrent spontaneous abortion. *Biosci. Trends* **2018**, *12*, 157–167. [CrossRef]
112. Chung, I.-B.; Yelian, F.; Zaher, F.; Gonik, B.; Evans, M.; Diamond, M.P.; Svinarich, D. Expression and Regulation of Vascular Endothelial Growth Factor in a First Trimester Trophoblast Cell Line. *Placenta* **2000**, *21*, 320–324. [CrossRef] [PubMed]
113. Poole, T.J.; Finkelstein, E.B.; Cox, C.M. The role of FGF and VEGF in angioblast induction and migration during vascular development. *Dev. Dyn.* **2001**, *220*, 1–17. [CrossRef]
114. Sherer, D.; Abulafia, O. Angiogenesis during Implantation, and Placental and Early Embryonic Development. *Placenta* **2001**, *22*, 1–13. [CrossRef] [PubMed]
115. Qian, D.; Lin, H.-Y.; Wang, H.-M.; Zhang, X.; Liu, D.-L.; Li, Q.-L.; Zhu, C. Involvement of ERK1/2 pathway in TGF-beta1-induced VEGF secretion in normal human cytotrophoblast cells. *Mol. Reprod. Dev.* **2004**, *68*, 198–204. [CrossRef] [PubMed]
116. Robertson, S.A.; Ingman, W.V.; O'Leary, S.; Sharkey, D.J.; Tremellen, K.P. Transforming growth factor β—A mediator of immune deviation in seminal plasma. *J. Reprod. Immunol.* **2002**, *57*, 109–128. [CrossRef]
117. Wahl, S.M.; Wen, J.; Moutsopoulos, N. TGF-beta: A mobile purveyor of immune privilege. *Immunol. Rev.* **2006**, *213*, 213–227. [CrossRef] [PubMed]
118. Nocera, M.; Chu, T.M. Characterization of Latent Transforming Growth Factor-β From Human Seminal Plasma. *Am. J. Reprod. Immunol* **1995**, *33*, 282–291. [CrossRef] [PubMed]

119. Loras, B.; Vételé, F.; El Malki, A.; Rollet, J.; Soufir, J.-C.; Benahmed, M. Seminal transforming growth factor-β in normal and infertile men. *Hum. Reprod.* **1999**, *14*, 1534–1539. [CrossRef] [PubMed]
120. Robertson, S.A.; Sharkey, D.J. The role of semen in induction of maternal immune tolerance to pregnancy. *Semin. Immunol.* **2001**, *13*, 243–254. [CrossRef] [PubMed]
121. Robertson, S.A.; Guerin, L.R.; Moldenhauer, L.M.; Hayball, J.D. Activating T regulatory cells for tolerance in early pregnancy-the contribution of seminal fluid. *J. Reprod. Immunol.* **2009**, *83*, 109–116. [CrossRef] [PubMed]
122. Robertson, S.A. Immune regulation of conception and embryo implantation—All about quality control? *J. Reprod. Immunol.* **2010**, *85*, 51–57. [CrossRef] [PubMed]
123. Sharkey, D.J.; MacPherson, A.M.; Tremellen, K.P.; Mottershead, D.G.; Gilchrist, R.B.; Robertson, S.A. TGF-β Mediates Proinflammatory Seminal Fluid Signaling in Human Cervical Epithelial Cells. *J. Immunol.* **2012**, *189*, 1024–1035. [CrossRef] [PubMed]
124. Sakaguchi, S. Regulatory T Cells. *Cell* **2000**, *101*, 455–458. [CrossRef]
125. Shevach, E.M. CD4+CD25+ suppressor T cells: More questions than answers. *Nat. Rev. Immunol.* **2002**, *2*, 389–400. [CrossRef]
126. Chen, W.; Jin, W.; Hardegen, N.J.; Lei, K.-J.; Li, J.; Marinos, N.J.; McGrady, G.; Wahl, S.M. Conversion of Peripheral CD4+CD25− Naive T Cells to CD4+CD25+ Regulatory T Cells by TGF-β Induction of Transcription Factor Foxp3. *J. Exp. Med.* **2003**, *198*, 1875–1886. [CrossRef]
127. Clark, D.A.; Fernandez, J.; Banwatt, D. ORIGINAL ARTICLE: Prevention of Spontaneous Abortion in the CBA × DBA/2 Mouse Model by Intravaginal TGF-β and Local Recruitment of CD4+ 8+ FOXP3+ Cells. *Am. J. Reprod. Immunol.* **2008**, *59*, 525–534. [CrossRef]
128. Robertson, S.A.; Guerin, L.R.; Bromfield, J.J.; Branson, K.M.; Ahlström, A.C.; Care, A.S. Seminal Fluid Drives Expansion of the CD4+CD25+ T Regulatory Cell Pool and Induces Tolerance to Paternal Alloantigens in Mice1. *Biol. Reprod.* **2009**, *80*, 1036–1045. [CrossRef]
129. Guerin, L.R.; Moldenhauer, L.M.; Prins, J.R.; Bromfield, J.; Hayball, J.D.; Robertson, S.A. Seminal Fluid Regulates Accumulation of FOXP3+ Regulatory T Cells in the Preimplantation Mouse Uterus Through Expanding the FOXP3+ Cell Pool and CCL19-Mediated Recruitment. *Biol. Reprod.* **2011**, *85*, 397–408. [CrossRef]
130. Chu, T.; Kawinski, E. Plasmin, Substilisin-like Endoproteases, Tissue Plasminogen Activator, and Urokinase Plasminogen Activator Are Involved in Activation of Latent TGF-β1 in Human Seminal Plasma. *Biochem. Biophys. Res. Commun.* **1998**, *253*, 128–134. [CrossRef]
131. Emami, N.; Diamandis, E.P. Potential role of multiple members of the kallikrein-related peptidase family of serine proteases in activating latent TGFβ1 in semen. *Biol. Chem.* **2010**, *391*, 85–95. [CrossRef]
132. Tomaiuolo, R.; Veneruso, I.; Cariati, F.; D'Argenio, V. Microbiota and Human Reproduction: The Case of Female Infertility. *High Throughput* **2020**, *9*, 12. [CrossRef] [PubMed]
133. Tomaiuolo, R.; Veneruso, I.; Cariati, F.; D'Argenio, V. Microbiota and Human Reproduction: The Case of Male Infertility. *High Throughput* **2020**, *9*, 10. [CrossRef] [PubMed]
134. Cariati, F.; D'Argenio, V.; Tomaiuolo, R. The evolving role of genetic tests in reproductive medicine. *J. Transl. Med.* **2019**, *17*, 1–33. [CrossRef] [PubMed]
135. Mignard, S.; Flandrois, J. 16S rRNA sequencing in routine bacterial identification: A 30-month experiment. *J. Microbiol. Methods* **2006**, *67*, 574–581. [CrossRef]
136. Bukin, Y.S.; Galachyants, Y.P.; Morozov, I.V.; Bukin, S.V.; Zakharenko, A.S.; Zemskaya, T.I. The effect of 16S rRNA region choice on bacterial community metabarcoding results. *Sci. Data* **2019**, *6*, 190007. [CrossRef]
137. Koot, Y.E.; Teklenburg, G.; Salker, M.; Brosens, J.; Macklon, N. Molecular aspects of implantation failure. *Biochim. Biophys. Acta Mol. Basis Dis.* **2012**, *1822*, 1943–1950. [CrossRef]
138. Ticconi, C.; Pietropolli, A.; Di Simone, N.; Piccione, E.; Fazleabas, A.T. Endometrial Immune Dysfunction in Recurrent Pregnancy Loss. *Int. J. Mol. Sci.* **2019**, *20*, 5332. [CrossRef]

nutrients

MDPI

*Review*

# Prevention and Management with Pro-, Pre and Synbiotics in Children with Asthma and Allergic Rhinitis: A Narrative Review

**Lien Meirlaen †, Elvira Ingrid Levy † and Yvan Vandenplas ***

KidZ Health Castle, UZ Brussel, Vrije Universiteit Brussel, 1090 Brussels, Belgium; lienmeirlaen@hotmail.com (L.M.); elvira.levy9@gmail.com (E.I.L.)
* Correspondence: yvan.vandenplas@uzbrussel.be; Tel.: +32-475748794
† These authors contributed equally to this work.

**Abstract:** Allergic diseases including allergic rhinitis and asthma are increasing in the developing world, related to a westernizing lifestyle, while the prevalence is stable and decreasing in the industrialized world. This paper aims to answer the question if prevention and/or treatment of allergic rhinitis and asthma can be achieved by administrating pro-, pre- and/or synbiotics that might contribute to stabilizing the disturbed microbiome that influences the immune system through the gut–lung axis. We searched for relevant English articles in PubMed and Google Scholar. Articles interesting for the topic were selected using subject heading and key words. Interesting references in included articles were also considered. While there is substantial evidence from animal studies in well controlled conditions that selected probiotic strains may offer benefits in the prevention of wheezing and asthma, outcomes from clinical studies in infants (including as well pre- and postnatal administration) are disappointing. The latter may be related to the multiple confounding factors such as environment, strain selection and dosage, moment of administration and genetic background. There is little evidence to recommend administration of pro, pre- or synbiotics in the prevention of asthma and allergic rhinitis in children.

**Keywords:** probiotics; prebiotics; synbiotics; microbiome; children; allergic rhinitis; asthma

**Citation:** Meirlaen, L.; Levy, E.I.; Vandenplas, Y. Prevention and Management with Pro-, Pre and Synbiotics in Children with Asthma and Allergic Rhinitis: A Narrative Review. *Nutrients* **2021**, *13*, 934. https://doi.org/10.3390/nu13030934

Academic Editors: Sonia González, Nuria Salazar and Silvia Arboleya

Received: 16 January 2021
Accepted: 11 March 2021
Published: 14 March 2021



## 1. Introduction

### 1.1. Prevalence of Asthma and Allergic Rhinitis

The global prevalence of atopic diseases such as asthma, allergic rhinitis and atopic dermatitis is remarkable and has been expanding over the years [1]. Allergic rhinitis occurs in 10 to 30% of adults and up to 40% in children and its prevalence is increasing [2]. With around 339 million people affected globally, asthma is one of the most common long-term non-transmissible diseases [3]. The worldwide prevalence of doctor-diagnosed asthma in adults is 4.3% (95% confidence interval (CI) 4.2–4.4), with a wide variation between countries: the highest occurrence is found in developed countries such as Australia (21%) and the lowest in third world countries such as Ethiopia (2%) [4]. In children, asthma is more frequent in boys than in girls due to their smaller airways relative to their lung size, with a turnaround during puberty, as the prevalence in women is 20% higher than in men [5]. Asthma prevalence is steady or even shrinking in many developed countries, but as lifestyles become more westernized in developing countries, there is a fast increase in its prevalence in these parts of the world [6]. The interaction between the genomic background, changing environmental conditions such as more pollution [7], increasing obesity, the "hygiene hypothesis" and less breastfeeding [8] is likely to play a crucial part. Parental reduction in smoking has proven to reduce asthma [9]. Important to mention is that in less developed countries, the detection rate of allergic disease is likely to be lower, which may result in an underestimation of its prevalence [10]. By identifying and characterizing more of these conditions and the involved lifestyle factors, epidemiologic studies try to deduce potential approaches for prevention of allergic diseases [11]. Asthma causes impaired life quality, substantial disability and preventable deaths in children and adolescents, combined with important health care costs [6]. As a consequence, the increased social and economic burden of asthma makes asthma prevention an important public health goal [12].

### *1.2. Pathophysiology Asthma and Allergic rhinitis*

Atopic diseases like asthma and allergic rhinitis are complex multifactorial conditions of which the outcome is strongly influenced by a complex interplay between genetic background, the state of the body's defenses, gut microbiota and the environment. There are different mechanisms and typical pathological characteristics of asthma immunopathology, which can be divided in three groups: non-eosinophilic (neutrophilic type 1 and type 17 and pauci-granulocytic), eosinophilic (allergic and non-allergic), and mixed granulocytic inflammation [6]. The eosinophilic group represents 50% of all asthma patients. In this process, allergen or trigger factor exposure stimulates local inflammatory responses mediated by immunoglobulin E (IgE) release. This leads to allergen sensitization and the forming of an atopic response. Type 2 T helper (Th2) cells play a crucial part in this inflammatory process by producing cytokines that control fabrication of allergen-specific immunoglobulin E and inflammation of tissue characterized by the invasion of eosinophils, mast cells and activated CD4+ T-cells. Regulatory T-cells (Treg) are involved in preventing the sensitization to allergens by the production of anti-inflammatory cytokines such as IL-10, by secreting transforming growth factor B, and by possibly suppressing the production of immunoglobulin E and proliferation of Type 1 T helper (Th1)/Type 2 T helper (Th2) balance. The mechanisms of tolerance induction are complex [13]. The intestinal microbiome contributes to the pathological process of allergic diseases because of its notable effect on mucosal immunity. A healthy microbiome at a young age changes the balance between T helper 1 T helper 2, shifting towards a T helper 1 cell response. About 60–70% of the immune cells are located within the gastrointestinal tract. On the other hand, atopic diseases involve Type 2 T helper reactions to allergens. Unusual allergic responses are believed to occur in cases of intestinal dysbiosis during the development of the immune system, causing a shift of the Th1/Th2 cytokine balance towards a Th2 response, a consequent activation of Th2 cytokines and increased production of IgE [14]. Additionally, there is increasing evidence that a balanced gut microbiome is needed for the proper formation of T-regulatory cells, which are important for tolerance induction [13].

### *1.3. Definitions Pro-, Pre- and Synbiotics*

Probiotics are live microorganisms that, when administered in sufficient quantities, give a health improvement of the host. Probiotics induce immunomodulatory mechanisms in many different ways, including skewing of the Th1/Th2 balance towards Th1 by inhibiting Th2 cytokines or indirectly expanding IL-10 and Treg formation via either dendritic cell development or Toll-like receptors, although the exact mechanism remains to be clarified [15]. Prebiotics are substrates that are selectively utilized by host microorganisms conferring a health benefit. Synbiotics are defined as a mixture comprising live microorganisms and substrate(s) selectively utilized by host microorganisms that confer a health benefit on the host.

### *1.4. Rationale for Using Pro-, Pre- and Synbiotics in Atopic Diseases*

Living circumstances in the industrialized world such as a decreased fermented food consumption, increased intake of antibiotics and other drugs, and improved hygiene are according to data from epidemiologic studies associated to the increase in allergic diseases. More or less exposure to microbial stimuli during infancy is associated to more or less allergic disease. The association has been described as the "hygiene hypothesis". A lack of exposure to microbial stimuli early in childhood is a major factor involved in the steep increase in allergy [16]. In those who spend their childhood on a farm, allergic diseases are less common [16]. The comparison between the composition of microbiota of farm children and the microbiota of children with other lifestyles shows a significant difference [16]. Children living on farms are exposed to a wider range of microbes than children not living on a farm, and this exposure explains a substantial fraction of the inverse relation between asthma and growing up on a farm [17]. The gastrointestinal microbiota composition differs between allergic and healthy infants, independent of the prevalence of allergic disease in the region [13]. In contrary to what has been believed for a long time, an amniotic microbiome has been reported, and as a consequence, the fetal intestine may not be sterile since there is the presence of microbial deoxyribonucleic acid in meconium [18]. Early life is characterized by a rapid change in gastrointestinal microbiota composition. The first altering factor of the neonatal microbiome is the contact with vaginal, fecal and skin bacteria of the mother. In caesarean section-born babies, a less diversified microbiome is observed. The second altering factor is feeding. Human milk is rich in oligosaccharides which have prebiotic properties (a substrate that is selectively utilized by host microorganisms conferring a health benefit [19]) and promote the growth of selected species of bacteria. Human milk is also a natural bacterial inoculum. The third altering factor is environmental influenced alterations, which may undo the first two beneficial gut alterations: environments like neonatal intensive care units and medication such as antibiotics or proton pump inhibitors administered perinatal or during early life [20,21].

During early life, a balanced gastrointestinal microbiota is of major importance for the balanced skewing of the developing of the immune system and also determines the gut–lung communication of the gut–lung axis. Therefore, dysbiosis of the intestinal microbiome during early life will contribute to immune-mediated diseases later in life [14]. However, these associations between gut microbiota and allergic disease cannot provide a satisfactory explanation for all observations and does not result in evidence to decrease the rise in allergic disorders. However, the microbiota hypothesis does provide a rationale for using pro-, pre- and synbiotics, to alter the microbiota composition in the intestine to result in a more balanced development of the immune system [13]. Since a child's microbiota does not reflect adult patterns until they are two years old, the infant microbiota may be more susceptible to manipulation [22].

More knowledge is needed on the mechanisms behind dysbiosis, translocation of microbiota from the intestine to the respiratory tract through various mechanisms and for a better evaluation of the therapeutic possibilities to correct this dysbiosis, which in turn can be used to manage various respiratory diseases [23].

In this paper, we will try to answer the question if probiotics or prebiotics and/or synbiotic supplementation can alter the microbiome sufficiently to have an efficacious prevention and/or management of allergic rhinitis and asthma.

## 2. Materials and Methods

A search was performed in PubMed, EMBASE, Google Scholar, Web of Science and Cochrane Library. We included preferably meta-analyses, systematic reviews and clinical trials from 1990 up until October 2020 published in the English language. The following keywords in the respective language were used: "asthma", "wheezing", "respiratory disease", "allergic rhinitis", "allergic coryza", "probiotics", "prebiotics", "synbiotics", "prevention", "therapy", "therapeutics", "child". These keywords were combined with the Boolean command "OR" and were linked by the Boolean command "AND". Records were screened based on the titles and abstracts. Articles were extracted using subject heading and key words of interest to the topic. A second selection was made by reading the abstract. Interesting references in included articles were also considered. Records were excluded if the abstract or full text was not available, if the topic was not relevant, if non-English or if the study design was not adequate. Duplicates were removed.

Search strategy for human studies in the results section: In PubMed, the following search string was used: ("Asthma"[MeSH Terms] OR "respiratory disease"[Title/Abstract] OR "wheezing"[Title/Abstract] OR "recurrent wheeze"[Title/Abstract] OR "rhinitis, allergic, seasonal"[MeSH Terms] OR "allergic coryza"[Title/Abstract]) AND ("Probiotics"[MeSH Terms] OR "Prebiotics"[MeSH Terms] OR "Synbiotics"[MeSH Terms]) AND "Child"[MeSH Terms].

## 3. Results

### 3.1. Probiotics for Prevention of Asthma

#### 3.1.1. Animal Studies

A beneficial effect of the administration of probiotics was suggested by showing that oral administration of *Lactococcus lactis* NZ9000 to rats resulted in a decrease in infiltration of pro-inflammatory leucocytes, mainly eosinophils and decreased lung IL-4 and IL-5 expression in the broncho-alveolar lavage and a reduced level of serum allergen-specific IgE [24]. Another study conducted in mice using *Lactobacillus rhamnosus* GR-1 significantly prevented airway hyperreactivity development and prevented microbiome disturbance in the asthmatic animals, supporting the existence of the gut–lung axis [25]. An interesting aspect is that most probiotics are given orally; however, a new approach was tested by giving probiotics (*Lactobacillus paracasei* NCC2461 [26] and *Lactobacillus rhamnosus* GG [27] in mice through the nose and showed benefits in reducing inflammation of the lungs [28]. The probiotic *Bifidobacterium breve* M-16V administered to pregnant mice was shown to be effective in lowering eosinophils in the broncho-alveolar lavage fluid of neonatal mice and reduced allergic lung inflammation in mice exposed to air pollution [29]. In another animal study, the intranasal administration of *Lactobacillus rhamnosus* GG (LGG), but not *Lactobacillus rhamnosus* GR-1, suppressed airway hyper-reactivity and reduced the counts of eosinophils, IL-13 and IL-5 in broncho-alveolar fluid [27]. In addition to inhibiting inflammatory cell infiltration in lung tissue, *Lactobacillus* GG was shown to decrease MMP9 expression, a class of enzymes that are involved in the degradation of the extracellular matrix and of which levels were significantly increased in asthma [30]. *Lactobacillus* GG and *Bifidobacterium lactis* were shown to increase natural regulatory T cells in the lungs of asthmatic mice in another animal study [31]. Lee et al. mentioned that four *Lactobacillus* species used in animal studies had different immunomodulatory effects [32] against allergy *Lactobacillus planetarum* had shown some beneficial effect, but this was not the case for *Lactobacillus salivarius* and *fermentum* [33]. Probiotic strain-specific induction of Foxp3þ T regulatory cells was found in mouse allergy models [34].

#### 3.1.2. Human Studies

In humans, evidence of the use of probiotics as a preventive agent for respiratory allergies in children was reported to be low [35] (Table 1). A meta-analysis of 2013 showed that by giving the most frequently used probiotics (*Lactobacillus* spp. and/or *Bifidobacteria* spp.) to prenatal mothers plus continued after birth versus only postnatally, no difference in IgE levels were seen. Less atopy was seen if the probiotics were given to pregnant women and continued after birth. Probiotics given after birth only decreases the risk of atopic sensitization in young children but not of asthma or wheeze [14]. This supports the theory that probiotics that have colonized the mothers' intestine will be transferred at birth during vaginal delivery. Further administration of pro- and prebiotics to the pregnant mother results in the potential transmission of tolerogenic mediators such as regulatory cytokines, antibodies and growth factors across the placenta, stimulating the development of the fetal immune system [36]. This could help to prevent asthma or allergic rhinitis. Like mentioned above, the findings in pregnant mice are of human interest since up to now, knowledge was restricted to the fact that *Bifidobacterium breve* M-16V in infants can suppress T-helper type 2 immune responses and modulate the systemic Type 1 T helper/Type 2 T helper balance. Exposure of the pregnant mother to air pollution increases asthma susceptibility of the newborn and later on. Therefore, *Bifidobacterium breve* M-16V might contribute to reducing asthma in a population living in highly polluted areas [29]. In 2014, the Panda Study showed that giving a probiotic mixture postnatally (two *Bifidobactera* spp. and *Lactococcus lactis*) for one year does not have a beneficial effect on the development of allergic diseases after six years [37]. After five years follow-up, the negative outcome persisted [38]. Furthermore, no association (relative risk (RR) 0.59, 95% CI 0.36–0.96, $p = 0.059$) was found in a study with a follow-up of 11 years. This study was a two-center RCT using *Lactobacillus rhamnosus* HN001 or *Bifidobacterium lactis* HN019 daily taken from 35-week gestation to six months postpartum in mothers while breastfeeding and from birth to the age of two years in infants [39]. Consistent with the previously mentioned studies, a more recent meta-analysis including 19 RCTs involving 5157 children showed no association as well in lowering the incidence of asthma and wheezing if probiotics were given to pregnant mothers or postnatally. However, in infants with atopic diseases, probiotics seem to reduce the wheezing incidence significantly (RR 0.61, 95% CI 0.42–0.90; $p < 0.05$). No association was found between probiotics and a subgroup analysis of asthma (RR 0.94, 95% CI 0.82–1.09). Important to mention is that due to the small sample size in the subgroup analysis, the information should be interpreted carefully. The question "Do infants with atopic disease benefit from probiotics (Lactobacillus spp. and/or Bifidobacteria spp., Propionibacterium freudenreichii ssp. shermanii JS)?" should be tested in more heterogenetic, well-designed RCTs. Beneficial effects of specific strains might become lost by pooling probiotic strains together, since the effects are strain-specific. As a consequence, meta-analysis should be strain-specific. Due to the wide heterogeneity of strains, mixture and doses administered, the efficacy of specific probiotic strains has been difficult to analyze. Therefore, further research is needed to optimize the selection of probiotic strains and the configuration of intervention regimens [12].

### 3.2. *Probiotics for the Treatment of Asthma*

The curative effects of probiotics in asthmatics are not well established [40] (Table 2). A recent study in baby mice indicated that *Bifidobacterium infantis* could reduce the infiltration of inflammatory cells by promoting Th1 immune responses and oppositely suppressing Th2 immune responses [41]. In a 2008 systematic review, probiotic administration showed no positive effect in the treatment of asthma [42]. A later meta-analysis from Das et al., which included 12 studies, showed no enhancement in quality-of-life scores in asthmatic patients. However, probiotics were found to be efficacious in diminishing the amount of asthma attacks [43]. Altogether, the present evidence does not support use of probiotics in the treatment of asthma, although some studies suggest some benefit while harm was not reported [40].

### 3.3. *Probiotics for Prevention of Allergic Rhinitis*

The occurrence of perennial allergic rhinitis and seasonal allergic rhinitis has been rising globally and their management is costly [44] (Table 3). Currently, there is no strong proof that probiotics are successful in preventing allergic rhinitis [45]. Surprisingly, some studies suggest that there may even be an increased prevalence of allergic rhino-conjunctivitis in patients taking probiotics in the perinatal period and in childhood [46]. In a systematic review published in 2014, five RCTs that have studied the preventive role of probiotics in allergic rhinitis were assessed. Combining data from adults and children, no difference in incidence of allergic rhinitis between the probiotic and control groups (odds ratio (OR) 1.07, 95% CI, 0.81–1.42, $p = 0.64$, fixed-effects model), and no significant difference in the prevention of allergic rhinitis have been found [47]. A 2019 meta-analysis of seventeen RCTs including 5264 children could not identify a clear advantage of probiotic supplementation during pre- and postnatal periods in the prevention of allergic rhinitis [48]. In follow-up research of a previous study investigating the pre- and postnatal usage of probiotics in high-risk children between five and ten years of age, Peldan et al. sent surveys to their parents to investigate if atopic diseases, including allergic rhinitis, were present. The lifetime prevalence of allergic rhinitis was equal in both probiotic and placebo groups (35.2% vs. 41.7%, adjusted OR 0.74, 95% CI 0.55–1.00, $p < 0.05$); nevertheless, the prevalence of allergic rhino-conjunctivitis at five to ten years of age was greater in the probiotic than in the placebo group (36.5% vs. 29.0%, OR 1.43, 95% CI 1.06–1.94, $p = 0.03$) [46]. Following the authors of this study, the question form may be biased since manifestations of viral rhinitis may be mistaken for allergic rhinitis [46]. After a follow-up of 11 years, the same negative outcome of no association (RR 0.85, 95% CI 0.65–1.1, $p = 0.24$) was found for probiotics *Lactobacillus rhamnosus* HN001 and *Bifidobacterium lactis HN019* taken by mothers every day from 35-week gestation to six months postnatally while breastfeeding and by infants from birth to two years of age [39]. However, similar to the prevention of asthma, the absence of evidence for a potential benefit may be due to shortcomings in study designs and the presence of multiple confounding variables. Probiotic intervention may have a favorable role in the prevention and additional treatment of allergic rhinitis, although results up to now are disappointing [49].

### 3.4. Probiotics for Treatment of Allergic Rhinitis

Avoidance of contact with allergens, medications to reduce symptoms to decrease inflammation and immunotherapy are standard approaches in the management of allergic rhinitis [50]. The question raised is if oral probiotics might modulate the microbiome in such a way that they result in an alteration of the immune system which would contribute to the treatment of allergic rhinitis [51] (Table 4). The development of allergic inflammation in a murine house dust mite asthma model is suppressed by synbiotic mixtures of non-digestible oligosaccharides and *Bifidobacterium breve* M-16V [52].

A review from 2010 (including seven trials, n = 616, children and adults mixed) suggested that probiotics (*Lactobacillus* spp. and *Bifidobacterium* spp.) contribute to a decrease in allergic rhinitis symptoms, quality of life and decrease the need for drug intake (standard mean difference (SMD) −1.17, 95% CI −1.47–0.86; $p < 0.00001$) [53]. Another meta-analysis performed in 2014 including 11 RCTs reported similar conclusions, as probiotics significantly improved both quality of life and nasal symptom scores (SMD −2.97, 95% CI, −4.77–1.16, $p = 0.001$). However, this was not associated with an improvement in immunologic variables [47]. This meta-analysis was criticized for its methodology [47,54]. A 2016 meta-analysis of 22 RCTs also came up with evidence of a potential benefit of probiotics, once more demonstrating improvement in quality of life. A clinically significant benefit was reported for at least one outcome in 17 studies, while no benefit could be shown in six trials. Improvement was mainly regarding quality of life (SMD −2.30, 85% CI −3.93 to −0.67, $p = 0.006$), while no effect was shown on rhinitis symptoms (SMD −0.34, 95% CI −0.62–0.07; $p = 0.13$) or total IgE levels (SMD 0.01, 95% CI −0.17–0.19, $p = 0.88$), and for antigen-specific IgE (SMD 0.09, 95% CI −0.44–0.62, $p = 0.74$) in the placebo group compared to the probiotic. Studies are characterized by a high degree of heterogeneity in probiotic strains tested, inclusion criteria and outcomes [55].

In 212 children under five-years-old from Pakistan, a probiotic product administered as a chewable tablet, containing two x $10^9$ CFU of *Lactobacillus* Paracasei (LP-33), was administered once a day for six weeks while the control group was treated with cetirizine tablet 2.5 mg (<two years) or 5 mg (two-five years) once daily. Significant improvement from baseline symptoms (rhinorrhea, sneezing, nasal blocking, coughing, feeding difficulties and sleeping difficulties) was reported equally in both groups in almost all children [56]. Although the title of the paper mentions probiotics, the study was in fact performed with postbiotics since it was lyophilized extracts of bifidobacteria which were shown to suppress allergic rhinitis in mice via inducing IL-10-producing B cells [57]. Another study (with mice) showed that Clostridium butyricum extracts—again, postbiotics—can efficiently inhibit experimental allergic rhinitis by increasing IL-10 expression in B cells [58].

A pilot study in only 20 adult (18–65-years-old) patients with allergic rhinitis caused by house dust mite allergy suggests that probiotics-impregnated bed linen with five natural genetically unmodified bacterial probiotic strains of Bacillus species (strains of Bacillus subtilis, Bacillus amyloliquefaciens and Bacillus pumilus) reduces symptoms and increases quality of life [59]. A large-scale study is recommended to further investigate all these findings [59].

**Table 1.** Studies examining probiotics for prevention of asthma in humans.

| # | Author and Publication Date | Country | Type of Study | Number of Trials and or Patients | Age | Type of Probiotic and Dose(s) (cfu) | Administration Duration of Probiotics | Follow-Up | Effect |
|---|---|---|---|---|---|---|---|---|---|
| 1 | Elazab et al., 2013 [14] | United States of America | Metanalysis including: Double-blinded, randomized, placebo-controlled trials | 25 studies of 20 cohorts ($n = 4031$) *Only* 10 trials were included ($n = 3143$) to look at probiotics and risk of asthma/wheeze | Birth up to 6 years. | *Lactobacillus* spp. and *Bifidobacterium* spp. or mixed probiotics *Dose:* 1–550 × $10^8$ cfu | 1–13.5 months 8 trial probiotics administrated also prenatal | 0–70 months | Probiotics did not significantly reduce asthma/wheeze (RR 0.96, 95% CI 0.85–1.07), no evidence of publication bias ($p = 0.25$) |
| 2 | Gorissen et al., 2014 [37] | The Nether-lands | Prospectively in a single-blinded (investigator blinded) design. | $n = 123$ and 83 at age of 6 years | Birth up to 6 years | Probiotic mixture consisting of *B bifidum*, *B lactis* and *Lactococcus lactis* *Dose:* not mentioned | 0–24 months Administration started prenatal. | 0–24 months and once at 6 years of age | Did not lead to prevention of asthma at 1 and 6 years of age |
| 3 | Wickens et al. 2018 [39] | New Zealand | two-center randomized placebo-controlled trial | $n = 407$ | Birth–11 years | *L. rhamnosus* HN001 or *B. lactis* HN019 *Dose*: 6 × $10^9$ cfu or 9 × $10^9$ cfu | Daily from 35-week gestation to 6 months' post-partum in mothers while breastfeeding and birth to age 1 years in infants | Birth to 11 years | No association with the development of allergic disease was found RR 0.59, 95% CI 0.36–0.96, $p = 0.059$ |
| 4 | Wei et al., 2020 [12] | China | meta-analysis included randomized, double blind, placebo-controlled trials | Total: 19 RCTs ($n = 5157$) 14 RCTs $n = 4021$ for the analysis of asthma (of which 10 RCTs already used in Elazab et al. (2013)) | Birth-8 years | 10 RCTs *L.* spp. 1 RCT *B.* spp. 6 RCTs probiotic mixtures *Dose*: daily ranged from $10^8$ to $10^{11}$ cfu | 3–24 months | 1–8 years | No significant association of probiotics with risk of asthma (RR 0.94, 95% CI, 0.82–1.09) or wheeze (RR 0.97, 95% CI, 0.88–1.06) compared with placebo. Subgroup analysis by asthma risk showed that probiotics significantly reduced wheeze incidence among infants with atopy disease (RR 0.61, 95% CI, 0.42–0.90), |
| 5 | Davies et al., 2018 [38] | United Kingdom | randomized, double-blind, placebo-controlled, parallel group trial | $n = 318$ | Birth up to 5 years | *L.* spp. and *B.* spp. or mixed probiotics *Dose*: 1 × $10^{10}$ cfu per day | from 36 weeks' gestation, and then administered to their infants during their first 6 months of life | Follow up at 2 years and 5 years | No reduction in asthma after 2 or 5 years |

Legend. #= number; RCT, randomized controlled trial; spp., species; cfu, colony-forming unit; CI, confidence interval; RR, relative risk; L: Lactobacillus; B: Bifidobacterium.

**Table 2.** Studies examining probiotics for the treatment of asthma.

| # | Author and Publication Date | Country | Type of Study | Number of Patients | Age | Type of Probiotic | Administration Duration of Probiotics | Follow-Up | Effect |
|---|---|---|---|---|---|---|---|---|---|
| 1 | Sharma et al., 2018 [40] | Korea | Review | 1 RCT examining probiotics for treatment of asthma with $n$ = 105 | 6 up to 2 years | *L. gasseri* PM-A0005 (A5; $2 \times 10^9$cells/capsule) twice a day | 8 weeks | Observation period of 10 weeks | Pulmonary function and PEFR increased significantly and the clinical symptom scores for asthma decreased in the probiotic group |
| 2 | Vliagoftis et al., 2008 [42] | Multi-center (Greece, The United States of America, Canada) | Systematic review of random-ized controlled trials | 4 RCTs ($n$= 257) | 2–13 years (1 RCT included patients up to 45 years) | 1 RCT *L. casei* ($10^{10}$ cfu); 1 RCT *Enterococcus faecalis* ($18 \times 10^7$ cfu); 1 RCT *L. rhamnosus* ($10^{10}$ cfu); 1 RCT *L.acidophilus* ($7.6 \times 10^8$ cfu) | 4 weeks up to 1 year | 22 up to 56 weeks | No effect of probiotics on asthma treatment |
| 3 | Das et al., 2013 [43] | India | Systematic review | 10 RCTs ($n$ = 860) and 2 Randomized crossover designs ($n$ = 39) | 2 up to 16 years (7 RCTs also included adults up to 57 years) | Different strains (*L. salivarius, gasseri, acidophilus, paracasei, rhamnosus, Bulgaricus; Streptococcus thermophilus; B. longum* 5 36) | 1 month up to 1 year | Not mentioned | No improvement in quality-of-life score in asthmatics. Longer time free from episodes of asthma (mean (95% CI 3.5, 2.7–4.3) versus 2.1 (1.5–2.7) months) ($p$ = 0.027)) |

Legend: #= number; RCT, randomized controlled trial; spp., species; cfu, colony-forming unit; CI, confidence interval; RR, relative risk; L: Lactobacillus; B: Bifidobacterium.

**Table 3.** Studies examining probiotics for prevention of allergic rhinitis.

| # | Author and Publication Date | Country | Type of Study | Number of Patients | Age | Type of Probiotic | Administration Duration of Probiotics | Follow-up | Effect |
|---|---|---|---|---|---|---|---|---|---|
| 1 | Zuccotti et al., 2015 [45] | Italy | Systematic review and meta-analysis | 17 RCTs ($n$ = 4755) | Children, not otherwise specified | ## | ## | 2 months up to 7 years | No significant difference in terms of prevention of rhino-conjunctivitis (RR 0.91, 95% CI 0.67–1.23, $p$ = 0.53) was documented. |
| 2 | Peng et al., 2015 [47] | China | Systematic review | A total of 11 RCTs of which 5 addressed the preventive role of probiotics in AR ($n$ = 1527) | Mothers from 36 weeks of gestation; Infants from birth up to adults | *L.* spp. and *B.* spp. or mixed probiotics *Dose*: wide range of probiotic doses applied | ## | ## | No difference in the incidence of AR between probiotic and placebo groups. Improvement in overall quality of life and nasal symptom scores (MD—2.97 95% CI, −4.77–1.16; $p$ = 0.001). |
| 3 | Du et al., 2019 [48] | China | Meta-analysis | 17 RCTs ($n$ = 5264) | Children | Variable strains | ## | ## | No clear benefit of probiotics in the prevention of allergic rhinitis |
| 4 | Wickens et al., 2018 [39] | New-Zealand | Two-center randomized placebo-controlled trial | $n$ = 407 | Birth up to 11 years | *L. rhamnosus* HN001 or *B. lactis* HN019 *Dose*: 6 × $10^9$ cfu or 9 × $10^9$ cfu | Daily from 35-week gestation to 6 months post-partum in mothers while breastfeeding and birth to age 1 years in infants | Birth up to 11 years | Children taking HN001 had a notable reduction in the risk of rhinitis (RR 0.79, 95% CI 0.59–1.05, $p$ = 0.1). Among *Bifidobacterium lactis* HN019 children, there was no notable reduction in allergic rhinitis prevalence. |

Legend: #= number; RCT, randomized controlled trials; spp., species; cfu, colony-forming unit; CI, conventional interval; RR, relative risk, ## not mentioned or not applicable; L: Lactobacillus; B: Bifidobacterium

**Table 4.** Studies examining probiotics for treatment of allergic rhinitis.

| # | Author and Publication Date | Country | Type of Study | Number of Patients | Age | Type of Probiotic | Administration Duration of Probiotics | Long-Term Follow Up | Effect |
|---|---|---|---|---|---|---|---|---|---|
| 1 | Das et al., 2010 [53] | India | systematic reviews | 7 RCTs $n = 616$ | Any age | *L.* spp. and *B.* spp. Additionally, one RCT Streptococcus thermophilus *Dose:* Huge variation | 1 up to 2 months | none | Decrease in allergic rhinitis symptoms, quality of life and need for drug intake |
| 2 | Peng et al., 2015[47] | China | systematic review and meta-analysis | 11 RCTs $n = 1527$ | During pregnancy up to 9 months of age | Different: *Propionibacterium freudenreichii* ssp. shermanii JS, *S. thermophiles; L.* spp. and *B.* spp. *Dose:* Huge variation | 4–9 months Some started during pregnancy | none | Significantly improved both quality of life and nasal symptom scores |
| 3 | Guvenc et al., 2016 [55] | Turkey | Systematic review and metanalysis | 22 RCTs $n = 2242$ ($n =$ 1953 after losses to follow-up) | 2 up to 65 years of age | Huge variance in probiotics and doses | From 1 up to 12 months | none | 16 RCTs had significant benefits of probiotics on clinical parameters; 9 RCTs had significant improvement in immunologic parameters compared with placebo. Meta-analysis significant ameliorated nasal and ocular symptoms and QoL scores |
| 4 | Ahmed et al., 2019 [56] | Pakistan | RCT | $n = 212$ | 6 to 60 months | *L. Paracasei* (LP-33) *Doses:* $2 \times 10^9$ cfu once daily | 6 weeks | none | Probiotic (LP-33) was equally effective as cetirizine in under five years children for the treatment of perennial allergic rhinitis |
| 5 | Berings et al., 2017 [59] | Belgium | pilot double-blind, randomized, placebo-controlled, crossover trial | $n = 24$ | 18–65 years | Purotex® textile treatment contains five different probiotic and natural (not genetically modified) bacterial strains of *Bacillus* species. *Doses*: unknown | | | significant improvement in symptoms and QoL |

Legend. #: number; RCT randomized controlled trials, spp. species cfu: colony-forming unit, CI confidence interval, RR relative risk; L: Lactobacillus; B: Bifidobacterium; S: Stretococcus.

### 3.5. Prebiotics for Prevention/Treatment of Asthma or Allergic Rhinitis

Inulin, fructo-oligosaccharides and galacto-oligosaccharides are well known examples of prebiotics. Table 5 provides an overview of the literature. These substrates will contribute to the growth of two common bacteria in the gut-bifidobacteria and lactobacilli [60]. Some of the substrates interacting with the infant's gut microbiome are human milk oligosaccharides (HMOs) [61], which form the third biggest fraction in human milk [36]. In a mouse model, 2′-fucosyllactose and 6′-sialyllactose decrease the symptoms of food allergy due to the induction of IL-10(+) T regulatory cells and indirect stabilization of mast cells [62]. Prebiotics such as non-human galacto- and fructo-oligosaccharides have been added to infant formula to try to mimic the results of HMOs. However, these non-human prebiotics are less structurally diverse than HMOs [47]. An 18-year follow-up of high-allergy-risk breastfed infants was conducted to evaluate the relation between HMO profiles of the mother and the risk of developing asthma, eczema and sensitization. One HMO profile, namely the acidic Lewis HMOs, showed an increased risk of developing allergic disease and asthma in youth (OR 5.82, 95% CI 1.59–21.23) compared to the neutral Lewis HMO profile. Another finding of the study is that the acidic-predominant profile was associated with a lower risk of food sensitization (OR 0.08, 95% CI 0.01–0.67, $p < 0.05$). HMOs have only been recently available on the market; nevertheless, there are some studies investigating their effect on allergies [63]. A meta-analysis with two studies reporting early respiratory symptoms as outcome (n = 249) has examined if these non-human oligosaccharides have effects on allergy. The study found that infants who received prebiotics (non-human oligosaccharides) had reduced asthma or recurrent wheezing (RR 0.37, 95% CI 0.17–0.80, $p < 0.01$) [64]. Another double blinded RCT (n = 461) compared Chinese toddlers drinking standard milk formula with those drinking a formula containing bioactive proteins and/or the HMO 2′-fucosyllactose and/or milk fat, for a period of six months. In this study, however, no difference was found in the occurrence of upper respiratory infections. No analysis for allergy was conducted [65]. Concluding, there is still little evidence to use prebiotics for the prevention of asthma and none for allergic rhinitis to our knowledge on rhinitis. No studies have been conducted to analyze the effects of prebiotics as a treatment for asthma or allergic rhinitis.

### 3.6. Synbiotics for Prevention/Treatment of Asthma or Allergic Rhinitis

#### 3.6.1. Asthma

The literature on synbiotics regarding prevention and/or treatment of allergic manifestations is still limited (Table 6). Some analyses do not differentiate between pre-, pro- and synbiotics. [66]. Ninety infants with atopic dermatitis were managed with a formula with extensively hydrolyzed protein and were included in a double-blind, placebo controlled multicenter trial for 12 weeks, randomized to the formula with or without synbiotics over a period of seven months. One year later, information regarding respiratory symptoms and asthma medication was collected with a questionnaire. The significant reduced prevalence of "frequent wheezing" and "wheezing and/or noisy breathing apart from colds" was observed in the synbiotic group (13.9% vs. 34.2%, absolute risk reduction (ARR) −20.3%, 95% CI −39.2% to −1.5%, and 2.8% vs. 30.8%, ARR −28.0%, 95% CI −43.3% to −12.5%, respectively). Additionally, the use of asthma medication was significantly lower (5.6% vs. 25.6%, ARR −20.1%, 95% CI −35.7% to −4.5%). However, total IgE levels did not differ. Increased specific cat-IgE levels were noticed in five children (15.2%) in the placebo group versus none in the synbiotic group (ARR −15.2%, 95% CI −27.4% to −2.9%). The outcome of this trial suggests that synbiotics may prevent asthma in infants presenting with atopic dermatitis [67]. However, the limited number of children included in this trial is a major limitation. Cabana et al. [68] performed an RCT in 92 infants with a mixture of LGG and inulin as synbiotic (in the study mentioned as probiotics) between birth and the age of six months of life in infants with mixed breast and formula feeding [68]. Asthma at the age of five years was a secondary outcome, but was not statistically different in both groups with an incidence of 17.4% in the control prebiotic and 9.7% in the symbiotic [68].

**Table 5.** Studies examining prebiotics for prevention/treatment of asthma or allergic rhinitis.

| # | Author and Publication Date | Country | Type of Study | Number of Patients | Age | Type of Prebiotic | Administration Duration of Prebiotics | Long-Term Follow Up | Effect |
|---|---|---|---|---|---|---|---|---|---|
| 1 | Lodge et al., 2020 [63] | Australia | randomized controlled trial of the effects of infant formulas weaning on allergic disease risk, then continued as an observational birth cohort | $n = 145$ | 0–18 years | HMO | 0–12 months breastfeeding | 18 times in the first 2 years, then yearly until 7 years, then at 12 and 18 years | some profiles of HMOs were associated with increased and some with decreased allergic disease risks over childhood |
| 2 | Cuello-Garcia et al. 2016 [64] | International ?? | Meta-analysis 2 RCTs | $n = 249$ | Infant | HMO | ## | ## | HMO reduced asthma or recurrent wheezing |
| 3 | Leung et al., 2020 [65] | China | randomized, controlled, double-blind, parallel-group clinical trial | $n = 461$ | 1–2.5 years | standard formula milk or containing bioactive proteins and/or the HMO 2′-fucosyllactose and/or milk fat | 6 months | 6 months | No reduction for respiratory and gastrointestinal infections in toddlers with HMO |

Legend. #: number; HMO, human milk oligosaccharide; RCT, randomized controlled trial; spp., species; cfu, colony-forming unit; CI, confidence interval; RR, relative risk; ## not mentioned or not applicable.

**Table 6.** Studies examining synbiotics for prevention/treatment of asthma.

| # | Author and Publication Date | Country | Type of Study | Number of Patients | Age | Type of Synbiotic | Administration Duration of Syniotics | Long-Term Follow Up | Effect |
|---|---|---|---|---|---|---|---|---|---|
| 1 | van der Aa et al., 2011 [67] | The Netherlands | double-blind, placebo-controlled multicenter trial | $n$ = 90 follow up $n$ = 75 | mean age 17.3 months | extensively hydrolyzed formula with *B. breve* M-16V and a galacto/fructo-oligosaccharide mixture *Dose*: Not mentioned | 4 weeks | 1 year | synbiotic mixture prevents asthma-like symptoms in infants with atopic dermatitis. Additionally, less started with medication |
| 2 | Cabana et al., 2017 [68] | United states of America | randomized, double-blind controlled trial | $n$ = 92 | 4 days old–6 years | $10^{10}$ cfu *L. rhamnosus* GG and 225 mg of inulin for first 6 months of life | 6 months | Up to 6 years | No significant reduction in asthma with synbiotics |
| 3 | Hassanzad et al., 2019 [69] | Iran | double-blinded, randomized, placebo-controlled clinical trial | $n$ = 100 | 6.9 ± 2.7 years | *S. thermophiles; L.* spp. and *B.* spp., zinc and FOS (prebiotic) | 6 months | Not mentioned | Less outpatient visits, no significant frequency of asthma attacks and hospitalization due to asthma being exacerbated |

Legend. #:number; RCT, randomized controlled trial; spp., species; cfu, colony-forming unit; CI, confidence interval; RR, relative risk; L: Lactobacillus; B: Bifidobacterium; S: Streptococcus; FOS: fructo-oligosaccharide.

A double-blinded, placebo-controlled RCT performed in Iranian children younger than 12 years tested the efficacy of synbiotic (Kidilact®: Streptococcus thermophilus, *Bifidobacterium* spp., *Lactobacillus* spp. zinc and fructo-oligosaccharide) asthma management. Multiple outcomes did not show a difference between both groups; the number of outpatient visits, 19 in the synbiotic versus 55 in the control arm ($p = 0.001$), was the only statistically significant difference [69].

3.6.2. Allergic Rhinitis

The effect of synbiotics on prevention of allergic rhinitis will remain unanswered because no RCTs have been conducted yet to our knowledge (Table 7). Clinical symptoms and quality of life improve with immunotherapy, but synbiotics do not contribute to this improvement.

**Table 7.** Studies examining synbiotics for prevention/treatment of allergic rhinitis.

| # | Author and Publication Date | Country | Type of Study | Number of Patients | Age | Type of Synbiotic | Administration Duration of Synbtics | Long-Term Follow Up | Effect |
|---|---|---|---|---|---|---|---|---|---|
| 1 | Dehnavi et al., 2019 [70] | Iran | placebo-controlled, double-blind RCT | $n = 20$ | 9 up to 53 years | *S. thermophilus*, *B.* spp., *L.* spp., FOS *Doses*: not written | 2 months | Total of 6 months | Significant reduction in IL-17 gene expression following administration of symbiotic. Clinical symptoms and quality of life were improved with immunotherapy. Synbiotics did not have additional effects |
| 2 | Jalali et al., 2019 [51] | Iran | Crossover RCT | $n = 152$ | Adults; 30.1 ± 7.6 years | seven different Gram-positive organisms: $9 \times 10^9$ cfu/g lyophilized lactobacilli (*L. acidophilus*, *L. casei*, *L. delbrueckii* subsp. L. bulgaricus, and *L. rhamnosus*), $1.25 \times 10^{10}$ of bifidobacteria (*B. longum*, and *B. breve*), and $1.5 \times 10^{-10}$ of *S. salivarius* subsp. thermophilus and 38.5 mg FOS | 4 months | Total 2 months | Addition of probiotics to budesonide significantly improved QoL in persistent AR patients |

Legend. #: number; RCT, randomized controlled trial; spp., species; cfu, colony-forming unit; CI, confidence interval; RR, relative risk; L: Lactobacillus; B: Bifidobacterium; S; Stretococcus; FOS; fructo-oligosaccharide.

Synbiotics in the treatment of allergic rhinitis are also poorly studied, although some of the trials reporting on the efficacy of probiotics, in fact, concern synbiotics [37]. A placebo-controlled, double-blind RCT in a small number of children and adults ($n = 20$, age nine-53 years) in Iran showed that immunotherapy and a synbiotic (Streptococcus thermophilus, *Bifidobacterium* spp., *Lactobacillus* spp., fructo-oligosaccharide) reduced the gene expression of IL-17 after two and six months ($p = 0.001$, $p = 0.0001$) more compared to the group receiving immunotherapy and a placebo [70]. Other probiotics [71] were also shown to reduce cytokine IL-17 by directly and indirectly downregulating and suppressing the T helper 17 subset. A 2019 crossover RCT ($n = 152$ subjects (30.1 ± 7.6 years) in adults in Iran showed that adding synbiotics (however, in the study, mentioned as probiotics) to budesonide significantly ameliorated quality of life in persistent allergic rhinitis patients ($p < 0.05$ for social functioning and $p < 0.001$ for mental health and vitalism) [37]. The patient population used in this study may not be representative for allergic rhinitis patients in the overall population, since symptoms did not respond to usual therapy with antihistamines, antileukotrienes, decongestants and nasal steroids [51].

More well-designed studies, investigating only the effects of synbiotics for allergy prevention and/or treatment, are needed [36].

## 4. Conclusions

Meta-analyses have showed marked heterogeneity as well in inclusion criteria, studied products and primary outcomes between studies, making direct comparison hazardous. Today, the American Academy of Pediatrics, the European Academy of Allergy and Clinical Immunology, the National Institute of Allergy and Infectious Disease, and the European Society for Pediatric Gastroenterology, Hepatology and Nutrition do not recommend the use of probiotics for primary prevention of allergic disease [13]. The lack of evidence is the consequence of large heterogeneities between study designs, differences in strains, and dosages and duration of probiotics administered. Future research may clarify these issues [35]. Data from laboratory research in well-controlled conditions demonstrate an important role for gastrointestinal microbiota composition on the development of allergic disease in the respiratory tract, suggesting even a causal relation. Data from clinical human studies remain disappointing. The multiple confounding variables in the clinical situation, therefore, illustrate the impact of environmental and other variables on the development of allergic disease. Overall, we have to conclude that the evidence is insufficient to recommend administration of pro-, pre- or synbiotics in the prevention or treatment of respiratory tract allergies. However, adverse effects are not reported. Additionally, data obtained in controlled situations suggest benefits. Future research requires thoughtful development of appropriate study design according to internationally set standards to ensure uniformity [72]. The modes of action of pro-, pre- and synbiotics need to be further clarified in health and disease [40].

**Author Contributions:** Conceptualization, L.M., E.I.L. and Y.V.; methodology, L.M., E.I.L.; software, L.M., E.I.L.; validation, L.M., E.I.L., Y.V.; formal analysis, L.M., E.I.L.; investigation, L.M., E.I.L.; resources, L.M., E.I.L.; data curation, L.M., E.I.L.; writing—original draft preparation, L.M., E.I.L.; writing—review and editing, L.M. E.I.L.; visualization, L.M., E.I.L.; supervision, E.I.L, Y.V.; project administration, L.M., E.I.L.; funding acquisition, none. Please turn to the CRediT taxonomy for the term explanation. Authorship must be limited to those who have contributed substantially to the work reported. All authors have read and agreed to the published version of the manuscript.

**Funding:** This research received no external funding.

**Institutional Review Board Statement:** Not applicable.

**Informed Consent Statement:** Not applicable.

**Data Availability Statement:** Not applicable.

**Conflicts of Interest:** L.M. and E.I.L. declare no conflict of interest. Y.V. has participated as a clinical investigator, and/or advisory board member, and/or consultant, and/or speaker for Abbott Nutrition,, ByHeart, CHR Hansen, Danone, ELSE Nutrition, Friesland Campina, Nestle Health Science, Nestle Nutrition Institute, Nutricia, Mead Johnson Nutrition, Phathom Pharmaceuticals, United Pharmaceuticals, Wyeth.

## References

1. Asher, M.I.; Montefort, S.; Björkstén, B.; Lai, C.K.; Strachan, D.P.; Weiland, S.K.; Williams, H. Worldwide time trends in the prevalence of symptoms of asthma, allergic rhinoconjunctivitis, and eczema in childhood: ISAAC Phases One and Three repeat multicountry cross-sectional surveys. *Lancet* **2006**, *368*, 733–743. [CrossRef]
2. Meltzer, E.O. Allergic rhinitis: Burden of illness, quality of life, comorbidities, and control. *Immunol. Allergy Clin. N. Am.* **2016**, *36*, 235–248. [CrossRef]
3. Global, regional, and national incidence, prevalence, and years lived with disability for 328 diseases and injuries for 195 countries, 1990–2016: A systematic analysis for the Global Burden of Disease Study 2016. *Lancet* **2017**, *390*, 1211–1259. [CrossRef]
4. To, T.; Stanojevic, S.; Moores, G.; Gershon, A.S.; Bateman, E.D.; Cruz, A.A.; Boulet, L.-P. Global asthma prevalence in adults: Findings from the cross-sectional world health survey. *BMC Public Health* **2012**, *12*, 204. [CrossRef]

5. Leynaert, B.; Sunyer, J.; Garcia-Esteban, R.; Svanes, C.; Jarvis, D.; Cerveri, I.; Dratva, J.; Gislason, T.; Heinrich, J.; Janson, C.; et al. Gender differences in prevalence, diagnosis and incidence of allergic and non-allergic asthma: A population-based cohort. *Thorax* **2012**, *67*, 625–631. [CrossRef] [PubMed]
6. Papi, A.; Brightling, C.; Pedersen, S.E.; Reddel, H.K. Asthma. *Lancet* **2018**, *391*, 783–800. [CrossRef]
7. Kuang, H.; Li, Z.; Lv, X.; Wu, P.; Tan, J.; Wu, Q.; Li, Y.; Jiang, W.; Pang, Q.; Wang, Y.; et al. Exposure to volatile organic compounds may be associated with oxidative DNA damage-mediated childhood asthma. *Ecotoxicol. Environ. Saf.* **2021**, *210*, 111864. [CrossRef] [PubMed]
8. Enilari, O.; Sinha, S. The global impact of asthma in adult populations. *Ann. Glob. Health* **2019**, *85*. [CrossRef] [PubMed]
9. Molero, Y.; Zetterqvist, J.; Lichtenstein, P.; Almqvist, C.; Ludvigsson, J.F. Parental nicotine replacement therapy and offspring bronchitis/bronchiolitis and asthma—A nationwide population-based cohort study. *Clin. Epidemiol.* **2018**, *10*, 1339–1347. [CrossRef]
10. Arokiasamy, P.; Uttamacharya; Kowal, P.; Capistrant, B.D.; Gildner, T.E.; Thiele, E.; Biritwum, R.B.; Yawson, A.E.; Mensah, G.; Maximova, T.; et al. Chronic noncommunicable diseases in 6 Low- and middle-income countries: Findings from wave 1 of the world health organization's study on global ageing and adult health (SAGE). *Am. J. Epidemiol.* **2017**, *185*, 414–428. [CrossRef] [PubMed]
11. Von Mutius, E.; Matsui, E.C. Prevention is the best remedy: What can we do to stop allergic disease? *J. Allergy Clin. Immunol. Pract.* **2020**, *8*, 890–891. [CrossRef]
12. Wei, X.; Jiang, P.; Liu, J.; Sun, R.; Zhu, L. Association between probiotic supplementation and asthma incidence in infants: A meta-analysis of randomized controlled trials. *J. Asthma* **2020**, *57*, 167–178. [CrossRef]
13. Kalliomäki, M.; Antoine, J.M.; Herz, U.; Rijkers, G.T.; Wells, J.M.; Mercenier, A. Guidance for substantiating the evidence for beneficial effects of probiotics: Prevention and management of allergic diseases by probiotics. *J. Nutr.* **2010**, *140*, 713s–721s. [CrossRef] [PubMed]
14. Elazab, N.; Mendy, A.; Gasana, J.; Vieira, E.R.; Quizon, A.; Forno, E. Probiotic administration in early life, atopy, and asthma: A meta-analysis of clinical trials. *Pediatrics* **2013**, *132*, E666–E676. [CrossRef] [PubMed]
15. Dicksved, J.; Flöistrup, H.; Bergström, A.; Rosenquist, M.; Pershagen, G.; Scheynius, A.; Roos, S.; Alm, J.S.; Engstrand, L.; Braun-Fahrländer, C.; et al. Molecular fingerprinting of the fecal microbiota of children raised according to different lifestyles. *Appl. Environ. Microbiol.* **2007**, *73*, 2284–2289. [CrossRef]
16. Braun-Fahrländer, C.; Gassner, M.; Grize, L.; Neu, U.; Sennhauser, F.H.; Varonier, H.S.; Vuille, J.C.; Wüthrich, B. Prevalence of hay fever and allergic sensitization in farmer's children and their peers living in the same rural community. SCARPOL team. Swiss study on childhood allergy and respiratory symptoms with respect to air pollution. *Clin. Exp. Allergy* **1999**, *29*, 28–34. [CrossRef]
17. Ege, M.J.; Mayer, M.; Normand, A.C.; Genuneit, J.; Cookson, W.O.; Braun-Fahrländer, C.; Heederik, D.; Piarroux, R.; von Mutius, E. Exposure to environmental microorganisms and childhood asthma. *N. Engl. J. Med.* **2011**, *364*, 701–709. [CrossRef]
18. Valdes, A.M.; Walter, J.; Segal, E.; Spector, T.D. Role of the gut microbiota in nutrition and health. *Br. Med. J.* **2018**, *361*, k2179. [CrossRef]
19. Swanson, K.S.; Gibson, G.R.; Hutkins, R.; Reimer, R.A.; Reid, G.; Verbeke, K.; Scott, K.P.; Holscher, H.D.; Azad, M.B.; Delzenne, N.M.; et al. The International Scientific Association for Probiotics and Prebiotics (ISAPP) consensus statement on the definition and scope of synbiotics. *Nat. Rev. Gastroenterol. Hepatol.* **2020**, *17*, 687–701. [CrossRef] [PubMed]
20. Goulet, O. Potential role of the intestinal microbiota in programming health and disease. *Nutr. Rev.* **2015**, *73*, 32–40. [CrossRef] [PubMed]
21. Buccigrossi, V.; Nicastro, E.; Guarino, A. Functions of intestinal microflora in children. *Curr. Opin. Gastroenterol.* **2013**, *29*, 31–38. [CrossRef] [PubMed]
22. Sepp, E.; Mikelsaar, M.; Salminen, S. Effect of administration of Lactobacillus casei strain GG on the gastrointestinal microbiota of newborns. *Microb. Ecol. Health Dis.* **1993**, *6*, 309–314. [CrossRef]
23. Trivedi, R.; Barve, K. Gut microbiome a promising target for management of respiratory diseases. *Biochem. J* **2020**, *477*, 2679–2696. [CrossRef] [PubMed]
24. Cervantes-García, D.; Jiménez, M.; Rivas-Santiago, C.E.; Gallegos-Alcalá, P.; Hernández-Mercado, A.; Santoyo-Payán, L.S.; Loera-Arias, M.J.; Saucedo-Cardenas, O.; Montes de Oca-Luna, R.; Salinas, E. Lactococcus lactis NZ9000 prevents asthmatic airway inflammation and remodelling in rats through the improvement of intestinal barrier function and systemic TGF-β production. *Int. Arch. Allergy Immunol.* **2020**, 1–15. [CrossRef]
25. Spacova, I.; Van Beeck, W.; Seys, S.; Devos, F.; Vanoirbeek, J.; Vanderleyden, J.; Ceuppens, J.; Petrova, M.; Lebeer, S. Lactobacillus rhamnosus probiotic prevents airway function deterioration and promotes gut microbiome resilience in a murine asthma model. *Gut Microbes* **2020**, *11*, 1729–1744. [CrossRef] [PubMed]
26. Pellaton, C.; Nutten, S.; Thierry, A.-C.; Boudousquié, C.; Barbier, N.; Blanchard, C.; Corthésy, B.; Mercenier, A.; Spertini, F. Intragastric and intranasal administration of -Lactobacillus paracasei- NCC2461 modulates allergic airway inflammation in mice. *Int. J. Inflamm.* **2012**, *2012*, 686739. [CrossRef] [PubMed]
27. Spacova, I.; Petrova, M.I.; Fremau, A.; Pollaris, L.; Vanoirbeek, J.; Ceuppens, J.L.; Seys, S.; Lebeer, S. Intranasal administration of probiotic Lactobacillus rhamnosus GG prevents birch pollen-induced allergic asthma in a murine model. *Allergy* **2019**, *74*, 100–110. [CrossRef] [PubMed]

28. Jamalkandi, S.A.; Ahmadi, A.; Ahrari, I.; Salimian, J.; Karimi, M.; Ghanei, M. Oral and nasal probiotic administration for the prevention and alleviation of allergic diseases, asthma and chronic obstructive pulmonary disease. *Nutr. Res. Rev.* **2020**, 1–16. [CrossRef] [PubMed]
29. Terada-Ikeda, C.; Kitabatake, M.; Hiraku, A.; Kato, K.; Yasui, S.; Imakita, N.; Ouji-Sageshima, N.; Iwabuchi, N.; Hamada, K.; Ito, T. Maternal supplementation with Bifidobacterium breve M-16V prevents their offspring from allergic airway inflammation accelerated by the prenatal exposure to an air pollutant aerosol. *PLoS ONE* **2020**, *15*, e0238923. [CrossRef] [PubMed]
30. Mennini, M.; Dahdah, L.; Artesani, M.C.; Fiocchi, A.; Martelli, A. Probiotics in asthma and allergy prevention. *Front. Pediatr.* **2017**, *5*, 165. [CrossRef]
31. Feleszko, W.; Jaworska, J.; Rha, R.D.; Steinhausen, S.; Avagyan, A.; Jaudszus, A.; Ahrens, B.; Groneberg, D.A.; Wahn, U.; Hamelmann, E. Probiotic-induced suppression of allergic sensitization and airway inflammation is associated with an increase of T regulatory-dependent mechanisms in a murine model of asthma. *Clin. Exp. Allergy* **2007**, *37*, 498–505. [CrossRef]
32. Lee, J.; Bang, J.; Woo, H.J. Immunomodulatory and anti-allergic effects of orally administered Lactobacillus species in ovalbumin-sensitized mice. *J. Microbiol. Biotechnol.* **2013**, *23*, 724–730. [CrossRef]
33. Huang, R.; Ning, H.; Shen, M.; Li, J.; Zhang, J.; Chen, X. Probiotics for the treatment of atopic dermatitis in children: A systematic review and meta-analysis of randomized controlled trials. *Front. Cell Infect. Microbiol.* **2017**, *7*, 392. [CrossRef]
34. Lyons, A.; O'Mahony, D.; O'Brien, F.; MacSharry, J.; Sheil, B.; Ceddia, M.; Russell, W.M.; Forsythe, P.; Bienenstock, J.; Kiely, B.; et al. Bacterial strain-specific induction of Foxp3+ T regulatory cells is protective in murine allergy models. *Clin. Exp. Allergy* **2010**, *40*, 811–819. [CrossRef] [PubMed]
35. Wang, H.T.; Anvari, S.; Anagnostou, K. The role of probiotics in preventing allergic disease. *Children* **2019**, *6*, 24. [CrossRef]
36. West, C.E.; Dzidic, M.; Prescott, S.L.; Jenmalm, M.C. Bugging allergy; role of pre-, pro- and synbiotics in allergy prevention. *Allergol. Int.* **2017**, *66*, 529–538. [CrossRef] [PubMed]
37. Gorissen, D.M.; Rutten, N.B.; Oostermeijer, C.M.; Niers, L.E.; Hoekstra, M.O.; Rijkers, G.T.; van der Ent, C.K. Preventive effects of selected probiotic strains on the development of asthma and allergic rhinitis in childhood. The Panda study. *Clin. Exp. Allergy* **2014**, *44*, 1431–1433. [CrossRef] [PubMed]
38. Davies, G.; Jordan, S.; Brooks, C.J.; Thayer, D.; Storey, M.; Morgan, G.; Allen, S.; Garaiova, I.; Plummer, S.; Gravenor, M. Long term extension of a randomised controlled trial of probiotics using electronic health records. *Sci. Rep.* **2018**, *8*, 7668. [CrossRef]
39. Wickens, K.; Barthow, C.; Mitchell, E.A.; Kang, J.; van Zyl, N.; Purdie, G.; Stanley, T.; Fitzharris, P.; Murphy, R.; Crane, J. Effects of Lactobacillus rhamnosus HN001 in early life on the cumulative prevalence of allergic disease to 11 years. *Pediatr. Allergy Immunol.* **2018**, *29*, 808–814. [CrossRef]
40. Sharma, G.; Im, S.H. Probiotics as a potential immunomodulating pharmabiotics in allergic diseases: Current status and future prospects. *Allergy Asthma Immunol. Res.* **2018**, *10*, 575–590. [CrossRef]
41. Wang, W.; Luo, X.; Zhang, Q.; He, X.; Zhang, Z.; Wang, X. Bifidobacterium infantis relieves allergic asthma in mice by regulating Th1/Th2. *Med. Sci. Monit.* **2020**, *26*, e920583. [CrossRef]
42. Vliagoftis, H.; Kouranos, V.D.; Betsi, G.I.; Falagas, M.E. Probiotics for the treatment of allergic rhinitis and asthma: Systematic review of randomized controlled trials. *Ann. Allergy Asthma Immunol.* **2008**, *101*, 570–579. [CrossRef]
43. Das, R.R.; Naik, S.S.; Singh, M. Probiotics as additives on therapy in allergic airway diseases: A systematic review of benefits and risks. *BioMed Res. Int.* **2013**, *2013*. [CrossRef] [PubMed]
44. Brozek, J.L.; Bousquet, J.; Baena-Cagnani, C.E.; Bonini, S.; Canonica, G.W.; Casale, T.B.; van Wijk, R.G.; Ohta, K.; Zuberbier, T.; Schünemann, H.J. Allergic rhinitis and its impact on asthma (ARIA) guidelines: 2010 revision. *J. Allergy Clin. Immunol.* **2010**, *126*, 466–476. [CrossRef]
45. Zuccotti, G.; Meneghin, F.; Aceti, A.; Barone, G.; Callegari, M.L.; Di Mauro, A.; Fantini, M.P.; Gori, D.; Indrio, F.; Maggio, L.; et al. Probiotics for prevention of atopic diseases in infants: Systematic review and meta-analysis. *Allergy* **2015**, *70*, 1356–1371. [CrossRef]
46. Peldan, P.; Kukkonen, A.K.; Savilahti, E.; Kuitunen, M. Perinatal probiotics decreased eczema up to 10 years of age, but at 5–10 years, allergic rhino-conjunctivitis was increased. *Clin. Exp. Allergy* **2017**, *47*, 975–979. [CrossRef] [PubMed]
47. Peng, Y.; Li, A.; Yu, L.; Qin, G. The role of probiotics in prevention and treatment for patients with allergic rhinitis: A systematic review. *Am. J. Rhinol. Allergy* **2015**, *29*, 292–298. [CrossRef] [PubMed]
48. Du, X.; Wang, L.; Wu, S.; Yuan, L.; Tang, S.; Xiang, Y.; Qu, X.; Liu, H.; Qin, X.; Liu, C. Efficacy of probiotic supplementary therapy for asthma, allergic rhinitis, and wheeze: A meta-analysis of randomized controlled trials. *Allergy Asthma Proc.* **2019**, *40*, 250–260. [CrossRef] [PubMed]
49. Dimitri-Pinheiro, S.; Soares, R.; Barata, P. The microbiome of the nose-friend or foe? *Allergy Rhinol.* **2020**, *11*, 2152656720911605. [CrossRef]
50. Juniper, E.F.; Ståhl, E.; Doty, R.L.; Simons, F.E.; Allen, D.B.; Howarth, P.H. Clinical outcomes and adverse effect monitoring in allergic rhinitis. *J. Allergy Clin. Immunol.* **2005**, *115*, S390–S413. [CrossRef] [PubMed]
51. Jalali, M.M.; Soleimani, R.; Alavi Foumani, A.; Ganjeh Khosravi, H. Add-on probiotics in patients with persistent allergic rhinitis: A randomized crossover clinical trial. *Laryngoscope* **2019**, *129*, 1744–1750. [CrossRef]
52. Verheijden, K.A.; Willemsen, L.E.; Braber, S.; Leusink-Muis, T.; Jeurink, P.V.; Garssen, J.; Kraneveld, A.D.; Folkerts, G. The development of allergic inflammation in a murine house dust mite asthma model is suppressed by synbiotic mixtures of non-digestible oligosaccharides and Bifidobacterium breve M-16V. *Eur. J. Nutr.* **2016**, *55*, 1141–1151. [CrossRef]

53. Das, R.R.; Singh, M.; Shafiq, N. Probiotics in treatment of allergic rhinitis. *World Allergy Organ. J.* **2010**, *3*, 239–244. [CrossRef] [PubMed]
54. Turner, J.H.; Adams, A.S.; Zajac, A. Probiotics in prevention and treatment of allergic rhinitis. *Am. J. Rhinol. Allergy* **2015**, *29*, e224. [CrossRef]
55. Güvenç, I.A.; Muluk, N.B.; Mutlu, F.S.; Eşki, E.; Altintoprak, N.; Oktemer, T.; Cingi, C. Do probiotics have a role in the treatment of allergic rhinitis? A comprehensive systematic review and meta-analysis. *Am. J. Rhinol. Allergy* **2016**, *30*, e157–e175. [CrossRef] [PubMed]
56. Ahmed, M.; Billoo, A.G.; Iqbal, K. Efficacy of probiotic in perennial allergic rhinitis under five year children: A randomized controlled trial. *Pak. J. Med. Sci.* **2019**, *35*, 1538–1543. [CrossRef] [PubMed]
57. Xue, J.M.; Ma, F.; An, Y.F.; Suo, L.M.; Geng, X.R.; Song, Y.N.; Mo, L.H.; Luo, X.Q.; Zhang, X.W.; Liu, D.B.; et al. Probiotic extracts ameliorate nasal allergy by inducing interleukin-35-producing dendritic cells in mice. *Int. Forum Allergy Rhinol.* **2019**, *9*, 1289–1296. [CrossRef] [PubMed]
58. Zeng, X.H.; Yang, G.; Liu, J.Q.; Geng, X.R.; Cheng, B.H.; Liu, Z.Q.; Yang, P.C. Nasal instillation of probiotic extracts inhibits experimental allergic rhinitis. *Immunotherapy* **2019**, *11*, 1315–1323. [CrossRef] [PubMed]
59. Berings, M.; Jult, A.; Vermeulen, H.; De Ruyck, N.; Derycke, L.; Ucar, H.; Ghekiere, P.; Temmerman, R.; Ellis, J.; Bachert, C.; et al. Probiotics-impregnated bedding covers in house dust mite allergic rhinitis patients: A double-blind, randomised, placebo-controlled, crossover clinical trial. *Allergy* **2017**, *72*, 23. [CrossRef]
60. Gibson, G.R.; Hutkins, R.; Sanders, M.E.; Prescott, S.L.; Reimer, R.A.; Salminen, S.J.; Scott, K.; Stanton, C.; Swanson, K.S.; Cani, P.D.; et al. Expert consensus document: The international scientific association for probiotics and prebiotics (ISAPP) consensus statement on the definition and scope of prebiotics. *Nat. Rev. Gastroenterol. Hepatol.* **2017**, *14*, 491–502. [CrossRef]
61. Bode, L. The functional biology of human milk oligosaccharides. *Early Hum. Dev.* **2015**, *91*, 619–622. [CrossRef]
62. Castillo-Courtade, L.; Han, S.; Lee, S.; Mian, F.M.; Buck, R.; Forsythe, P. Attenuation of food allergy symptoms following treatment with human milk oligosaccharides in a mouse model. *Allergy* **2015**, *70*, 1091–1102. [CrossRef] [PubMed]
63. Lodge, C.J.; Lowe, A.J.; Milanzi, E.; Bowatte, G.; Abramson, M.J.; Tsimiklis, H.; Axelrad, C.; Robertson, B.; Darling, A.E.; Svanes, C.; et al. Human milk oligosaccharide profiles and allergic disease up to 18 years. *J. Allergy Clin. Immunol.* **2020**. [CrossRef] [PubMed]
64. Cuello-Garcia, C.A.; Fiocchi, A.; Pawankar, R.; Yepes-Nuñez, J.J.; Morgano, G.P.; Zhang, Y.; Ahn, K.; Al-Hammadi, S.; Agarwal, A.; Gandhi, S.; et al. World allergy organization-McMaster university guidelines for allergic disease prevention (GLAD-P): Prebiotics. *World Allergy Organ. J.* **2016**, *9*, 10. [CrossRef] [PubMed]
65. Leung, T.F.; Ulfman, L.H.; Chong, M.K.C.; Hon, K.L.; Khouw, I.; Chan, P.K.S.; Delsing, D.J.; Kortman, G.A.M.; Bovee-Oudenhoven, I.M.J. A randomized controlled trial of different young child formulas on upper respiratory and gastrointestinal tract infections in Chinese toddlers. *Pediatr. Allergy Immunol.* **2020**, *31*, 745–754. [CrossRef] [PubMed]
66. Azad, M.B.; Coneys, J.G.; Kozyrskyj, A.L.; Field, C.J.; Ramsey, C.D.; Becker, A.B.; Friesen, C.; Abou-Setta, A.M.; Zarychanski, R. Probiotic supplementation during pregnancy or infancy for the prevention of asthma and wheeze: Systematic review and meta-analysis. *Br. Med. J.* **2013**, *347*, 15. [CrossRef] [PubMed]
67. Van der Aa, L.B.; van Aalderen, W.M.; Heymans, H.S.; Henk Sillevis Smitt, J.; Nauta, A.J.; Knippels, L.M.; Ben Amor, K.; Sprikkelman, A.B. Synbiotics prevent asthma-like symptoms in infants with atopic dermatitis. *Allergy* **2011**, *66*, 170–177. [CrossRef]
68. Cabana, M.D.; McKean, M.; Caughey, A.B.; Fong, L.; Lynch, S.; Wong, A.; Leong, R.; Boushey, H.A.; Hilton, J.F. Early probiotic supplementation for eczema and asthma prevention: A randomized controlled trial. *Pediatrics* **2017**, *140*. [CrossRef]
69. Hassanzad, M.; Mostashari, K.M.; Ghaffaripour, H.; Emami, H.; Limouei, S.R.; Velayati, A.A. Synbiotics and treatment of asthma: A Double-blinded, randomized, placebo-controlled clinical trial. *Galen Med. J.* **2019**, *8*. [CrossRef]
70. Dehnavi, S.; Azad, F.J.; Hoseini, R.F.; Moazzen, N.; Tavakkol-Afshari, J.; Nikpoor, A.R.; Salmani, A.A.; Ahanchian, H.; Mohammadi, M. A significant decrease in the gene expression of interleukin-17 following the administration of synbiotic in patients with allergic rhinitis who underwent immunotherapy: A placebo-controlled clinical trial. *J. Res. Med. Sci.* **2019**, *24*, 51. [CrossRef]
71. Tanabe, S. The effect of probiotics and gut microbiota on Th17 cells. *Int. Rev. Immunol.* **2013**, *32*, 511–525. [CrossRef] [PubMed]
72. Huang, Y.J.; Marsland, B.J.; Bunyavanich, S.; O'Mahony, L.; Leung, D.Y.; Muraro, A.; Fleisher, T.A. The microbiome in allergic disease: Current understanding and future opportunities-2017 PRACTALL document of the American Academy of Allergy, Asthma and Immunology and the European Academy of Allergy and Clinical Immunology. *J. Allergy Clin. Immunol.* **2017**, *139*, 1099–1110. [CrossRef] [PubMed]

nutrients

*Article*

# Levels of Predominant Intestinal Microorganisms in 1 Month-Old Full-Term Babies and Weight Gain during the First Year of Life

Sonia González [1,2], Marta Selma-Royo [3], Silvia Arboleya [2,4], Cecilia Martínez-Costa [5,6], Gonzalo Solís [7,8], Marta Suárez [7,8], Nuria Fernández [2,9], Clara G. de los Reyes-Gavilán [2,4], Susana Díaz-Coto [10], Pablo Martínez-Camblor [11], Maria Carmen Collado [3,*] and Miguel Gueimonde [2,4,*]

1 Department of Functional Biology, University of Oviedo, 33004 Oviedo, Spain; soniagsolares@uniovi.es
2 Diet, Microbiota and Health Group, Instituto de Investigación Sanitaria del Principado de Asturias (ISPA), 33004 Oviedo, Spain; silvia.arboleya@ipla.csic.es (S.A.); nuriajmhd@gmail.com (N.F.); greyes_gavilan@ipla.csic.es (C.G.d.l.R.-G.)
3 Institute of Agrochemistry and Food Technology (IATA-CSIC), 46980 Paterna, Spain; mselma@iata.csic.es
4 Department of Microbiology and Biochemistry of Dairy Products, Instituto de Productos Lácteos de Asturias (IPLA-CSIC), 33300 Asturias, Spain
5 Department of Pediatrics, School of Medicine, University of Valencia, 46010 Valencia, Spain; cecilia.martinez@uv.es
6 Pediatric Gastroenterology and Nutrition Section, INCLIVA Research Center, Hospital Clínico Universitario Valencia, 46010 Valencia, Spain
7 Pediatrics Service, Hospital Universitario Central de Asturias, SESPA, 33004 Oviedo, Spain; gsolis@telefonica.net (G.S.); msr1070@hotmail.com (M.S.)
8 Pediatrics Research Group, Instituto de Investigación Sanitaria del Principado de Asturias (ISPA), 33004 Oviedo, Spain
9 Pediatrics Service, Hospital de Cabueñes, SESPA, 33201 Gijón, Spain
10 Department of Statistics, University of Oviedo, 33004 Oviedo, Spain; UO266718@uniovi.es
11 Department of Anesthesiology, Geisel School of Medicine at Dartmouth, Dartmouth, NH 03756, USA; Pablo.Martinez-Camblor@hitchcock.org
* Correspondence: mcolam@iata.csic.es (M.C.C.); mgueimonde@ipla.csic.es (M.G.)

**Citation:** González, S.; Selma-Royo, M.; Arboleya, S.; Martínez-Costa, C.; Solís, G.; Suárez, M.; Fernández, N.; de los Reyes-Gavilán, C.G.; Díaz-Coto, S.; Martínez-Camblor, P.; et al. Levels of Predominant Intestinal Microorganisms in 1 Month-Old Full-Term Babies and Weight Gain during the First Year of Life. *Nutrients* **2021**, *13*, 2412. https://doi.org/10.3390/nu13072412

Academic Editor: Nadja Haiden

Received: 15 June 2021
Accepted: 13 July 2021
Published: 14 July 2021



**Abstract:** The early life gut microbiota has been reported to be involved in neonatal weight gain and later infant growth. Therefore, this early microbiota may constitute a target for the promotion of healthy neonatal growth and development with potential consequences for later life. Unfortunately, we are still far from understanding the association between neonatal microbiota and weight gain and growth. In this context, we evaluated the relationship between early microbiota and weight in a cohort of full-term infants. The absolute levels of specific fecal microorganisms were determined in 88 vaginally delivered and 36 C-section-delivered full-term newborns at 1 month of age and their growth up to 12 months of age. We observed statistically significant associations between the levels of some early life gut microbes and infant weight gain during the first year of life. Classifying the infants into tertiles according to their *Staphylococcus* levels at 1 month of age allowed us to observe a significantly lower weight at 12 months of life in the C-section-delivered infants from the highest tertile. Univariate and multivariate models pointed out associations between the levels of some fecal microorganisms at 1 month of age and weight gain at 6 and 12 months. Interestingly, these associations were different in vaginally and C-section-delivered babies. A significant direct association between *Staphylococcus* and weight gain at 1 month of life was observed in vaginally delivered babies, whereas in C-section-delivered infants, lower *Bacteroides* levels at 1 month were associated with higher later weight gain (at 6 and 12 months). Our results indicate an association between the gut microbiota and weight gain in early life and highlight potential microbial predictors for later weight gain.

**Keywords:** infants; microbiota; *Staphylococcus*; *Enterococcus*; *Bifidobacterium*; weight gain

## 1. Introduction

From birth, and initially depending on the delivery mode, the neonatal gut is colonized by a rapidly diversifying microbiota, reaching an adult-type microbiota around 3–5 years of life. During early life, other perinatal factors, such as feeding practices, environment or antibiotic treatments, also contribute to shaping the microbiota development [1]. Current evidence supports the role of this early microbiota in promoting and maintaining a balanced immune response and adequate brain development and, subsequently, in the future health of the infant [2,3]. Induction of early microbiota alterations by antibiotics use has been linked to allergic diseases [4], obesity [5], risk of colorectal cancer [6] and other potential non-communicable diseases (NCDs) later in life [7,8]. These studies underline the importance of the early life microbiota as a key driver for adequate infant development and later health. Moreover, recent evidence indicates that altering this early microbiota may also have long-lasting effects on body weight and weight gain in childhood and on the later risk of obesity during adulthood [9–13]. Indeed, higher birth weight and rapid growth during early life have been linked to increased risk of overweight and obesity during childhood and adulthood [14–18]. Interestingly, a recent study reported the very early life microbiota which is present in meconium or first-pass neonatal samples as a predictor of infant overweight by the age of 2 years [19]. Early microbiota composition has also been linked to overweight and obesity at later infancy [20,21]. On the other side, other recent studies have highlighted the effect of antibiotic treatment on infant growth and development during the first 6 years of life [12].

In this context, the potential relationship between early microbiota and weight gain is of great interest since this relationship offers opportunities for the microbiota-mediated modulation of weight gain [22] and/or the prevention of growth impairment [23]. Recently, some studies have assessed the potential association between early microbiota and weight gain in preterm infants [24,25]; however, data on full-term babies are still scarce.

In this study, we evaluated the association between the early life microbiota and the later weight gain in both vaginally and C-section-delivered healthy full-term infants. We aimed at identifying if the levels of selected microorganisms at 1 month of age were related to later weight gain during the first year of life in these two groups of infants. With this goal, we used quantitative PCR (qPCR) for assessing specific microbial groups in the infant's feces at the age of 1 month and monitored weight gain during the first year of life.

## 2. Subjects and Methods

### 2.1. Recruitment and Sampling

A total of 124 infants born after a full-term, uncomplicated pregnancy by vaginal delivery ($n$ = 88) or by C-section ($n$ = 36) were recruited at the neonatology units of the University Central Hospital of Asturias (Oviedo, Northern area, Spain) and the University Clinic Hospital of Valencia (Valencia, Mediterranean area, Spain). Inclusion criteria were no metabolic (obesity, diabetes) or chronic diseases and no probiotics consumption by mothers during late pregnancy or infants' early life and no antibiotics administration to the infants. Clinical data such as gestational age or perinatal maternal antibiotics were recorded, as well as neonatal weight and height at birth, at 1, 6 and 12 months of life, the change in weight (weight gain), Z-scores for weight, height and weight-for-height were calculated for each infant at each time point. A fresh fecal sample was collected at 1 month of age and immediately frozen for later microbiota analyses.

Families received detailed study information and signed an informed consent form. The study recruitment and sampling have been approved by the Regional Ethics Committee of Clinical Research of Asturias (Ref. 12/16. 3 February 2016), the Ethics Committee of the Hospital Clínico Universitario de Valencia INCLIVA (Ref. 9 January 2015) and the Committee on Bioethics of CSIC. The procedures were performed in accordance with the fundamental principles set out in the Declaration of Helsinki, the Oviedo Bioethics Convention, the Council of Europe Convention on Human Rights and Biomedicine and the Spanish legislation on bioethics. The Directive 95/46/EC of the European Parliament and the Council of 24 October 1995, on the protection of individuals regarding the processing of personal data and on the free movement of such data were (and will be) strictly followed.

### 2.2. Fecal Microbiota Analyses

Total DNA was isolated from all fecal pellets as described previously by using the QiAGEN Stool Kit (QIAgen. Hilden. Germany) [26]. The extracted DNA was then used for quantifying fecal levels of the Enterobacteriaceae family, the *Bacteroides*-group and the genera *Bifidobacterium*, *Enterococcus*, *Staphylococcus* and *Lactobacillus* group by quantitative PCR using previously described primers, conditions and standard cultures (Table 1). PCRs were performed either in a LightCycler® 480 Real-Time PCR System (Roche®) or a 7500 Fast Real-Time PCR system (Applied Biosystems) by using the SYBR Green. A subgroup of samples ($n = 33$) was analyzed in both machines to ensure comparability, without detecting statistically significant differences between the data obtained in each of them (data not shown), with both machines showing high correlation (Pearson's correlation coefficients ranging between 0.785 and 0.942 depending in the primer pair used).

**Table 1.** Primers and PCR conditions for the different primer pairs used in this study.

| Bacterial Group | Primers | Tm | Reference |
|---|---|---|---|
| *Bacteroides*-group | F-GAGAGGAAGGTCCCCCAC<br>R-CGCKACTTGGCTGGTTCAG | 56·°C | [26] |
| *Bifidobacterium* genus | F-GATTCTGGCTCAGGATGAACGC<br>R-CTGATAGGACGCGACCCCAT | 60 °C | [27] |
| Enterobacteriaceae family | F-CATTGACGTTACCCGCAGAAGAAGC<br>R-CTCTACGAGACTCAAGCTTGC | 63 °C | [28] |
| *Enterococcus* genus | F-CCCTTATTGTTAGTTGCCATCATT<br>R-ACTCGTTGTACTTCCCATTGT | 61 °C | [29] |
| *Lactobacillus*-group | F-AGCAGTAGGGAATCTTCCA<br>R-CACCGCTACACATGGAG | 58 °C | [30,31] |
| *Staphylococcus* genus | F-ACGGTCTTGCTGTCACTTATA<br>R-TACACATATGTTCTTCCCTAATAA | 56 °C | [32] |

### 2.3. Anthropometrical Determinations

Child height and weight were recorded to the nearest 0.1 cm and 0.1 kg, respectively, through standardized procedures by pediatric nurse and registered at birth, 1, 6 and 12 months. With this information and the date of birth, Z-score was calculated by using WHO ANTHRO, Software for Calculating anthropometry, Version 3.2.2 (https://www.who.int/childgrowth/software/es/; accessed on 14 July 2021). The WHO Child Growth Standards provide child growth measures standardized by age and sex using Z-score.

### 2.4. Statistical Analyses

For statistical analysis, the free software R (https://www.r-project.org; accessed on 7 June 2021) was used. Variables are described by mean and standard deviations, median and percentiles or by counts and frequencies. Student–Welch and Chi-square tests were used for checking the mean and distributions equalities, respectively. Pearson correlation coefficients were used for studying the association between continuous variables. A heatmap, generated in R package using the heatmap.2 application in ggplots package [33] was employed for summarizing the analyses. Comparisons on bacterial levels among the different infants' groups were achieved by using a *t*-test with Bonferroni's correction. Multiple mixed linear models were used for studying the effect of the microbial levels at one month and the weight gain, weight, height and Z-scores at 1, 6 and 12 months. These models were used unadjusted and after adjusting for potential confounders in both groups of infants (vaginally or C-section-delivered). Backward stepwise analyses based on the Aikaike Information Criterion (AIC) were employed to determine whether the variables were included in a potential predictive model. A forest plot was used to show the effect sizes (with 95% confidence intervals) in both the so-labeled univariate and the multivariate models in both groups of infants (adjusting by infant gender and feeding type). Time variation in weight was determined according to the tertile classification of each of the microbial groups analyzed in this study. For this purpose, the cut-off points established were: 1) for vaginally delivered babies; *Bacteroides* group (T1 < 6.72; T2 6.72–8.59; T3 > 8.59); *Bifidobacterium* (T1 < 8.37; T2 8.37–8.98; T3 > 8.98); Enterobacteriaceae (T1 < 7.79; T2 7.79–8.60; T3 > 8.60); *Enterococcus* (T1 < 6.52, T2 6.52–7.60, T3 >7.60); *Lactobacillus* group (T1 < 5.60, T2 5.60–6.75, T3 > 6.75); *Staphylococcus* (T1 < 5.94, T2 5.94–6.75, T3 > 6.75) and 2) for C-section-delivered babies; *Bacteroides* group (T1 < 6.48; T2 6.48–7.30; T3 > 7.30); *Bifidobacterium* (T1 < 7.78; T2 7.78–8.79; T3 > 8.79); Enterobacteriaceae (T1 < 6.95; T2 6.95–8.42; T3 > 8.42); *Enterococcus* (T1 < 6.58, T2 6.58−7.92, T3 > 7.92); *Lactobacillus* group (T1 < 5.23, T2 5.23−6.54, T3 > 6.54); *Staphylococcus* (T1 < 5.40, T2 5.40–6.77, T3 > 6.77). *p*-values below 0.05 were considered statistically significant.

## 3. Results

### 3.1. General Description of the Population

The 124 full-term babies (55 males/69 females) included in this study were born at gestational ages ranging from 37 to 41 weeks (mean 39.6). Of these, 88 babies were delivered vaginally (birth weights between 2135 and 4800 g) and 36 by C-section (birth weights between 2215 and 4690 g). There were no statistically significant differences in mean weight according to delivery mode (mean weight of 3189 vs. 3215 for vaginal and C-section babies, respectively). Fifty-six of the infants born vaginally were exclusively breastfed, whereas 31 babies received formula or mixed feeding at the age of 1 month. In C-section babies, the proportion of children receiving each of these feeding types was 50 percent. Female babies showed a significantly higher rate of vaginal delivery than males (80% vs. 58%, $p = 0.019$), whereas no differences in feeding habits were observed between boys and girls.

### 3.2. Gut Microbiota Composition and Weight Gain Are Affected by Different Variables

In this study, the main microbial phyla representatives were quantified by using group-specific qPCR methods. As expected, the microbiota of vaginally delivered babies was different from that of C-section ones, with significantly ($P < 0.05$) higher levels of *Bacteroides*-group of microorganisms (7.65 $\pm$ 1.42 vs. 6.74 $\pm$ 0.98 Log n° cells/g, respectively) and *Bifidobacterium* (8.52 $\pm$ 0.76 vs. 8.05 $\pm$ 1.02) in the former group. No differences between both groups of infants were observed for any of the other microbial groups analyzed (Enterobacteriaceae, 8.07 $\pm$ 1.11 vs. 7.66 $\pm$ 1.38; *Enterococcus*, 6.96 $\pm$ 1.38 vs. 6.97 $\pm$ 1.56; *Lactobacillus*, 6.08 $\pm$ 1.52 vs. 5.79 $\pm$ 1.58; *Staphylococcus*, 5.94 $\pm$ 1.56 vs. 5.84 $\pm$ 1.60). These differences in the levels of some of the microbial groups analyzed between both delivery mode groups prompted us to consider them as two different groups and analyze them separately.

In both groups of 1 month-old infants, the genus *Bifidobacterium* was the bacterial group showing the highest levels, followed by members of the Enterobacteriaceae family and *Bacteroides*-group (Table 2). Interestingly, no differences in bacterial levels were observed between 1 month-old males and females neither in vaginally delivered nor in C-section-delivered babies. Concerning infant feeding practices, exclusive breastfeeding was found to be associated with reduced levels of enterococci at 1 month of age compared to formula/mixed feeding; the differences reaching statistical significance ($P < 0.05$) in vaginally delivered babies (Table 2).

**Table 2.** Levels (Log n° cells/g) of some relevant bacterial groups in fecal samples of the vaginally delivered or C-section-delivered infant population included in this study, categorized by feeding type, gender and mode of delivery.

| | | Vaginally Delivered Babies ($n$ = 88) | | | | C-Section-Delivered Babies ($n$ = 36) | | | |
|---|---|---|---|---|---|---|---|---|---|
| | | Gender | | Feeding Type | | Gender | | Feeding Type | |
| Phyla | Bacterial Groups | Male ($n$ = 33) | Female ($n$ = 55) | EBF ($n$ = 56) | MF ($n$ = 31) | Male ($n$ = 22) | Female ($n$ = 14) | EBF ($n$ = 18) | MF ($n$ = 18) |
| Bacteroidetes | *Bacteroides* group | 7.63 ± 1.57 $ | 7.67 ± 1.34 $ | 7.55 ± 1.57 | 7.79 ± 1.08 $ | 6.67 ± 0.97 $ | 6.85 ± 1.03 $ | 6.85 ± 0.88 | 6.63 ± 1.09 $ |
| Actinobacteria | *Bifidobacterium* genus | 8.45 ± 0.92 | 8.57 ± 0.65 $ | 8.48 ± 0.76 | 8.60 ± 0.78 $ | 8.16 ± 1.01 | 7.89 ± 1.06 $ | 8.05 ± 1.10 | 8.05 ± 0.98 $ |
| Proteobacteria | Enterobacteriaceae | 8.18 ± 1.11 | 8.01 ± 1.12 | 8.01 ± 1.00 | 8.17 ± 1.33 | 7.51 ± 1.38 | 7.89 ± 1.39 | 7.69 ± 1.31 | 7.62 ± 1.48 |
| Firmicutes | *Enterococcus* genus | 6.64 ± 1.64 | 7.15 ± 1.16 | 6.54 ± 1.31 * | 7.77 ± 1.13 * | 6.89 ± 1.38 | 7.10 ± 1.87 | 6.60 ± 1.68 | 7.34 ± 1.38 |
| | *Lactobacillus* group | 6.05 ± 1.63 | 6.10 ± 1.47 | 6.08 ± 1.55 | 6.10 ± 1.52 | 5.95 ± 1.63 | 5.55 ± 1.52 | 5.41 ± 1.66 | 6.17 ± 1.44 |
| | *Staphylococcus* genus | 6.00 ± 1.48 | 6.02 ± 1.41 | 6.15 ± 1.47 | 5.71 ± 1.34 | 5.43 ± 1.55 | 6.48 ± 1.51 | 5.98 ± 1.62 | 5.70 ± 1.61 |

All values are shown as mean ± standard deviation. EBF, exclusive breastfeeding; MF, formula/mixed feeding. There is a missing value in feeding type ($n$ = 87). * Denotes statistically significant differences ($p \leq 0.05$) between genders or feeding types within the same delivery group. $ Denotes statistically significant differences ($p \leq 0.05$) for infants from the same gender or feeding type between the two delivery groups (vaginally delivered or C-section-delivered).

As expected, when analyzing the anthropometric parameters in the sample (Table 3), statistically significant differences were found between both genders, with body weight and height being higher in males. Moreover, C-section-delivered infants on formula/mixed-feeding showed a significantly lower birth weight and weight and height by the age of 1 month ($P$ = 0.022) without noticing statistically significant differences at a later age. Z-scores showed statistically significant differences in weight for height at 1 and 6 months but not at 12 months of age, and no other statistically significant differences in Z-scores were obtained between vaginally delivered and C-section babies (Supplementary Table S1).

**Table 3.** Weight and weight gain during the first year of life in the infants included in this study as categorized by feeding type, gender and delivery mode.

| | Vaginally Delivered Babies ($n$ = 88) | | | | C-Section-Delivered Babies ($n$ = 36) | | | |
|---|---|---|---|---|---|---|---|---|
| | Gender | | Feeding Type | | Gender | | Feeding Type | |
| Variable | Male ($n$ = 33) | Female ($n$ = 55) | EBF ($n$ = 56) | MF ($n$ = 31) | Male ($n$ = 22) | Female ($n$ = 14) | EBF ($n$ = 18) | MF ($n$ = 18) |
| Weight birth (g) | 3358 ± 520 * | 3036 ± 579 * | 3202 ± 328 | 3087 ± 867 | 3317 ± 629 | 3055 ± 479 | 3504 ± 479 * | 2927 ± 508 * |
| Height birth (cm) | 49.9 ± 2.6 | 48.7 ± 1.8 | 49.2 ± 1.8 | 48.9 ± 2.9 | 49.7 ± 2.6 | 48.6 ± 2.1 | 50.0 ± 2.3 | 48.5 ± 2.4 |
| Weight 1 month (g) | 4310 ± 531 * | 3898 ± 497 * | 4117 ± 428 $ | 3954 ± 702 | 4235 ± 751 | 3887 ± 602 | 4415 ± 769 *,$ | 3784 ± 482 * |
| Height 1 month (cm) | 54.1 ± 2.0 | 53.2 ± 2.7 | 53.7 ± 2.2 | 53.3 ± 3.1 | 53.5 ± 2.8 | 52.6 ± 2.8 | 54.6 ± 2.2 * | 51.7 ± 2.6 * |
| Weight 6 month (g) | 8129 ± 712 * | 7106 ± 738 * | 7355 ± 860 | 7721 ± 877 | 8142 ± 972 * | 7206 ± 890 * | 7951 ± 1169 | 7606 ± 886 |
| Height 6 month (cm) | 68.4 ± 2.3 * | 65.9 ± 2.5 * | 66.6 ± 2.8 | 67.2 ± 2.6 | 67.4 ± 2.5 * | 65.7 ± 2.5 * | 67.2 ± 2.5 | 66.3 ± 2.7 |
| Weight 12 month (g) | 10369 ± 1003 * | 8921 ± 921 * | 9414 ± 1204 | 9515 ± 1142 | 9971 ± 1019 * | 9020 ± 937 * | 9838 ± 1227 | 9363 ± 885 |
| Height 12 month (cm) | 76.7 ± 2.9 * | 73.5 ± 2.9 * | 74.9 ± 3.3 | 74.4 ± 3.3 | 75.3 ± 2.5 | 73.5 ± 3.8 | 75.5 ± 2.8 | 73.7 ± 3.4 |
| Weight gain 1 month (g) | 952 ± 305 | 872 ± 511 | 914 ± 334 | 884 ± 609 | 917 ± 344 | 832 ± 292 | 911 ± 397 | 857 ± 235 |
| Weight gain 6 month (g) | 4717 ± 633 * | 4071 ± 876 * | 4208 ± 882 | 4499 ± 767 | 4824 ± 1058 * | 4151 ± 746 * | 4446 ± 905 | 4679 ± 1091 |
| Weight gain 12 month (g) | 7015 ± 882 * | 5886 ± 1030 * | 6271 ± 1200 | 6350 ± 950 | 6653 ± 1026 * | 5965 ± 831 * | 6334 ± 1007 | 6437 ± 1024 |

All values are shown as mean ± standard deviation. EBF, exclusive breastfeeding; MF, formula/mixed feeding. * Denotes statistically significant differences ($p \leq 0.05$) between gender or feeding types within the same delivery group. $ Denotes statistically significant differences ($p \leq 0.05$) for infants from the same gender or feeding type between the two delivery groups (vaginally delivered or C-section-delivered).

### *3.3. Gut Microbial Groups at 1 Month Are Associated with Weight Gain*

The analysis of Pearson correlation coefficients pointed out different associations between microbes and infant growth depending on the mode of delivery (vaginal and C-section-delivered babies). In vaginally delivered infants, the family Enterobacteriaceae was the microbial group showing more correlations with the infant's growth variables (Figure 1). A significant positive association was observed between the levels of these microorganisms at 1 month and Z-score birth weight, weight at 1 month, Z-score weight at 1 month, Z-score weight for height at 1 month and Z-score weight at 6 months (Figure 1). Similarly, in this group of infants, the levels of *Staphylococcus* showed a significant positive association with weight and Z-score for weight at 1 month of age. In C-section-delivered babies, the only significant correlations observed were the negative association between the levels of *Bacteroides* at 1 month and weight and height (as raw measures and as Z-scores) at the age of 6 months, and the direct association between levels of enterocci and weight gain at 6 months (Figure 1). Although no other statistically significant differences were obtained, the data indicate different interactions between bacteria and infant development depending on the delivery mode; the levels of some microorganisms at 1 month of age, such as *Bacteroides* or *Staphylococcus*, showed a clearly different pattern in vaginally and C-section-delivered infants.

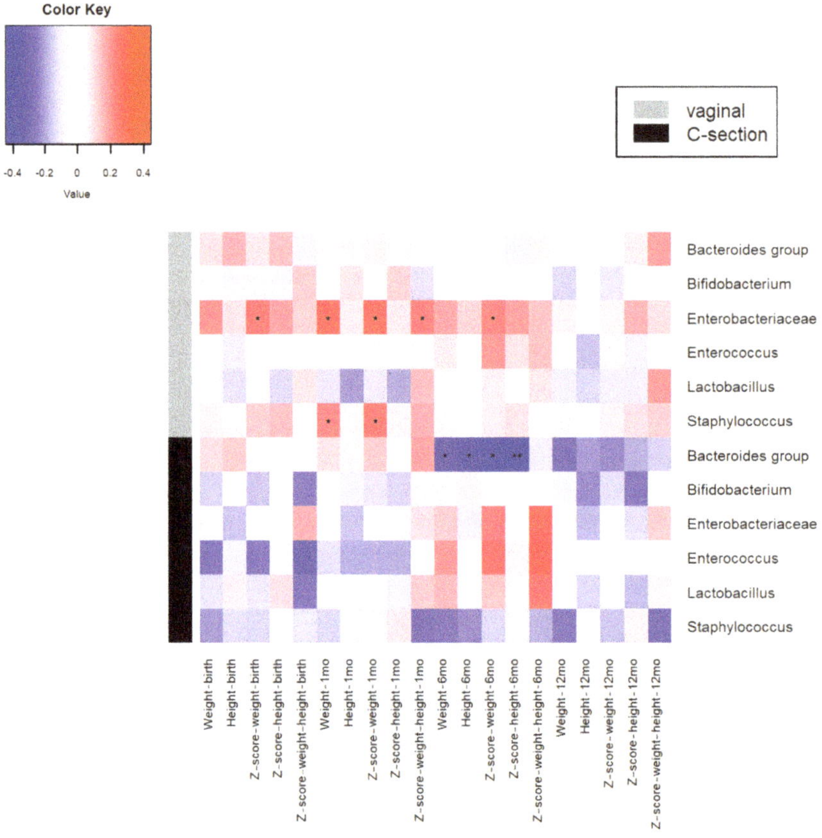


**Figure 1.** Heatmap showing the pairwise Pearson correlation coefficients between microbial groups at 1 month of age and the analyzed growth variables for both vaginally delivered and C-section-delivered babies. * $p < 0.05$.

To gain further insight into these associations, infants were classified according to the tertiles of the levels of the different microorganisms analyzed, and the variations on body weight, along the first year of life, were compared among these tertiles (Figure 2). No statistically significant differences were observed on the evolution of weight during the first 12 months of life among the tertiles for the fecal levels of *Bacteroides*, *Bifidobacterium*, Enterobacteriaceae, *Enterococcus* and *Lactobacillus*, neither in vaginally delivered nor in C-section babies. However, C-section children classified according to the tertiles obtained for the levels of *Staphylococcus* showed statistically significant differences in their weight trajectory (Figure 2). C-section infants harboring high levels of staphylococci at 1 month of age (upper tertile) displayed a significantly lower weight at 1 year of age, with this phenomenon not being observed in vaginally delivered babies.

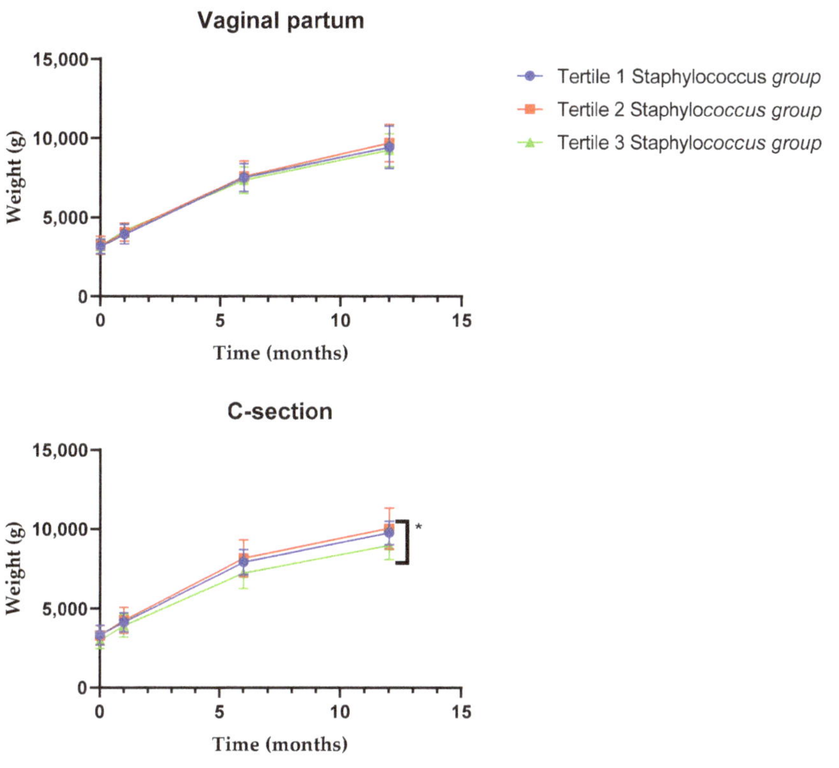


**Figure 2.** Long-term variations in body weight for the tertiles according to fecal levels of the *Staphylococcus* at 1 month of age in vaginally delivered or C-section-delivered babies ($n$ = 122) (T1, tertile 1; T2, tertile 2; T3, tertile 3). * $p \leq 0.05$.

Hereafter, uni- and multivariate regression models were performed for a deeper assessment of the association between early microbiota and weigh-gain in both groups of infants (Figure 3). To take into consideration the potential effects of gender and feeding type, the models were controlled for these two variables, and the relationship between microbiota at 1 month of age and infant weight gain at 1, 6 and 12 months of age was assessed (Figure 3). Different effects were observed between both groups of infants. A significant positive effect of the levels of *Staphylococcus* at 1 month of age on weight gain at 1 month was obtained in both the unadjusted ($p$ = 0.016) and adjusted ($p$ = 0.036) models in vaginally delivered babies, but these do not reach significance in C-section-delivered infants. On the contrary, a negative association of *Bacteroides* levels at 1 month of age with weight gain at 6 and 12 months was observed in the C-section group ($p$ = 0.007 in unadjusted and $p$ = 0.014 in the adjusted model at 6 months of age, and $p$ = 0.031 and $p$ = 0.052, respectively, at 12 months of age). The other microbial groups analyzed did not show any statistically significant effect.

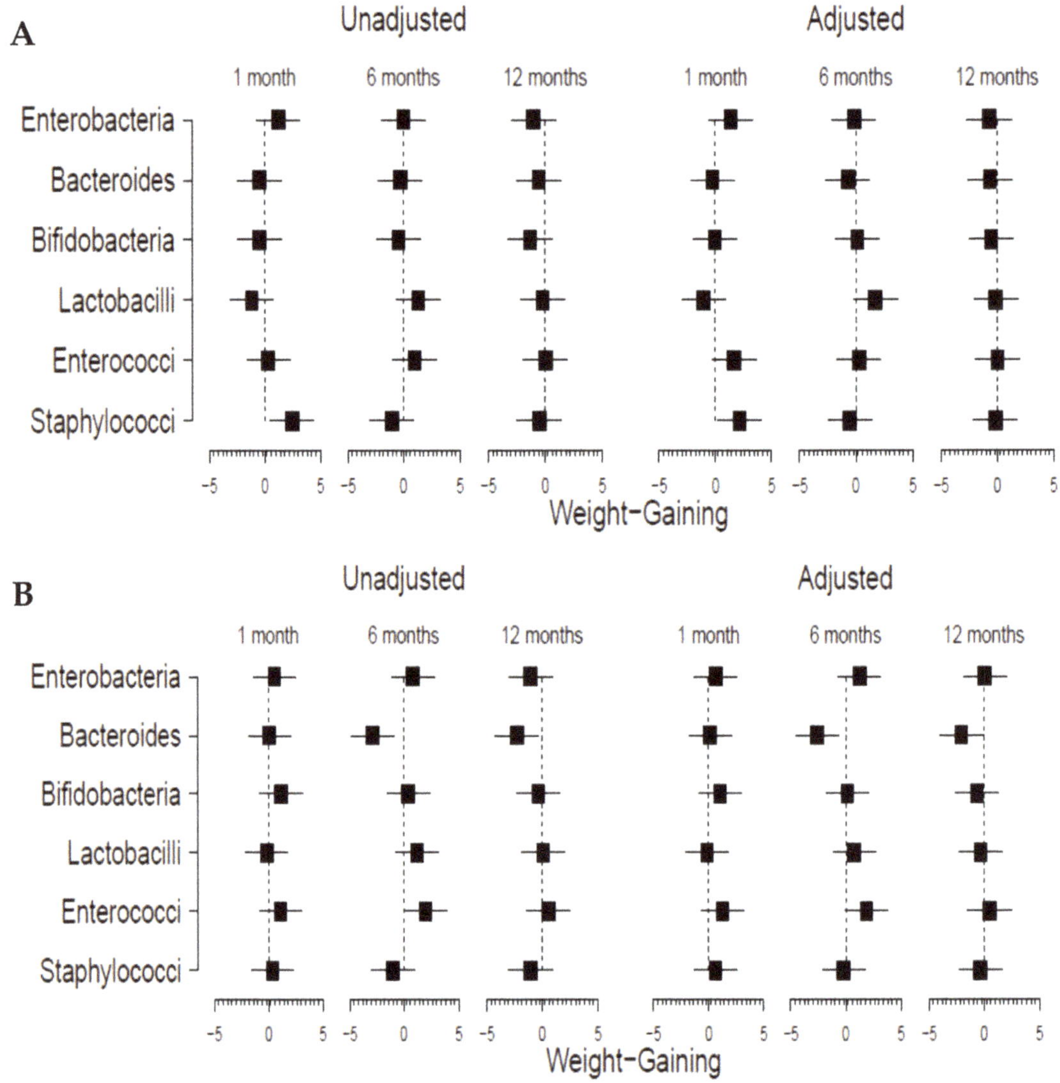


**Figure 3.** Forrest plots showing the effect sizes (with 95% confidence interval) of the association of microbiota-related variables with infant weight gain at 1, 6 and 12 months of age according to delivery mode (vaginal delivery (**A**) and C-section (**B**)). Results from unadjusted model and adjusted by gender and type of feeding. Dotted lines represent the zero value, with values on the left side indicating negative associations and those on the right side indicating associations with positive sign.

## 4. Discussion

The levels of the different microbial groups analyzed in this study were in line with those previously described for 1 month-old full-term infants [26,34,35]. Additionally, in accordance with previous studies, bifidobacteria was the bacterial group showing the highest levels, followed by enterobacteria, which is another of the dominant microbial groups in un-weaned infants [26,34]. Additionally, as expected [1], differences in the microbial levels were observed between vaginal and C-section-delivered babies.

Gender-associated differences in the infant microbiota composition and diversity have been previously reported [36,37]; however, in the present work, we did not notice any significant differences in the levels of the analyzed microbial groups between males and females, neither in vaginally nor in C-section-delivered babies. This is one aspect that deserves further attention since understanding the potential gender differences in the microbiome, the so-called microgenderome [38], and the role that these play in the risk of disease is of utmost importance for developing microbiota modulation strategies in early life. It must be taken into consideration that early life constitutes a critical moment. Some studies have reported an association between antibiotic treatments during the postnatal period microbiota [34,39] and an increased risk of obesity and related metabolic disorders [40,41]. This suggests a possible influence of microbiota alterations during this period in obesity risk later in life, as it has been demonstrated in animal models [42].

Moreover, some studies have also reported associations between early microbiota and weight gain [24,37,43]. However, due to the growing evidence linking the microbiota in early life to obesity risk, we consider that studies focused on full-term infants, as the present one, are especially relevant in this context. In this regard, previous studies demonstrated an altered microbiota during the first year of life in infants developing obesity later on [20], pointing out at the first months of life as the key moment for later metabolic homeostasis.

Interestingly, some microorganisms, such as *Staphylococcus* or *Enterococcus*, have been previously reported to be negatively associated with infant weight/weight gain in preterm infants during very early life [24]. The levels of these microorganisms were also found to be lower, at 5 and 9 months of life, in excessive weight gaining full-term infants than in those showing an appropriate weight gain [43]. However, some differences among studies are also present, likely due to the different methodologies and experimental designs used; for instance, we analyzed the microbiota at 1 month of age, whereas others analyzed it at a later stage (5 and 9 months of age) [43], and we segregated the analyses by delivery mode whilst other authors did not. Actually, our results indicate the existence of different interactions in vaginally delivered and in C-section-delivered babies. We observed that changes in the sign of the microbe-weigh association might occur along different sampling times, as evidenced by our data on staphylococci, showing a positive association with weight gain at 1 month of age but not at later ages when the interaction seems to be even negative. Interestingly, in C-section-delivered babies, but not in vaginal infants, the levels of the *Bacteroides*-group at 1 month of age correlated negatively with later weight, even when the model was adjusted by feeding mode. Delayed colonization by this microorganism has been often reported in C-section-delivered babies [44,45] and C-section delivery has been linked to increased risk of childhood obesity [46]. These observations point out at the levels of *Bacteroides* during early life as a potential early marker for the later risk of excessive weight gain in C-section-delivered babies, an aspect that should be the subject of further studies.

It is important to point out that different factors may influence infant growth trajectories. Among these, infant feeding habits may be of importance. Our results showed that exclusive breastfeeding was associated with significantly lower levels of *Enterococcus* in vaginally delivered babies and with a trend (non-statistically significant) also observed in the C-section-delivered group. In the former group, a trend towards higher *Staphylococcus* levels was also observed. These two microorganisms have been linked to the feeding pattern of the infant. Breastmilk has been previously described as a source of *Staphylococcus*, with increased levels of this microorganism being found in breastfed babies [47]. Other studies, in accordance with our results, reported lower levels of *Enterococcus* in breastfed infants [48]. Altogether, these results suggest that the observed differences in microbial groups and weight gain may be partly related to the feeding habit of the infant. However, although the feeding habit is likely an important factor, our multivariate models were corrected for this variable and some of the effects still remained significant, indicating a microbiota–weight association independent of the feeding type. Therefore, the microbiota–host relation needs to be considered in the analyses focused on infant growth trajectories in order to shed light on the influence of this relationship for child development. Once this relationship is fully understood, it may be possible to develop nutritional strategies, such as dietary probiotics or prebiotics targeting the infant, or perhaps the pregnant or lactating mother, for modulating early life microbiota and the later infant weight gain.

It is also important to underline that our sample size is still limited for establishing strong general conclusions, especially in a context where several potential confounder factors may be present, as is the case in infant microbiota studies. However, it is also true that the infants included originated not just from a unique hospital and geographical location, which could be a source of bias, but from two distant locations. It is worth pointing out as well that our microbiota data are restricted to defined microbial groups for which absolute levels were determined and the potential influence of other microorganisms may have been overseen.

## 5. Conclusions

This work is among the first ones assessing the relationship between the absolute levels of relevant early life intestinal microorganisms, such as bifidobacteria, enterobacteria, lactobacilli, enterococci or staphylococci, and the later weight gain in either vaginally or C-section-delivered full-term infants. The data point out the relationship between specific infant gut microbes and healthy infant development. Our results underline the interest in exploring the intestinal microbiota as a potential target for favoring proper growth and weight gain in infants with potential consequences in later health.

**Supplementary Materials:** The following are available online at https://www.mdpi.com/article/10.3390/nu13072412/s1, Table S1: General description of the WHO Z-scores in the sample across time by type of partum and feeding.

**Author Contributions:** S.G., M.S.-R., S.A., C.M.-C., G.S., M.S., N.F., C.G.d.l.R.-G., S.D.-C., P.M.-C., M.C.C. and M.G. wrote sections of the first draft, thoroughly edited the manuscript and approved the final draft. All authors have read and agreed to the published version of the manuscript.

**Funding:** This work was funded by the EU Joint Programming Initiative—A Healthy Diet for a Healthy Life (JPI HDHL. http://www.healthydietforhealthylife.eu/; accessed on 14 July 2021. Project EarlyMicroHealth) and the Project AGL2017-83653R funded by the Spanish "Ministerio de Ciencia, Innovación y Universidades (MCIU), Agencia Estatal de Investigación (AEI) and FEDER" and by the European Research Council under the European Union's Horizon 2020 research and innovation program (ERC starting grant, no. 639226). Silvia Arboleya is the recipient of a Juan de la Cierva Postdoctoral Contract from the Spanish Ministry of Science and Innovation (Ref. IJCI-2017-32156) funded by the Spanish Ministry of Science, Innovation and Universities.

**Institutional Review Board Statement:** The study was conducted according to the guidelines of the Declaration of Helsinki, and approved by the Institutional Review Board by the Regional Ethics Committee of Clinical Research of Asturias (Ref. 12/16. 3 February 2016), the Ethics Committee of the Hospital Clínico Universitario de Valencia INCLIVA (Ref. 9 January 2015) and the Committee on Bioethics of CSIC.

**Informed Consent Statement:** Informed consent was obtained from the paernts of all the infants involved in the study.

**Data Availability Statement:** The data are available upon reasoned request to the authors, under the restrictions established by the ethical approval of the study.

**Acknowledgments:** We thank all the families involved in the study.

**Conflicts of Interest:** The authors declare no conflict of interest.

## References

1. Milani, C.; Duranti, S.; Bottacini, F.; Casey, E.; Turroni, F.; Mahony, J.; Belzer, C.; Delgado Palacio, S.; Arboleya Montes, S.; Mancabelli, L.; et al. The First Microbial Colonizers of the Human Gut: Composition. Activities, and Health Implications of the Infant Gut Microbiota. *Microbiol. Mol. Biol. Rev.* **2017**, *81*, e00036-17. [CrossRef]
2. Gensollen, T.; Iyer, S.S.; Kasper, D.L.; Blumberg, R.S. How colonization by microbiota in early life shapes the immune system. *Science* **2016**, *352*, 539–544. [CrossRef]
3. O'Mahony, S.M.; Clarke, G.; Dinan, T.G.; Cryan, J.F. Early-life adversity and brain development: Is the microbiome a missing piece of the puzzle? *Neuroscience* **2017**, *342*, 37–54. [CrossRef] [PubMed]
4. Tsai, H.J.; Wang, G.; Hong, X.; Yao, T.C.; Ji, Y.; Radovick, S.; Ji, H.; Cheng, T.L.; Wang, X. Early Life Weight Gain and Development of Childhood Asthma in a Prospective Birth Cohort. *Ann. Am. Thorac. Soc.* **2018**, *15*, 1197–1204. [CrossRef] [PubMed]
5. Schulfer, A.F.; Schluter, J.; Zhang, Y.; Brown, Q.; Pathmasiri, W.; McRitchie, S.; Sumner, S.; Li, H.; Xavier, J.B.; Blaser, M.J. The impact of early-life sub-therapeutic antibiotic treatment (STAT) on excessive weight is robust despite transfer of intestinal microbes. *ISME J.* **2019**, *13*, 1280–1292. [CrossRef]
6. Jensen, B.W.; Bjerregaard, L.G.; Ängquist, L.; Gögenur, I.; Renehan, A.G.; Osler, M.; Sørensen, T.I.A.; Baker, J.L. Change in weight status from childhood to early adulthood and late adulthood risk of colon cancer in men: A population-based cohort study. *Int. J. Obes. (Lond.)* **2018**, *42*, 1797–1803. [CrossRef]
7. Crowther, N.J. Early determinants of chronic disease in developing countries. *Best Pract. Res. Clin. Endocrinol. Metab.* **2012**, *26*, 655–665. [CrossRef]
8. Ferraro, A.A.; Bechere Fernandes, M.T. Relationship between childhood growth and later outcomes. *Nestle Nutr. Inst. Workshop Ser.* **2013**, *71*, 191–197.
9. Cox, L.M.; Blaser, M.J. Antibiotics in early life and obesity. *Nat. Rev. Endocrinol.* **2015**, *11*, 182–190. [CrossRef] [PubMed]
10. Azad, M.B.; Moossavi, S.; Owora, A.; Sepehri, S. Early-Life Antibiotic Exposure. Gut Microbiota Development, and Predisposition to Obesity. *Nestle Nutr. Inst. Workshop Ser.* **2017**, *88*, 67–79.
11. Block, J.P.; Bailey, L.C.; Gillman, M.W.; Lunsford, D.; Daley, M.F.; Eneli, I.; Finkelstein, J.; Heerman, W.; Horgan, C.E.; Hsia, D.S.; et al. PCORnet Antibiotics and Childhood Growth Study Group. Early Antibiotic Exposure and Weight Outcomes in Young Children. *Pediatrics* **2018**, *143*, e20180290. [CrossRef]
12. Uzan-Yulzari, A.; Turta, O.; Belogolovski, A.; Ziv, O.; Kunz, C.; Perschbacher, S.; Neuman, H.; Pasolli, E.; Oz, A.; Ben-Amram, H.; et al. Neonatal antibiotic exposure impairs child growth during the first six years of life by perturbing intestinal microbial colonization. *Nat. Commun.* **2021**, *12*, 443. [CrossRef] [PubMed]
13. Wilkins, A.T.; Reimer, R.A. Obesity, Early Life Gut Microbiota, and Antibiotics. *Microorganisms* **2021**, *9*, 413. [CrossRef]
14. Sacco, M.R.; de Castro, N.P.; Euclydes, V.L.; Souza, J.M.; Rondó, P.H. Birth weight, rapid weight gain in infancy and markers of overweight and obesity in childhood. *Eur. J. Clin. Nutr.* **2013**, *67*, 1147–1153. [CrossRef] [PubMed]
15. Sutharsan, R.; O'Callaghan, M.J.; Williams, G.; Najman, J.M.; Mamun, A.A. Rapid growth in early childhood associated with young adult overweight and obesity-evidence from a community based cohort study. *J. Health Popul. Nutr.* **2015**, *33*, 13. [CrossRef]
16. Matthews, E.K.; Wei, J.; Cunningham, S.A. Relationship between prenatal growth, postnatal growth and childhood obesity: A review. *Eur. J. Clin. Nutr.* **2017**, *71*, 919–930. [CrossRef]
17. Geserick, M.; Vogel, M.; Gausche, R.; Lipek, T.; Spielau, U.; Keller, E.; Pfäffle, R.; Kiess, W.; Körner, A. N Acceleration of BMI in Early Childhood and Risk of Sustained Obesity. *N. Engl. J. Med.* **2018**, *379*, 1303–1312. [CrossRef] [PubMed]
18. Johnson, W.; Bann, D.; Hardy, R. Infant weight gain and adolescent body mass index: Comparison across two British cohorts born in 1946 and 2001. *Arch. Dis. Child.* **2018**, *103*, 974–980. [CrossRef] [PubMed]
19. Korpela, K.; Renko, M.; Vänni, P.; Paalanne, N.; Salo, J.; Tejesvi, M.V.; Koivusaari, P.; Ojaniemi, M.; Pokka, T.; Kaukola, T.; et al. Microbiome of the first stool and overweight at age 3 years: A prospective cohort study. *Pediatr. Obes.* **2020**, *15*, e12680. [CrossRef]
20. Kalliomäki, M.; Collado, M.C.; Salminen, S.; Isolauri, E. Early differences in fecal microbiota composition in children may predict overweight. *Am. J. Clin. Nutr.* **2008**, *87*, 534–538. [CrossRef]
21. Stanislawski, M.A.; Dabelea, D.; Wagner, B.D.; Iszatt, N.; Dahl, C.; Sontag, M.K.; Knight, R.; Lozupone, C.A.; Eggesbø, M. Gut Microbiota in the First 2 Years of Life and the Association with Body Mass Index at Age 12 in a Norwegian Birth Cohort. *mBio* **2018**, *9*, e01751-18. [CrossRef] [PubMed]
22. Dror, T.; Dickstein, Y.; Dubourg, G.; Paul, M. Microbiota manipulation for weight change. *Microb. Pathog.* **2017**, *106*, 146–161. [CrossRef] [PubMed]
23. Blanton, L.V.; Charbonneau, M.R.; Salih, T.; Barratt, M.J.; Venkatesh, S.; Ilkaveya, O.; Subramanian, S.; Manary, M.J.; Trehan, I.; Jorgensen, J.M.; et al. Gut bacteria that prevent growth impairments transmitted by microbiota from malnourished children. *Science* **2016**, *351*, aad3311. [CrossRef] [PubMed]
24. Arboleya, S.; Martinez-Camblor, P.; Solís, G.; Suárez, M.; Fernández, N.; de Los Reyes-Gavilán, C.G.; Gueimonde, M. Intestinal Microbiota and Weight-Gain in Preterm Neonates. *Front. Microbiol.* **2017**, *8*, 183. [CrossRef]
25. Yee, A.L.; Miller, E.; Dishaw, L.J.; Gordon, J.M.; Ji, M.; Dutra, S.; Ho, T.T.B.; Gilbert, J.A.; Groer, M. Longitudinal microbiome composition and stability correlate with increased weight and length of very-low-birth-weight infants. *mSystems* **2019**, *4*, e00229-18. [CrossRef]

26. Arboleya, S.; Binetti, A.; Salazar, N.; Fernández, N.; Solís, G.; Hernández-Barranco, A.; Margolles, A.; de Los Reyes-Gavilán, C.G.; Gueimonde, M. Establishment and development of intestinal microbiota in preterm neonates. *FEMS Microbiol. Ecol.* **2012**, *79*, 763–772. [CrossRef] [PubMed]
27. Gueimonde, M.; Tölkko, S.; Korpimäki, T.; Salminen, S. New real-time quantitative PCR procedure for quantification of Bibidobacteria in human fecal samples. *Appl. Environ. Microbiol.* **2004**, *70*, 4165–4169. [CrossRef]
28. Bartosch, S.; Fite, A.; Macfarlane, G.T.; McMurdo, M.E. Characterization of bacterial communities in feces from healthy elderly volunteers and hospitalized elderly patients by using real-time PCR and effects of antibiotic treatment on the fecal microbiota. *Appl. Environ. Microbiol.* **2004**, *70*, 3575–3581. [CrossRef]
29. Rinttila, T.; Kassinen, A.; Malinen, E.; Krogius, L.; Palva, A. Development of an extensive set of 16S rDNA-targeted primers for quantification of pathogenic and indigenous bacteria in faecal samples by real-time PCR. *J. Appl. Microbiol.* **2004**, *97*, 1166–1177. [CrossRef]
30. Walter, J.; Hertel, C.; Tannock, G.W.; Lis, C.M.; Munro, K.; Hammes, W.P. Detection of *Lactobacillus*, *Pediococcus*, *Leuconostoc*, and *Weissella* species in human feces by using group-specific PCR primers and denaturing gradient gel electrophoresis. *Appl. Environ. Microbiol.* **2001**, *67*, 2578–2585. [CrossRef]
31. Heilig, H.G.; Zoetendal, E.G.; Vaughan, E.E.; Marteau, P.; Akkermans, A.D.; de Vos, W.M. Molecular diversity of lactobacillus spp. and other lactic acid bacteria in the human intestine as determined by specific amplification of 16s ribosomal DNA. *Appl. Environ. Microbiol.* **2002**, *68*, 114. [CrossRef] [PubMed]
32. Matsuda, K.; Tsuji, H.; Asahara, T.; Kado, Y.; Nomoto, K. Sensitive quantitative detection of commensal bacteria by rRNA-targeted reverse transcription-PCR. *Appl. Environ. Microbiol.* **2007**, *73*, 32–39, Erratum in *Appl. Environ. Microbiol.* **2007**, *73*, 6695. [CrossRef]
33. Warnes, G.R.; Bolker, B.; Bonebakker, L.; Gentleman, R.; Huber, E.; Liaw, A.; Lumley, T.; Maechler, M.; Magnusson, A.; Moeller, S.; et al. gplots: Various R Programming Tools for Plotting Data. R Package Version 3.1.1. Available online: https://CRAN.R-project.org/package=gplots (accessed on 7 June 2021).
34. Arboleya, S.; Sánchez, B.; Milani, C.; Duranti, S.; Solís, G.; Fernández, N.; de los Reyes-Gavilán, C.G.; Ventura, M.; Margolles, A.; Gueimonde, M. Intestinal microbiota development in preterm neonates and effect of perinatal antibiotics. *J. Pediatr.* **2015**, *166*, 538–544. [CrossRef] [PubMed]
35. Echarri, P.P.; Graciá, C.M.; Berruezo, G.R.; Vives, I.; Ballesta, M.; Solís, G.; Morillas, I.V.; Reyes-Gavilán, C.G.; Margolles, A.; Gueimonde, M. Assessment of intestinal microbiota of full-term breast-fed infants from two different geographical locations. *Early Hum. Dev.* **2011**, *87*, 511–513. [CrossRef]
36. Cong, X.; Xu, W.; Janton, S.; Henderson, W.A.; Matson, A.; McGrath, J.M.; Maas, K.; Graf, J. Gut Microbiome Developmental Patterns in Early Life of Preterm Infants: Impacts of Feeding and Gender. *PLoS ONE* **2016**, *11*, e0152751. [CrossRef]
37. Kozyrskyj, A.L.; Kalu, R.; Koleva, P.T.; Bridgman, S.L. Fetal programming of overweight through the microbiome: Boys are disproportionately affected. *J. Dev. Orig. Health Dis.* **2016**, *7*, 25–34. [CrossRef]
38. Vemuri, R.; Sylvia, K.E.; Klein, S.L.; Forster, S.C.; Plebanski, M.; Eri, R.; Flanagan, K.L. The microgenderome revealed: Sex differences in bidirectional interactions between the microbiota, hormones, immunity and disease susceptibility. *Semin. Immunopathol.* **2019**, *41*, 265–275. [CrossRef] [PubMed]
39. Tanaka, S.; Kobayashi, T.; Songjinda, P.; Tateyama, A.; Tsubouchi, M.; Kiyohara, C.; Shirakawa, T.; Sonomoto, K.; Nakayama, J. Influence of antibiotic exposure in the early postnatal period on the development of intestinal microbiota. *FEMS Immunol. Med. Microbiol.* **2009**, *56*, 80–87. [CrossRef]
40. Azad, M.B.; Bridgman, S.L.; Becker, A.B.; Kozyrskyj, A.L. Infant antibiotic exposure and the development of childhood overweight and central adiposity. *Int. J. Obes.* **2014**, *38*, 1290–1298. [CrossRef] [PubMed]
41. Bailey, L.C.; Forrest, C.B.; Zhang, P.; Richards, T.M.; Livshits, A.; DeRusso, P.A. Association of antibiotics in infancy with early childhood obesity. *JAMA Pediatr.* **2014**, *168*, 1063–1069. [CrossRef]
42. Cox, L.M.; Yamanishi, S.; Sohn, J.; Alekseyenko, A.V.; Leung, J.M.; Cho, I.; Kim, S.G.; Li, H.; Gao, Z.; Mahana, D.; et al. Altering the intestinal microbiota during a critical developmental window has lasting metabolic consequences. *Cell* **2014**, *158*, 705–721. [CrossRef]
43. Laursen, M.F.; Larsson, M.W.; Lind, M.V.; Larnkjær, A.; Mølgaard, C.; Michaelsen, K.F.; Bahl, M.I.; Licht, T.R. Intestinal Enterococcus abundance correlates inversely with excessive weight gain and increased plasma leptin in breastfed infants. *FEMS Microbiol. Ecol.* **2020**, *96*, fiaa066. [CrossRef]
44. Mitchell, C.M.; Mazzoni, C.; Hogstrom, L.; Bryant, A.; Bergerat, A.; Cher, A.; Pochan, S.; Herman, P.; Carrigan, M.; Sharp, K.; et al. Delivery Mode Affects Stability of Early Infant Gut Microbiota. *Cell Rep. Med.* **2020**, *1*, 100156. [CrossRef] [PubMed]
45. Selma-Royo, M.; Calatayud Arroyo, M.; García-Mantrana, I.; Parra-Llorca, A.; Escuriet, R.; Martínez-Costa, C.; Collado, M.C. Perinatal environment shapes microbiota colonization and infant growth: Impact on host response and intestinal function. *Microbiome* **2020**, *8*, 167. [CrossRef] [PubMed]
46. Yuan, C.; Gaskins, A.J.; Blaine, A.I.; Zhang, C.; Gillman, M.W.; Missmer, S.A.; Field, A.E.; Chavarro, J.E. Association Between Cesarean Birth and Risk of Obesity in Offspring in Childhood, Adolescence, and Early Adulthood. *JAMA Pediatr.* **2016**, *170*, e162385. [CrossRef] [PubMed]

47. Jiménez, E.; Delgado, S.; Maldonado, A.; Arroyo, R.; Albújar, M.; García, N.; Jariod, M.; Fernández, L.; Gómez, A.; Rodríguez, J.M. *Staphylococcus epidermidis*: A differential trait of the fecal microbiota of breast-fed infants. *BMC Microbiol.* **2008**, *8*, 143. [CrossRef]
48. Ma, J.; Li, Z.; Zhang, W.; Zhang, C.; Zhang, Y.; Mei, H.; Zhuo, N.; Wang, H.; Wang, L.; Wu, D. Comparison of gut microbiota in exclusively breast-fed and formula-fed babies: A study of 91 term infants. *Sci. Rep.* **2020**, *10*, 15792. [CrossRef]

MDPI
St. Alban-Anlage 66
4052 Basel
Switzerland
Tel. +41 61 683 77 34
Fax +41 61 302 89 18
www.mdpi.com

*Nutrients* Editorial Office
E-mail: nutrients@mdpi.com
www.mdpi.com/journal/nutrients

www.ingramcontent.com/pod-product-compliance
Ingram Content Group UK Ltd.
Pitfield, Milton Keynes, MK11 3LW, UK
UKHW060123010826
14089UKWH00006B/129